Complete Accounts of Integrated Drug Discovery and Development: Recent Examples from the Pharmaceutical Industry

Volume 3

ACS SYMPOSIUM SERIES **1369**

Complete Accounts of Integrated Drug Discovery and Development: Recent Examples from the Pharmaceutical Industry

Volume 3

Jaan A. Pesti, Editor
Pharma Resource Group, Inc.
Drexel Hill, Pennsylvania, United States

Ahmed F. Abdel-Magid, Editor
Therachem Research Medilab, LLC
Chelsea, Alabama, United States

Rajappa Vaidyanathan, Editor
Medicines for All Institute
Virginia Commonwealth University
Richmond, Virginia, United States

Sponsored by the
ACS Division of Organic Chemistry

American Chemical Society, Washington, DC

Library of Congress Cataloging-in-Publication Data

Names: Pesti, Jaan A., 1954- editor. | Abdel-Magid, Ahmed F., 1947- editor. | Vaidyanathan, Rajappa, editor. | American Chemical Society, issuing body. | American Chemical Society. Division of Organic Chemistry, sponsoring body.
Title: Complete accounts of integrated drug discovery and development : recent examples from the pharmaceutical industry, vol. 3/ Jaan A. Pesti, editor, Pharma Resource Group, Inc., Drexel Hill, Pennsylvania, United States, Ahmed F. Abdel-Magid, editor, Therachem Research Medilab, LLC, Chelsea, Alabama, United States, Rajappa Vaidyanathan, editor, Medicines for All Institute, Virginia Commonwealth University, Richmond, Virginia, United States ; sponsored by the ACS Division of Organic Chemistry.
Description: Washington, DC : American Chemical Society, [2018-] | Series: ACS symposium series ; 1369 | Includes bibliographical references and index.
Identifiers: LCCN 2018048857 (print) | LCCN 2018050283 (ebook) | ISBN 9780841298637 (ebook) | ISBN 9780841298644 (print)
Subjects: LCSH: Drug development. | Pharmaceutical industry. | Pharmaceutical technology.
Classification: LCC RM301.25 (ebook) | LCC RM301.25 .C66 2018 (print) | DDC 615.1/9--dc23
LC record available at https://lccn.loc.gov/2018048857

The paper used in this publication meets the minimum requirements of American National Standard for Information Sciences—Permanence of Paper for Printed Library Materials, ANSI Z39.48n1984.

Foreword

The ACS Symposium Series is an established program that publishes high-quality volumes of thematic manuscripts. For over 40 years, the ACS Symposium Series has been delivering essential research from world leading scientists, including 36 Chemistry Nobel Laureates, to audiences spanning disciplines and applications.

Books are developed from successful symposia sponsored by the ACS or other organizations. Topics span the entirety of chemistry, including applications, basic research, and interdisciplinary reviews.

Before agreeing to publish a book, prospective editors submit a proposal, including a table of contents. The proposal is reviewed for originality, coverage, and interest to the audience. Some manuscripts may be excluded to better focus the book; others may be added to aid comprehensiveness. All chapters are peer reviewed prior to final acceptance or rejection.

As a rule, only original research papers and original review papers are included in the volumes. Verbatim reproductions of previous published papers are not accepted.

ACS Books

Contents

Indexes

Preface

Scientists owe an enormous debt of gratitude to the French biologist, chemist, and artist Louis Pasteur. His exploits in biology and chemistry are astounding, and his research is responsible for saving and enhancing literally millions of lives. As a young scientist, shortly after receiving his doctorate degree, he studied tartaric acid crystals forming on a wine vat and was able to separate mirror image crystals of the acid. He discovered chirality by comparing the effect of plane-polarized light on these mirror image crystals. The artist in him may have seen the mirror image relationship between the crystals, but the scientist in him concluded they must be similarly related at the molecular level—that is, their structures must also be chiral. This discovery eventually led to the realization that mirror image structures (enantiomers) not only interact differently with plane-polarized light but may also vary in their biological responses. This intuitive leap has ultimately changed the way we discover and develop pharmaceuticals today. The current FDA guidelines require the preparation, separation, and testing of enantiomers to determine which one has the desired biological effect. Because of Louis Pasteur's pioneering work and discoveries, we may consider him the father (or one of the founders) of modern pharmaceutical science. He placed science at the forefront of his work and laid the underpinnings of modern science and, in turn, the modern pharmaceutical establishment. His quote, "Science knows no country, because knowledge belongs to humanity, and is the torch which illuminates the world," is particularly relevant to this book as we highlight the contributions of scientists from multiple countries.

In this book we present stories of recent drug discovery and development from different pharmaceutical companies. Drug discovery and development is a long and meticulous process that starts with Discovery Chemists who quite logically "discover" promising novel lead compounds. Through a tedious process of lead optimization, they identify and select new molecular entities that become drug candidates. When a compound has been shown to possess sufficient novelty and utility toward a disorder or disease, it is selected to progress into development. The typical procedure is to hand it off to process chemists, who are trained to identify efficient, safe, and sustainable procedures for the synthesis of the drug candidates. The synthesis initiates at gram scale, progresses to multi-kilogram scale, and—if all goes well—ultimately to metric ton manufacturing scale. All of our stories have reached at least the multi-kilogram scale. The nexus of the two disciplines, Discovery Research and Process Research, is necessary to produce nearly all the new drugs that come into being today. Their combined efforts lead to the introduction of novel effective drugs that are produced by safe and cost-effective manufacturing processes to meet the needs of patients globally.

The success of our previous books (*Comprehensive Accounts of Pharmaceutical Research and Development: From Discovery to Late-Stage Development*, Volume 1 and Volume 2, and *Complete Accounts of Integrated Drug Discovery and Development: Recent Examples from the Pharmaceutical Industry*, Volume 1 and Volume 2) has encouraged us to further inform synthetic organic chemists as well as future researchers with real life success stories from the pharmaceutical world. This book is the result of that effort.

We hope that the reader will be able to take away new concepts that would enable their own research. Thus, not only should this book be read for the pleasure in seeing how discovery and development problems were solved, we also aim for it to have a material effect on modern pharmaceutical research.

The chapters in this book are written by a carefully chosen group of exceptional, exceedingly proficient medicinal and process chemists representing nine of the top pharmaceutical companies. In addition to stories of drug invention and development, the chapters also highlight new ideas, novel solutions, and extensive lists of primary journal articles and useful reviews.

The narrative of the discovery of each new drug is initially provided by the discovery chemist and is immediately followed by the process chemist's discussion of the development of the same drug, hopefully up to manufacturing scale. Unlike examples pulled from the primary literature, this enables a complete picture of drug discovery and development — the story is not fragmented among numerous journals. Furthermore, unlike the primary literature, the book format allows us to delve into the esoteric details that journal publication rarely affords. Occasionally, science simply goes wrong — how does one recover from a predicament when time or resources may be a factor? Our readers will establish a familiarity with the drug discussed and the practical barriers experienced en route to an advanced stage of development or to a marketed pharmaceutical. Narratives of approved drugs are undoubtedly the ultimate success stories, but they number very few. The majority of drug candidates fall by the wayside due to various reasons: toxicity, mutagenicity, market considerations, etc. However, there are many remarkable success stories even for drug candidates that did not make it to market, including new discoveries and innovative problem-solving ideas that contribute to the advancement of the science of drug discovery and development. Therefore, our book chapters chronicle stories from molecules that advanced into development—some all the way to commercialization. In addition, we aimed to augment the value of this book series by allowing the authors to include all pertinent details of the process. Few restrictions were placed on the authors — these are their own stories.

Our particular thanks go out to all the authors listed below. These extraordinary scientists have volunteered substantial time and effort to write this book. Their willingness to contribute the demanding effort expended in the writing of their chapters bears exquisite witness to their exceptional work. Some authors have written for us before; we owe them mightily.

David R. Witty, David T. MacPherson, Gerard M. P. Giblin, and Michael T. Williams of Convergence Pharmaceutical Ltd., Robbie Chen, Vinny Couming, John Guzowski, Jr., Erwin Irdam, William F. Kiesman, Daw-Iong Albert Kwok, Wenli Liang, Tamera L. Mack, Erin M. O'Brien, Suzanne M. Opalka, Daniel Patience, Donald G. Walker, and Frederick Osei-Yeboah of Biogen , Inc., and Stefan Sahli of Biogen International, describe the invention and development of vixotrigine, a use-dependent voltage-gated sodium channel blocker useful for the treatment of neuralgia (Chapter 1). The discovery and lead optimization strategies were designed to harness the proven mode of action of existing sodium channel blockers, while significantly improving the therapeutic index and side-effect profile. Medicinal chemistry routes were adapted for early scale-up but were far from optimal. Subsequent development of scalable syntheses and the careful application of green chemistry principles has led to a highly efficient production process suitable for commercial application.

Heather J. Finlay, Scott A. Savage, Jason M. Stevens, Steven R. Wisniewski, and Miao Yu of Bristol Myers Squibb describe the drug discovery and early development efforts to advance BMS-919373 as a selective I_{Kur} inhibitor for the potential treatment of atrial fibrillation (Chapter 2). Atrial fibrillation (AF) is the most commonly occurring cardiac arrhythmia; and despite recent success in the reduction of stroke risk with oral anticoagulants, AF patients remain in abnormal sinus rhythm. Development of an orally administered agent to reduce AF burden and maintain normal sinus rhythm in this patient population has been a significant challenge in the industry and continues to be an area of high, unmet medical need.

Bradley P. Morgan of Cytokinetics, and Seb Caille and Shawn D. Walker of Amgen, Inc., discuss the discovery and development of omecamtiv mecarbil, a first-in-class, selective activator of cardiac myosin for the potential treatment of heart failure with reduced ejection fraction (HFrEF) (Chapter 3). Omecamtiv mecarbil promotes the force-generating cycle of myosin and was designed to improve heart function in HFrEF without perturbation of intracellular calcium levels. Insights into the transition from candidate selection to early development are shared, as well as how aggressive timelines and continued optimization provided the foundation for the selection of the initial manufacturing route. Based on this early synthetic work, the switch from a "fit for purpose" manufacturing route to a highly efficient and robust manufacturing process that has afforded over a metric ton of drug substance (DS) to support large, multinational registration studies is reviewed. The process design minimized variability introduced from the DS powder properties to allow for the development of modified release oral formulations that preserved oral bioavailability while lowering C_{max} and maintained small and consistent peak-to-trough fluctuation at steady state. This chapter provides the reader with the molecule's journey from idea to candidate selection, from fit-for-purpose early clinical supply route to scalable factory process designed with performance of the final product in mind.

Fredrik Lehmann of Oncopeptides AB and Johan Wennerberg of DuPont Chemoswed AB relate the discovery and process development of melflufen, an alkylating agent for the treatment of multiple myeloma (Chapter 5). Melflufen has a distinctive background—it is directly related to the nitrogen alkylator chloromethine, the first substance used for chemotherapy and developed during the 1940s. Melflufen is a lipophilic agent that readily penetrates membranes which result in high intracellular concentrations of its metabolites. The result is a unique pharmaceutical for the treatment of cancer.

Christophe Hardouin of Oril Industrie and Jérôme-Benoît Starck of the Servier Research Institute describe the discovery and chemical development of S 44563, a potent inhibitor of anti-apoptotic proteins involved in programmed cell death (Chapter 6). S 44563 is an acylsulfonamide-based compound isolated as a sodium salt bearing two controlled stereocenters (a chiral pyrazinoquinolone and a chiral 1,3-diamine motif) and also a substituted biaryl moiety. Medicinal chemistry efforts relying on a restriction of conformations approach of an aryl piperazine moiety are briefly presented. The original racemic synthesis to provide the first quantities is described, and several topics that could be potential issues for the scale-up are also discussed and addressed to reach a kilogram-scale enantioselective approach.

Nathan B. Mantlo and Scott A. Frank of Eli Lilly and Company discuss the discovery and development of the cholesteryl transfer protein inhibitor evacetrapib. Evacetrapib is a potent and selective inhibitor of cholesteryl ester transfer protein and, favorably, does not induce aldosterone and cortisol biosynthesis through off-target biological pathways. Beginning with a practical medicinal chemistry route that provided multiple analogs as well as several hundred-gram quantities for preclinical toxicological evaluation, a new route was developed which enabled large-scale manufacture for Phase III clinical supply material demands. Insights gained from the medicinal chemistry route with regards to stable benzazepine configurational isomers at various intermediates greatly informed development of a practical and highly efficient redesigned synthetic process. The new route leveraged continuous processing unit operations at two key transformations that established a key stereocenter and isomeric fragment, which led to the production of well over 1500 kilograms of final material. Evolution of the medicinal chemistry strategy and process chemistry is discussed.

Feng Xu and John A. McCauley of Merck and Co., Inc. discuss the discovery and development of the hepatitis C virus NS3/4a protease inhibitor, grazoprevir, approved in combination with the NS5a inhibitor elbasvir as ZEPATIER,® for the treatment of HCV infection (Chapter 9). Structure-based drug design was a key element in multiple aspects of the discovery of grazoprevir, including the introduction of a novel macrocyclic framework. The chemical development of grazoprevir was a journey of evolution of chemistry and process. Of importance, mechanistic studies played a crucial role for understanding of the origin of process issues and therefore allowed the team to navigate through the process evolution and development. With innovative breakthroughs and thorough evaluation of various bond connection strategies, the most effective synthesis of this highly complex molecule with an optimal sequence of bond connection to assemble four building blocks has been established in the final manufacturing process.

Matthew R. Kaller, Robert P. Farrell, and Ted C. Judd of Amgen, Inc., and Paul Dransfield of H3 Biomedicine, Inc., discuss the discovery and development of AMG 986, a potent and efficacious APJ (a Gi coupled G-protein receptor closely related to the angiotensin II type 1 receptor) agonist for the treatment of heart failure (Chapter 4). Rapid identification of a lead compound was achieved from an initial screening hit with >500-fold improvement in potency. A critical synthetic challenge was overcome through a collaborative effort between Medicinal Chemistry and Process Development teams, enabling the implementation of a novel enantioselective hydrogenation for the synthesis of a key intermediate. Finally, the early Medicinal Chemistry route was replaced with a second-generation convergent approach well suited to process chemistry.

Bin Ma, Chaomin Li, Brian T. Hopkins, Lloyd Franklin, Robbie Chen, Tamera Mack, Daniel Patience, William F. Kiesman, and Erin M. O'Brien of Biogen, Inc. describe the discovery and development of BIIB068, a highly selective, reversible Bruton's tyrosine kinase inhibitor shown to exhibit favorable ADME properties and a preclinical safety profile to support advancing into clinical studies to evaluate as a potential small-molecule oral therapy for the treatment of systemic lupus erythematosus (Chapter 10). In addition, the medicinal chemistry strategy and Phase I clinical results are discussed.

Arnaud C. Tiberghien and Jeremy S. Parker of AstraZeneca discuss the discovery and development of tesirine, a cytotoxic antibody-drug conjugate drug-linker designed to destroy cancer cells (Chapter 7). The properties of tesirine and its advanced development made it a perfect candidate for conjugation to numerous anticancer antibodies. Loncastuximab tesirine is scheduled for commercial launch in 2021, and camidanlumab tesirine is undergoing Phase II registrational trials.

Feng Xu, Benjamin D. Sherry, and Timothy A. Blizzard of Merck and Co., Inc. relate the discovery and development of relebactam, a novel beta-lactamase inhibitor, recently approved by the FDA for the treatment of resistant Gram-negative bacterial infections (Chapter 8). The discovery of relebactam was enabled by a novel first-generation large-scale synthesis of bridged bicyclic urea intermediate. To support the development of relebactam, more efficient synthetic routes to access the unique urea key intermediate were then developed. Finally, the understanding and control of the formation of oligomeric impurities from the late-stage labile hydroxy urea intermediates played an important role in the development of the robust manufacturing process to relebactam.

We are forever grateful to this assembly of scientists who wrote these stories. This was truly a labor of love—as they took on this work for no reward other than sharing their science. Their willingness to tackle this difficult task, contribute numerous hours of writing, and freely offer their scientific expertise made this book possible.

We would be remiss if we did not thank the numerous referees who made innovative suggestions for tweaking and improving the narratives. We happily thank our friends at ACS Books who permitted and encouraged the assemblage of this book: Sara Tenney, Amanda Koenig, Alison Kreckmann, and Arlene Furman. Finally, we thank the Organic Chemistry Division of the American Chemical Society for their support of our symposium, which gave rise to this book.

Jaan A. Pesti
Pharma Resource Group, Inc.
Drexel Hill, Pennsylvania 19026, United States

Ahmed F. Abdel-Magid
Therachem Research Medilab, LLC
Chelsea, Alabama 35043, United States

Rajappa Vaidyanathan
Medicines for All Institute
Virginia Commonwealth University
Richmond, Virginia 23219, United States

Chapter 1

Discovery and Chemical Development of the Use-Dependent Sodium Channel Blocker Vixotrigine for the Treatment of Pain

David R. Witty,[*,1,4] David T. MacPherson,[1,5] Gerard M. P. Giblin,[1,6] Michael T. Williams,[1,7] Robbie Chen,[2] Vinny Couming,[2] John Guzowski, Jr.,[2] Erwin Irdam,[2] William F. Kiesman,[2] Daw-Iong Albert Kwok,[2] Wenli Liang,[2] Tamera L. Mack,[2] Erin M. O'Brien,[*,2] Suzanne M. Opalka,[2] Daniel Patience,[2] Stefan Sahli,[3] Donald G. Walker,[2] and Frederick Osei-Yeboah[2]

[1]Convergence Pharmaceuticals Ltd., a Biogen Company, Babraham Research Campus, Cambridge SG22 3AT, United Kingdom

[2]Biogen Inc., Product and Technology Development, Cambridge, Massachusetts 02142, United States

[3]Biogen International, Neuhofstrasse 30, 6340 Baar, Switzerland

[4]Current address: Witnet Ltd., St John's Innovation Park, Cambridge CB4 0DS, United Kingdom

[5]Current status: Retired

[6]Current address: Giblin Consultancy, Hertfordshire, United Kingdom

[7]Current address: CMC Consultant, Deal, Kent CT14 9TY, United Kingdom

[*]Emails: david.r.witty@gmail.com; erin.obrien@biogen.com (Biogen author)

This chapter describes the research program that led to the discovery of vixotrigine, a use-dependent voltage-gated sodium channel blocker designed to deliver an improved safety profile. The properties that led to the selection of this candidate molecule are discussed. The choice of initial synthetic strategy and the development of an early scalable synthesis are examined. Further development of a more efficient process and the incorporation of green chemistry principles to successfully prepare for commercial utility are described.

Introduction

Voltage-gated sodium channels (*1*) (Na_Vs) are ubiquitous proteins found in all known classes of organisms, indicating their essential role in the basic functions of life. There is a long history of sodium channel blockers showing utility in the treatment of human diseases. For example, carbamazepine (Figure 1) was introduced for the treatment of epilepsy and trigeminal neuralgia in the 1960s (2), and lamotrigine for epilepsy and bipolar disorder in the 1990s (3, 4). More recently, driven by discoveries in human sodium channel genetics, there has been a resurgence of effort aimed at discovering novel sodium channel blockers for the treatment of pain (5).

Sodium channels are trans-membrane proteins, consisting of a large tetrameric pore-forming α-subunit and smaller auxiliary β-subunits. There are nine known subtypes designated by their respective α-subunits, $Na_V1.1$ - 1.9, plus the more structurally distinct Na_x which arise from genes SCN1A–SCN11A (6). Sodium channels at resting membrane potential are closed to sodium ion transport but respond to external stimuli by detecting small changes in membrane potential. These trigger activation of the channel via a conformational change. This results in transient channel opening permitting selective cellular influx of sodium and thus depolarization resulting in neuronal activation and excitation. Channels remain in the open state for a short period before entering an inactivated state that is non-conducting and resistant to activation. An increase in membrane potential known as hyperpolarization causes the channel to return to the resting state. In order to intervene pharmacologically it is preferable to stabilize the inactivated state rather than blocking the open state which has important roles in homeostasis (7).

Clinically there have been several human studies investigating the efficacy of sodium channel blockers in neuropathic pain. Local anaesthetics such as lidocaine (Figure 1) are non-selective sodium channel blockers. The analgesic activity of the local anaesthetics is clinically well characterised although the therapeutic index is low, limiting use to local administration. Non-selective blockade of voltage-gated sodium channels is part of the mechanism of action of several drugs in current clinical practice. Both lamotrigine (8) and carbamazepine (9) are characterised as non-selective sodium channel blockers. However, their utility is limited by side effects and multiple drug-drug interactions that require careful and prolonged titration to the safe, effective clinical dose. Hence there is a need for sodium channel blockers with improved safety profiles.

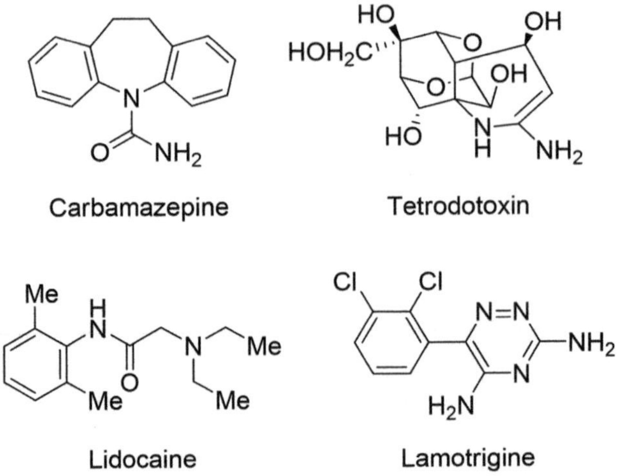

Figure 1. Compounds with a sodium channel blockade mode of action.

Both site-directed mutagenesis studies and ligand competition experiments indicate channel blocking compounds can bind at many differing sites within the ion channel assembly (10). Certain compounds, such as tetrodotoxin (Figure 1), block the outer pore which is wide enough to accommodate hydrated sodium ions. Marketed drugs, including lidocaine, are believed to bind at the 'local anaesthetic' site on the inside of the pore; this is a broad cavity accessible to small molecules capable of penetrating the cell membrane (11, 12, 13).

Our goal was to identify a Na_V blocker with potential utility in the treatment of neuropathic pain, such as trigeminal neuralgia, and other central nervous system (CNS) disorders (*14*). Selective stabilization of the inactivated state can reduce the propensity for channel opening in response to high frequency firing. This is important to pain function as pain is typically associated with a high frequency of nerve pulses - often in the 100 Hz region. In contrast many other physiological functions, such as heartbeat, also modulated by sodium channel action, are associated with much lower frequency activity. Achieving a sufficient degree of inactivated-state block of sodium channels was targeted for optimal safety and efficacy. Measurement in the change in conduction after a rapid sequence of action potentials indicates the 'use-dependence' of an agent which blocks the inactivated state of the channel (*15*). Furthermore, based on genetic association studies (*16*) through state dependence, we were aiming for enhanced Na_V blockade in neurons with pathologically elevated activity, while sparing channels associated with low frequency activities such as cardiac function (*17*).

Importantly, we chose to optimize for CNS penetration on the basis that nociceptors (*18*) express a variety of Na_V subtypes, and blockade of brain sodium channels (Na_Vs 1.1, 1.2, 1.3) is highly precedented (*19*) for efficacious Na_V blockers. Consequently, we sought compounds showing use-dependence by preferential binding to the inactivated state of relevant sodium channels. We also wanted selectivity over the hERG potassium channel which is known to modulate heart function (*20*), a profile suitable for oral dosing, and low potential for drug-drug interactions.

Clinical Target

Trigeminal neuralgia (*21*) is a rare disease linked to entrapment of the trigeminal nerve. The disease is associated with intense, unilateral facial pain characterised by electric shock-like episodes There are about 150,000 diagnosed cases in the United States each year. Treatment options for this debilitating condition include microvascular decompression surgery, which risks facial numbness and paralysis, and drug treatment with carbamazepine or oxcarbazepine. These agents are non-selective voltage-gated sodium channel blockers but are poorly tolerated and their use is hampered by multiple clinically relevant drug-drug interactions and the active pharmacology of their metabolites (*22*).

Strategy

We were keen to find a chemotype structurally differentiated from currently marketed drugs to give an improved safety profile, avoiding the need for dose titration (*23*). The first approach was to conduct a high throughput fluorescence imaging screen using cells expressing the brain sodium channel $Na_V1.2$ to find general chemotypes with sodium channel blocking activity. Hits were assessed for their tonic (initial) block and use-dependent block by automated parallel electrophysiology using the $Na_V1.3$ channel as a surrogate for brain sodium channels. Blockade of the open form of the ion channel (tonic block) was assessed by measuring the effect of test compounds on the peak current elicited by the first pulse in a 10 Hz pulse train at a range of concentrations, from which a pIC_{50} was determined (*24*). The use-dependent block was assessed on the final pulse, and data presented here show the negative log of the concentration that caused a 15% inhibition of the sodium current (pUD_{15}) – a level reported to show analgesic effects. The ratio of pUD_{15} to pIC_{50} is reported as the Tonic Selectivity (TS).

The pharmacophore for a local anaesthetic agent such as lidocaine can be described by an aromatic group linked to an amide with a pendant amine. This has been a starting point for the design of some classes of systemic analgesic agents. However, we demonstrated use-dependent compounds from a variant of the pharmacophore in which the amino and amide groups are reversed such that a lipophilic aromatic group is linked to a basic amine or polar heterocycle, which in turn is linked to an amide or amide isostere (Figure 2).

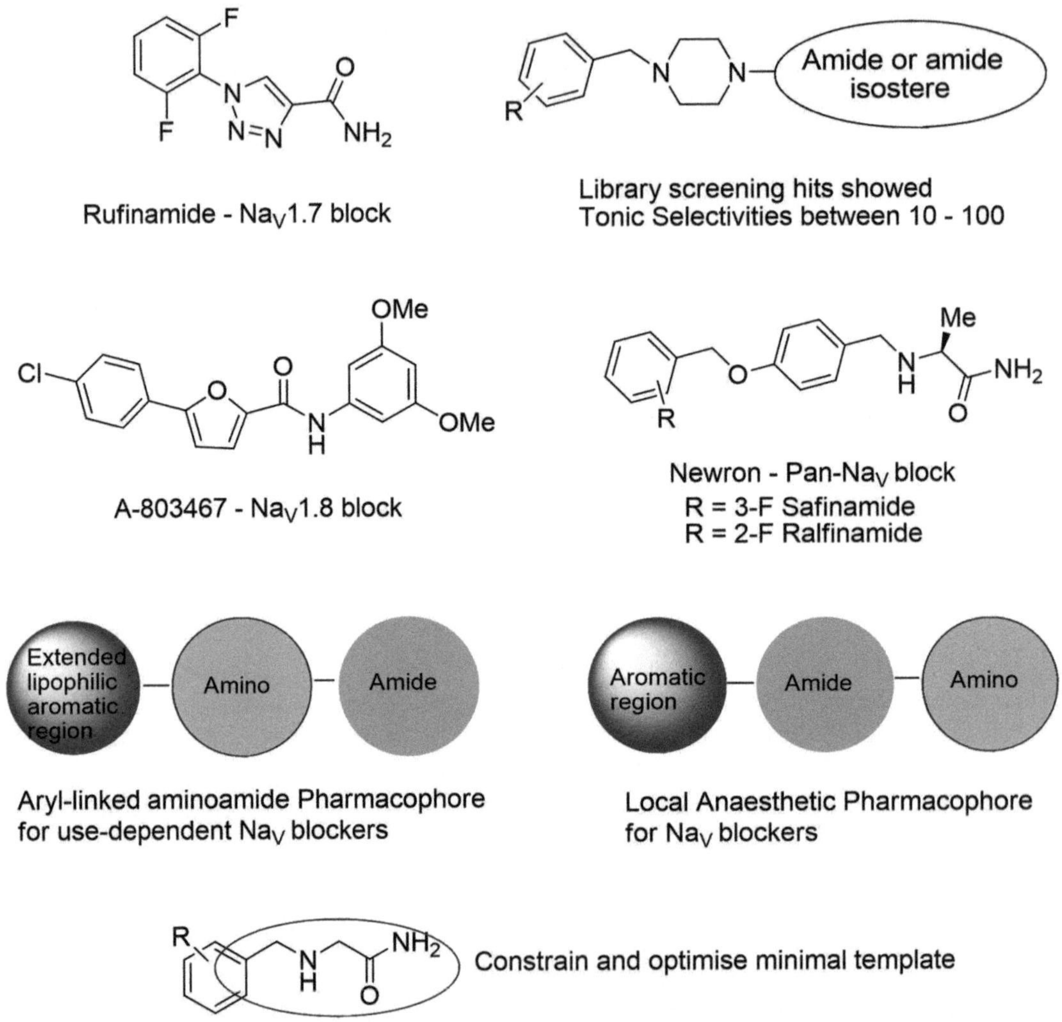

Figure 2. Na_V Pharmacophore generation and selected published Na_V blockers.

This pharmacophore was present in high throughput screening hits. It is also exemplified in specific compounds in the published literature, with ralfinamide and safinamide reported to be state dependent blockers by Newron (25). In addition, the Abbott compound A-803467 and Eisai's rufinamide, mapping to this pharmacophore, are reported as $Na_V1.8$ and $Na_V1.7$ blockers, although for rufinamide no channel selectivity or state dependence was claimed (26, 27).

We chose to explore the *N*-benzyl-glycinamide scaffold as a minimum template which demonstrated both tonic and use-dependent block of $Na_V1.3$. Our strategy was to optimize through substitution and derivatization. Amide **1**, an unsubstituted *N*-benzyl glycinamide incorporating the *para*-2-fluorobenzyloxy group present in published NaV blockers (5), was synthesized and showed modest tonic and use-dependent block of $Na_V1.3$ (pIC_{50} 4.3, pUD_{15} 5.5, TS 11, clogP 2.4). We hypothesized that conformational restriction around the benzyl-amino-amide feature could be associated with enhanced TS. Compounds of this class, all small amino amides, had the additional advantage of good solubility and reasonable synthetic accessibility, which helped us design a lead optimization campaign.

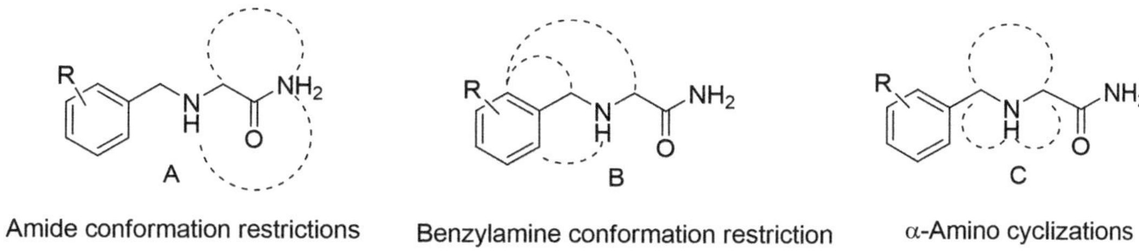

Figure 3. Targeted constraints of benzyl glycinamides.

We undertook systematic conformational constraint of this template by introducing cyclization along the benzyl glycinamide backbone: between the amide nitrogen and either the α-position or the basic nitrogen (**A**), from the aromatic ring on to the glycinamide backbone (**B**), and between positions α-to the amino nitrogen (**C**) (Figure 3). Alkyl-substituted benzyl glycinamides such as ralfinamide were brain penetrant. In order to maintain developability properties within a range suitable for a brain penetrant drug, we targeted constrained analogs with clogP values between 2 and 3.5 and with a molecular weight below 400.

Cyclization between the amide nitrogen and the glycine amine nitrogen gave compound **2** (Figure 4) which showed no significant sodium channel blockade (pIC_{50} <4.1). In contrast cyclization through α-carbon gave analog **3** which retained some use-dependent block (pUD_{15} = 4.8), but at a significantly lower level than **1** and with very low tonic block (pIC_{50} <4.1).

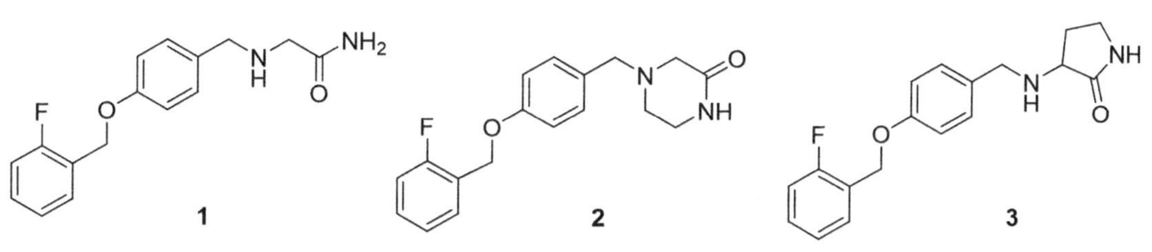

Figure 4. Initial benzyl-amino-amide analogs explored.

Amides **4-10** demonstrate constraint of the benzylic position onto the central aromatic ring, either through nitrogen or benzylic carbon, while retaining an unsubstituted amide (Figure 5). Ring systems incorporating the amino nitrogen with branching at nitrogen such as tetrahydroisoquinoline **4** (pIC_{50} <4.1), benzoxepine **5** (pIC_{50} 4.2) and isoindoline **6** (pIC_{50} <4.1) showed markedly reduced tonic block at $Na_V1.3$. Similar compounds such as **7** (pIC_{50} = 4.7, pUD_{15} 4.8) in which the amide group was attached to the ring α to the nitrogen showed more potent tonic block but had significantly reduced use-dependent block relative to **1**. Exocyclic amines in which the benzylic position alone was constrained such as indane **8** (pIC_{50} 4.5, pUD_{15} 5.5) and tetrahydronaphthalene **9** (pIC_{50} 4.2, pUD_{15} 5.6, clogP 3.8) were comparably potent tonic blockers of $Na_V1.3$ and showed significant use-dependent block, but for **9** this was at the expense of increasing logP beyond the target range. However, efforts to reduce log P, for example by introduction of an oxygen atom as a heterocyclic constraint, as in benzopyran **10** (pIC_{50} <4.1), led to the loss of sodium channel blockade.

Further evaluation of analogs showed that while indanes and tetrahydronapthalenes related to **8** and **9** had a low potential for hERG blockade and showed low microsomal clearance, they proved to be inhibitors of cytochrome P450 isoforms. Modification with additional alkyl substituents typically led to much higher hERG inhibition, high logPs or increased rat microsomal clearance and therefore this class was not pursued further.

An alternative constraint incorporated the amino nitrogen into a discrete pyrrolidine ring. Methyl substituted compound **11**, incorporating a tertiary nitrogen, had higher use-dependent block (pIC_{50} 4.2, pUD_{15} 5.7), while the proline derivative **12** gave the highest use-dependent block in the series (pIC_{50} 4.1, pUD_{15} 5.8), with a TS of 50 (Figure 6). The isomeric pyrrolidines **13-16** all maintained good use-dependent blockade of $Na_V1.3$ (pIC_{50} 4-4.2, pUD_{15} 5.4-5.7).

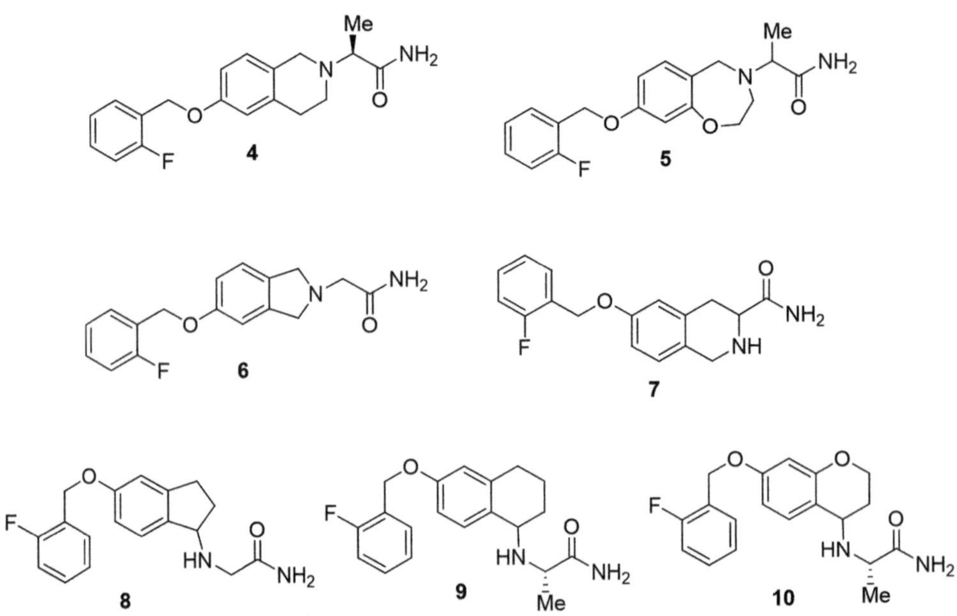

Figure 5. Analogs that constrained the benzylic position onto the central aromatic ring (undefined stereocenters indicate compounds tested as mixtures).

Further investigation revealed that pyrrolidines **11**, **12** and their related analogs, all possessing a tertiary pyrrolidine ring, were generally associated with high clearance in rat microsomes. Attempts to block metabolically vulnerable positions adjacent to the basic nitrogen by alkyl substitution typically gave rise to higher hERG inhibition and compounds of this class were reported to have other off-target activities (*28*). In contrast, the individual 2-5-substituted pyrrolidine isomers **13-16**, while showing broadly similar use-dependent block of $Na_V1.3$, demonstrated differential effects in their other properties. *Cis*-isomers **13** and **16** displayed a trend towards lower blockade of hERG, while, interestingly, there was a clear diastereomeric preference for the 2-(*S*) configuration alpha to the carbonyl, compounds **14** and **16** giving the lowest rat microsomal clearance. Of these, 2*S*,5*R* isomer **16** had the most attractive overall profile ($Na_V1.3$ pIC_{50} 4.2, pUD_{15} 5.5, TS 16, hERG pIC_{50}

4.9 and rat CLi = 1.2 mL/min/g, clogP 3.1) so an analog programme was initiated around **16** while maintaining clogP values between 2.5 and 3.5.

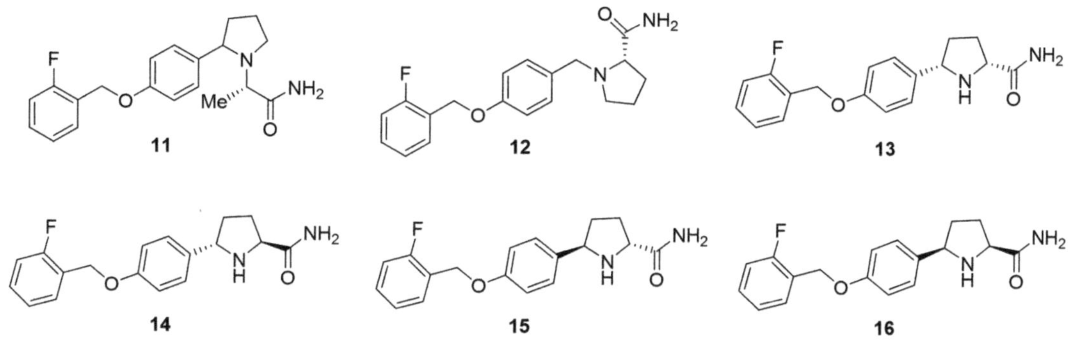

Figure 6. Analogs that incorporated the amino nitrogen into a pyrrolidine ring.

Moving the position of the fluoro substituent on the terminal ring to *meta* **17** (Figure 7) caused the hERG inhibition to rise markedly, but the *para* analog **18** had a similar in vitro profile to **16**. A nitrile group in **19** was also tolerated as a replacement for the fluorine of **16**. Expanding the ring from a pyrrolidine to piperidine **20** increased the hERG inhibition. Phenyl ethers proved comparably potent sodium channel blockers to benzyl ethers, with compound **21** showing use-dependent blockade equivalent to **16**. Substitution on the linker in compound **22** increased hERG inhibition while substitution in the central aromatic ring such as **23**, or its replacement with a heterocycle **24**, reduced use-dependent blockade. Compound **25** without the fluorine had higher hERG inhibition.

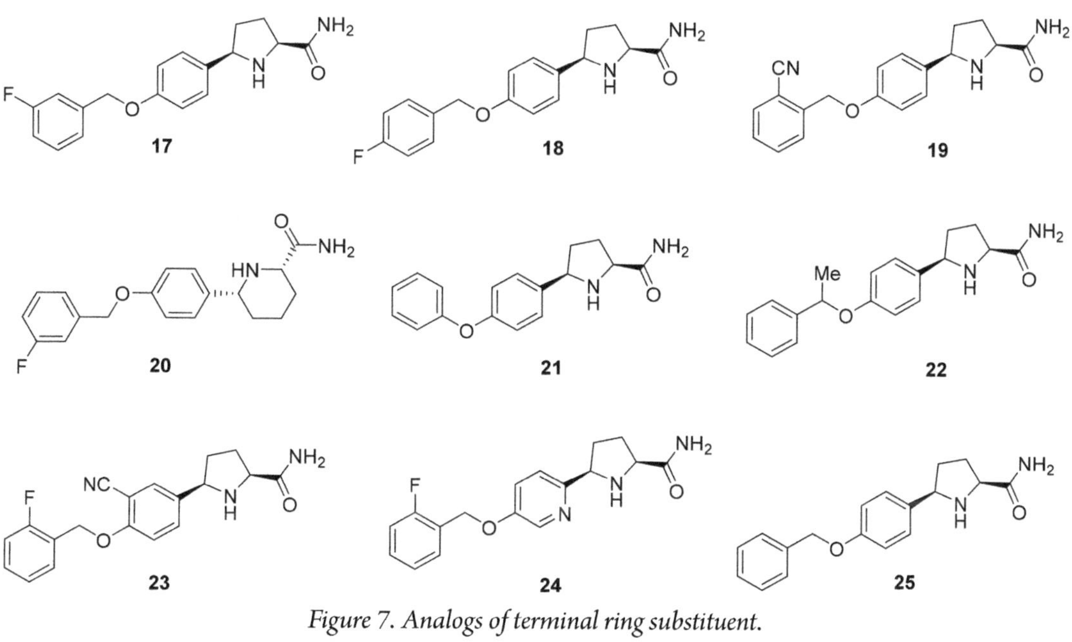

Figure 7. Analogs of terminal ring substituent.

Drug Metabolism and Pharmacokinetics (DMPK)

Four compounds, **16**, **18**, **19** and **21**, meeting the criteria of TS greater than 10 fold, rat microsomal clearance less than 2.0 mL/min/g and hERG pIC_{50} less than 5.0, were progressed into rat in vivo exposure studies, to determine clearance, bioavailability and in vitro tissue distribution (Table 1).

Table 1. Rat Exposure

Compound	*Blood Clb (mL/min/kg)*	*Brain:Blood ratio*	*% Oral Bioavailability*
16	39	6.8	43
18	91	3.5	31
19	37	0.8	39
21	40	4.8	25

Only compounds **16, 18** and **19** showed oral bioavailability greater than 30% and all three were analysed for brain penetration. The more polar analog **19** showed a low brain:blood ratio but both 2-fluorobenzyloxy- and 4-fluorobenzyloxy analogs, **16** and **18,** achieved high brain penetration. However *para*-fluoro analog **18** was highly cleared in vivo. Compound **16** had a good oral bioavailability of 43%, moderate clearance of 39 mL/min/kg and a brain to blood ratio of 6.8. This compound was therefore selected for further studies.

The interactions of *ortho*-fluoro analog **16** with binding sites on 130 receptors, ion channels, transporters, and enzymes were examined in a radioligand binding cross screening panel. This compound had a pan-sodium channel state dependent inhibition profile (29) and the most significant off-target interaction was inhibition of MAO-B. This activity was confirmed in human liver microsomes where **16** inhibited MAO-B enzyme activity with a pIC_{50} of 8.4. In contrast **16** had no measurable inhibitory effect on MAO-A enzyme activity (pK_i <5).

Pharmacology

Compound **16** was evaluated in a model of neuropathic pain by oral administration at 0.5 or 5 mg/kg, BID. It significantly reversed mechanical allodynia in the chronic constriction injury (CCI) model after 6 and 8 days of dosing (Figure 8) starting at day 22 post surgery. In the same study, gabapentin (30 mg/kg orally BID) was dosed as a positive control and significantly inhibited CCI induced mechanical allodynia on days 1, 6 and 8 of dosing period.

Compound **16** was also evaluated in the Complete Freund's Adjuvant (CFA) in vivo model of inflammatory pain. After single oral doses, it produced a dose-related reversal of the established CFA-induced hypersensitivity to pain, as measured by weight bearing. The ED_{50} for this effect was 0.91 mg/kg. A dose of 1 mg/kg po was associated with a mean free plasma concentration of 7.8 ng/mL at 1 h post-dose. A second experiment was conducted in the CFA-induced knee joint model of sub-chronic inflammation. At an oral dose of 1 mg/kg BID, **16** showed a reversal of hypersensitivity that was equipotent to the effect produced by the non-steroidal anti-inflammatory drug celecoxib over the 5-day dosing period.

In order to assess the centrally mediated side-effects such as sedation or ataxia, **16** was tested in a model of spontaneous locomotor activity (sLMA) of male rats. It was administered orally at doses of 60, 100 or 300 mg/kg, 30 minutes before the test session. No significant reduction in sLMA was observed with a dose of 60 mg/kg. The mean plasma concentration of **16**, 60 min post-dose was 13.1 μg/mL. Higher doses of 100 or 300 mg/kg of **16** significantly reduced locomotor activity, typical CNS side effects associated with Na_V channel blockers. We concluded that **16** has a substantial therapeutic index (TI) between plasma exposures associated with central side effects and exposures required to demonstrate efficacy in the in vivo pain models.

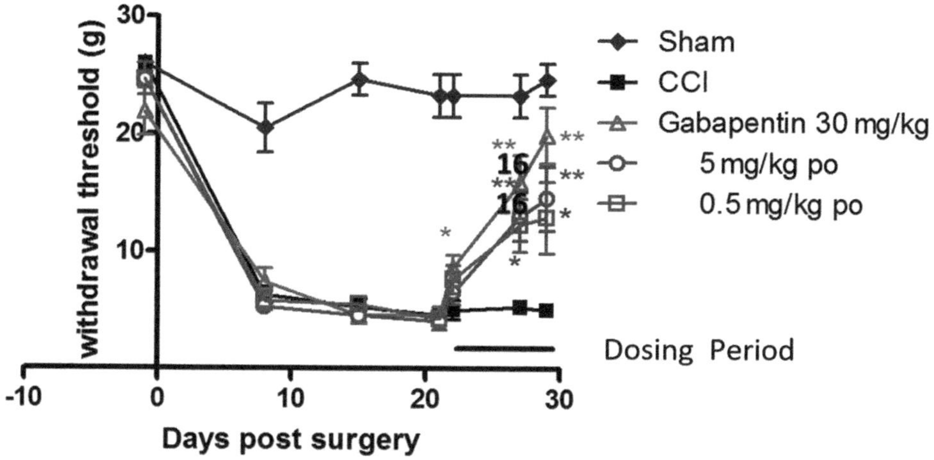

Figure 8. Compound 16 CCI Induced static allodynia (0.5 h post dose). Reproduced from ref (24). Copyright 2020 American Chemical Society.

The pharmacokinetics of **16** following a single intravenous and an oral dose were investigated in several species. Blood clearance of **16** was moderate in rat and dog (46% and 39% of liver blood flow, respectively) and low (11% of liver blood flow) in monkey. The volume of distribution at steady state (Vss) was in excess of total body water in all species, suggesting moderate to extensive tissue distribution and resulting in half lives of 0.8, 2.2 and 5 h in rat, dog and monkey, respectively.

The physicochemical properties of **16** lie within CNS drug-like space (MW 314.4 Da; Log P 1.6 and pKa 7.6). These properties confer excellent aqueous solubility (solubility of HCl salt:water 16.3 mg/mL; FeSSIF 6.9 mg/mL) and high membrane permeability (Papps = 386 nm/s at pH 7.4) in the MDCK cell assay. Compound **16** has been assigned to BCS Class I (high solubility, high permeability).

Compound **16** has a low potential for clinically relevant drug-drug interactions. Against a panel of CYP450 enzymes, it showed low levels of inhibition (pIC_{50}: 1A2, 13 μM; 2A6, >100 μM; 2B6, 31 μM; 2C8, >100 μM; 2C9, >100 μM; 2C19, 79 μM; 2D6, >100 μM; 3A4, >100 μM (vs atorvastatin, midazolam and nifedipine)). Furthermore, in the same panel of CYP450 enzymes, it displayed no time/mechanism dependent inhibition. Up to concentrations of 100 μM, **16** is not an inhibitor of the transporters, P-glycoprotein, OATP1B1, BCRP, OATP1B3, OAT1 or OCT2.

After successful first time-in-human enabling studies (genetic toxicology, safety pharmacology and 28-day repeat dose toxicology in rat and dog), **16** was progressed into clinical Phase I studies in healthy volunteers. A single ascending dose study demonstrated that **16** had a good tolerability profile and was well absorbed with a dose-related increase in exposure (Figure 9) (*30*). In a repeat dose study, steady state blood levels were achieved rapidly and **16** exhibited a 12 h half-life consistent with twice-daily dosing (*31*).

The HCl salt of compound **16**, now named vixotrigine (*32*), is a novel and use-dependent voltage-gated sodium channel blocker, that possesses broad spectrum in vivo anti-hyperanalgesic efficacy. It has successfully completed ICH mandated toxicology, genotoxicity and safety pharmacology studies and, subsequently, a suite of Phase I human volunteer studies. Vixotrigine is well tolerated in humans at predicted pharmacologically active exposures and shows rapid absorption and high oral bioavailability, and a half-life sufficient to support BID or TID dosing in the clinic. A Phase II study has been conducted showing efficacy in reducing the intensity and frequency of paroxysms of pain associated with trigeminal neuralgia (*33*).

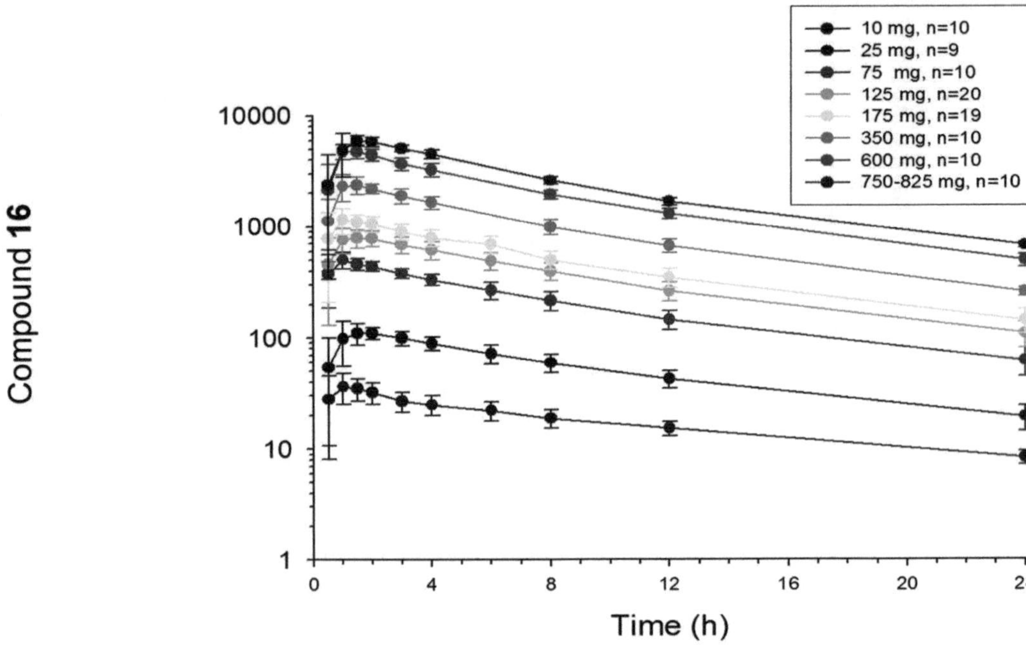

*Figure 9. Plasma concentration-time profile for **16**. Reproduced from ref (24). Copyright 2020 American Chemical Society.*

Medicinal Chemistry Synthetic Route

The initial medicinal chemistry route to vixotrigine (Scheme 1) was designed to permit access to a range of analogs and isomers (*34*). Chiral Boc protected (*S*)-pyroglutamate (*35, 36*) **26** was treated with the organolithium reagent formed from 4-benzyloxy-1-bromobenzene **27** at low temperature to furnish the ketone adduct **28**. Deprotection under acidic conditions formed the cyclic 2H-pyrrolidine **29**. Stereoselective reduction of the double bond over a platinum catalyst installed predominantly the 5(*R*)-stereogenic center in **30** but the epimer **31** could be separated chromatographically. Hydrolysis of the ester and re-protection of the nitrogen with Boc anhydride gave acid **32** which could be amidated using hexamethyl disilazane as a source of ammonia to provide **33**. The terminal benzyl group was removed by a second hydrogenation, this time over palladium, to give a versatile phenolic intermediate **34**. Alkylation of this material with 2-fluorobenzyl bromide **35** furnished **36**, which upon Boc removal with HCl generated in situ afforded the crystalline hydrochloride salt of carboxamide **16**, vixotrigine. Purifications for all intermediates were achieved using silica gel chromatography.

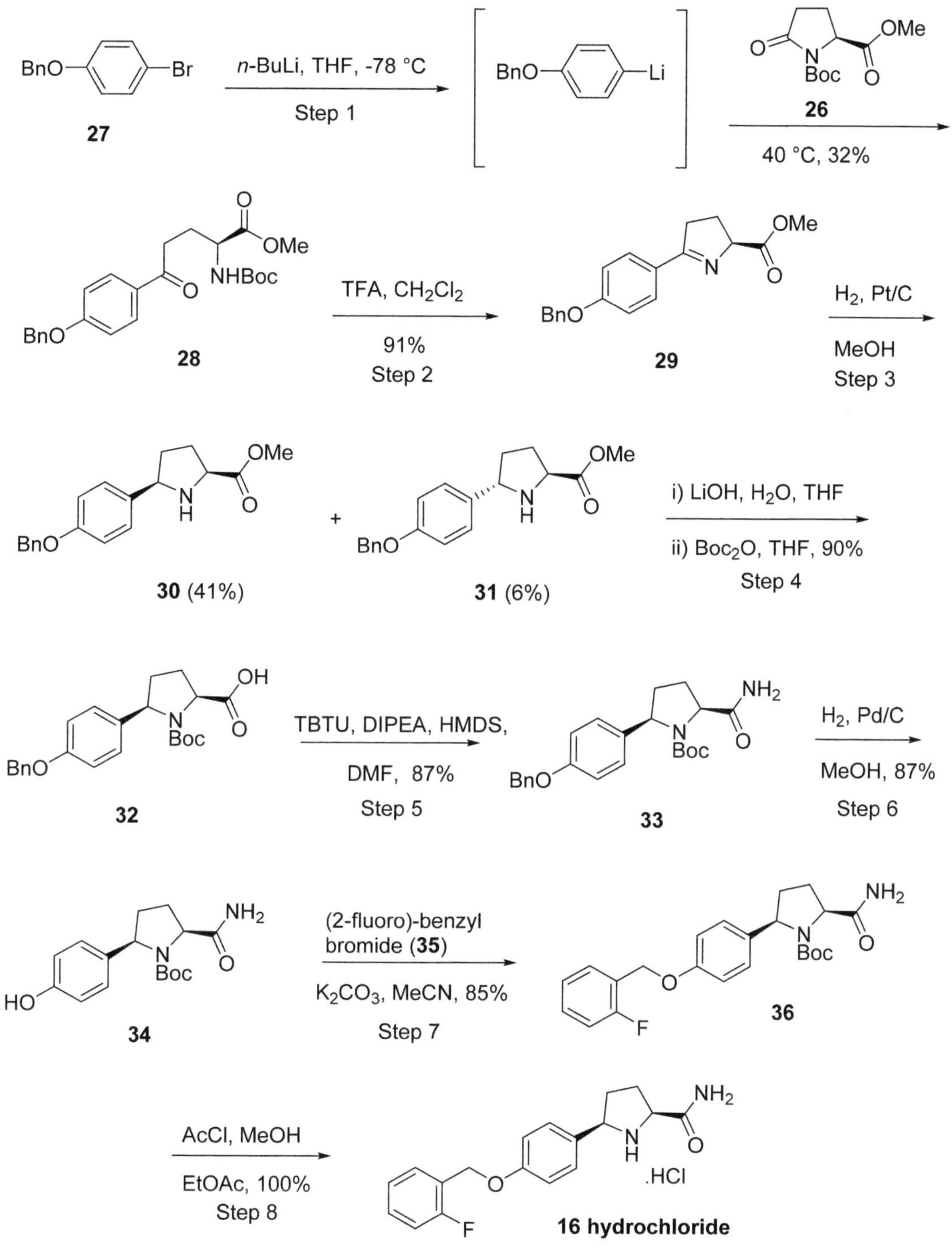

Scheme 1. Enantioselective synthesis of vixotrigine.

Evaluation of Scale Up Routes

Route A

A shorter and more scalable route to vixotrigine **16** was subsequently developed in which the chromatographic operations were removed. This synthetic route, designated **Route A,** was employed to furnish drug substance for preclinical studies and early Phase I clinical studies. It is a substantially shorter and more efficient synthesis than the discovery route. Scheme 2 shows **Route A** as adapted for early production. (2-Fluorobenzyoxy)-4-bromobenzene **37** was obtained through a Williamson ether synthesis from commercial precursors 2-fluorobenzylbromide **35** and 4-bromophenol **38**. This was transformed to Grignard reagent **37b** which was subsequently added to pyroglutamate **26**. Acid hydrolysis of the resulting Boc protected ketone **39** gave the cyclic imine **40** which was stereoselectively hydrogenated to ester **41**. Aminolysis of this ester to **16**, which was converted to the hydrochloride salt, provided vixotrigine (*37, 38*). Variants of the route were used to prepare more than 15 kg of API, however, several changes to the process were necessary as it was scaled up.

Route A Step 1 – Ether Synthesis

The key issue in the ether formation from commercially available 2-fluorobenzyl bromide (**35**) and 4-bromophenol (**38**)**,** was the purity of **35**. Commercial supplies of **35** contained no more than 0.1% (*39*) of the 4-fluoro isomer, but up to 4% of the 3-fluoro isomer. Unfortunately, fate and purge studies demonstrated poor purging of the isomeric 3-fluoro-impurities in the downstream process. However, use tests demonstrated that a 3-fluoroisomer level of ≤ 0.32% gave acceptable impurity levels and batches of **35** containing greater than this level were rejected.

A simple precipitation method was developed for the isolation of **37** to reduce the Step 1 cycle time. Thus, the filtered Step 1 reaction mixture was concentrated and the product precipitated by addition of 2% aqueous NaOH. Filtration of the precipitated solid followed by water washing afforded **37** in high yield and purity. With these improvements, we carried out pilot plant scale batches to furnish up to 58 kg of **37** in 91% yield as a white solid containing less than 0.2% of the 3-fluoro isomer and no more than 0.3% of the other impurities.

Route A Step 2 – Grignard Addition

A chemoselective addition of the Grignard reagent derived from **37** to the commercially available chiral pyroglutamate **26** provided the ketone **39** establishing the (2*S*) stereogenic center of vixotrigine. Control of this cryogenic reaction (-50 to -60 °C), and of process related impurities, became a major challenge on scale-up. The process was demonstrated at 30-75 kg scale, but several deficiencies were identified, which resulted in significant cost and/or quality issues.

The major issues relate to the formation of substantial levels of impurities **42** and **43**, which are likely formed via a benzyne intermediate (*40*), as shown in Scheme 3. A biaryl impurity **44**, is also formed by the Wurtz homo-coupling of the Grignard reagent (*41*).

A detailed analysis of the deficiencies in **Route A** was undertaken to see if they could be overcome with minor modifications, in parallel to the design of entirely different routes (**Routes B** and **C**). Compounds **42** and **43** are very difficult to purge at Step 2 and are carried through the route to related impurities in the final API. Substantial effort was therefore invested to improve the profile and selectivity of the Grignard reaction. A range of alternative ethereal solvents was evaluated (2-

methyl THF, cyclopentyl methyl ether and a toluene-THF mixture), but THF remained the solvent of choice. Addition of dibromoethane was required to sustain Grignard formation.

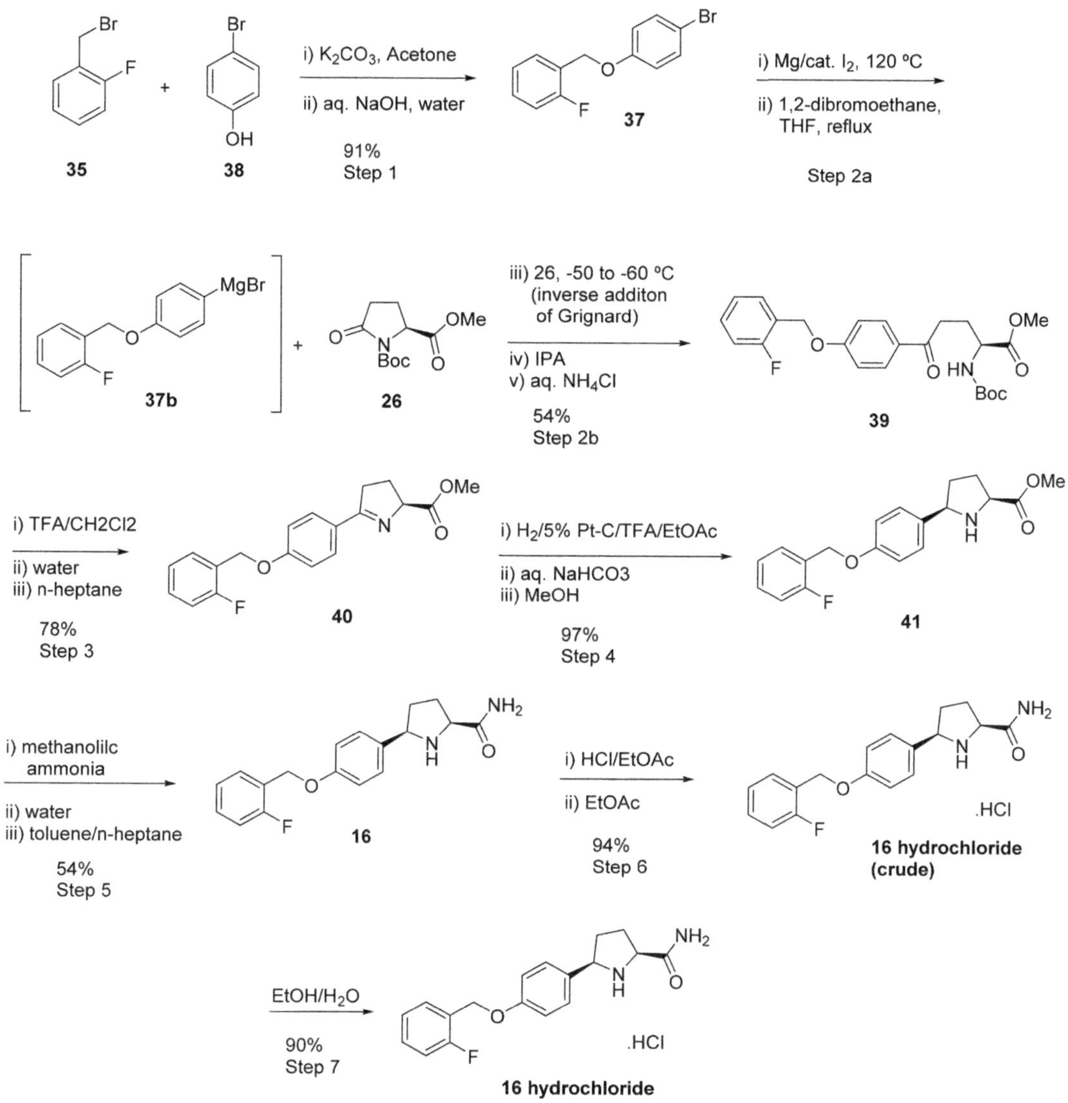

Scheme 2. Route A: The first scalable synthesis of ***16***.

Addition of the Grignard reagent **37b** to a -55 °C solution of **26** was preferred to the reverse addition of **26** to the similarly cooled Grignard reagent. Reverse addition of **26** to the Grignard reagent **37b** led to a considerable increase in the level of the homo-coupled impurity **44**. Alternative methods to generate a Grignard or a Grignard-related reactant were tried including Knochel's method of catalysis with lithium salts via *i*PrMgCl/LiCl exchange (*42*); formation of the arylzinc reagent (*43*); use of the iodoaryl substrate; formation of the organomagnesium "ate complex (*44, 45, 46*);" and a range of magnesium activating methods such as DIBAL (*47*) and TMSCl (*48*). All these methods either failed to promote the formation of the Grignard reagent or resulted in low step yields and/or inferior impurity profiles.

The isolation procedure for ketone **39** was also evaluated and extensive screening of solvents for precipitation and isolation resulted in identification of a ternary mixture of ethyl acetate:IPA:*n*-

hexane 2:2:6. This resulted in the most effective purge of the major benzyne-derived impurity **42** to below 4%, as compared with the previous conditions which resulted in levels of **42** at 5%. The pilot plant scale batches were run implementing these improvements providing up to 28 kg of ketone **39** in 54% yield with the level of **42** below 3%. While we established modest yield improvements, the reaction still generated significant levels of benzyne-derived reaction impurities.

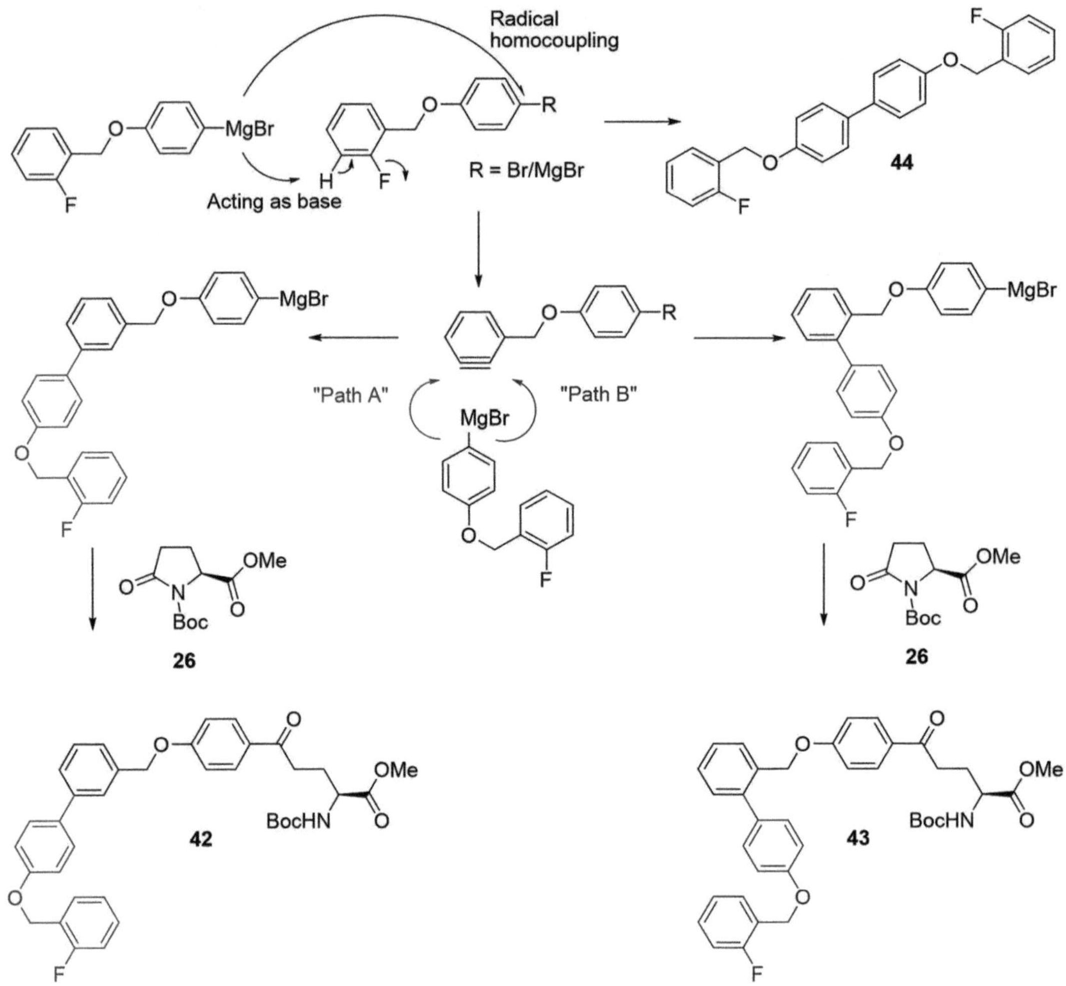

Scheme 3. Putative mechanism of formation of homo-coupled and benzyne-derived impurities.

Route A Step 3 – Cyclization to Imine

The Step 3 acid-catalysed removal of the Boc protecting group with trifluoroacetic acid (TFA) in dichloromethane (DCM) also resulted in cyclization to the pyrrolidine-imine **40**. After completion of the reaction and neutralization, the DCM layer was concentrated. In the first bulk campaign *n*-hexane was added and the precipitated solid product was isolated in 82–85% yield. Process development of this step focused on evaluating the optimal control of pH during workup to avoid emulsion formation, and replacement of *n*-hexane as an antisolvent with a safer alternative. Neutralization with aqueous $KHCO_3$ to pH 6.0–7.0 eliminated the emulsion and maintained the chemical yield above 80%, while *n*-heptane successfully replaced *n*-hexane as an antisolvent.

These changes were implemented on a pilot plant scale for two batches using 25–30 kg inputs of ketone **39**, and furnished pyrrolidine-imine **40** in 78% and 88% yields as off-white solids. Levels of the cyclized versions of impurities **42** and **43** were only modestly reduced by the Step 3 crystallization process. For example, impurity **42** in the stage 2 product was present at 2.7% and its cyclised derivative was present at levels of 2.5% in the stage 3 product. Remaining issues included the prolonged neutralization time (up to 19 h) and the large volumes of TFA and DCM, which were undesirable from a safety and environmental perspective.

Route A Step 4 – Reduction of Imine

A detailed evaluation of the hydrogenation was conducted with the objective to increase its chemo- and stereoselectivity. The reduction of the imine was prone to high and variable levels (5–10%) of the (5*S*) *trans*-diastereomer **45**. The sub-optimal chemoselectivity in this reaction also resulted in formation of impurity **46**, resulting from the ring-opening over-reduction (Figure 10). A range of different catalysts and loadings, solvents and reaction conditions were evaluated. Replacing platinum catalyst with Pd/C resulted in much higher levels of the hydrogenolysis impurities **46** and **47**, while Raney nickel led to an unacceptably slow reaction, with 72% of unreacted pyrrolidine imine substrate **40** remaining after 18 h. A selection of different grades of 5% Pt/C was screened under standard conditions with only minor impact on yield and the level of *trans*-isomer **45**.

The effect of hydrogen pressure was evaluated with no significant difference in impurity profile between 15 and 60 psig, but higher pressures gave faster reaction times. There was a tendency for levels of hydrogenolysis products to increase at 80 psig, and, specifically, an increased Pt/C catalyst loading of 10% increased levels of the ring opened hydrogenolysis product **47**. A solvent screen did not reveal a solvent that improved this transformation.

Thus Step 4 remained sub-optimal, and hydrogenation of pyrrolidine-imine **40** to introduce the (5*R*) stereogenic center was carried out using 5% Pt/C (5% w/w) under 50-60 psi hydrogen pressure in ethyl acetate and TFA (1.0 equiv). After workup, the pyrrolidine **41** was isolated in consistently high yields (95 - 97%) as an oil, but the stereoselectivity of the reaction remained difficult to control. The content of the unwanted (5*S*) *anti*-diastereomer **45** varied from approximately 6% to 10% with some tendency for further erosion of stereoselectivity as scale increased. In addition, varying levels of hydrogenolysis of the 2-fluorobenzyl ether (leading to phenol **46**) and/or cleavage of the benzylic C-N bond (leading to primary amine **47**) were produced (see Figure 10).

On pilot plant scale, the reaction and workup conditions developed yielded **41** in 97 - 100% chemical yield for inputs of **40** of 8.3 kg. Levels of *trans*-isomer **45** were between 7 and 10%, and cyclised pyrrolidine-containing species formed from benzyne-derived impurities **42** and **43** persisted at levels of 1.8–2.6%.

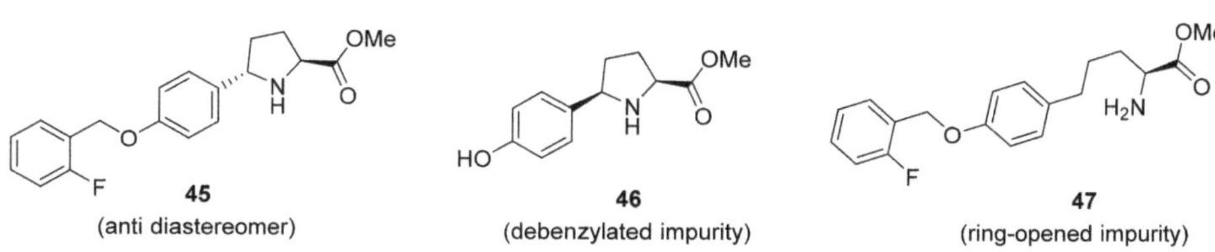

45 (anti diastereomer) **46** (debenzylated impurity) **47** (ring-opened impurity)

Figure 10. Impurities in Route A Step 4 process.

Route A Step 5 – Ester Amidiation

In Step 5, the ester **41** in methanol was treated with 7 M methanolic ammonia solution at 3–4 kg/cm^2 ammonia pressure for a period of 10-12 h. The original workup involved a prolonged cycle time of about 50 h, comprising multiple washings and distillations, a 16 h crystallization period and prolonged drying to remove toluene. Yields of crystallized free base **16** were variable, in the range of 50-65%. Laboratory optimization of this step focussed on simplification of the post reaction processing and development of an improved crystallization procedure. Resulting changes included the use of toluene rather than ethyl acetate to remove methanol via azeotropic distillation, elimination of brine washes, and replacement of toluene washes of the filter cake with *n*-heptane to improve drying and reduce material loss. Through this optimization, the overall process cycle time was reduced to 32 h, and chemical yields were consistent in the range of 60 - 63% in the laboratory. Levels of the unwanted *trans*-isomer **14** in **16** were reduced from 9% to 2% by crystallization. On a pilot plant scale, **16** was isolated in more variable yields (54 to 63%) as a light brown solid having a *trans*-isomer **14** content of between 0.8 and 3.8%, and benzyne-derived impurity levels between 0.2% and 0.8%.

Route A Steps 6/7 – Salt Formation and Crystallization

For Step 6 salt formation, a filtered ethyl acetate solution of free base **16** was treated with ethyl acetate/HCl, and slurried at 0-5° C to furnish vixotrigine (**16·HCl**). Kilo lab batches produced vixotrigine of ~97% purity in 94–96% yield. The recrystallization of **16·HCl** in Step 7 was performed by suspending the crude solid in 5% aqueous ethanol, heating to 65–72°C to effect dissolution, seeding at 50–52° C, and then aging at 5--10° C before isolation by filtration. The resulting vixotrigine API was typically isolated in 90–95% yield, with 99% purity. The final product was colored (yellow/brown), and attempts to reduce color with charcoal treatment (various grades), or with alternative solvents were all unsuccessful. This inability to control the color of API manufactured using **Route A** necessitated the use of color masking in the drug product manufacture, with yellow dye coatings applied to both active and placebo tablets.

Route A Summary

Using the optimized **Route A** process, scale up and production provided batches of API between 1.5 kg and 15.2 kg with typical purity levels of 98.5–99.0%. API was typically pale yellow to light brown in color. Assays were developed for **13**, the enantiomer of **16** (typical level in API <0.05%), **48**, the acid degradant of vixotrigine (typical level <0.05–0.15%), the *trans*-isomer **14** (typical level 0.17–0.69%), and both **50** (typical levels 0.14–0.41%) and **49** (typical levels <0.05%) from downstream intermediates of the benzyne-derived impurities (Figure 11).

Therefore, **Route A** performed adequately to provide API for preclinical toxicology studies and early clinical studies. However, this route had several deficiencies that precluded its use for later stage development and commercialization. The yield of the Grignard addition reaction (Step 2) was poor and produced 'benzyne' reaction derived impurities that were difficult to purge in the downstream process. The stereoselectivity of the catalytic hydrogenation to reduce the imine **39** to introduce the (5*R*) stereocenter was inadequate. Moreover, levels of impurities arising from hydrogenolysis side reactions (forming **46** and **47**) were difficult to control. The overall yield was affected by the crystallization in Step 5 which was required to ultimately reduce levels of the trans-isomer in drug substance to acceptable levels. Impurities **42** and **43**, formed in Step 2, were transformed, and the downstream impurities were carried through to the final API. Purging these impurities was challenging, leading to a high burden for purification in the latter steps of the process that was operationally cumbersome. The appearance of the final, isolated API was also variable with respect to color.

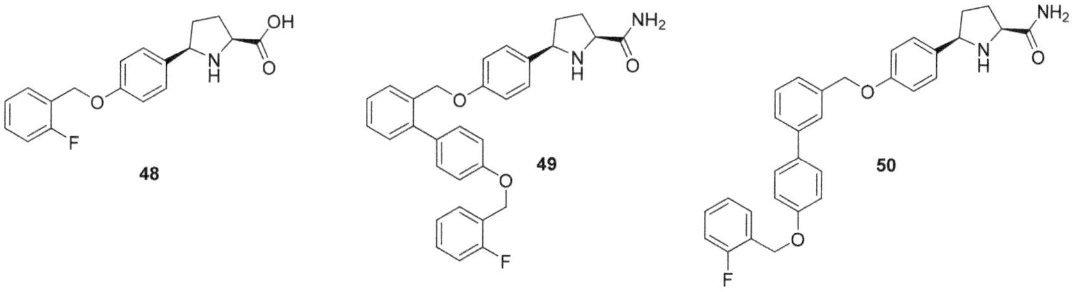

Figure 11. Key impurities in Route A vixotrigine.

These deficiencies led us to investigate two alternative routes that avoided the use of the troublesome Grignard reaction used in **Route A**. Both routes targeted the same imine **40** as **Route A** as this was the most suitable intermediate for introduction of the desired 5-(*R*) stereochemistry via stereoselective hydrogenation. It was recognized that further work to improve the stereoselectivity of the imine reduction would be required (see **Route D** for this discussion).

Route B

The key step in this route (Scheme 4) is an asymmetric Michael addition of the glycine benzophenone imine **57** to the Michael acceptor **56** (*49, 50*). It starts with the etherification of phenol **51** with 2-fluorobenzyl bromide **35** to produce the 2-fluorobenzyloxy benzoate ester **52**. Saponification of ester **52** gave carboxylic acid **53** which in turn was readily converted to the Weinreb amide **55** via intermediate acyl chloride **54**. Addition of vinyl magnesium bromide to **55** provided the Michael acceptor **56** in high yield. An asymmetric copper catalyzed Michael addition of the glycine derivative **57** to enone **56** provided adduct **58**. Several ligands were screened for the asymmetric Michael addition and the most suitable in terms of performance and availability was the FOXAP ligand shown in Scheme 4. Acidic workup resulted in the hydrolysis of the benzophenone imine intermediate **58** and subsequent condensation to form the cyclic imine **40** in high yield and good enantioselectivity. The product **40** was obtained as a mixture of the (*S*) and (*R*) enantiomers in 97.7:2.3 ratio when the reaction was carried out at -78 °C. However, the enantioselectivity diminished as the reaction temperature increased (-30 °C furnished 95.4 (*S*):4.6 (*R*); -20 °C gave 94.1 (*S*):5.9 (*R*)).

While limited optimization was carried out, **Route B** was ultimately dropped. This was primarily due to cost of purchasing or preparing the FOXAP ligand (*51*), and the challenges associated with synthesizing enone **56** at scale. The Weinreb amide precursor represented a significant cost and involved a subsequent cryogenic reaction, while alternative chemistries to introduce the α,β-unsaturated ketone appeared unattractive for scale-up.

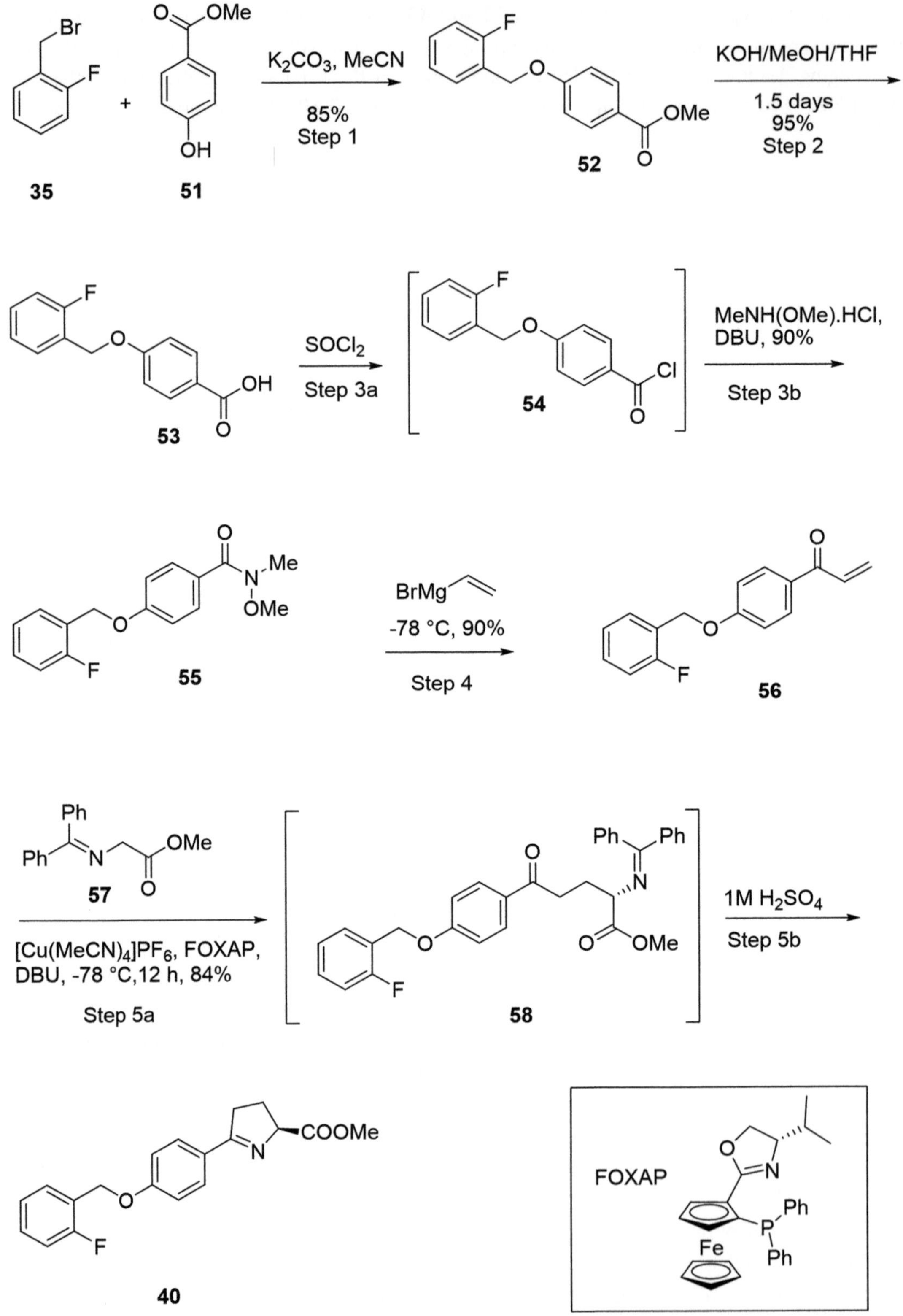

*Scheme 4. Route B to imine intermediate **40**.*

Route C

A route which entirely avoided cryogenic steps was investigated and is shown in Scheme 5. This approach employs the *N*-Boc-methyl ester of (*S*)-propargylglycine **61** as the source of chirality (*4, 52*). **Route C** has two catalytic steps in which the catalyst cost is critical to the economics: a Sonogashira coupling (Step 2) and a silver triflate catalyzed cyclization reaction (Step 4), after which this route converges on the imine **40** common to **Route A** and **Route B**.

Alkylation of 4-iodophenol **59** with 2-fluorobenzyl bromide **35** provides the iodide partner **60** for a Sonogashira reaction with **61**. A wide variety of conditions has been reported for the Sonogashira reaction (*53*). A brief survey indicated that use of an excess of iodide **60** (1.2 equiv) and slow addition of **61** to the reaction mixture were required to give good yields of **62** and to minimise Glaser dimerization of the alkyne component (*54, 55*). Removal of the Boc protecting group followed by silver triflate catalysed cyclization provided imine **40** in high yield.

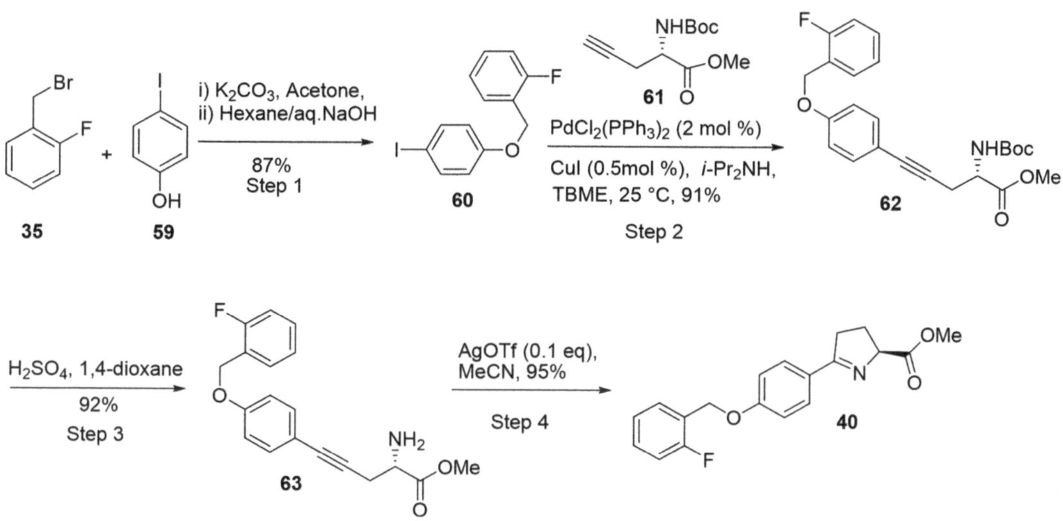

Scheme 5. Route C to Imine 40.

Although we carried out limited optimization of **Route C**, the requirement for two metal-catalyzed transformations and the high cost of (*S*)-propargylglycine starting material discouraged us against continuing with this approach. Therefore, we switched our focus to the development of **Route D** to provide material for clinical studies.

Route D. Development of the Improved Second Pilot Plant Process

In addition to the deficiencies of **Route A** already described, the starting material **35** is expensive. Considering that the overall yield of **Route A** is only 18 %, **35** contributed about 30% to the cost of **Route A** raw material. The pursuit of an improved process to supply vixotrigine API therefore considered the introduction of the 2-fluorobenzyl group later in the process, necessitating the selection of an appropriate phenol protecting group for the Grignard step. This modification would remove the opportunity for the difficult-to-purge benzyne impurities observed in Route A. Following a laboratory evaluation, *O*-benzyl protection was selected, with the route depicted in Scheme 6 chosen for development and scale-up.

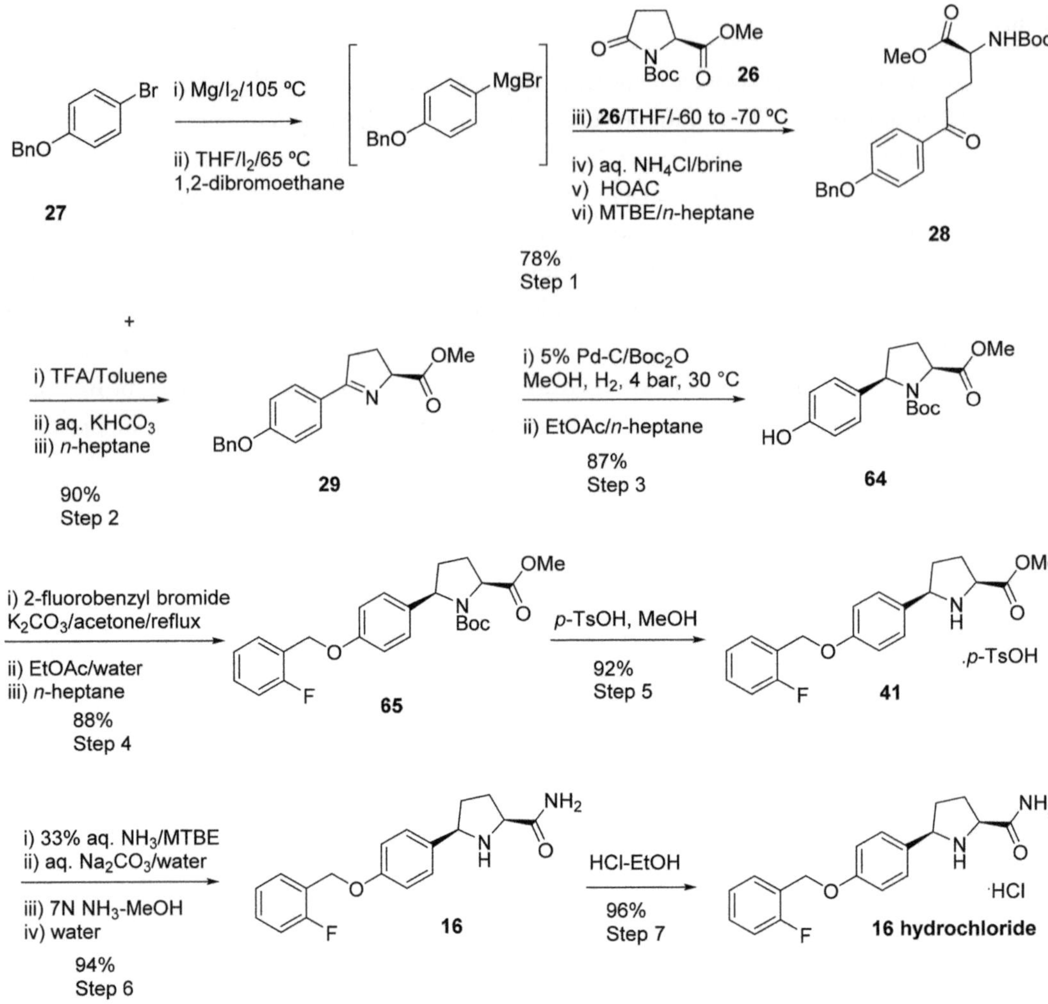

Scheme 6. Route D manufacturing process for vixotrigine.

Route D Step 1 – Grignard Addition

High quality commercial 4-benzyloxy bromobenzene (**27**) is available with less than 0.15% of either the 2-Br or 3-Br isomer. Thus, the first step in the **Route D** process became the addition of 4-benzyloxyphenyl magnesium bromide to *N*-Boc-methyl-(*S*)-pyroglutamate (**26**). The conditions used for the Grignard reagent formation and reaction in **Route A** were retained in the first **Route D** pilot plant campaign. The product crystallization solvent was changed to MTBE/*n*-heptane. Two batches provided 24.6 kg of the pyroglutamate adduct **28**, with both an improved yield (78% vs 54%) and HPLC purity (98.7% vs 95.5%) relative to the analogous step in the **Route A** process. In the absence of benzyne-derived side-products, the major impurity in **28** was the homo-coupled byproduct **66** at about 0.7%, with no other impurities above 0.2%. The level of **66** was reduced during the isolation of **29** in Step 2, and the downstream impurity **67** was readily purged in the Step 3 isolation.

For the second campaign, a 75% excess of 4-benzyloxy-bromobenzene was required for the complete consumption of the pyroglutamate **26** in Step 1. The reaction was scaled up over four-fold to 37 kg per batch, with either 5 or 7 batches combined for product isolation. The additional benzyl ether side-products **68**, from Grignard oxidation (56), and **69**, from unreacted Grignard quenching, were identified as low-level impurities that were purged in subsequent steps (Figure 12). The product quality was maintained across these campaigns, with 900 kg of **28** produced in ~73% yield for campaigns two, three and four.

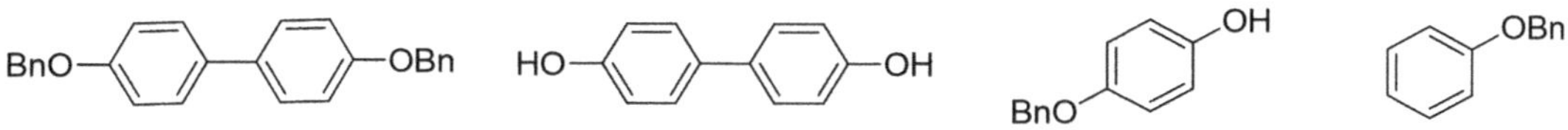

66 **67** **68** **69**

Figure 12. Impurities Derived from Route D Step 1.

Route D Step 2 – Cyclization to Imine

The cleavage of the Boc protecting group of pyroglutamate adduct **28** effects an in situ cyclization to the pyrrolidine-imine **29**. A comparison with the equivalent **Route A** step processing showed that in **Route D**, both the excess of TFA and the volume of DCM were reduced (from 11 to 8.9 equivalents and from 10 to 6.5 mL/g, respectively). In the pilot plant this step scaled smoothly, with inputs of **28** up to 21 kg producing the crystalline imine **29** in ~90% yield. The more environmentally acceptable solvent toluene, also at 6.5 mL/g, was used rather than DCM for the remaining five campaigns in comparable quality and yield.

This revised process scaled smoothly, with inputs of **28** from 117 kg up to 562 kg. Apart from the homo-coupled impurity **66** (levels 0.1 - 0.5%) which is purged at Step 3, the major impurity was the acid **70** (present at levels at 0.30 - 0.76%), formed by hydrolysis during the workup. The fate of acid **70** in the process was studied, and the acid analog of the API is a specified impurity. As the imine **29** might be prone to epimerization, a chiral assay was developed, but the enantiomer **71** (see Figure 13) was consistently controlled to levels below 0.1%.

70 **71**

Figure 13. Impurities in Route D Step 2 Process.

Route D Step 3 – Reduction, Boc Protection and Debenzylation

The key third step of the process was developed to sequentially accomplish three distinct transformations in the same reaction vessel (Scheme 7) (57). Firstly to reduce the imine to generate the *cis*-pyrrolidine intermediate; secondly to Boc protect to give **72**; and thirdly hydrogenolysis to cleave the *O*-benzyl group to provide **64** at such a rate that debenzylation occurs after imine reduction and Boc protection. Remarkably, these transformations were successfully achieved to provide ester **64** in high yield and stereoselectivity.

The aryl C-N bond of the pyrrolidine intermediate is susceptible to hydrogenolysis, leading to a ring-opened impurity **73**, but Boc protection of the nitrogen atom inhibits benzylic C-N bond cleavage. Careful choice of the grade of Pd/C catalyst ensured that ring opening prior to Boc addition was minimized, while achieving a good *cis/trans*-selectivity in the imine reduction. The *cis/trans*-selectivity of the **Route A** Pt/C-catalyzed imine reduction reaction was only about 7:1 (Scheme 1), but with 5% Pd/C type 487 paste a *cis/trans*-selectivity of greater than 30:1 was achieved in methanol in the **Route D** process, minimizing the level of *trans*-isomer **74** formed. A fate and purge study showed that a 0.4% level of **72** in the step product **64** taken through the rest of the process as an impurity, accounted for 0.08% of analog **25**, lacking the benzylic fluorine, present in the API.

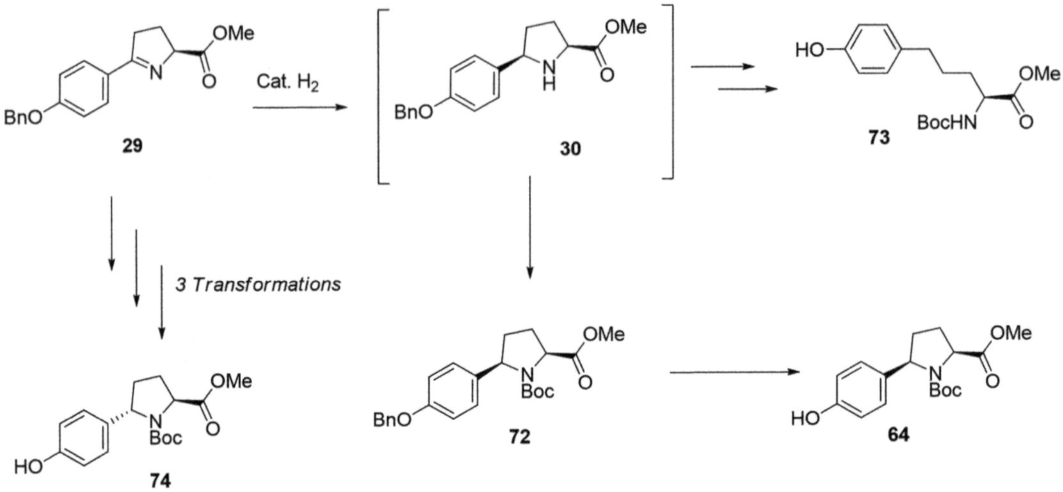

*Scheme 7. Intermediates and by-products in the conversion of **29** to **64** in Step 3.*

A range of crystallization solvents was examined to optimize the purging of impurities. An ethyl acetate/*n*-heptane solvent system achieved the best results, and enabled efficient purging of impurities due to the highly crystalline character of **64**. Six pilot plant batches contained less than 0.05% each of **72**, **73** and **74**, and no detectable biphenyl **67** arising from the Step 1 process impurity dibenzyl **66** occurred. Phenol **64** was formed in yields between 82.3% and 90.6% with an input between 13.7 and 26 kg of imine **29**. The high crystallinity and particle characteristics of **64** isolated from its crystallization resulted in quick filtration and drying of the solid, establishing it as a key intermediate and point of purity control for this and future routes.

Route D Step 4 – Fluorobenzyl Ether Formation

Alkylation of the phenol **64** with 2-fluorobenzyl bromide (**35**) in Step 4 was performed in acetone with potassium carbonate as base. The levels of the regioisomers of 2-fluorobenzyl bromide were closely monitored, due to the placement of this reaction later in the process for **Route D**, as compared with **Route A**. Trials found that using **35** spiked with 0.4% each of 3-fluorobenzyl bromide and 4-fluorobenzyl bromide led to the coupled Step 4 product **65** in which the levels of corresponding 3- and 4-fluorobenzyl regioisomers had been reduced two to four-fold.

The six **Route D** manufacturing campaigns produced over 1000 kg of **65**, with a yield range of 85–89%. The only impurities present in crystalline **65** at levels above 0.1% in these campaigns were phenol **64** at up to 0.6% and the 3-fluorobenzyl analog at less than 0.2%. In the first campaign we discovered that unreacted **64** did not purge well in the crystallization of **65**, but the 0.6% level seen in one batch was shown to be efficiently purged in downstream processing.

Route D Step 5 – Deprotection

The first campaign for the cleavage of the Boc protecting group of **65** in Step 5 of **Route D** process was carried out using 85% aqueous phosphoric acid in MTBE at 15-20 °C (*58*). However, the process suffered from several drawbacks including using relatively large workup solvent volumes, generating significant amounts of inorganic salts from the neutralization of the large excess of phosphoric acid, and producing the free base **41** as an oil by concentrating to dryness. In addition, there was poor plant volume yield. A new process was therefore sought that enabled the formation of the Step 5 intermediate **41** as a filterable solid salt in a "direct drop process." Studies of the best acid and solvent combination resulted in the selection and use of *p*-toluenesulfonic acid (*p*-TsOH) in methanol. Yields of between 88.6 and 96.1 % were achieved with inputs going from 14 to 328 kg and levels of the acid **48** kept below 0.3%

The procedure used for the first **Route D** campaign provided **41** as the free base, which was isolated as an oil with 99.15% purity, with unreacted **65** as the main impurity at 0.58%. For campaigns 2 to 6, **41** was isolated as its *p*-TsOH salt in ~99.7% purity, with acid **48** present at about 0.2%, and the 3-F regioisomer present at a level below 0.05%.

Route D Step 6 – Ester Amidation

Isolation of the Step 5 product **41** as the *p*-TsOH salt led to advantages in handling and impurity control. The extra operations in Step 6, including the salt break required additional studies to demonstrate the extent of *p*-TsOH ester formation and their purge in subsequent process steps. The developed Step 6 amidation process also employed a large excess of ammonia, and was one of the steps that subsequently required attention to improve the volume productivity and green metrics of the vixotrigine manufacturing route.

The amino ester **41** was converted to amino amide **16** using a 7N solution of ammonia in methanol. For the first campaign using the free base of **41** (an oil) as the input, the product was isolated by a replacement solvent swap with toluene, and the addition of *n*-heptane antisolvent to crystallize **16**. In the subsequent campaigns, the *p*-TsOH salt of **41** was converted to its free base in MTBE, which was replaced by methanol for the amidation process, and water was added to crystallize the amino amide **16**. In six campaigns with inputs ranging from 10.3 to 362 kg, all batches of product **16** were formed with at least 99.7% HPLC area purity. Acid **48** and the *trans*-amide **14** were the only impurities that exceeded the 0.10% level in any of these batches and yields for the transformation ranged between 88 and 93%.

Route D Step 7 – Salt Formation

For the first three campaigns for Step 7 using **Route D**, conversion of free base **16** to the HCl salt was carried out by addition of 1.25N ethanolic HCl to an ethanol solution of the free base at 25–35 °C. This process provided excellent yield and quality, but the initially formed ethanol solvate of the HCl salt required extended drying time to break the solvate and achieve the required residual solvent content. In addition, the mean particle size distribution (PSD) was lower than preferred at that time for the formulation process, mainly due to the presence of fine particles.

An alternative process was therefore developed for campaigns 4 and 5 in which the solution of free base **16** in IPA was treated with aqueous HCl at 60–65 °C then cooled to 0–5 °C before collection of the salt. This second process gave a product that dried more rapidly, with higher bulk density and a d_{50} measurement in the required range of 50—100 μm. Following a reassessment of the PSD requirement to smaller particle size, campaign 6 reverted to the process run in ethanol solvent, giving four batches with d_{50} measurements in the 15 to 20 μm range. The drying times for these batches were significantly reduced by a steeper temperature ramp and more intensive agitation in a Nutsche filter drier.

Route D Summary

The six **Route D** campaigns produced over 660 kg of **16·HCl** with a Step 7 yield range of 91%–95%. The purities of all batches of API, were at least 99.8% by HPLC with the acid impurity **48** reduced in the salt formation below 0.07%, all unknown impurities less than 0.10%, and no detectable levels of the *trans*-amide **14** or of the enantiomer **13**. In addition, material prepared by **Route D** was white, obviating the need for color masking required for API produced using **Route A** in the manufacture of drug product.

Late-Stage Chemical Development

While **Route D** was adequate to supply earlier phases of development, a more robust process was required to provide the quantities of vixotrigine needed to support Phase III studies and late-stage manufacturing (*59*). Further improvements were deemed necessary to address commercial process considerations including reductions in cost of goods, PMI (process mass intensity measures total waste including water relative to the product) (*60*), cycle time, and designing the synthesis for a projected high-volume demand for vixotrigine.

The first step of the synthesis was deemed unsuitable for the long-term needs of commercial manufacturing given the necessity for cryogenic (-70 °C) reaction conditions, an issue for metric ton scale manufacturing based on capacity. As such, an engineering solution was chosen to address these challenges, and we decided to pursue development of a flow process.

An additional area deemed critical for optimization was increasing the selectivity and throughput of Step 3, a selective one-pot catalytic imine reduction/benzyl group hydrogenolysis/Boc protection of the pyrrolidine formed, to afford intermediate **64**. Telescoping the remaining alkylation and amide bond formation steps to vixotrigine was also considered advantageous for reducing cycle time and hence the cost of goods. The powder flow properties of the solid API were found to be a critical quality attribute for the formulation process. For this reason, wet milling was chosen for development to enable the manufacture of batches of vixotrigine larger than 100 kg with the desired physical properties.

The calculated PMIs for the final **Route D** methodology are shown for each step in Table 2.

The original process for **Route D** Step 1 to generate **28** had several issues including volume inefficiency (PMI=32) and moderate yield. The batch process was therefore optimized to provide adequate supply of lower volume needs for clinical development.

Table 2. Step Efficiencies of Route D

Step Yield	*Reaction*	*PMI*
Step 1 78%	**27** + **26** → **28**	32
Step 2 90%	**28** → **29**	35
Step 3 87%	**29** → **64**	20

Table 2. (Continued). Step Efficiencies of Route D

Step Yield	*Reaction*	*PMI*
Step 4 88%	**64 → 65**	39
Step 5 92%	**65 → 41**	16
Step 6 94%	**41 → 16**	51
Step 7 96%	**16 → 16·HCl API**	20

Late Process Step 1 – Batch Processing Optimization

The original process used to prepare 37 kg batches of **28** commenced with the heating of dry magnesium with iodine. The Grignard reagent was then formed by reaction of the magnesium after activation with iodine and 1,2-dibromoethane with aryl bromide **27.** The Grignard addition was performed in THF at -60 °C. In this process 1.75 equiv of **27** relative to pyroglutamate **26** were needed to effectively consume **26**. The reaction was quenched directly with IPA at -60 °C. Following the IPA quench, aqueous ammonium chloride/sodium chloride was added, followed by additional water and 50% acetic acid in water to adjust the pH to 6-7. After extraction and a solvent swap from THF to MTBE, the product was crystallized from MTBE/*n*-heptane to provide pyroglutamate adduct **28** in 78% yield, with 98.7% purity and PMI=32. The major process impurity, at 0.7%, was the Wurtz homo-coupled biphenyl impurity **66** (Figure 12).

The initial round of process optimization revealed that the excess aryl bromide **27** could be reduced from 1.75 equiv to 1.21 equiv. The addition of aqueous ammonium chloride/sodium chloride was also removed. The quench was modified to IPA/water and 50% aqueous acetic acid. Extensive work was undertaken to understand the mass-balance of the reaction. The traditional Grignard impurities associated with oxygen and moisture exposure were noted. At least four identified impurities were formed in this step and it was important to control their formation. During production, the known THF-opened-adduct **76** was observed at higher than expected levels in **28** and forward processing of this adduct lead to a cyclized downstream impurity 77, (Scheme 8). More stringent control of oxygen was necessary to avoid air introduction in both the Grignard formation phase and cooling phase, prior to the addition of **26**, so reactor and vent line check valves were installed.

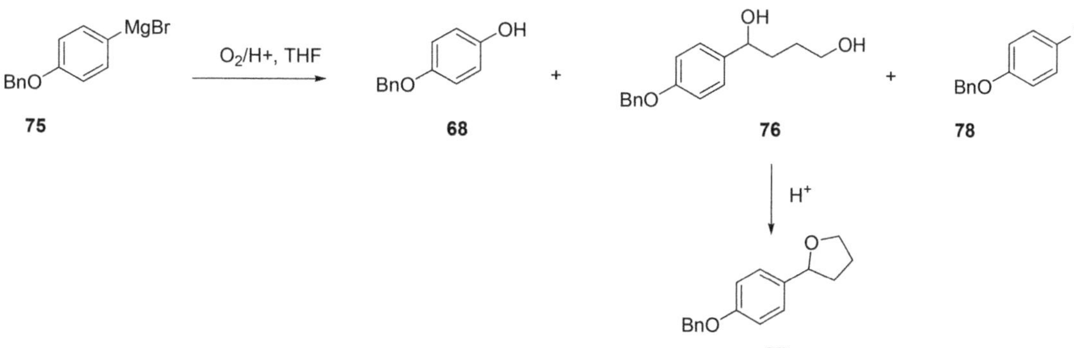

Scheme 8. Potential process impurities in the Grignard formation promoted by oxygen.

Reaction of **26** with itself to form **79**, and multiple addition products as shown in Scheme 9 were another source of yield loss. To address this issue, careful control of the Grignard reagent **75** stoichiometry was implemented. The Grignard reagent was stable when under a nitrogen atmosphere. However, upon switching off the nitrogen bubbler and exposing the solution to air, an immediate and sustained increase in phenol impurity **68** and THF adduct impurity **76** were noted (Scheme 8).

Optimization of reactor utilization resulted in a four-reactor setup that utilized dedicated reactors for specific unit operations to minimize cleaning between batches. In the original process, the same tank was used to prepare the THF solution of the Grignard reagent **75**, the second quench solution, the solvent swap to MTBE, and crystallization using *n*-heptane as antisolvent. Another tank was used for the Grignard addition and first part of the quench–addition of IPA, then aqueous ammonium chloride/brine. The improved process had dedicated tanks for each of the THF solution of **75**, the Grignard formation, the Grignard addition to **26**, and the reaction quench/crystallization.

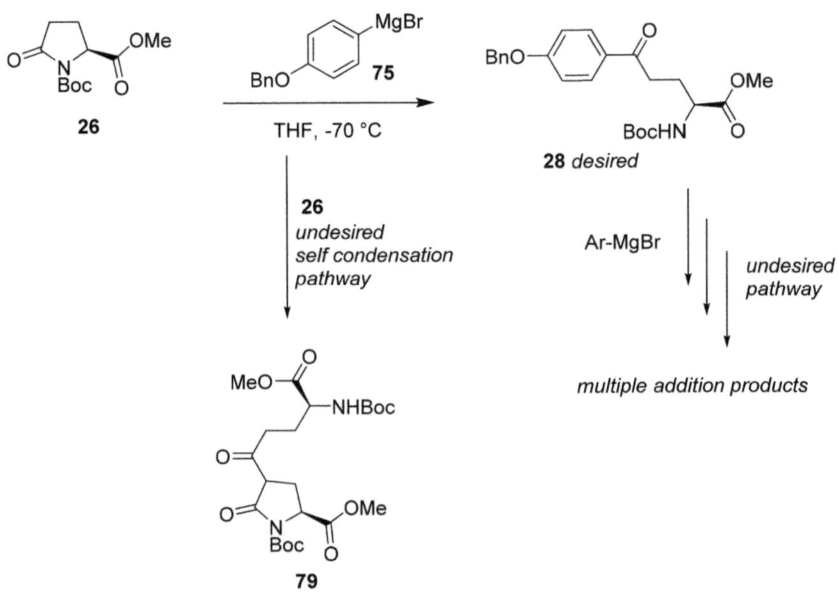

Scheme 9. Over-addition process impurities from Step 1.

To accomplish this change, the quench was changed from a direct quench with IPA at -60 °C to a reverse quench whereby THF/IPA was prepared in a separate reactor and the -60 °C reaction mixture solution was added. This design kept the Grignard formation and addition vessels free of water so they could be re-used directly in the next batch.

Another key area for immediate cost reduction and improved sustainability was replacement of the high-volume MTBE crystallization with a THF-*n*-heptane crystallization that more efficiently purged impurities and took advantage of the THF previously utilized in the reaction. Finally, we discovered that the Step 1 drying step could be eliminated, and the wet pyroglutamate adduct **28** could be forward processed directly by implementing an acetonitrile wash of the wet cake. These changes resulted in a 36% increase in batch size using the same reactors and approximately two-fold reduction in time to run a batch. In turn, the PMI of the process was reduced to 16.

Late Process Step 1 – Flow Processing

We envisioned that **28** could be prepared in a series of transformations in a flow setup. We anticipated that the Grignard reagent formation, Grignard addition, quench, and phase separation could be executed before proceeding into a batch crystallization. This would be like the currently implemented THF-*n*-heptane crystallization for batch manufacturing campaigns. Various equipment configurations were considered: a flow setup utilizing a CSTR (continuous stirred tank reactor) for Grignard formation, plug flow reactors for Grignard addition and quench followed by centrifugal extractors for phase separations and washes was designed (Figure 14).

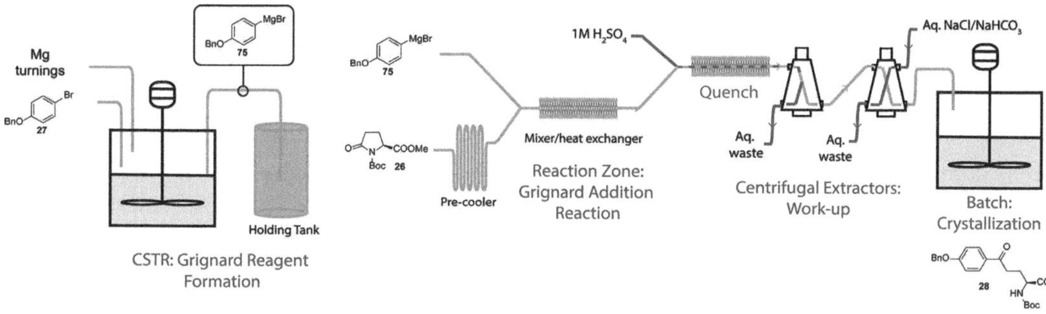

Figure 14. Overview of flow process design for Grignard addition and workup. Reproduced from ref (59). Copyright 2020 American Chemical Society.

Our learnings from the batch process development provided insight into developing successful flow chemistry conditions. The use of flow chemistry for the synthesis of Grignard reagents was published previously. We capitalized on the previously published CSTR strategy for formation of the Grignard while making some modifications to specifically suit the conversion of benzyloxy-4-bromobenzene (**27**) to the Grignard reagent **75** and our specific metric ton manufacturing constraints (*61*). The main technical challenges surrounded the solid magnesium handling and charging, magnesium sequestration, and analytical monitoring. The key challenge for the Grignard addition was designing a process that avoided clogging from the metal particles and implementing a streamlined quench amenable to a flow process. Finally, given the good phase separation properties of the reaction, a two-stage centrifugal extraction process was readily implemented.

The flow process was demonstrated on a 1 kg scale suitable for scale-up. The kilo-lab equipment featured a two-tank, one-stage CSTR setup for Grignard formation. The first tank was a CSTR setup where the Grignard formation would take place. The 500 mL vessel was fitted with a solid-addition device for magnesium delivery, an inlet for delivery of the bromobenzene **27**, a sampling port, and a specially designed product removal tube to transfer the material to the second tank to settle before transfer into a storage vessel. The entire reaction and storage vessels were under an inert atmosphere. The advantage of a CSTR setup was that fresh **27** and magnesium could be charged continuously to an already activated tank of Grignard, removing the need for initiation with each charge of fresh reactants.

Once initiation was confirmed by measurement of an exotherm as well as observation of a change in color, both the starting feed of **27** in THF and the magnesium addition device were turned on to provide a 60 min residence time for the Grignard formation. Upon reaction execution, magnesium was continuously added at a rate matching the aryl bromide stoichiometry. Extra equivalents of magnesium used during the initiation underlined the importance of including a suitable way to withdraw the Grignard solution without entrainment and transfer of magnesium particles. A specially designed product removal tube was created to facilitate the transfer of the supernatant Grignard solution to a settling tank to remove particulates before transfer to a storage tank. A feed solution of **75** in THF at 0.83 M was prepared and kept under inert conditions.

For the reaction of the Grignard reagent **75** with Boc-pyroglutamate **26**, a flow setup was used that consisted of a mixer plus heat exchanger. All three reagent streams: **75**, **26** and the quench solution were charged from pressurized stainless-steel reservoir containers and controlled by mass-flow metering valves. The Boc-pyroglutamate stream **26** was passed through a heat exchanger to adjust the temperature of the stream to -8 °C. The mixer was directly attached to a heat exchanger that removed heat from the setup.

In comparison to the quench process in batch, the flow process was streamlined. Among the by-products generated during the quench of the Grignard addition were magnesium salts. The potential to clog the flow reactor during quenching was mitigated by choice of the quench solution. Sulfuric acid was found to perform well given the high solubility of $MgSO_4$ in water (360 g/L at 20 °C), and its compatibility with the materials of construction of the flow reactor. Ultimately, centrifugal extractors were used to perform the phase separation, which was followed by a brine wash. This allowed direct collection of the product in THF, which was taken directly into the THF-*n*-heptane batch crystallization. This material was dried and further processed to imine **29** (see Step 2 section for batch processing details) to produce a total of 1032 g of the target product in 74% isolated yield based on pyroglutamate adduct **28**. While the PMI for the flow process was comparable to the batch process (15 versus 17 respectively), this flow processing enabled us to avoid cryogenic conditions at scale.

Late-Stage Process Step 2 – Cyclization to Imine

The next step involved deprotection of **28** and cyclization to form imine **29**. The original process used 8.9 equiv of TFA in toluene stirred at ambient temperature. The workup procedure required multiple operations, including a neutralizing quench with 8.5 vol of aqueous potassium bicarbonate solution, phase separation with water (3 vol) and distillation of toluene to 1.2 vol and addition of *n*-heptane (2 vol) before filtration of the product. This workup resulted in a long cycle time and generation of an excessive amount of waste.

The focus of further optimization was placed on finding a cheaper alternative acid to TFA and reduction of the reaction solvent volume. Acetonitrile was identified as a suitable solvent, allowing for increased volume efficiency, from 6.5 volumes of toluene to 3 volumes of acetonitrile. The additional benefit resulting from the change in solvent to acetonitrile was that the Wurtz impurity biphenyl **65** carried over from the previous step could be removed in a polish filtration. TFA was replaced with a more economically viable acid, methanesulfonic acid (MsOH) and the acid stoichiometry required for efficient conversion of **28** to **29** was reduced to 2.9 equiv.

Changes were also made to further improve the workup by eliminating off-gassing due to bicarbonate neutralization and minimizing aqueous waste during the quench by replacing the potassium bicarbonate with 1.5 volumes of ammonium hydroxide solution. Crystallization of imine **29** was modified such that it was performed in an acetonitrile/IPA mixture, using water as the antisolvent with seeding, for a controlled cooling crystallization.

Compared to **Route D**, the newly developed process had greatly improved volume efficiency, a reduction in the number of unit operations, and a shortened cycle time. The process changes implemented consistently produced the compound at commercial manufacturing scale ($\geq$750 kg), in

88% yield and ≥99.5% HPLC area purity. Beyond the improvements in process efficiency described for Step 2, the PMI was reduced from 35 to 20.

Late-Stage Process Step 3 – Reduction, Boc Protection, and Debenzylation

The conversion of imine **29** to phenol **64** involved three distinct chemical reactions, as described in Scheme 7. Further by-products beyond the 2,5-*trans* impurity **74** and pyrrolidine ring-opened impurity **73** occurred during process optimization and scale up, and are shown in Figure 15. These include **79**, the impurity arising from reaction of the phenolic OH group in **64** with Boc anhydride. The phenol **80**, arising from reductive cleavage of the benzyl group in **30** prior to formation of the *N*-Boc intermediate **72** was observed. Additionally, the amino ester **81**, from ring-opening reduction of the C-N benzylic bond in **80**, and the phenol **82**, from reductive cleavage of the benzyl group in the imine **29**, were noted. Impurities **80** and **81** were significant in experiments that omitted Boc anhydride and were also observed when Boc anhydride was not used in excess.

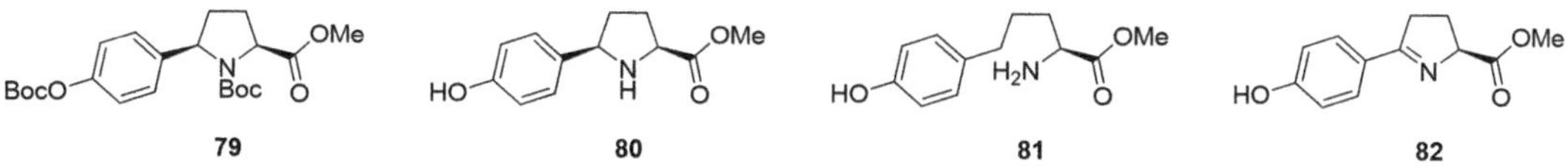

Figure 15. Further Impurities formed in Step 3.

After initial evaluation, the following parameters were targeted for additional studies:

- lowering the catalyst charge (currently 7.5 wt%) or changing to an alternative catalyst
- reducing the Boc anhydride charge
- decreasing the reaction time and
- eliminating multiple distillations to accomplish the solvent swap from methanol to ethyl acetate–*n*-heptane for isolation of the product.

A series of small-scale experiments was performed in hopes of finding a more active catalyst than 5% Pd/C type 487 paste, in order to attempt to address both the quantity of the catalyst charge and the long reaction time required for Step 3. A wide variety of commercially available Pd/C and Pt/C catalysts from Johnson Matthey was screened.

In general, Pt-based catalysts produced **72** (Scheme 7), without final debenzylation, as the major reaction product and therefore, these catalysts were not pursued further. Of the Pd/C catalysts screened, the use of Pearlman's catalyst (20 wt% $Pd(OH)_2/C$) at a loading of 3 wt% versus the use of 5% Pd/C type 487 at a loading of 7.5 wt% gave encouraging results. Pearlman's catalyst resulted in consistent completion of the reaction within 2–5 h. In providing a shortened reaction time, it is hypothesized that, under the reaction conditions, Pearlman's catalyst produces a more active Pd(0) catalytic species in situ in comparison to the use of the 5% Pd/C type 487 catalyst (*62, 63*). After optimisation, 1.01 equiv of Boc anhydride were found to afford the best impurity profile for reactions that went to completion (≤0.15% area purity of intermediate **72** with respect to the desired phenol **64**).

Temperature and agitation rate were not found to have a significant impact on conversion to phenol **64** or on the impurity profile of the reaction.

For the previous process, **Route D**, the procedure for isolation of **64** required three distillations to accomplish a solvent swap from MeOH to EtOAc, which preceded a charge of *n*-heptane to beget crystallization. The conversion of imine **29** to phenol **64** was studied in EtOAc, in order to streamline operations for the **Step 3** workup and isolation. Since results from these experiments were not promising, crystallization of phenol **64** directly from MeOH was studied. It was pleasing to discover that phenol **64** could be isolated in 90–92% yield from 4 volumes of 1:1 (v/v) MeOH–water. In all experiments, phenol impurities **74** and **73**, and Boc-phenol impurity **79** were purged to levels below 0.05% in isolated phenol **64**.

In readying the process for an initial scale up to 30 kg in a 500 L hydrogenation reactor, additional process development work was performed. Minor changes to the procedure included:

- correcting the charge of imine **29** for its weight percent purity
- pre-dissolution of imine **29** in MeOH (5 vol) at 30–40 °C prior to reaction initiation to eliminate variations in reaction time based on the morphology/PSD dependence of its solubility
- lowering of the starting reaction temperature from 35 °C to 25 °C to better control the exotherm associated with conversion of imine **29** to ether **72**
- increasing the agitation rate from 500 rpm to 1000 rpm to enable more efficient heat removal and mass transfer efficiencies.

A 30 kg run was performed and the expected yield of phenol **64** was obtained with the optimised process giving a yield of 89% and purity of 99.8%, and residual **79** below 0.2%.

While previous scale ups were performed in headspace reactors, subsequently a BUSS Loop® reactor was employed which provided improved heat removal and mass transfer efficiencies (*64, 65*). Since previous studies were performed at pressures of no more than 4 barg, additional experiments were evaluated at pressures of 10–40 barg. Versus the use of 4 barg, a lower percent of impurity **73** occurred at 10–20 barg, while at greater than 25 barg, significant amounts of **72** arose in the impurity profile, indicating a slowed rate of reaction. Based on these results, an operating pressure of 14–16 barg, with a target of 15 barg, was selected for use.

Careful monitoring of the impurity profile of isolated solids from an initial crystallization experiment revealed that impurity **73** precipitated during an extended aqueous methanol slurry hold time at -7 to -3 °C. Therefore, a study of the impurity profile as a function of MeOH – water composition and volumes, and slurry hold temperature and time, was performed. The results showed that the use of 1.5 vol water and an extended slurry hold time at 3–7 °C did not result in significant precipitation of phenol **73** during the crystallization of **64** and an excellent impurity profile of **64** was obtained.

These combined changes resulted in a PMI reduction from 20 to 9 for the step and permitted the reaction to be run twice daily at 320 kg scale. Both filtrates were combined for isolation of phenol **64** as one batch. Table 3 below summarizes the results for a completed registration stability campaign.

Table 3. Step 3 Process Summary

*Phenol **64** / batch*	*% yield*	*HPLC Area%*	*Assay*	*Impurities Area%*	*Residual Pd*
575–606 kg	86–92%	≥99.8	≥99.6 wt%	≤0.19 **71** ≤0.08 **72**	<5 ppm

Late-Stage Process Step 4 – Ether Formation and Amidation

The **Route D** process converted phenol **64** to benzyl ether **65** using 2-fluorobenzyl bromide (**35**), with K_2CO_3 as base in acetone at reflux. Of concern was that the reaction time for this step was greater than 50 h. After workup and multiple co-distillations from *n*-heptane, the desired product was crystallized. This process resulted in a high PMI of 39. The Boc group was deprotected in MeOH using *p*-TsOH to afford **41** (*p*-TsOH salt).The *p*-TsOH salt was converted to its free base utilizing aqueous NH_4OH in MTBE. The workup involved multiple separations, followed by MTBE extraction, which was replaced by MeOH via a solvent swap. **41** was converted to vixotrigine free base using 7N NH_3 in MeOH.

The **Route D** amidation reaction required 24 h to reach completion with a poor volume efficiency (PMI = 51). Final salt formation from HCl/EtOH in Step 6 gave an EtOH solvate after filtration. Ethanol was removed during drying, but it was difficult to dry the final solid to meet the residual EtOH limits. In addition, the PSD of the HCl salt included a significant proportion of fines, and resulted in poor powder flow properties which caused problems during tableting. Finally, the volume efficiency of this step was also poor (PMI=20).

For the early phase process, the next three steps were alkylation, de-Boc and amidation. The reaction conditions of the step sequence were basic to acidic to basic. It was hypothesized that if the basic reactions were in sequence rather than interspersed, i.e. alkylation and amidation using the same solvent and base, the complete E-factor and efficiency could be improved (Scheme 10).

Amidation reactions can be accomplished with the use of formamide and an alkoxide base (*66*). Since alkylation using K_2CO_3 was the original process, we tested the possibility to continue using the same base for the amidation and found that at 90 °C it reached completion. One reason to use mild K_2CO_3 as base was to prevent the epimerization at the relatively acidic C-2 proton. With optimization, an alkylation-amidation one-pot process using acetonitrile as the solvent and K_2CO_3 as the base was developed and scaled up to more than 100 kg.

The drawbacks of this process are that a filtration is required after alkylation and fresh K_2CO_3 required charging for the amidation reaction. In addition, this heterogeneous process did not scale up consistently in terms of reaction time for both alkylation and amidation reactions. Analysis also revealed increased levels of impurities (Figure 16): acid **84** (2.0%), *trans*-amide epimer **83** (2.2%), (*2R, 5R*) and ester **65** (1.5%). The increased levels of these impurities for production batches compared to laboratory batches was attributed to longer reaction time. The reaction time was attributed to the heterogeneous nature of the process.

It was found that the commercially available 25% NaOMe in MeOH was a convenient and inexpensive alternative to potassium carbonate and using this base, the alkylation reaction reached completion in under 2 h at 25 °C. However, it was crucial to have slightly more benzyl bromide **35** (0.05 equiv in excess) versus NaOMe to keep the reaction medium from attaining basic conditions and leading to C-2 center epimerization. It is reported that sodium formamide (formed by reaction of NaOMe with formamide) is an effective amidation reagent (*67*). The pKa of MeOH (in DMSO) is 27.9 while that of formamide is about 4 pKa units lower. We discovered that the C-2 center epimerized completely in minutes using alkoxide base without the use of formamide, however, if formamide was charged before the addition of NaOMe, the basicity of the reaction mixture was lowered, and epimerization was minimized.

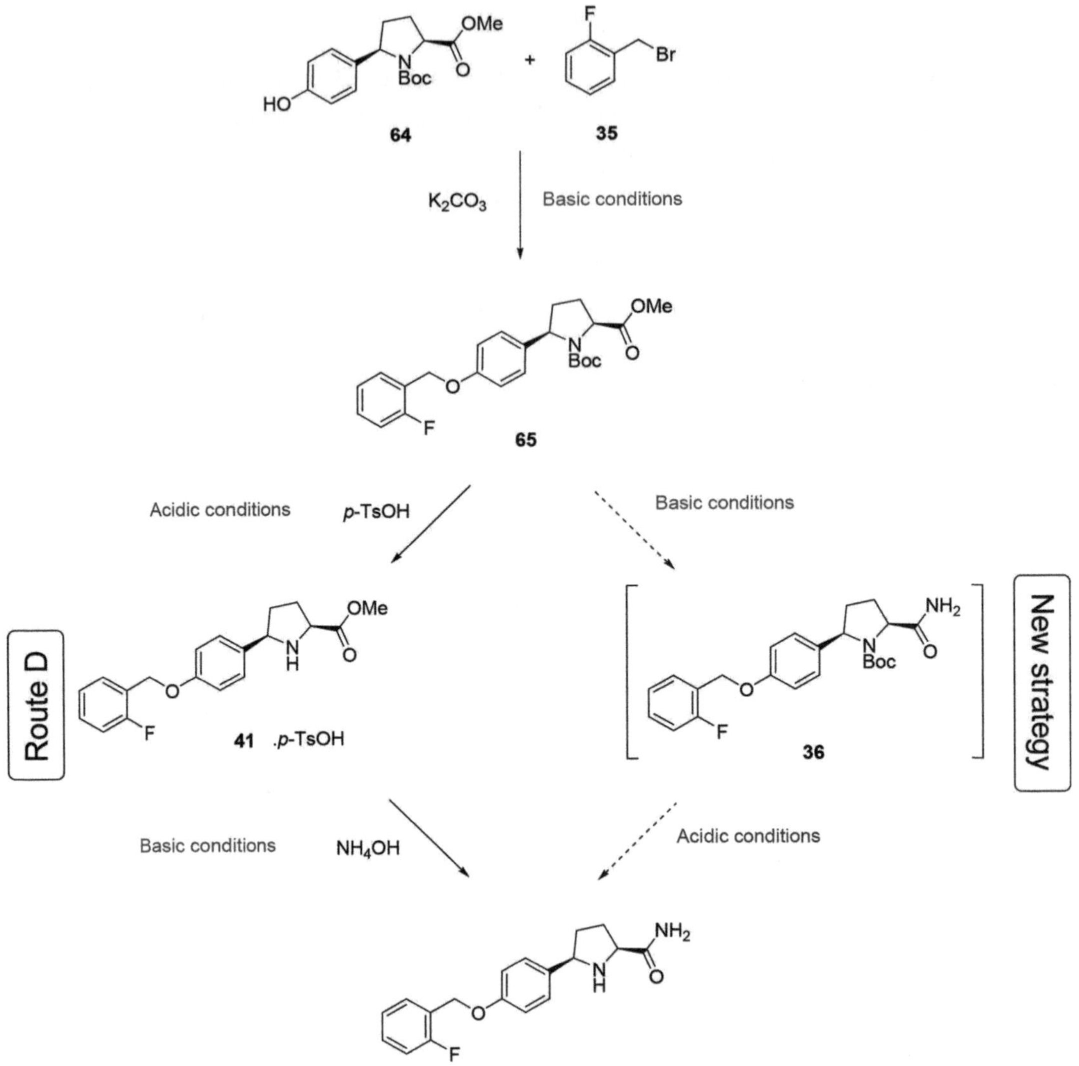

Scheme 10. Proposed telescoped strategy combining Steps 4 and 5.

The rate of epimerization is dependent on the number of added formamide equivalents (Figure 17). The higher the ratio of formamide to NaOMe the lower the basicity. For the final telescoped process, 24 equiv of formamide and 1.8 equiv of NaOMe were used resulting in less than 0.6% of *trans* isomer **83**.

There is also a clear advantage on reaction completion time (Figure 18) in using a higher formamide/NaOMe ratio. When 16 equiv of formamide was used, it took more than 6 h to reach 1% unreacted benzyl **65** but less than 3 h if 20 or 24 equiv was used. The amount of unreacted **65** reached a steady state for all curves with time. The lowest amount of unreacted **65** that can be reached using 8 equiv of formamide is 2.7% whereas only 0.6% of unreacted **65** remains using 24 equiv. Thus, the yield was improved by 2% by increasing formamide from 8 equiv to 24 equiv.

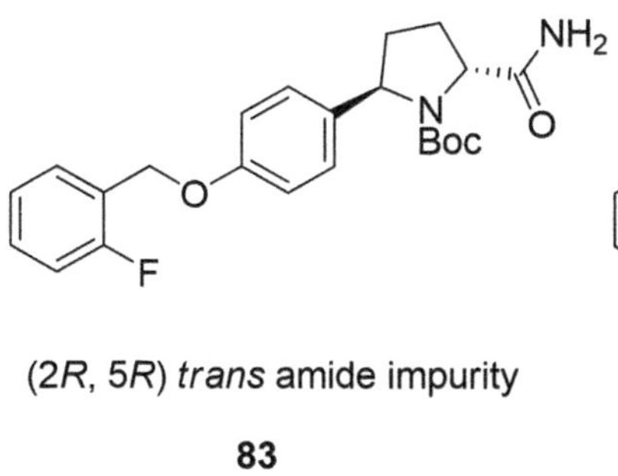

Figure 16. Impurities observed in process combining phenol alkylation and ester amidation with K_2CO_3 as base.

This new NaOMe telescoped process converting phenol **64** to **36** was demonstrated at 600 kg scale. Compared to K_2CO_3 process, this new NaOMe one-pot process improved the yield from 91% to 96% and purity (assay) from 96.4% to 99.4%. Also, the maximum volume for this NaOMe process was reduced by 50% from the original K_2CO_3 process hence doubling the volumetric efficiency, while the production time was reduced by more than 50%.

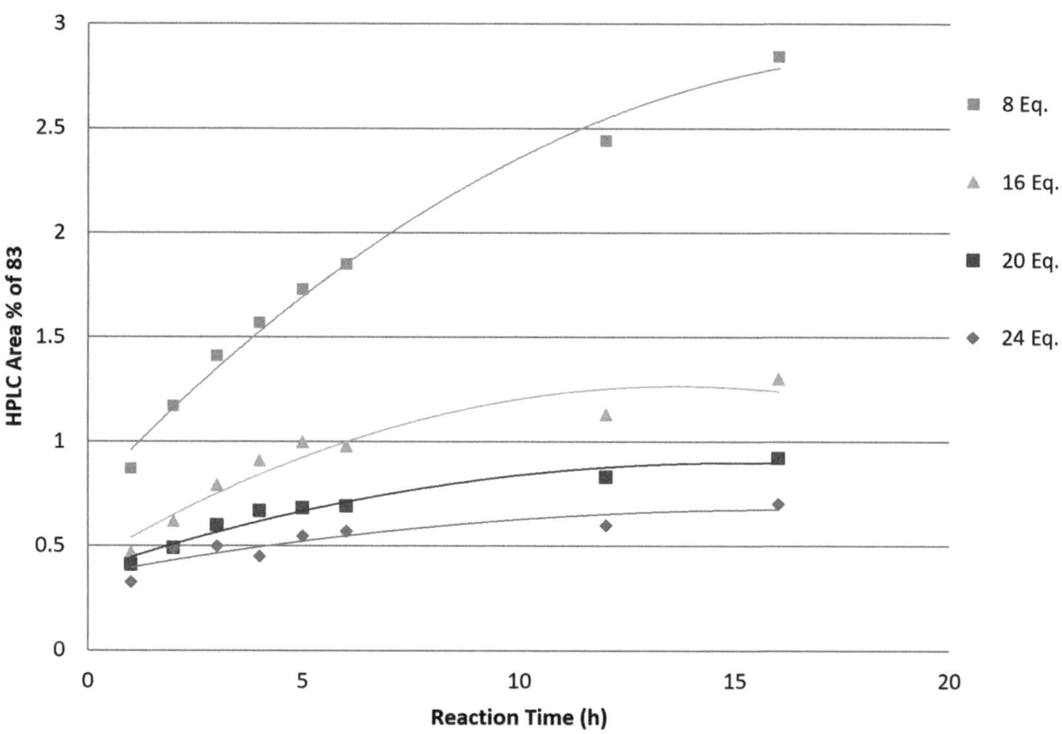

*Figure 17. Dependence of % of epimerization to **83** vs equiv of formamide. Reproduced from ref (59). Copyright 2020 American Chemical Society.*

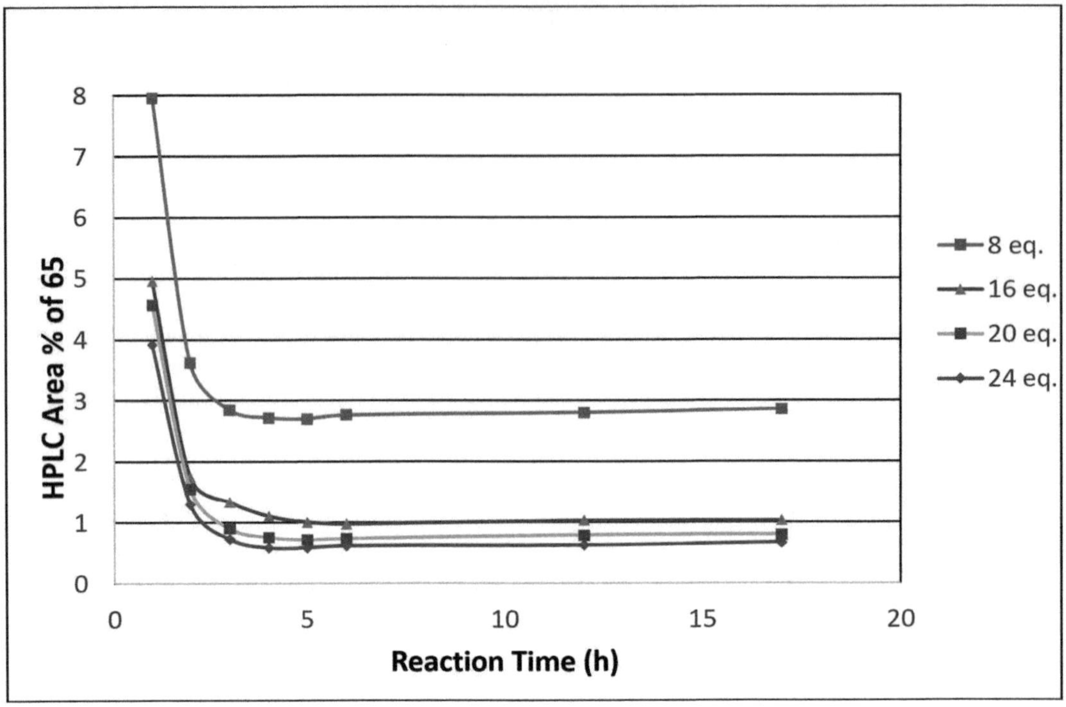

*Figure 18. Proportion of unreacted **65** vs equiv of formamide (8eq, 16 eq, 20 eq, 24 eq) versus time.*

Late-Stage Process Step 5 – Boc Deprotection

Since **16** is ultimately isolated as an HCl salt, using HCl to accomplish both Boc removal from **36** and HCl salt formation in one step was expected to be viable. A process using HCl/IPA to cleave the Boc group of **36** and form **16** HCl salt in one pot was developed and scaled to 160 kg. Although the molar yield was 98%, the process did not purge process impurities consistently.

Use of 1.3 equivalents of MsOH, in 5 volumes of acetonitrile at a reaction temperature of 45°C was effective for Boc cleavage to provide the **16** MsOH salt. The reaction was complete in 2-5 h at 45°C. Crystallization of the MsOH salt was observed during the reaction. Although the MsOH salt of **16** could be isolated, it required conversion to the free base prior to forming the final API HCl salt. To streamline the process, the MsOH salt was neutralized in situ at 50°C to **16** free base by charging 5 vol. of 1 N NH_4OH. A seed bed was formed toward the end of addition and the mixture was then cooled to 0°C, filtered, washed with H_2O and dried to afford free base **16** of 96% yield and more than 99.8% purity (Scheme 11).

Late-Stage Process Step 6 – Final Salt Formation and Particle Engineering

In the **Process D** route, the pilot scale process initially generated the metastable ethanol solvate of the HCl salt of **16.** The desired anhydrous form was formed after filtration and drying of the solid, however this temperature-driven removal of EtOH was time consuming for larger scale manufacturing. The de-solvation process caused crystal structure to collapse, resulting in a product with small particles and a wide PSD. As a result, the tableting process was complicated by poor powder flow. The solid properties were addressed through pre-blending, and dry granulation processes, which enabled powder flow properties compatible with the tableting process. It was clear that improvement of the bulk solid properties of the drug substance would benefit the drug product manufacturing, by allowing for a direct compression (DC) process.

Solvent-mediated removal of EtOH through introduction of an ethyl acetate cake wash was attempted. This was successful in reducing the drug substance drying time, but still produced undesirable particle properties (*68*). As such, an alternative solvent, IPA, was selected to replace ethanol in the salt formation step. Fortunately, salt formation in IPA did not produce a solvate and the anhydrous HCl salt of **16** (solvent free) was obtained. In addition, we found that salt formation with aqueous HCl produced larger particles than when HCl gas was used. The increase in particle size of the final isolated drug substance are shown in the micrographs in Figure 19.

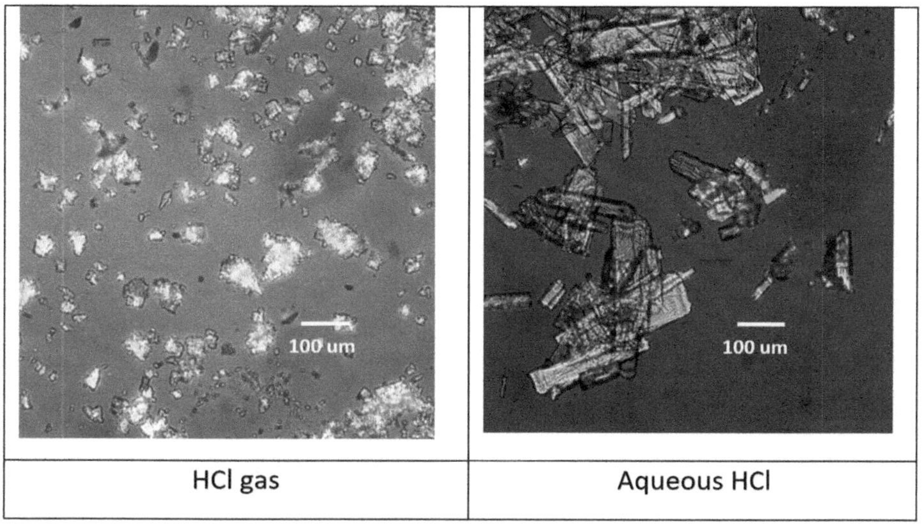

Figure 19. PLM images showing differences of ***16·HCl*** *produced by HCl gas and by ~12% aqueous HCl. Reproduced from ref (59). Copyright 2020 American Chemical Society.*

In expectation of generating particles within the target d_{50} range of 80-110 μm, an IPA/aq. HCl salt formation and drug substance isolation protocol was developed (75°C, IPA and 20% aq. hydrochloric acid solution). Due to low drug substance solubility in IPA at ambient temperature, the salt formation was performed by direct precipitation of the HCl salt at a higher temperature after dissolving free base **16**. Concerns about amide hydrolysis to form the corresponding carboxylic acid impurity **84** were addressed by limiting the HCl stoichiometry to 1.0 equiv. Under these conditions the carboxylic acid impurity did still form but within manageable levels (0.6% in final isolated API) when the product slurry formed was held for 5.5 h at 75°C. The target particle size in the precipitation process was achieved by controlling the acid addition and agitation rates. Laboratory studies suggested that a rapid rate of aq. HCl addition, combined with vigorous agitation resulted in particle sizes of 80-100 μm, within the target d_{50} range.

Early development work found particles from the IPA/HCl (aq.) precipitation method to be compatible with direct compression for drug product manufacturing. Unfortunately, a difficulty was encountered during initial drug product manufacture using direct compression. The powder blend resulting from the IPA/HCl precipitation/isolation had inadequate flow properties when held prior to compression. It required immediate compression after blending to avoid flow problems. Further examinations identified that increased cohesiveness of the API upon standing may likely be due to the presence of large amounts of fines.

Although the IPA/HCl (aq) precipitation method produced larger particles, fines persisted. As such, a wet milling process was selected. For the milling process, the slurry was recirculated through a high-shear mixer, consisting of rotor and stator screens with interchangeable configurations, as depicted in Figure 20. Milling completion was determined by there being no further change in particle size by in-line particle tracking or the number of volume turnovers.

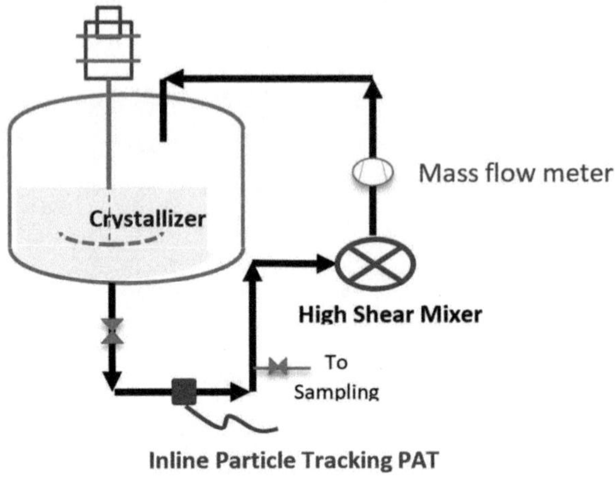

Figure 20. Wet mill configuration. Reproduced from ref (59). Copyright 2020 American Chemical Society.

The wet milling operation consisted of three stages:

- first the salt was formed and crystallized out of solution to produce large crystals
- second the crystals were circulated through a high-shear mixer
- third the milled slurry was exposed to thermal cycling in the reactor prior to isolation.

Material properties were obtained by systematic characterization using shear cell and FT4 powder rheometer measurements, and tableting feasibility was conducted using a compaction simulator. Process conditions that resulted in optimum powder properties were selected for manufacturing.

The selected conditions incorporated salt formation in an aqueous IPA mixture at 65°C and crystallization during cooling to 0-10°C (Step 1). The second stage was slurry milling using an IKA high-shear mixer with a tip speed of 25 ms^{-1} (Step 2). Milling was completed after the in-line particle tracking showed no more size reduction. The final stage was temperature cycling by heating to 60°C and cooling to 20°C (Step 3). As shown by comparison of images B and C in Figure 21, this effectively reduced the amounts of fines generated to obtain the final API **16·HCl**.

These conditions were effectively demonstrated in large-scale campaigns with batch sizes ranging from 120 to 584 kg and produced drug substance meeting the desired target PSD and other key bulk solid attributes allowing for implementation of a direct compression formulation Table 4.

A EtOH/HCl gas	B IPA/HCl aq. precipitation	C IPA/HCl aq. crystallization with wet milling and temperature cycling. Typical d_{50}: 80-125 µm; note absence of fines
Typical d_{50}: 15-20 µm	Typical d_{50}: 70-110 µm; note presence of fines around larger particle	

Figure 21. SEM images of API produces from 3 different methods. Reproduced from ref (59). Copyright 2020 American Chemical Society.

Table 4. Average PSD Obtained from Large-Scale Manufacturing Campaigns

	Batch Size (kg)	$d_{10}(\mu m)$	$d_{50}(\mu m)$	$d_{90}(\mu m)$
Campaign I	120–174	32	102	217
Campaign II	473–584	30	108	256

Summary – Late-Stage Process

Process improvements included optimization of a large-scale cryogenic Grignard reaction to increase throughput, and in preparation for future manufacturing, kilo-scale demonstration of the process in flow. Intermediate steps were streamlined to reduce unit operations, including a key hydrogenation reaction that incorporated three chemical transformations in one-pot. Additionally, two late-stage steps were telescoped in a one-pot process to improve throughput. The PMI dropped by 64% yet maintained drug substance purity above 99.8%. The flow properties of the drug substance were improved from the previous process by choosing optimal salt formation conditions, wet milling and temperature cycling. This new process was scaled up to generate 2.5 tons of drug substance for registration batches with an average yield of 90% for each reaction step (Scheme 11).

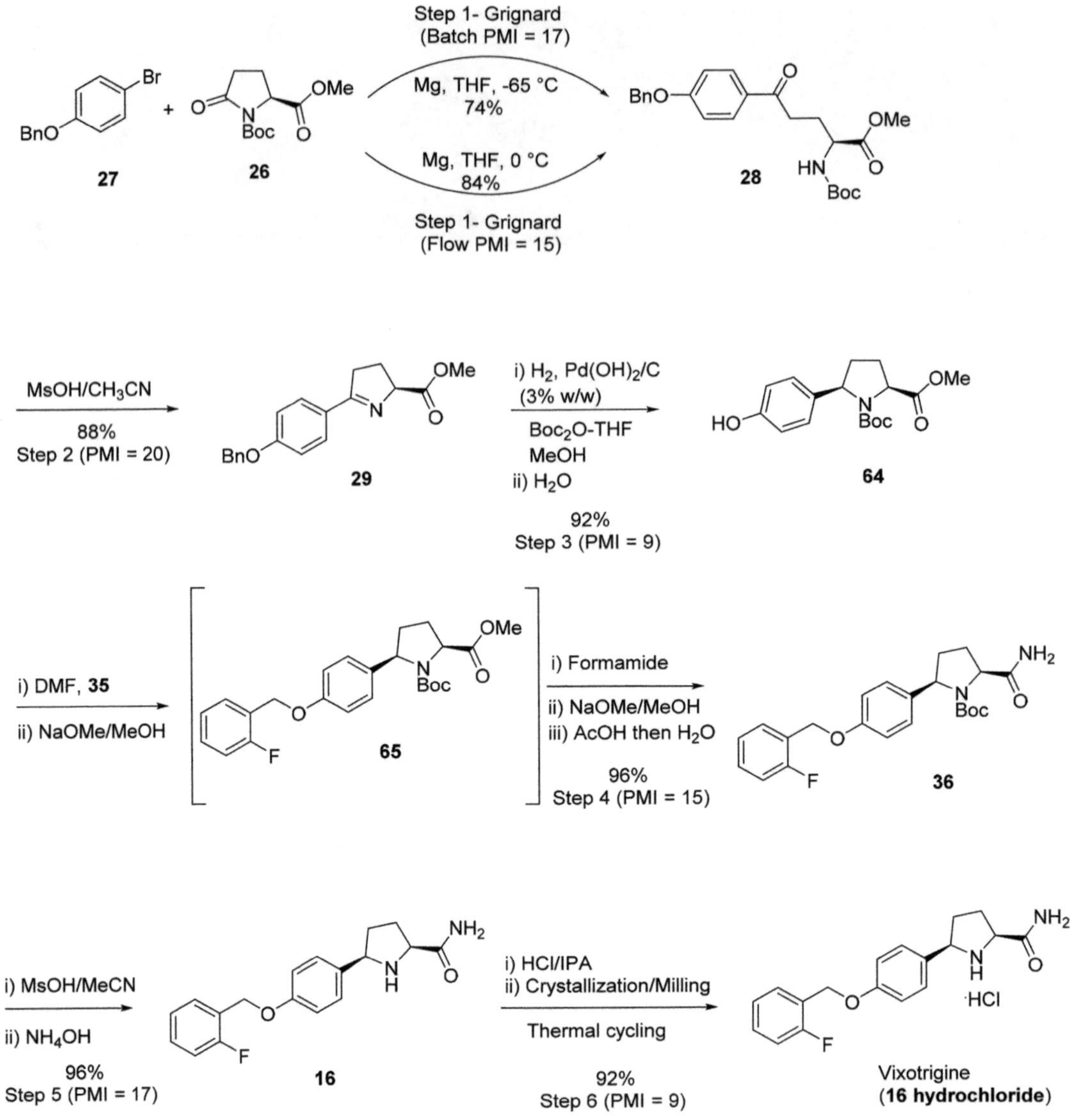

Scheme 11. Late-stage process for vixotrigine 16.

Conclusions

This case history of drug discovery and development chemistry illustrates the differing challenges faced by chemists, chemical engineers and formulators in the process of designing a new medicine. They range from the rapid synthesis of compounds for hit identification and lead optimization, to the more stringent purity and scalability attributes necessary for provision of GMP quality material for clinical trials, and finally to the efficiency and environmental requirements for a successful commercial manufacture. It is a process that depends on the scientific insight and inventiveness of dozens of individual scientists collaborating over many years. Creating a new medicine is for many of us the most exciting challenge a pharmaceutical scientist can face.

Disclaimer

This work was funded by Biogen Inc. and by Convergence Pharmaceuticals Ltd., which has since been acquired by Biogen Inc.

David R Witty, David T MacPherson, Gerard M P Giblin hold stock in Biogen and were employees of Convergence Pharmaceuticals Ltd. and Biogen Inc. at the time of the conduct of the study.

Robbie Chen, Vinny Couming, John Guzowski Jr., and Tamera Mack were employees of Biogen at the time of the conduct of the study and were employees during the creation of the manuscript.

Erwin Irdam, William F. Kiesman, Daw-Iong Albert Kwok, Wenli Liang, Erin M. O'Brien, Suzie Opalka, Daniel Patience, Stefan Sahli, Donald G. Walker, Frederick Osei-Yeboah are employees of Biogen and hold stock in Biogen.

References and Notes

1. For a recent review of the field see: Witty, D. R.; MacPherson, D. T.; Giblin, G. M. Sodium Channel Blockers for the Treatment of Pain. In *Comprehensive Medicinal Chemistry III*; Chackalamannil, S., Rotella, D., Ward, S. E., Eds.; Elsevier: Amsterdam, 2017; Vol. 7, pp 131–175. http://doi.org/10.1016/b978-0-12-409547-2.12439-4.
2. Scott. D. F. Carbamazepine. In *The History of Epileptic Therapy: An Account of How Medication Was Developed*; The Parthenon Publishing Group: Carnforth, U.K., 1993.
3. Lees, G.; Leach, M. J. *Brain Res.* **1993**, *612*, 190–199.
4. Goldberg, J. F.; Calabrese, J. R.; Saville, B. R.; Frye, M. A.; Ketter, T. A.; Suppes, T.; Post, R. M.; Goodwin, F. K. *J. Clin. Psychiatry* **2009**, *70*, 1273–1280.
5. Bagal, S. K.; Chapman, M. L.; Marron, B. E.; Prime, R.; Storer, R. I.; Swain, N. *Bioorg. Med. Chem. Lett.* **2014**, *24*, 3690–3699.
6. Catterall, W. A.; Goldin, A. L.; Waxman, S. G. *Pharmacol. Rev.* **2005**, *57*, 397–409.
7. Curia, G.; Biagini, G.; Perucca, E.; Avoli, M. *CNS Drugs* **2009**, *23*, 555–568.
8. Blaszczyk, B.; Czuczwar, S. J. *Neuropsychiatr. Dis. Treat.* **2010**, *6*, 145–150.
9. Mahesh Kumar, S.; Suresh, I.; Satish Kumar, C. H.; Swaroop Kumar, S. L. V. V. S. N. K.; Vijey Aanandhi, M. *J. Adv. Drug Res.* **2011**, *1*, 68–81.
10. Mike, A.; Lukacs, P. *Curr. Mol. Pharmacol.* **2010**, *3*, 129–144.
11. Payandeh, J.; Gamal El-Din, T. M.; Scheuer, T.; Zheng, N.; Catterall, W. A. *Nature* **2012**, *486*, 135–139.
12. Bagnéris, C.; DeCaen, P. G.; Naylor, C. E.; Pryde, D. C.; Nobeli, I.; Clapham, D. E.; Wallace, B. A. *Proc. Natl. Acad. Sci.* **2014**, *111*, 8428–8433.
13. Bagnéris, C.; Naylor, C. E.; McCusker, E. C.; Wallace, B. A. *J. Gen. Physiol.* **2015**, *145*, 5–16.
14. Wu, M.-T.; Huang, P.-Y.; Yen, C.-T.; Chen, C.-C.; Lee, M.-J. *PLoS One* **2013**, *8*, e55212. https://doi.org/10.1371/journal.pone.0055212.
15. *Ion Channels from structure to function*; Kew, J. N. C., Davies, C. H., Eds.; Oxford University Press: Oxford, U.K., 2010.
16. Bennett, D. L. H.; Woods, C. G. *Lancet Neurol.* **2014**, *13*, 587–599.
17. For a description of cardiac channelopathies associated with $Na_V1.5$ see: Remme, C. A. *J. Physiol.* **2013**, *591*, 4099–4116.

18. A nociceptor ("pain receptor") is a sensory neuron that responds to damaging or potentially damaging stimuli by sending "possible threat" signals to the spinal cord and the brain. If the brain perceives the threat as credible, it creates the sensation of pain to direct attention to the body part, so the threat can hopefully be mitigated; this process is called nociception.
19. Sitges, M.; Chiu, L. M.; Guarneros, A.; Nekrassov, V. *Neuropharmacology* **2007**, *52*, 598–605.
20. Johannesen, L.; Vicente, J.; Gray, R. A.; Galeotti, L.; Loring, Z.; Garnett, C. E.; Florian, J.; Ugander, M.; Stockbridge, N.; Strauss, D. G. *Clin. Pharmacol. Ther.* **2014**, *95*, 501–508.
21. Zakrzewska, J. M. *BMJ* **2014**, *348*, g474. https://www.aans.org/Patients/Neurosurgical-Conditions-and-Treatments/Trigeminal-Neuralgia.
22. Bu, H.-Z.; Kang, P.; Deese, A. J.; Zhao, P.; Pool, W. F. *Drug Metab. Dispos.* **2005**, *33*, 1920–1924.
23. Lorberg, B.; Youssef, N. A.; Bhagwagar, Z. *Int. J. Neuropsychopharmacol.* **2009**, *12*, 257–265.
24. Witty, D. R.; Alvaro, G.; Derjean, D.; Giblin, G. M. P.; Gunn, K.; Large, C.; Macpherson, D. T.; Morisset, V.; Owen, D.; Palmer, J.; Rugiero, F.; Tate, S.; Hinckley, C. A.; Naik, H. *ACS Med. Chem. Lett.* **2020**in press. http://dx.doi.org/10.1021/acsmedchemlett.0c00263.
25. Pevarello, P.; Bonsignori, A.; Dostert, P.; Heidempergher, F.; Pinciroli, V.; Colombo, M.; McArthur, R. A.; Salvati, P.; Post, C.; Fariello, R. G.; Varasi, M. *J. Med. Chem.* **1998**, *41*, 579–590.
26. Jarvis, M. F.; Honore, P.; Shieh, C.-C.; Chapman, M.; Joshi, S.; Zhang, X.-F.; Kort, M.; Carroll, W.; Marron, B.; Atkinson, R.; Thomas, J.; Liu, D.; Krambis, M.; Liu, Y.; McGaraughty, S.; Chu, K.; Roeloffs, R.; Zhong, C.; Mikusa, J. P.; Hernandez, G.; Gauvin, D.; Wade, C.; Zhu, C.; Pai, M.; Scanio, M.; Shi, L.; Drizin, I.; Gregg, R.; Matulenko, M.; Hakeem, A.; Gross, M.; Johnson, M.; Marsh, K.; Wagoner, P. K.; Sullivan, J. P.; Faltynek, C. R.; Krafte, D. S. *Proc. Natl. Acad. Sci.* **2007**, *104*, 8520–8525.
27. Suter, M. R.; Kirschmann, G.; Laedermann, C. J.; Abriel, H.; Decosterd, I. *Anesthesiology* **2013**, *118*, 160–172.
28. Caccia, C.; La Porta, E.; Maestroni, S.; Melloni, P.; Sabido, D. C.; Salvati, P. *Prolinamide derivatives as sodium and/or calcium channel blockers or selective MAO-B inhibitors*. Patent EP 1524265, 2005.
29. Characterisation of Vixotrigine, a Broad Spectrum Voltage-Gated Sodium Channel Blocker Hinkley, C.; Hajos, M.; Kuryshev, Y.; Sers, A.; Barre, A.; Buisson, B.; Naik, H. Submitted for publication.
30. Naik, H.; Steiner, D.; Versavel, M.; Palmer, J.; Fong, R. *Safety, Tolerability, and Pharmacokinetics of Single Doses of the Nav1.7 Selective Sodium channel Blocker Vixotrigine in Healthy Volunteers*. Presented at American Society for Clinical Pharmacology and Therapeutics 119th Annual Meeting, Orlando FL, March 22–24, 2018.
31. Safety, Tolerability, and Pharmacokinetics of Repeat Doses of the Nav1.7 Selective Sodium channel Blocker Vixotrigine in Healthy Volunteers. Naik H., Steiner D., Versavel M., Palmer J., Fong R. *American Society for Clinical Pharmacology and Therapeutics 119th Annual Meeting*, 2018
32. Also reported as CNV1014802, BIIB074
33. Zakrzewska, J. M.; Palmer, J.; Morisset, V.; Giblin, G. M. P.; Obermann, M.; Ettlin, D. A.; Cruccu, G.; Bendtsen, L.; Estacion, M.; Derjean, D.; Waxman, S. G.; Layton, G.; Gunn, K.; Tate, S. *Lancet Neurol.* **2017**, *16*, 291–300.

34. Alvaro, G.; Andreotti, D.; Bergauer, M.; Giovannini, R.; Marasco A. *Prolinamide Derivatives as Sodium Channel Modulators*. Patent WO 2007042250, 2007.
35. Nájera, C.; Yus, M. *Tetrahedron Asymmetry* **1999**, *10*, 2245–2303.
36. Ezquerra, J.; Pedregal, C.; Rubio, A.; Valenciano, J.; García Navio, J. L.; Alvarez-Builla, J.; Vaquero, J. J. *Tetrahedron Lett.* **1993**, *34*, 6317–6320.
37. Giblin, G. M. P.; Heseltine, A.; Kiesman, W.; MacPherson, D. T.; Ramsden, J.; Williams, M. T.; Witty, D. R. *Org. Process Res. Dev.* **2020**submitted for publication.
38. Alvaro, G.; Bergauer, M.; Giovannini, R.; Profeta R. *Prolinamide Derivatives as Sodium Channel Modulators*. Patent WO 2007042239, 2007.
39. All purities are by HPLC area % unless otherwise stated.
40. García-López, J.-A.; Greaney, M. F. *Chem. Soc. Rev.* **2016**, *45*, 6766–6798.
41. Behloul, C.; Guijarro, D.; Yus, M. Formation of Biaryls by Homocoupling of Grignard Reagents. In *7th Electronic Conference on Synthetic Organic Chemistry*, Nov 1–30, 2003. https://www.mdpi.org/ecsoc/ecsoc-7/papers/a019/a019.htm.
42. Krasovskiy, A.; Knochel, P. *Angew. Chem., Int. Ed.* **2004**, *43*, 3333–3336.
43. Piller, F. M.; Metzger, A.; Schade, M. A.; Haag, B. A.; Gavryushin, A.; Knochel, P. *Chem. Eur. J.* **2009**, *15*, 7192–7202.
44. Hawkins, J. M.; Dube, P.; Maloney, M. T.; Wei, L.; Ewing, M.; Chesnut, S. M.; Denette, J. R.; Lillie, B. M.; Vaidyanathan, R. *Org. Process Res. Dev.* **2012**, *16*, 1393–1403.
45. Inoue, A.; Kitagawa, K.; Shinokubo, H.; Oshima, K. *J. Org. Chem.* **2001**, *66*, 4333–4339.
46. Iida, T.; Wada, T.; Tomimoto, K.; Mase, T. *Tetrahedron Lett.* **2001**, *42*, 4841–4844.
47. Tilstam, U.; Weinmann, H. *Org. Process Res. Dev.* **2002**, *6*, 906–910.
48. Ace, K. W.; Armitage, M. A.; Bellingham, R. K.; Blackler, P. D.; Ennis, D. S.; Hussein, N.; Lathbury, D. C.; Morgan, D. O.; O'Connor, N.; Oakes, G. H.; Passey, S. C.; Powling, S. C. *Org. Process Res. Dev.* **2001**, *5*, 479–490.
49. Zajac, M. A. *Process for preparing α-carboxamide pyrrolidine derivatives*. Patent WO 2011029762, 2011
50. Strohmeier, M.; Leach, K.; Zajac, M. A. *Angew. Chem., Int. Ed.* **2011**, *50*, 12335–12338.
51. Miyake, Y.; Nishibayashi, Y.; Uemura, S. *Synlet* **2008**, 1747–1758.
52. Van Esseveldt, B. C. J.; Vervoort, P. W. H.; van Delft, F. H.; Rutjes, F. P. J. T. *J. Org. Chem.* **2005**, *70*, 1791–1795.
53. Chinchilla, R.; Nájera, C. *Chem. Soc. Rev.* **2011**, *40*, 5084–5121.
54. Taylor, I. N.; Brown, R. C.; Bycroft, M.; King, G.; Littlechild, J. A.; Lloyd, M. C.; Praquin, C.; Toogood, H. S.; Taylor, S. J. C. *Biochem. Soc. Trans.* **2004**, *32*, 290–292.
55. Adimurthy, S.; Malakar, C. C.; Beifuss, U. *J. Org. Chem.* **2009**, *74*, 5648–5651.
56. He, Z.; Jamison, T. F. *Angew. Chem., Int. Ed.* **2014**, *53*, 3353–3357.
57. MacPherson, D.; Witty, D.; Giblin, G.; Williams, M. T.; Walker, D.; Kiesman, W.; Mack, T. *Process for Preparing Alpha-Carboxamide Pyrrolidine Derivatives*. Patent WO 2016102967, 2016.
58. Li, B.; Berliner, M.; Buzon, R.; Chiu, C. K-F.; Colgan, S. T.; Kaneko, T.; Keene, N.; Kissel, W.; Le, T.; Leeman, K. R.; Marquez, B.; Morris, R.; Newell, R.; Wunderwald, S.; Witt, M.; Weaver, J.; Zhang, Z.; Zhang, Z. *J. Org. Chem.* **2006**, *71*, 9045–9050.
59. Chen, R.; Couming, V.; Guzowski, J., Jr.; Irdam, E.; Kiesman, W.; Kwok, D.-l.; Liang, W.; Mack, T. L.; O'Brien, E. M.; Opalka, S. M.; Patience, D.; Sahli, S.; Walker, D. G.; Osei-

Yeboah, F.; Gu, C.; Zhang, X.; Stöckli, M.; Stucki, T.; Matzinger, H.; Kuhn, R.; Thut, M.; Grohmann, M.; Nonnenmacher, M.; Cerea, P. *Org Process Res. Dev.* **2020**. DOI: https://pubs.acs.org/doi/10.1021/acs.oprd.0c00427.

60. Roschangar, F.; Zhou, Y.; Constable, D. J. C.; Colberg, J.; Dickson, D. P.; Dunn, P. J.; Eastgate, M. D.; Gallou, F.; Hayler, J. D.; Koenig, S. G.; Kopach, M. E. D. K.; Mergelsberg, I.; Scholz, U.; Smith, A. G.; Henry, M.; Mulder, J.; Brandenburg, J.; Dehli, J . R.; Fandrick, D. R.; Fandrick, K. R.; Gnad-Badouin, F.; Zerban, G.; Groll, K.; Anastas, P. T.; Sheldon, R. A.; Senanayake, C. H. *Green Chem.* **2018**, *20*, 2206–2211.
61. Wong, S.-W.; Changi, S. M.; Shields, R.; Bell, W.; McGarvey, B.; Johnson, M. D.; Sun, W.-M.; Braden, T. M.; Kopach, M. E.; Spencer, R. D.; Flanagan, G.; Murray, M. *Org. Process Res. Dev.* **2016**, *20*, 540–550.
62. Itaya, T.; Morisue, M.; Takeda, M.; Kumazawa, Y. *Chem. Pharm. Bull.* **1990**, *38*, 2656–2661.
63. Walker, D. G. *Private conversation with Johnson Matthey precious metal catalyst personnel.*
64. For an animated presentation of a BCT-Loop reactor's operation, see Pratteln-Basel, K. Hydrogenation – Buss Chemtech AG, 2020. www.buss-ct.com/buss_loop_reactor.html (accessed 16 September 2020)
65. For a white paper describing the operating features of a BCT-Loop reactor, see buss-ct.com, 2020. http://www.buss-ct.com/up/files/PDFs_RT/Brochure_RT_Advanced_Loop_Reactor_operating_principle.pdf (accessed 16 September 2020).
66. Le Roux, J.; Leriche, C.; Chamiot-Clerc, P.; Feutrill, J.; Halley, F.; Papin, D.; Derimay, N.; Mugler, C.; Grépin, C.; Schio, L. *Bioorg. Med. Chem. Lett.* **2016**, *26*, 454–459.
67. Sony, J.; Prasenjit, D.; Bindu, S.; Hashim, N.; Mohan, P. *Tetrahedron Lett.* **2013**, *54*, 929–931.
68. Vasudevan, K. V.; Osei-Yeboah, F.; Tran, K. K.; Patience, D.; Irdam, E.; Kwok, D.-l.; Peterson, M. L. *Org Process Res. Dev.* **2018**, *22*, 535–541.

Chapter 2

The Discovery and Chemical Development of BMS-919373: A Selective I_{Kur} Inhibitor for the Potential Treatment of Atrial Fibrillation

Heather J. Finlay,[*,1] Scott A. Savage,[2] Jason M. Stevens,[2] Steven R. Wisniewski,[*,2] and Miao Yu[2]

[1]Department of Discovery Chemistry, Bristol Myers Squibb, Princeton, New Jersey 08540, United States

[2]Department of Chemical Process Development, Bristol Myers Squibb, New Brunswick, New Jersey 08903, United States

[*]Emails: Heather.Finlay@bms.com; Steven.Wisniewski@bms.com

This chapter describes identification and characterization of the clinical candidate from the phenylquinazoline series of I_{Kur} inhibitors as a potential treatment for atrial fibrillation (AF). An introduction to the I_{Kur} target in the context of current standard of care options for AF and the discovery approach taken to optimize the series for potency, ion channel selectivity, metabolic stability, pharmacodynamic and safety profile leading to clinical candidate, BMS-919373 are described. The transition to process chemistry, the initial scale-up route, commercial route exploration, and commercial route development is also detailed herein.

Introduction

Atrial fibrillation (AF) is a disease which contributes to the mortality for 160,000 patients annually (*1*) and is responsible for 450,000 hospitalizations in the United States (*2*). The occurrence is steadily increasing in prevalence with an estimated global incidence of 0.5% of the population in 2020 (*3*). There has been a revolution in the treatment of AF patients with the approval of a number of direct acting oral anti-coagulants for stroke prevention in this patient population without underlying valve dysfunction. Anti-coagulant treatment has significantly reduced the risk of stroke and thrombo-embolism in AF patients with the most widely prescribed and effective oral anti-coagulant being Eliquis® (*4*). Despite the success in stroke risk reduction, occurrence of the arrhythmia and AF burden are still an issue in AF patients. Key goals of therapy in this population continue to be reduction of AF burden and maintenance of normal sinus rhythm to confer additional benefit in terms of treatment of symptoms and further reduce risk of stroke.

A hallmark of AF is a rapid and irregular heartbeat with a shortened effective refractory period (*5*). One approach to reducing AF burden is to prolong the effective refractory period by inhibiting ion channels that contribute to the cardiomyocyte repolarization. Current antiarrhythmic agents for AF target ion channels that are expressed in both atrium and ventricle. While these drugs are effective to varying degrees for treatment of AF, they all carry significant risk of causing potentially life-threatening arrhythmias in the ventricle (*6*). The ultrarapid-delayed rectifying potassium current (I_{Kur}) channel is encoded by the $hK_V1.5$ gene and is a repolarizing current functionally expressed in atrial myocytes, but not in the ventricular myocytes in humans (*7*). Selective inhibition of I_{Kur} increases atrial refractory period and has the potential to be a safer approach to maintaining normal sinus rhythm than currently used multi-ion inhibitors (*8*) or b-blockers (*9*) to treat arrhythmia. Multiple pharmaceutical companies have sought to identify selective I_{Kur} inhibitors to treat AF over the last 30 years. BMS-919373 (Figure 1) (*10*) is the first selective compound with oral bioavailability in this class to enter clinical studies in AF patients.

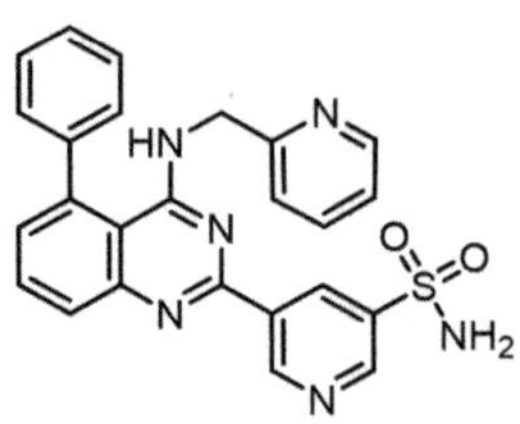

1

Figure 1. Structure of BMS-919373. Reproduced with permission from ref. (10). Copyright 2019 American Chemical Society.

Phenylquinazolines as I_{Kur} Inhibitors

Our initial efforts to identify selective I_{Kur} inhibitors, leveraged the calcium ion channel inhibitor dihydropyridine template (e.g. nifedipine, Figure 2) as a starting point. Structural modification of the C6 ester to an amide and optimization of the C7 aryl group and amide groups led to our first clinical compound BMS-394136 (Figure 3, **2**) (*11*). In Phase 1 normal healthy volunteers, the observed exposure of BMS-394136 was consistent with the predicted human exposure and was dose proportional. However, as the metabolite profile was fully elucidated and the potential for reactive intermediates identified, clinical studies were halted. In addition, the chemical stability with respect to aromatization of the dihydropyrimidine core led to challenges with scale up of the material and an extensive effort was undertaken to evaluate alternate chemotypes which led to the identification of the phenylquinazoline series.

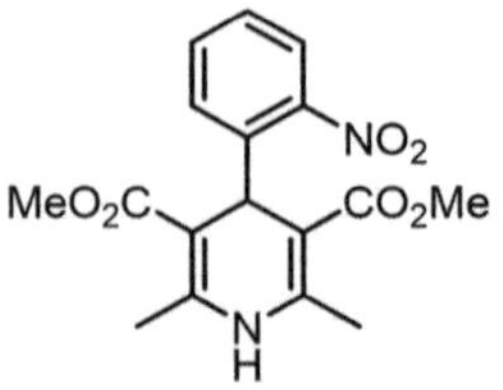

Figure 2. Structure of Nifedipine.

Discovery of BMS-919373

We initially screened templates with 5,6 membered fused heterocyclic rings based on the dihydropyrimidine and then rapidly expanded our efforts to 6,6 membered fused rings and observed $K_V1.5$ activity in analogs in the quinazoline series (*12*). The quinazoline template is an established and well known kinase template; however, observational SAR from published patent applications and disclosures led to an initial hypothesis that substituents at the C5 position could attenuate off-target kinase activity. We first evaluated positional isomers of the phenyl group on the quinazoline template. To measure compound activity in vitro, patch clamp techniques were used to evaluate inhibition of $K_V1.5$, (which encodes the I_{Kur} channel) in L929 cells at increasing concentrations. Phenyl isomers at C6, C7 and C8 had weak potency for inhibition of $K_V1.5$; however, we were encouraged that the C5 phenyl isomer (**3**) had potency for $K_V1.5$, selectivity versus *h*ERG, Ca and Na channels as measured in functional patch clamp assays and did not inhibit any kinase below 5 μM in the BMS primary kinase panel screen of 20 kinases. As a benchmark, quinazoline **3** was advanced to our rabbit pharmacodynamic (PD) model and demonstrated an increase in atrial effective refractory period (AERP) although high plasma exposures (9 μM) were required.

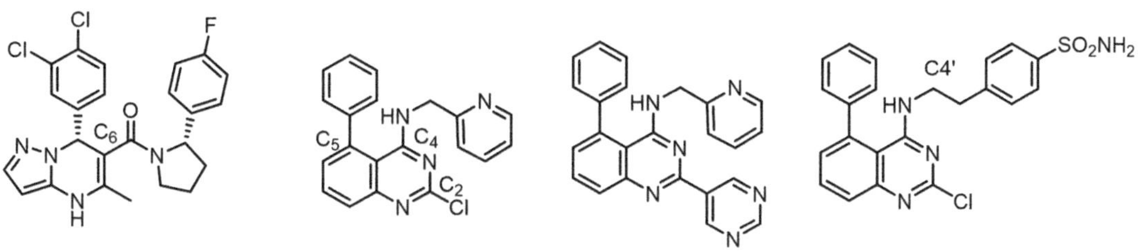

2
BMS-394136
3
4
5

Figure 3. Structure of BMS-394136 and key compounds in the phenylquinazoline series.

With the C5 phenyl group in place, we focused on developing SAR at the C4 and C2 positions to optimize for $K_V1.5$ potency and reduce clearance in an effort to improve the PK/PD profile in this series. Through transition metal-mediated cross-coupling reactions, we were able to introduce a diverse set of 5 and 6 membered heterocycles at C2. All analogues were assessed for potency, selectivity and in vitro microsomal stability (human, rat and mouse). Oxazoles, thiazoles and pyridines at C2 had potency and acceptable microsomal stability, however, selectivity vs *h*ERG was insufficient to advance the compound. Incorporating pyrimidine at C2 addressed potency, microsomal stability and ion channel selectivity and Compound **4** was advanced to the rabbit PD model where it demonstrated a robust and dose-dependent increase in AERP at a plasma concentration 3-fold lower than **3**. As we further advanced pyrimidine **4** through the in vivo testing tree, we observed an unacceptable level of brain penetration for the compound in the rat model. At a dose of 3 mg/kg, administered by intravenous bolus, the brain to plasma concentration of **4** was >1 and seizures were observed in our rat model of brain penetration as we increased to higher doses. We established a target brain to plasma ratio of 0.1 or less in this model to reduce the potential for centrally mediated effects.

Balancing potency and the polarity required to reduce brain penetration proved challenging. In our earlier SAR studies at C4 and C4', we observed that the primary sulfonamide group maintained low micro-molar potency while reducing brain penetration (**5**). Benzenesulfonamide **5** had low microsomal stability across species and was not advanced further. In addition, when we combined the C4' sulfonamide with the C2 pyrimidine, the $K_V1.5$ potency decreased significantly. Leveraging the primary sulfonamide to increase polar surface area (PSA) as a strategy to decrease brain penetration, we sought to incorporate this key functional group at alternate positions in the molecule. Exploring pyridines, pyrimidines and pyridazines at C2 with polar group substitutions such as primary amide, urea, sulfonylurea and sulfonamide led to a series of potent compounds. The best combination for metabolic stability, solubility, selectivity and increased $K_V1.5$ potency, was the sulfonamide group incorporated at the meta- position of the C2 pyridine resulting in the identification of BMS-919373 (**1**).

BMS-919373 demonstrated potency for $K_V1.5$ (IC_{50}: 50 nM), acceptable selectivity versus *h*ERG (37-fold), sodium and calcium channels and an acceptable in vitro safety profile with respect to cytochrome P450 inhibition, Pregnane X receptor transactivation and off target glycoprotein coupled receptors and nuclear hormone activity. BMS-919373 demonstrated low clearance in all species (mouse, rat, cynomolgus monkey, dog and rabbit) and had acceptable bioavailability when dosed as a crystalline suspension. We utilized two models of efficacy in vivo. The pharmacodynamic effects of BMS-919373 were first demonstrated in an anesthetized slow infusion rabbit model. In the rabbit pharmacodynamic model, BMS-919373 demonstrated a dose-dependent increase in AERP with a 10% increase in AERP at a plasma exposure of 0.2 – 0.4 μM. There were no significant effects on VERP in this model consistent with a selective molecule and atrial specific target. In addition to the rabbit model, BMS-919373 was dosed orally in the canine inducible AF heart failure model. The percent reduction in AF duration was compared to baseline measurements and, at a dose of 0.1 mg/Kg (C_{plasma} 0.4 μM), a reduction in AF burden of 50% was observed which was comparable to non-selective ion channel blockers that were efficacious in human (*13*).

The human projected clearance calculated from allometric scaling of the preclinical species PK, together with the exposure required for efficacy in the rabbit and dog models, led to a projected human dose for efficacy of 30-80 mg BID for BMS-919373. With an acceptable safety, pharmacokinetic profile and efficacious dose range in humans, BMS-919373 was advanced to longer term toxicology studies in rats and dogs. BMS-919373 was well tolerated in these long-term studies.

One observation in the rodent and dog PK studies was pH dependent absorption and the potential for food effects and drug-drug interactions. In addition, the solubility of BMS-919373 at pH 7 was low, leading to a pro-drug effort to increase solubility and to mitigate pH dependent absorption. Prodrugs attached at C4 amine and at C2' sulfonamide were initially explored, however, C2' attachments were readily prepared and more stable in solution. The phosphoroamide pro-drug (Figure 4, **6**) was identified with an excellent crystalline suspension rat PK profile and no circulating pro-drug at high dose (10). Further, when dosed in a monkey study, **6** demonstrated improved bioavailability and reduced pH effect with famotidine pre-treatment as compared to BMS-919373.

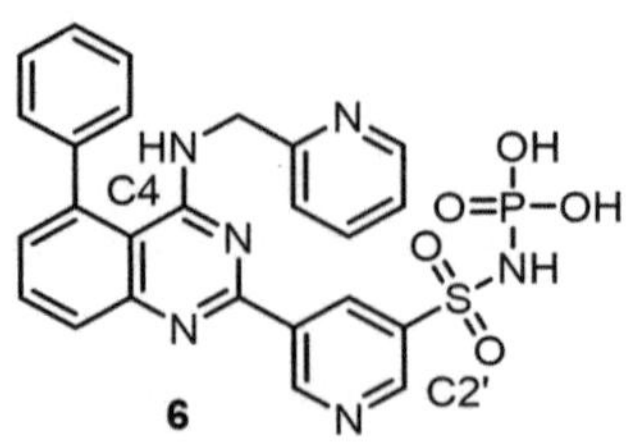

6

Figure 4. Structure of BMS-919373 phosphoramide pro-drug.

Given the low projected dose for efficacy for BMS-919373 and the ability to give an oral dose as a suspension, BMS-919373 was selected to advance to Phase I clinical studies in 2011. A total of 169 normal healthy volunteers were administered BMS-919373 in a single dose up to 100 mg and in multiple doses up to 30 mg for 2 weeks with QD dosing. The compound was rapidly absorbed with a T_{max} of from 1.5 - 3 h and generally dose proportional exposure. The observed human half-life was longer than predicted at 120 h with 5-7 fold accumulation observed following QD dosing. No clinically meaningful drug-drug interactions were observed with midazolam (Cytochrome P450 3A4), warfarin (Cytochrome P450 2C9) and statins. Subsequently, BMS-919373 was advanced to Phase II studies in 2014 (*14*).

Medicinal Chemistry Synthesis of BMS-919373

The first-generation synthetic route for BMS-919373 was designed to enable analogue synthesis by building diversity at the C5, C4 and C2 positions (Scheme 1) (10). Starting with readily available bromoisatin **7**, oxidative cleavage followed by condensation to the pyrimidinedione **9** led to the C5 bromide which could be cross-coupled to build in diversity at this position. Cross-coupling with phenylboronic acid generated 5-phenylpyridmidinedione **10**. Chlorination using phosphorous oxychloride gave the dichloroquinazoline **11**, which upon displacement with aminomethylpyridine at the more labile C4 position led to phenylquinazoline, **12**. Compound **12** could then undergo cross-coupling under Suzuki-Miyaura conditions with a diverse set of boronic acids or esters to yield C2 analogues. The yields for cross-coupling heterocyclic boronic esters were typically low and products were challenging to purify, a key issue in the first-generation synthesis of BMS-919373. In addition, a multi-step synthesis of the sulfonamido pyridine boronate ester **15** was required from the commercially available 3-pyridylsulfonic acid **13** via bromination of the corresponding sulfonyl chloride, conversion to the primary sulfonamide **14**, and subsequent installation of the boronate ester from the bromide.

Transition to Process Chemistry and Initial Kilogram Synthesis of BMS-919373

With BMS-919373 as the clear clinical candidate for the I_{Kur} inhibitor program, we began exploring routes that would enable rapid delivery of multi-kilogram quantities of the active pharmaceutical ingredient (API) to support both registrational toxicology studies and Phase I clinical trials. No longer needing the flexibility of the discovery route to install various groups at the C4 and C5 positions, we focused our efforts on a condensation approach where the sulfonamidopyridine group at C4 could be incorporated into the quinazoline ring formation. This disconnection would enable us to avoid both the large-scale chlorination to make 2,4-dichloroquinazoline **11** and the challenging Suzuki-Miyaura coupling in the first-generation route.

Identical to the Discovery Chemistry route, 4-bromoisatin **7** was chosen as the starting material for the second-generation route (Scheme 2) due to its commercial availability. It provided access to the desired 1,2,3-substituted arene core in two steps via oxime formation using hydroxylamine followed by oxidative ring opening using PCl_5 and $POCl_3$, which generated the desired cyanoaniline **16**. A Suzuki-Miyaura coupling with phenylboronic acid gave good yield of the target intermediate **17**.

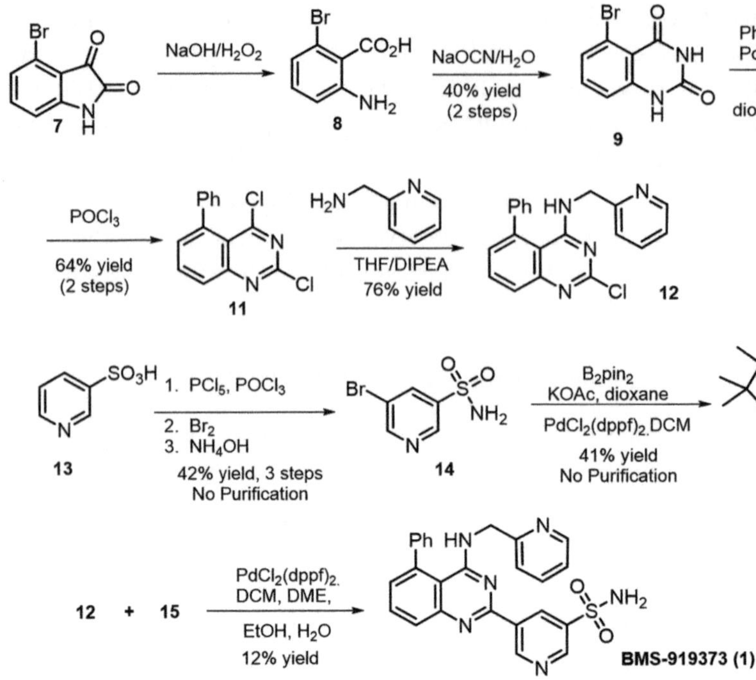

Scheme 1. Medicinal chemistry synthesis of BMS-919373.

The other condensation partner, sulfonamidopyridine **20**, was prepared in two steps from 3,5-dibromopyridine **18**. The first step was a challenging Grignard process to sequentially metallate, sulfonate, and amidate to yield the *tert*-butyl protected sulfonamide **19**. The cyanation of this compound using palladium catalysis with potassium ferrocyanide allowed conversion to the corresponding nitrile substituted pyridine **20**. The centerpiece for this route is the efficient and robust condensation reaction between the cyanoaniline **17** and cyanopyridine **20** using sodium *tert*-butoxide to form **21**. Having the core of BMS-919373 assembled, the last two transformations were alkylation of the primary amine with 2-chloromethylpyridine to form **22** followed by deprotection of the sulfonamide with hydrochloric acid. Three campaigns were completed using this effective route, the last of which supplied 7 kg of API for Phase II studies.

Route Scouting for Commercial Route of BMS-919373

Initial deliveries centered on speed to patient as BMS-919373 moved through the early stages of clinical development. The start of Phase II clinical studies brought a change in strategy in overall route design (*15, 16*). As the quantities of material needed to support the program increased along with the likelihood of a commercial drug, the requirements of the route shifted to support large-scale manufacturing. This new route would need appropriate characteristics such as low cost of goods, process mass intensity (PMI), and decreasing step count while improving process greenness, process robustness, and ensuring intrinsically safe reactions. To this point, several approaches to assemble the quinazoline core had been demonstrated. While there were advantages of some routes over others, none provided the disruptive innovation desired as every route utilized the relatively expensive 4-bromoisatin (7) to provide the ability to install the phenyl group at C5 (Scheme 3, paths A-C). The only exception was a C-H arylation approach to installing this group via a proposed directed C-H activation utilizing the picolylamine at C-4 (Scheme 3, path D).

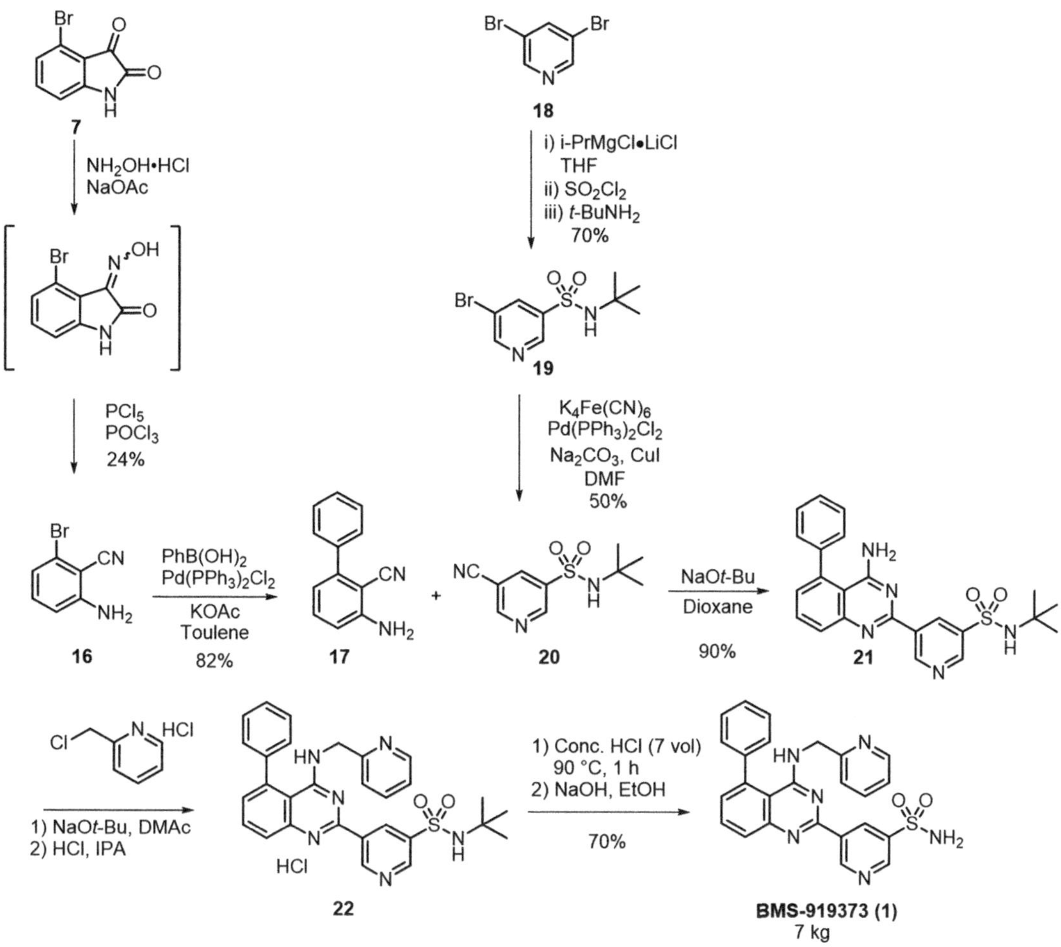

Scheme 2. Initial kilogram production of BMS-919373.

Route Strategy: Direct Arylation

Of the routes considered to BMS-919373, the shortest route to API that we envisioned employed a series of palladium-catalyzed transformations (Scheme 4). The initial discovery route utilized a des-phenyl Suzuki-Miyaura coupling similar to the one proposed in this route; however, it only afforded the cross-coupled product in 12% yield. We believed that optimization of the des-phenyl coupling would provide us with conditions also effective for the cross-coupling in the discovery route as the phenyl ring should not significantly impact the Suzuki-Miyaura coupling. This decision allowed us to focus on the key bond-forming reactions by starting from the commercially available 2,4-dichloroquinazoline **23**.

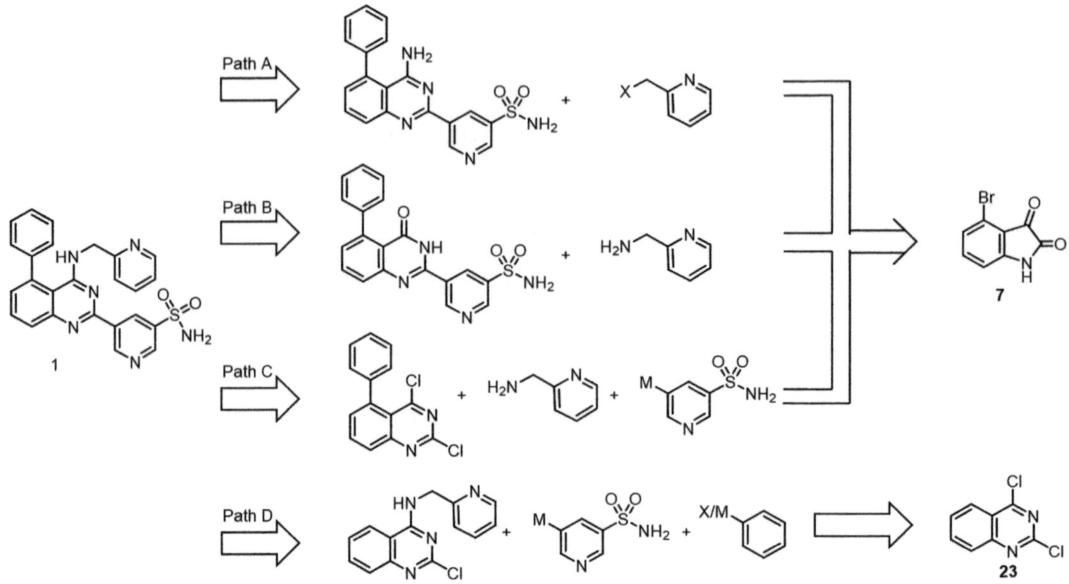

Scheme 3. Retrosynthetic Analysis of BMS-919373. Reproduced with permission from ref. (15). Copyright 2019 American Chemical Society.

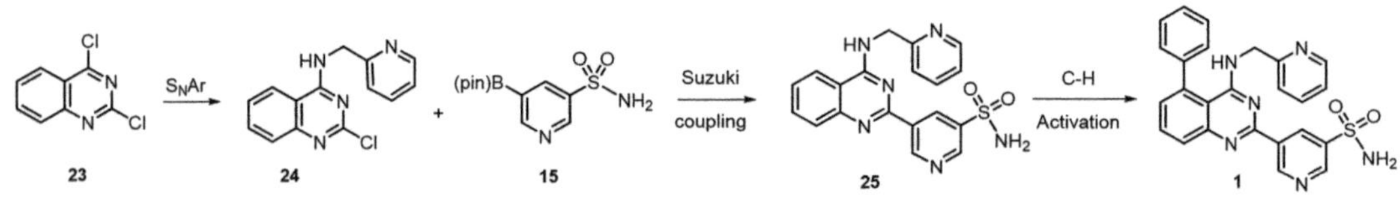

Scheme 4. Route to API from 2,4-Dichloroquinazoline.

Installation of the picolylamine subunit was rather straightforward by performing an S_NAr reaction with DIPEA as the base and IPA as the solvent. After reaction completion, the desired product **24** was isolated in 93% yield through a direct drop process by the addition of water (Eq 1 - Reproduced with permission from ref. (*15*); Copyright 2019 American Chemical Society).

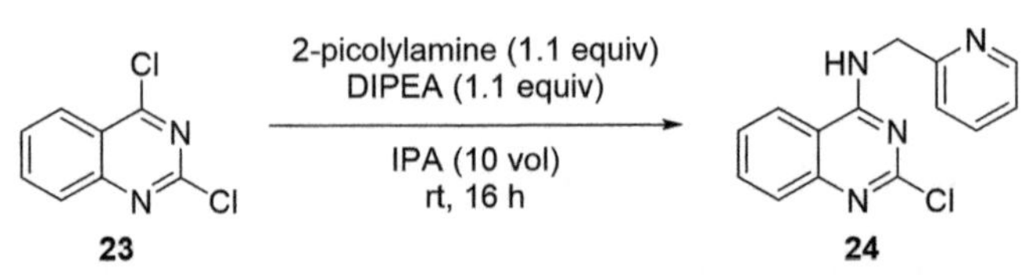

(1)

This heteroaryl chloride would serve as the electrophilic partner in the Suzuki-Miyaura cross-coupling. The nucleophilic partner, the boronate pinacol ester **15**, was prepared through a palladium-catalyzed borylation with B_2pin_2. Earlier work on the project suggested that the sulfonamide may have to be protected to increase the solubility of the late stage intermediates. Therefore, the free and *tert*-butyl protected bromo-substituted sulfonamides could be subjected to the same set of reaction conditions to afford the desired heteroaryl boronate esters **15** and **26** in yields of 89% and 83%, respectively (Eq 2). With the goal of having the shortest route without a protecting group, we chose to initially develop the cross-coupling with the free sulfonamide as we believed any conditions could also be applied to the *tert*-butyl protected analogue.

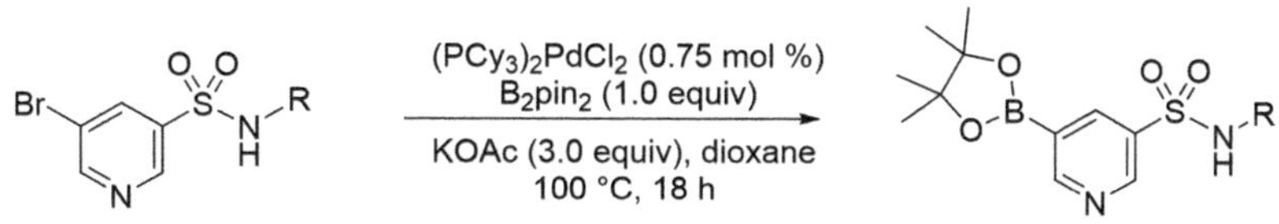

(2)

15 (R= H), 89%
26 (R = *t*-Bu) 83%

The Suzuki-Miyaura reaction (Eq 3) was optimized via High-Throughput Experimentation (HTE) where several rounds of screening investigated the impact of ligand, solvent, base, and temperature on the desired transformation. The corresponding potassium organotrifluoroborate was also prepared and tested in a series of cross-coupling reactions. However, lower conversion was observed, and therefore, the coupling with the boronate pinacol ester was investigated further.

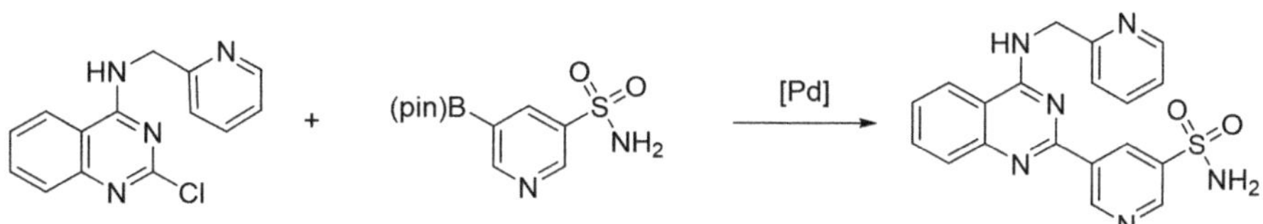

(3)

24 **15** **25**

A summary of the results of coupling between the quinazoline chloride **24** and boronate pinacol ester **15** can be seen in Figure 5. Several classes of ligands were investigated in the coupling, including triarylphosphines, bidentate ligands, and Buchwald ligands; however, the only class of ligands that consistently provided high levels of conversion to the desired product were ferrocenyl ligands. The top ligand was 1,1′-bis(di-tert-butylphosphino)ferrocene (DtBPF), providing 75% (*17*) of product in 4:1 dioxane/water with K_2HPO_4 as the base.

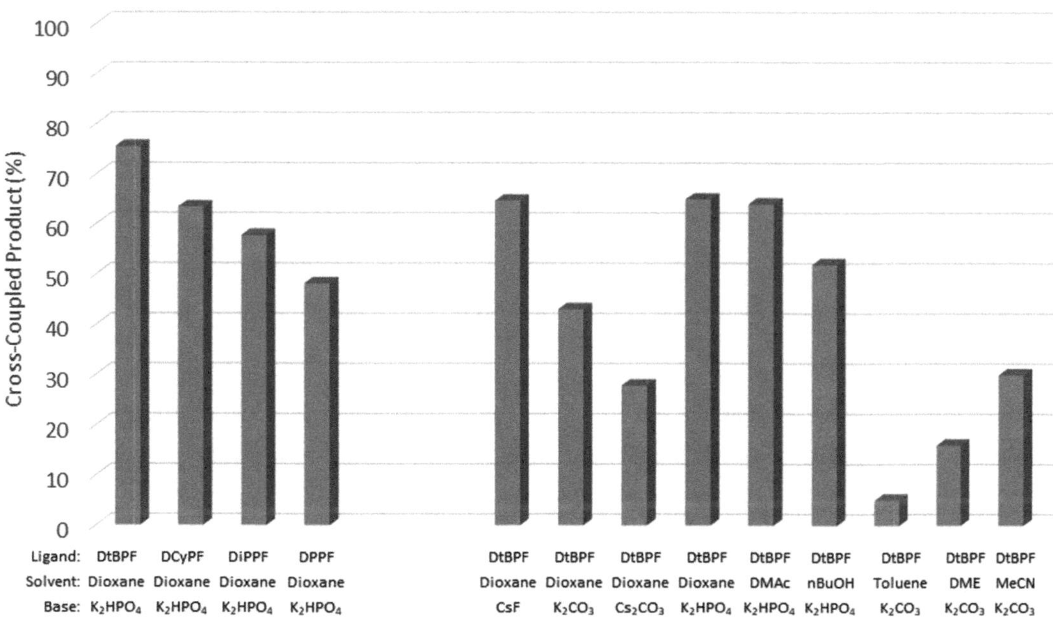

Figure 5. Results of the HTE screening for the Suzuki-Miyaura cross-coupling. Reaction conditions: $Pd(OAc)_2$ (5 mol %), Ligand (5 mol %), Base (3.0 equiv), Solvent (4:1 solvent/water, 0.1 M), 95 °C, 16 h.

Under this optimized system, other ferrocene ligands afforded the product in ≤60% yield. A class-variable screen of bases showed that K_2HPO_4 was uniquely effective at the desired transformation as more common bases provided the product in lower yield. A solvent screen (4:1 solvent/water) utilizing DtBPF as the ligand also demonstrated that dioxane was superior to other solvents by at least 10% in yield.

Although further optimization could have led to improved reaction conditions, downstream chemistry led us to rethink our approach with the free sulfonamide. The inherent insolubility of the Suzuki-Miyaura coupled product resulted in poor reactivity in subsequent steps. We therefore subjected the *tert*-butyl protected sulfonamide substrate **26** to the same set of reaction conditions, which resulted in the desired cross-coupled product in 95% **27** (Eq 4). However, the work-up and isolation of the cross-coupled product from this reaction stream proved to be difficult, presumably from residual ligand in the organic stream. We were unable to either effectively crystallize the product without a series of manipulations that would lead to a tedious process or to extract the product into the aqueous with strong acid.

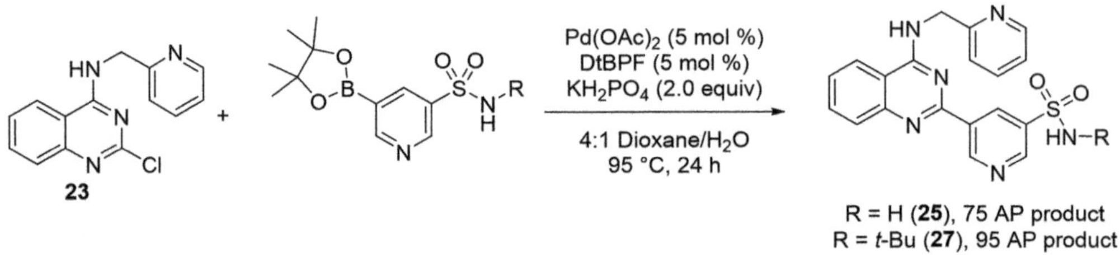

(4)

Based on the high conversion observed under the DtBPF conditions, we were hopeful that incorporation of the *tert*-butyl protecting group would allow for a wider range of reaction conditions to afford the desired product. We designed one HTE screen to rapidly identify a second set of conditions for the desired cross-coupling (Figure 6). Unlike the free sulfonamide where the only class of ligands that worked well were ferrocene ligands, several different ligands provided high conversion to the desired product. We were most intrigued by PCy_3●HBF_4, as the same ligand was used in the borylation reaction. This ligand provided the des-phenyl API **27** in 85% yield.

Further optimization of this cross-coupling led to the use of a preformed Pd(II) complex with K_2CO_3 as the base in 1:1 MeTHF/H_2O. The low solubility of the product resulted in a reactive crystallization, affording the desired product in 82% yield (Eq 5). If this route was selected for development, a telescoped borylation/cross-coupling process would have been investigated to avoid isolation of the pinacol boronate ester.

An important aspect of our route scouting efforts was to define the protecting group strategy on the sulfonamide because of the improved reactivity and isolability of the *tert*-butyl protected sulfonamide For the IND-Tox and FIH routes, the *tert*-butyl group was selected as the protecting group. This was introduced by reaction of sulfonyl chloride with *tert*-butyl amine, and was removed by conc. HCl at 90 °C. To avoid the harsh conditions to remove the *tert*-butyl group on the pyridine fragment, various other protecting groups were investigated (Scheme 5). Diphenylmethyl, the first one evaluated, could be introduced by using diphenylmethylamine in the amidation. The substrate **28** was underwent Pd-catalyzed borylation and Suzuki-Miyaura coupling to provide product in 76-80% yield, however, deprotection of diphenylmethyl group appeared more challenging with conc. HCl than the *tert*-butyl group (Scheme 5a).

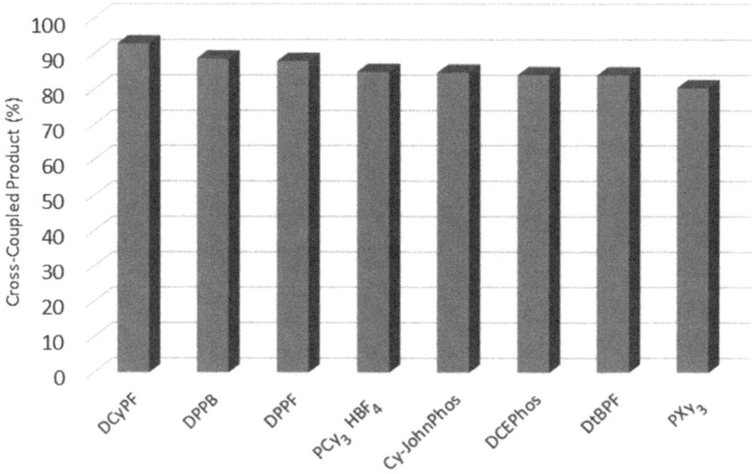

Figure 6. Results of the HTE screening for the Suzuki-Miyaura cross-coupling. Reaction conditions: $Pd(OAc)_2$ (10 mol %), Ligand (21 mol %), K_2CO_3, (3.0 equiv), Solvent (3.33:1 solvent/water, 0.1 M), 75 °C, 16 h.

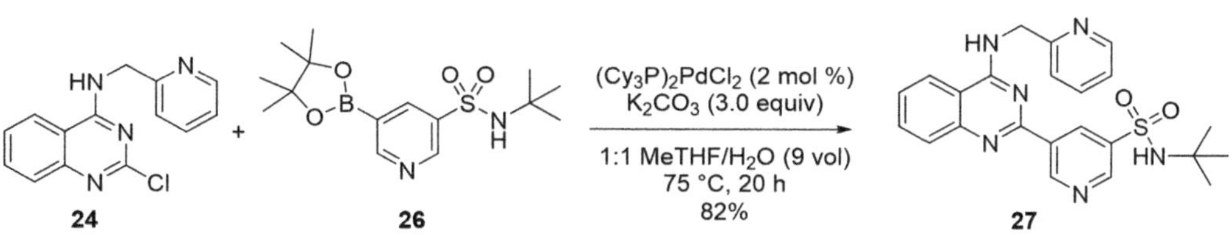

(5)

The more acid-labile triphenylmethyl (trityl) group was then evaluated. Substrate **29** was tested with the borylation/Suzuki-Miyaura sequence to afford the coupled product in 84% yield. The trityl group can be removed completely by heating with conc. HCl, but any milder conditions, such as 4 M HCl or TFA resulted in partial deprotection (Scheme 5b). Several more labile protecting groups were tested, including dimethoxytrityl, Boc, and TBS, but these were either not stable to the Suzuki-Miyaura reaction conditions or were deprotected when exposed to moisture. Therefore, we decided to use the *tert*-butyl group for the commercial route.

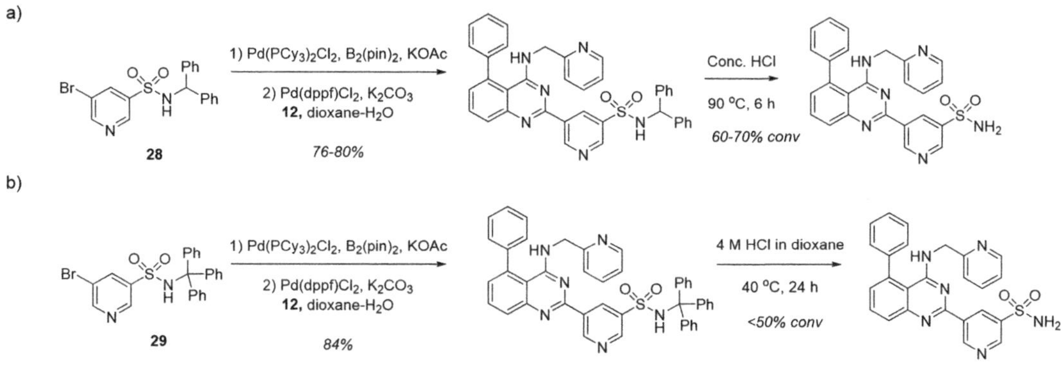

Scheme 5. Evaluation of alternative protecting groups of sulfonamide.

The optimization and implementation of the Suzuki-Miyaura reaction afforded us the opportunity to start looking at the key retrosynthetic disconnection: the C-H activation (*18, 19, 20*). We were optimistic about the proposed direct arylation as literature precedent showed the direct arylation of a similar aromatic bicyclic compound, installing an arene at the C8 position of a substituted naphthalene (Scheme 6) (*21*, *22*). Further, several methodologies have been developed over the last few years to selectively functionalize the C-8 position of substituted naphthalenes (*23*, *24*, *25*, *26*, *27*, *28*).

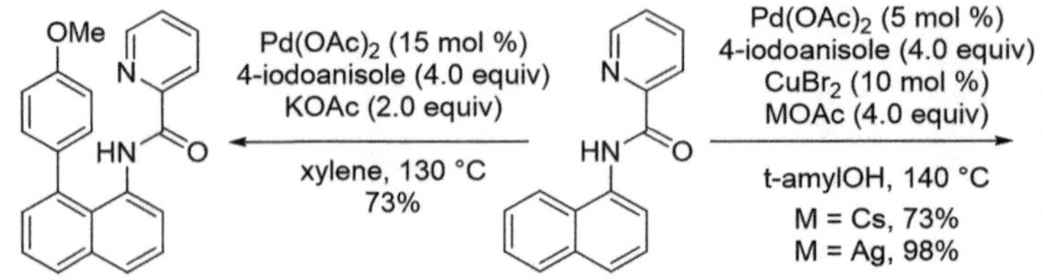

Scheme 6. Literature Precedent for the C-H activation. Reproduced with permission from ref. (15). Copyright 2019 American Chemical Society.

During the route ideation phase of this program, four different substrates for the C-H activation were proposed (Figure 7). The Schiff base **30** was quickly ruled out as it was unable to be synthesized under an array of reaction conditions. The amide **31**, utilizing the same directing group as the naphthalene system reported by Qi and Daugulis (*21*, *22*), would require a subsequent reduction step to afford the picolylamine subunit. A model system with the picolinamide subunit was investigated in the C-H activation and resulted in moderate conversion to the desired product. As mentioned previously, the free sulfonamide Suzuki-Miyaura product **25** was highly insoluble, even at reaction temperatures above 100 °C, resulting in no conversion under a variety of reaction conditions. Therefore, the *tert*-butyl protected sulfonamide **27** was the top substrate for optimization of the direct arylation.

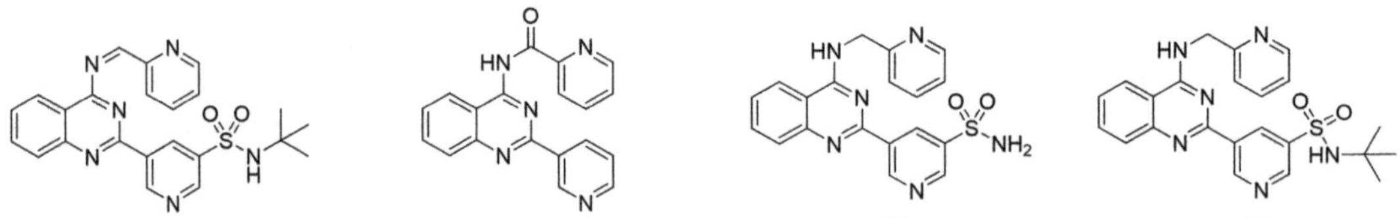

Figure 7. Proposed substrates for the C-H activation. Reproduced with permission from ref. (15). Copyright 2019 American Chemical Society.

Although there were several metals (Ru, Rh, Pd) and arene sources (boronates, halides, iodonium salts) capable of performing this type of C-H activation, we focused on a palladium-catalyzed direct arylation with an iodoarene (*29*, *30*, *31*, *32*, *33*, *34*, *35*, *36*, *37*) The direct arylation was optimized via HTE. The goal of the first screen was to determine if the C-H activation was even possible. Therefore, 4.0 equiv iodobenzene in various co-solvents (*p*-xylene, DMAc) as well as neat iodobenzene and bromobenzene were tested along with different bases (KOAc, KOPiv, TMAOAc, Cs_2CO_3) at 118 °C. Based on the literature precedent, the impact of CuBr as an additive was also investigated.

The results of this screen can be seen in Figure 8. Bromobenzene as a solvent was ineffective across all conditions tested. The top result utilized KOAc as a base in neat iodobenzene, providing proof of concept on the desired transformation in 93% yield. Changing the base to KOPiv resulted in slightly lower product formation. The addition of CuBr was not advantageous as several impurities were observed under these reaction conditions. Decreasing the equivalents of iodobenzene greatly impacted the reaction, and resulted in <50% product when 4.0 equiv were employed in *p*-xylene. Interestingly, DMAc as a solvent was not effective across all conditions tested. Lastly, TMAOAc and Cs_2CO_3 both provided low levels of desired product.

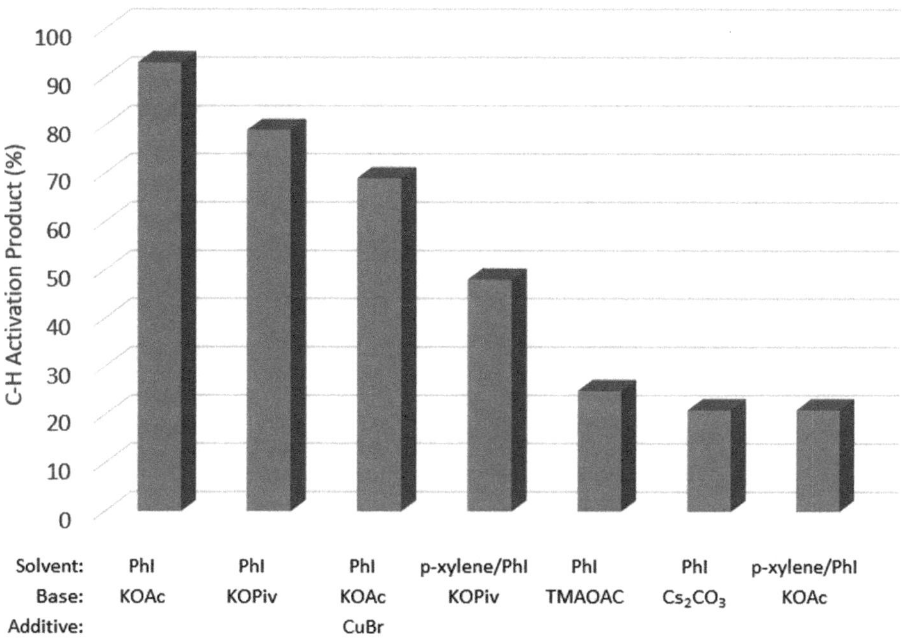

Figure 8. Results of the first direct arylation screen.

Although further optimization of the direct arylation was needed to afford a robust, scalable process, the conditions from this first screen (Eq 6 - Equation 6 is reproduced with permission from ref. (*15*). Copyright 2019 American Chemical Society.) provided support that a direct arylation route to the I_{Kur} API was feasible and should be considered for the commercial route.

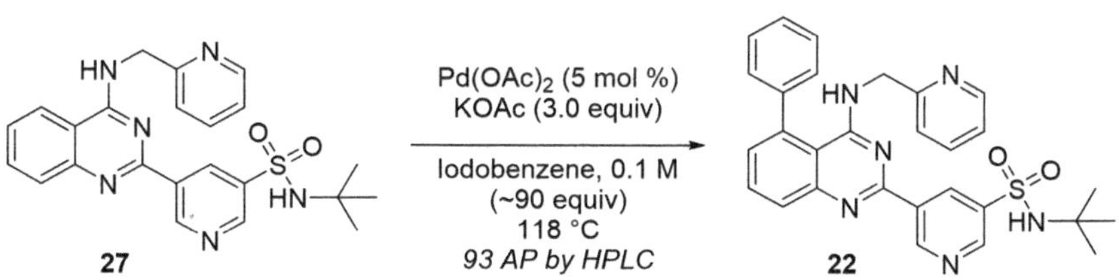

(6)

Route Strategy: Modified Niementowski Reaction

Another attractive disconnection strategy to BMS-919373 would be to install the 2-picolylamine fragments through an S_NAr type reaction utilizing an in situ activation of the corresponding cyclic amide (Scheme 7). This would circumvent issues with the existing alkylation chemistry that provided deleterious over-alkylation side products. The quinazolinone would then derive from a condensation reaction between two core intermediates that would be accessible from the existing starting materials from the previous synthesis.

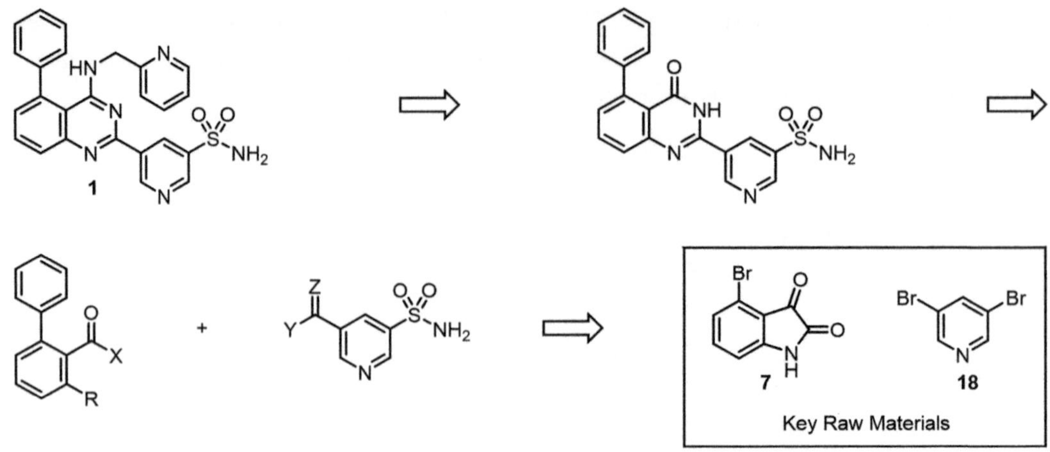

Scheme 7. Disconnection strategy for cyclization reaction.

A significant advantage of the condensation approach is that numerous iterations of condensation reactions could be evaluated from a common starting material (Scheme 8). It was envisioned that no less than four potential condensation partners for the biaryl fragment could be delivered from a common isatoic anhydride intermediate **32**. In a similar fashion, it was expected that existing intermediate 3-bromopyridine **15** would be able to deliver up to six different coupling fragments.

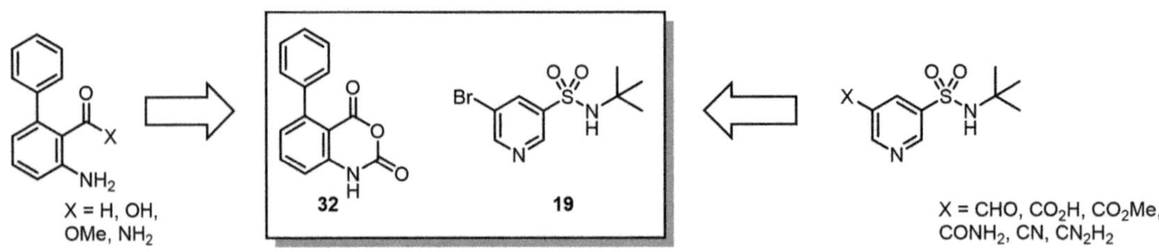

Scheme 8. Key intermediates in the cyclization route.

The effort to access our desired isatoic anhydride **32** began with a Suzuki-Miyaura reaction of 4-bromoisatin and phenyl boronic acid (Eq 7).

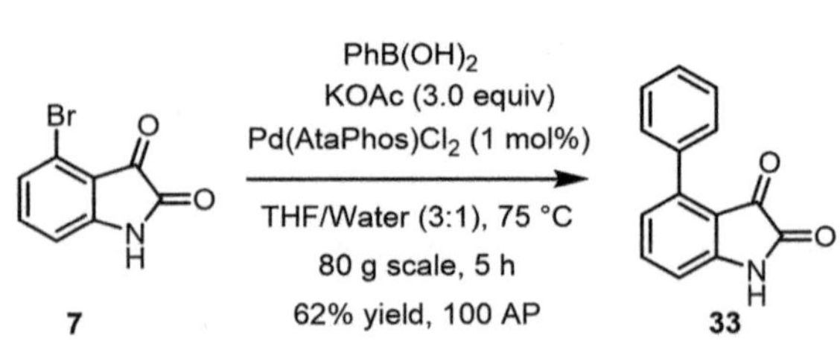

(7)

A rapid screen of Suzuki-Miyaura conditions (Table 1) revealed that the AtaPhos system, developed previously by BMS, gave optimal performance, affording a 62% isolated yield on 80 g scale.

Table 1. Optimization of the Suzuki-Miyaura Reaction of 4-Bromoisatin

Pd Catalyst	*Base*	*% Product*	*Pd Catalyst*	*Base*	*% Product*[a]
Pd(dppf)Cl_2	Cs_2CO_3	63	Pd(AtaPhos)$_2Cl_2$	Na_2CO_3	90
Pd(dppf)Cl_2	K_3PO_4	68	Pd(AtaPhos)$_2Cl_2$	Cs_2CO_3	85
Pd(dppf)Cl_2	NaO*t*Bu	30	Pd(AtaPhos)$_2Cl_2$	K_3PO_4	84
Pd(XantPhos)Cl_2	Cs_2CO_3	15	Pd(AtaPhos)$_2Cl_2$	K_2CO_3	86
Pd(XantPhos)Cl_2	K_3PO_4	43	Pd(AtaPhos)$_2Cl_2$	KOAc	92
Pd(XantPhos)Cl_2	NaO*t*Bu	4	Pd(AtaPhos)$_2Cl_2$[b]	KOAc	95

[a] Major impurities were unreacted boronic acid and biphenyl. [b] Reaction was run with 1 mol% Pd catalyst.

The subsequent oxidation of biaryl **33** to afford isatoic anhydride **32** was studied (Table 2). The survey found that persulfate oxidants were ineffective and that *m*-CPBA afforded higher conversions. The unoptimized process was scaled to 10 g to afford the desired isatoic anhydride **32** in 65% isolated yield. From a process development perspective, the use of *m*-CPBA was problematic but the chemistry in its present state was acceptable at this point to push material forward for downstream route investigation.

Table 2. Optimization of the Oxidation to the Isatoic Anhydride 32

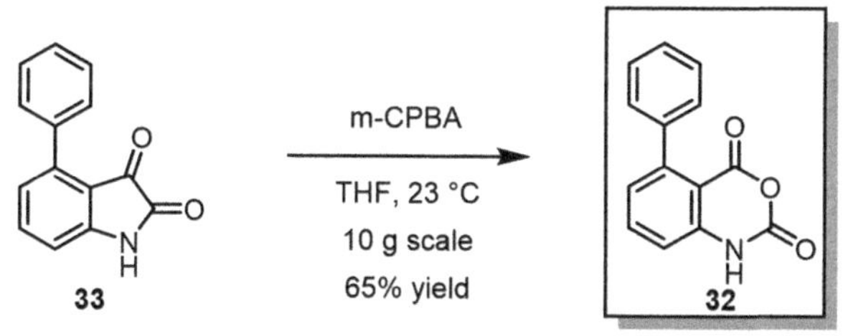

Entry	Conditions	Result[a]
1	$K_2S_2O_8$, NaOMe, MeOH	No reaction
2	Oxone, CH_3CN/Water	No reaction
3	1.5 equiv *m*-CPBA, CH_2Cl_2, 2 days	71% conversion
4	1.5 equiv *m*-CPBA, EtOAc, 24 h	85% conversion
5	1.1 equiv *m*-CPBA, THF, 24 h	90% conversion

[a]Conversion was measured by ^1H NMR

The first condensation to be targeted was based on literature precedent (39) that demonstrated the condensation of an aryl nitrile with an aryl anthranilate. The ester fragment **34** was prepared through DMAP catalyzed methanolysis of isatoic anhydride **32**. Intermediate pyridyl nitrile **20** from the previous synthesis was subjected to sulfonamide deprotection to give primary sulfonamide **35** (Scheme 9).

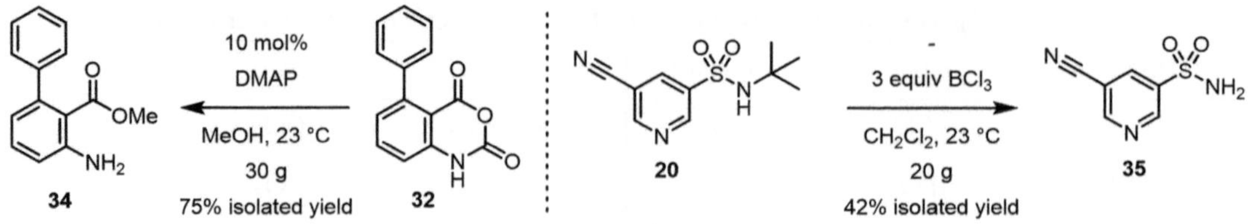

Scheme 9. Initial conditions for condensation starting materials.

With the two starting materials prepared, the first coupling iteration was explored. We used established literature conditions, employing anhydrous HCl in dioxane at 90 °C, which afforded the desired quinazolinone product **36** in 40% yield on 1 g scale (*39*). The aggressive conditions were not likely to be considered a serious option for commercialization, although the demonstrated proof of concept for a condensation approach allowed further investigation of the proposed endgame. Therefore, the final coupling was examined using PyBrop and 2-picolylamine and found to deliver the API with 90% conversion on 500 mg scale (Scheme 10). The initial investigation demonstrated the viability of a condensation strategy and delivered a potential protecting group-free synthesis of the API in seven total steps with the longest linear steps being five reactions.

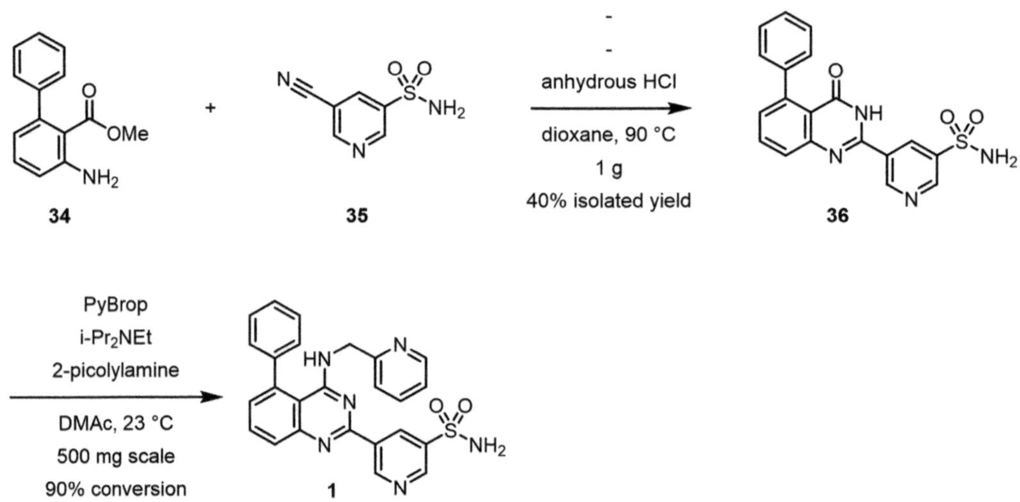

Scheme 10. Acid-catalyzed cyclization route.

The existing condensation route provided a solid foundation to build a potential commercial process although several critical factors required a remedy. The intermediates featuring the primary sulfonamide functionality were poorly soluble in most organic solvents and presented a liability for delivering high quality API. A similar risk to API quality was the presence of the key 2-picolylamine fragment in the final step. As opposed to a near quantitative deprotection reaction, utilizing a key bond-forming reaction in the API step leaves minimal opportunity to purge any potential impurities related to this fragment. Therefore, in addition to relieving the harsh conditions needed to accomplish the condensation, the next iteration of this route would seek to employ a sulfonamide protecting group to both improve solubility of intermediates and to utilize a near quantitative deprotection reaction in the API step instead of a key bond-forming reaction.

Investigation of a second-generation condensation route began with the search for a reaction that would utilize more reactive condensation partners to mitigate the need for forcing conditions. A variant of the Niementowski reaction (Scheme 11) (*40*) that had been shown to effect the condensation of amidines with isatoic anhydrides was particularly intriguing (*41*). There remained a question though: utilizing amidines without substitution at the amidine nitrogens, a necessity for accessing quinazolin-4(3*H*)-ones, had not been demonstrated. Nevertheless, this strategy was advantageous as the requisite isatoic anhydride was already in hand and effort could be directed toward the synthesis of the desired amidine.

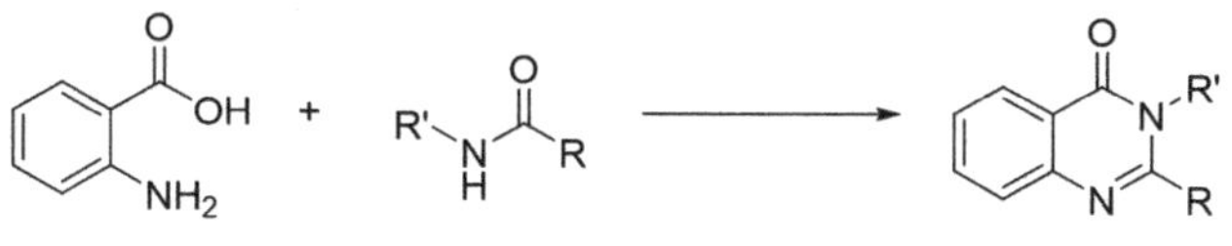

Scheme 11. General Niementowski reaction.

It was quickly found that treatment of aryl nitrile **20** with catalytic sodium methoxide generated intermediate imidate **37**, which upon treatment with ammonium chloride afforded desired amidine **38** (Scheme 12). The amidine was expected to impart higher reactivity than the corresponding nitrile in prospective condensation reactions and the *tert*-butyl protected sulfonamide was expected to improve solubility for downstream intermediates. Importantly, the amidine hydrochloride salt was readily isolable as a crystalline solid in high purity.

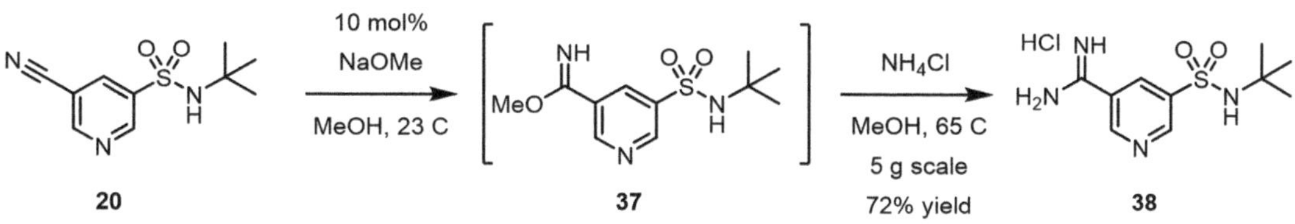

Scheme 12. Conditions for the amidine formation.

The preliminary work on the condensation focused on the use of isatoic anhydride as a model system (Table 3). We discovered that the reaction occurred in a variety of solvents, including acetonitrile, *tert*-amyl alcohol, dioxane, THF, and toluene at 65 °C. The reaction in many cases proceeded spontaneously without the need for the DMAP catalyst. The exceptionally mild reaction conditions that accommodated a variety of solvents was encouraging.

The newly discovered condensation conditions were explored using the desired isatoic anhydride **32.** The reaction proceeded with 84% conversion on 1 g scale and delivered a 65% isolated yield on the first attempt. The previously established S_NAr conditions were then applied to the condensation product giving 90% conversion on 500 mg scale to complete a formal synthesis in a six longest linear step sequence (Scheme 13).

Route Strategy: Alternative Cyclizations

With proof of concept on a promising condensation approach utilizing a modified Niementowski reaction, we investigated an alternative cyclization to form the quinazoline core (Scheme 14). In a model system, 2-amino-6-phenylbenzonitrile (**40**), an intermediate used in the first-generation synthesis, was coupled with nicotinic acid to generate amide **41** in 52% yield. The product was then heated in sulfolane in the presence of excess PCl_5 to form chloroimidate **42**, which was in situ quenched with 2-picolylamine to afford the desired compound **43** in 60% yield over two steps. Although the cyclization was successful in the model study, it was never applied to the actual system, due to the concern of the hazardous reagent PCl_5 and the harsh cyclization conditions that could decompose the protecting group on the sulfonamide.

Table 3. Optimization of the Condensation Reaction Using Isatoic Anhydride as a Model

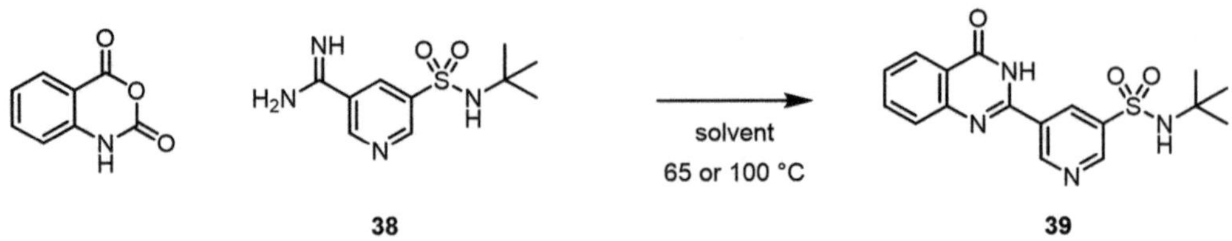

Entry	Solvent	Additive	Conv.
1	Acetonitrile	-	94%
2	Acetonitrile	DMAP	95%
3	t-Amyl Alcohol	-	83%
4	t-Amyl Alcohol	DMAP	99%
5	Dioxane	-	94%
6	Dioxane	DMAP	91%
7	Tetrahydrofuran	-	100%
8	Tetrahydrofuran	DMAP	96%
9	Toluene	-	20%
10	Toluene	DMAP	92%

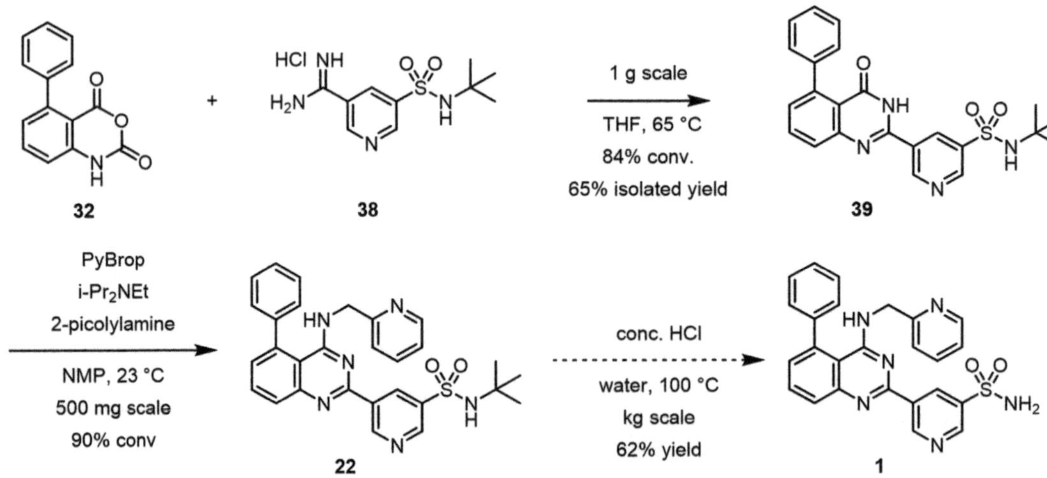

*Scheme 13. Route to I_{Kur} API (**1**) utilizing condensation route.*

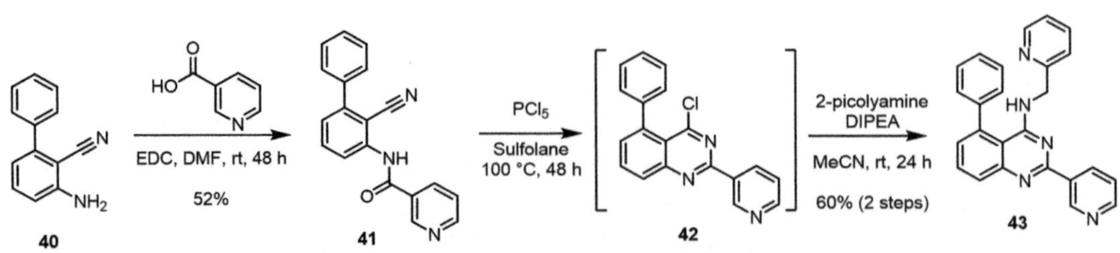

Scheme 14. Model system to study alternative cyclization.

Another type of condensation reaction that enabled further evaluation of the Suzuki-Miyaura route identified by discovery team was investigating an alternative approach to the key intermediate, 5-phenyl-2,4-dichloroquinazoline **11**. This can be converted to the API by an S_NAr reaction using 2-picolylamine and subsequent Suzuki-Miyaura coupling to form the pendant pyridine fragment.

We initially used the reaction of triphosgene with 2-amino-6-phenylbenzonitrile **40** to prepare the cyclized product **11** in 54% yield (Scheme 15a). This one-step preparation was then replaced with a more effective two-step procedure to avoid the usage of the toxic triphosgene. Thus, heating a mixture of **40** and DBU in DMF under 1 atm of CO_2, followed by direct drop crystallization afforded the quinazolinone **10** in 80% yield. Treatment of **10** with $POCl_3$ led to the formation of dichloroquinazoline **11** in 82% yield (Scheme 15b). We assumed this approach could also be used to make des-phenyl dichloroquinazoline for the C-H activation route in order to avoid sourcing the heterocycle.

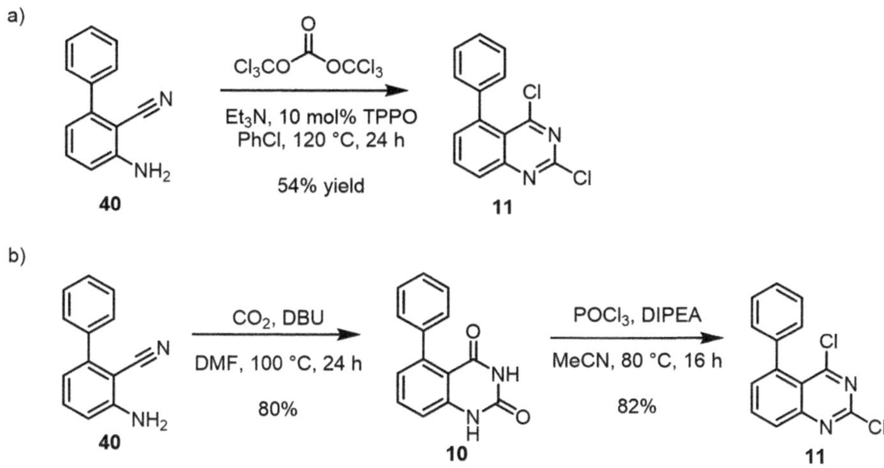

Scheme 15. Two routes to synthesis of 5-phenyl-2,4-dichloroquinazoline.

Route Strategy: Path Forward

We determined at this point that all of the routes that achieved proof of concept had suffered from several challenges. The C-H arylation route and the alternative cyclization route required the use of 2,4-dichloroquinazoline, which is a potential sensitizer and is hampered by a high cost. Furthermore, preparing 2,4-dichloroquinazoline would require the use of hazardous reagents. The use of palladium catalysts in three out of the six reactions was concerning not only because of cost but due to the ligand-like nature of the intermediates as well as their poor solubility in organic solvents that could stymie palladium remediation.

Similarly, the condensation approach (modified Niementowski reaction) utilized a high cost/kg starting material, utilized a palladium-catalyzed Suzuki-Miyaura reaction in the first step, and a high-energy oxidation in the second step with *m*-CPBA. However, an interesting opportunity presented itself with the combination of the direct arylation and modified Niementowski routes (Scheme 16). The key challenge to C-H arylation route involving a high-cost and hazardous starting material would be eliminated if the key intermediate could be synthesized by the condensation route. Similarly, the key challenge to the condensation route was the need to use the highly functionalized 4-bromoisatin to install the aryl fragment en route to corresponding isatoic anhydride. The new combined condensation, C-H activation strategy would utilize isatoic anhydride as the starting material, which greatly reduces the cost to \$10/kg, and nitrile **20** to access quinazolinone **27**. An S_NAr reaction would then furnish the desired quinazoline needed for the direct C-H arylation.

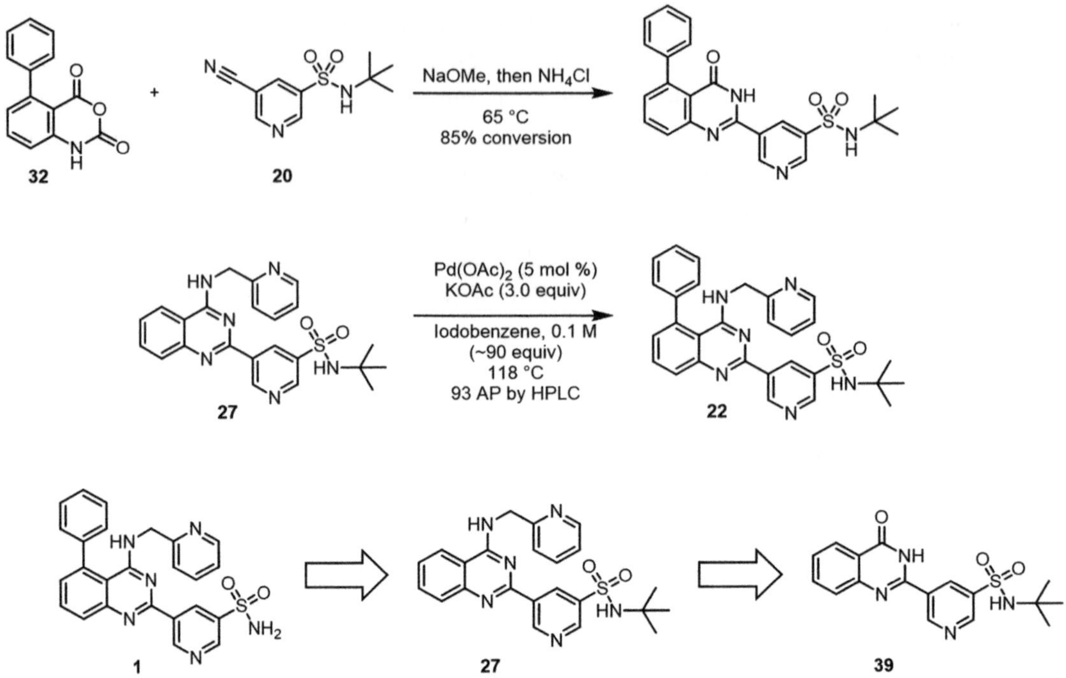

Scheme 16. Realization of hybrid condensation, C-H activation route.

The new route was investigated by scaling the conditions from our previous model system (Scheme 17). The existing S_NAr chemistry was run using the condensation product **39** and proceeded in higher yield than previously observed. The C-H activation starting material **27** was subjected to the C-H arylation conditions, and the condensation material performed identical to the material prepared from the 1,4-dichloroquinazoline route. Deprotection of the sulfonamide yielded BMS-919373 and achieved proof of concept for a six step longest linear synthesis with nine steps total. The benefits of the new chemistry were the use of less expensive and less hazardous raw materials.

Process Development of Proposed Commercial Route

The synthesis of 5-bromo-*N*-(*tert*-butyl)pyridine-3-sulfonamide **19** was initially accomplished with a three-step procedure for the first-generation route (Scheme 18). The reaction of *i*-PrMgCl·LiCl with 3,5-dibromopyridine (**18**) generated the heteroaryl Grignard reagent **44** through a magnesium-halogen exchange (*42, 43*). Subsequent addition of sulfuryl chloride to the heteroaryl Grignard reagent **44** provided sulfonyl chloride **45**. Condensation of *tert*-butylamine with crude **45** afforded **19**. The process behaved adequately on laboratory scale; however, incomplete Mg-Br exchange was encountered on multi-kilogram scale, which could not be resolved by kicker charges of *i*-PrMgCl·LiCl. Attempted crystallization of the batch failed due to the low yield of **19** in the crude product mixture. The root cause of the stalled reaction was not apparent as all lab-scale experiments had achieved complete conversion at the first stage. Stability studies by React-IR indicated decomposition of Grignard intermediate **44** over time, attributed to degradation in the presence of moisture.

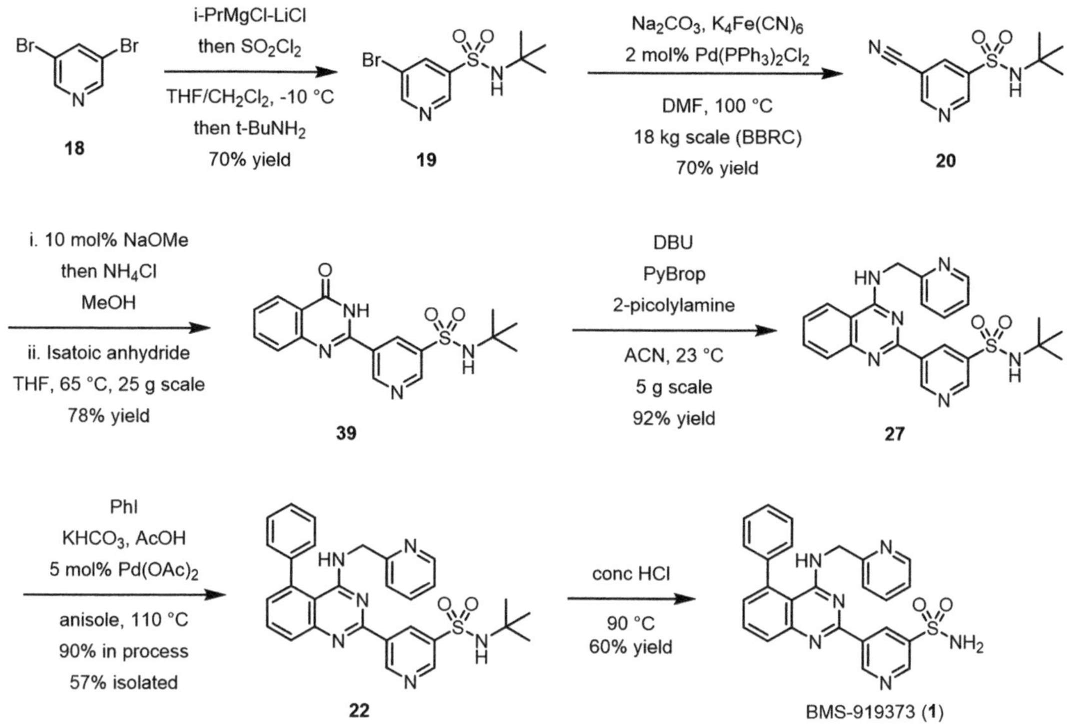

Scheme 17. Status of the project at route selection.

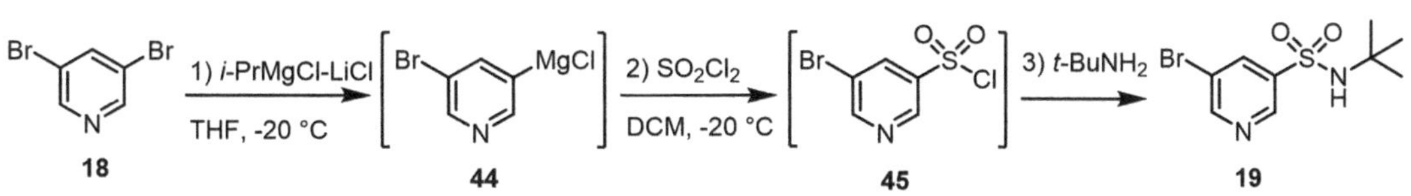

Scheme 18. Batch process for the preparation of arylsulfonamide 19.

As the program moved to later stage development, it was important to establish a robust process to meet the growing demand for sulfonamide **19**. Because of the significant chance of failure of the previous process, we evaluated the three alternative routes to the synthesis of **19** highlighted in Scheme 19 (*44, 45, 46*):

- **Route A** (scheme 19A): 3-Pyridinesulfonic acid **13** was chlorinated, and then the pyridine ring was brominated, followed by reaction with t-BuNH$_2$ to generate sulfonamide **19** in 52% yield. However, this approach suffered from new challenges. For example, instability of the reaction stream prevented scale-up to greater than a few kilograms per batch. In addition, the bulk material **13** was contaminated with 5000 ppm of Hg, which led to concerns about possible residual levels of the toxic metal I **19**.
- **Route B** (scheme 19B): The S$_N$Ar/sulfide-oxidation (*47*) was evaluated in parallel, but low yield of the oxidation of sulfide **42** made us to deprioritize the option.
- **Route C** (Scheme 19C): A Sandmeyer reaction (*48, 49*) was carried out with 3-amino-5-bromopyridine (**47**) followed by reaction with t-BuNH$_2$ afforded **19** in a moderate yield.

After assessment of all three alternative approaches, we refocused our efforts on optimizing the above described Grignard reaction (Scheme 18) and resolving the issues associated with the failures on larger scale. Monitoring the reaction progress by React-IR provided valuable mechanistic information. Intermediate **44** formed rapidly upon introduction of *i*-PrMgCl·LiCl to 3,5-dibromopyridine, but there was a significant shift of the diagnostic IR signal (*50*) when holding the reaction that often resulted in lower product yields. Switching the order of reagents for Grignard formation also led to a different IR spectrum that further supported concerns of the emergence of multiple metallated intermediates (*51*).

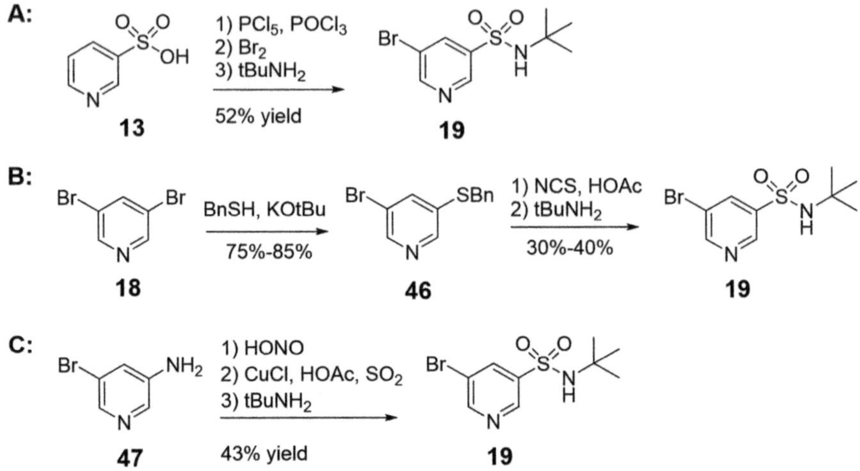

Scheme 19. Alternative routes to arylsulfonamide 2. Reproduced with permission from ref. (52). Copyright 2019 American Chemical Society.

Therefore, we started to evaluate continuous processing (*52*). We expected this to resolve concerns associated with the short lifetime of arylmagnesium intermediate **44** by minimizing changes in heat transfer and mass transfer environment across different scales, ultimately aiming to improve process robustness (*53, 54, 55, 56, 57, 58, 59, 60, 61, 62, 63*). We envisioned that large-scale continuous processing would bring additional benefits from perspectives of safety, robustness, and operation:

1) The continuous process allowed for close monitoring of the reaction outcome in real-time, significantly reducing the probability of failing acceptance criteria during operation. The output of a continuous process could be collected in small fractions, thereby any portions which did not meet the acceptance criteria would be discarded.
2) The continuous process would accurately control reaction time, while still assuring full conversion, minimizing the risk of degradation of unstable intermediates.
3) The hazard evaluation of the previous process showed noticeable adiabatic temperature increases for all three steps. Particularly, the adiabatic temperature rise of the sulfonylation was 51 °C, and could result in thermal decomposition (*64, 65*). The continuous process allows more efficient removal of heat due to the increased surface area to volume ratio, diminishing risks inherent to a highly exothermic transformation. Moreover, because of its design, the flow reactor decreased the chance of over-pressurization due to volatile solvents (THF and DCM).

4) The nature of efficient mixing and rapid heat transfer of a continuous process could lead to avoidance of the cryogenic (-20 °C) conditions that had been used for the batch process, saving energy and infrastructure costs.
5) The potential safety hazard and chemical reactivity associated with the Grignard reagent and sulfuryl chloride (*66, 67*) could be controlled by the fully enclosed continuous process, compared to the batch process where open handling often occurs.

The laboratory scale continuous process design, shown in Figure 9, involved three stages. A stream of **18** in THF (10 mL/g or 0.42 M) and a stream of *i*-PrMgCl·LiCl in THF (1.3 M), operated with syringe pumps, were mixed into a glass static mixer placed in a cold bath (-20 °C), and proceeded through the coil to afford the Grignard intermediate **44**. Then a DCM solution of SO_2Cl_2 was introduced by pumping through a separate static mixer and mixed with the Grignard solution to afford sulfonyl chloride **45**. Although it is known that the Grignard potentially reacts with DCM, and SO_2Cl_2 is incompatible with THF, the fast kinetics of the second step granted the desired reactivity.

At the end of the coil, the reaction mixture was pumped into a flask containing an agitated DCM solution of *t*-butylamine at 0 °C to afford *N*-*tert*-butyl sulfonamide **19**. Based on our experiences with the batch processes, we expected fast kinetics of all three steps, a requisite for continuous processing. Due to only minimal safety and robustness concerns, the first two steps were executed under flow mode, whereas the amidation step was conducted by dropwise addition of the sulfonyl chloride **45** stream to a vessel containing *tert*-butylamine.

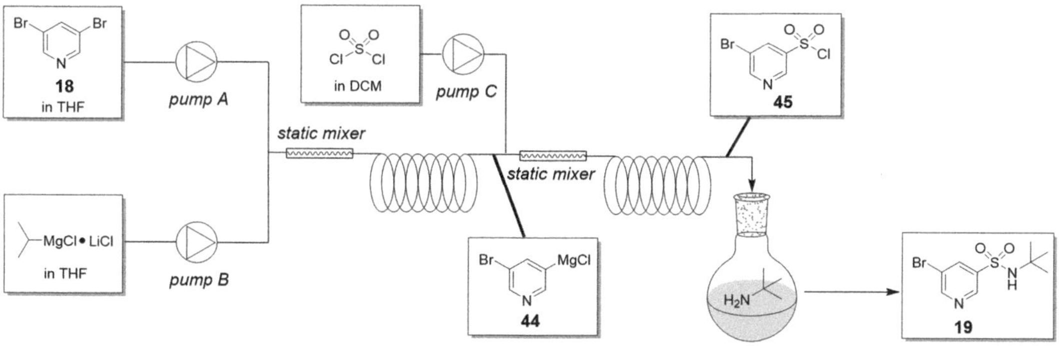

Figure 9. Equipment design for the development of a continuous process to arylsulfonamide ***19****. Reproduced with permission from ref.* (*52*)*. Copyright 2019 American Chemical Society.*

In an effort to garner an understanding of the magnesium-bromine exchange under flow mode (*68*), studies were initiated on sub-gram scale, utilizing a 2 mL Uniqsis glass chip as both the static mixer and the reactor. The chip was placed in a bath that was cooled to -20 °C prior to the reaction. The two stock solutions, **18** in THF (10 mL/g or 0.42 M) and *i*-PrMgCl·LiCl (1.3 M in THF), were introduced by two syringe pumps to mix together at the rate to maintain the corresponding stoichiometry. At the end of reaction, the reaction mixture was collected into a vial containing methanol to protonate intermediate **44** to generate 3-bromopyridine **48**, and then analyzed by HPLC for the reaction composition.

The relative equivalents of *i*-PrMgCl·LiCl and the residence time of the reaction were evaluated by varying the flow rate of the two components and the temperature of the bath (Table 4). We achieved a proof of concept quickly, as the conversion reached >98% under a residence time of 120 seconds using 1.5 equiv Grignard (entries 1-3,). We noticed, however, because of the solubility limit of the substrate in THF at -20 °C, it sometimes crashed out at the inlet of the glass chip and caused a blockage.

To resolve the solubility challenge at -20 °C while not increasing the volume, dimethoxyethane (DME) was identified as a co-solvent with THF that would keep the substrate in solution. Interestingly, reactions in the presence of DME displayed slower reaction rates, as higher levels of residual starting material were observed with even up to eight minutes of residence times (entries 4-6). Meanwhile, higher reaction temperatures were investigated as no sign of thermal instability was observed. Maintaining a residence time of 120 seconds, reactions conducted at 0 °C or 20 °C afforded similar results as reactions conducted at -20 °C (entries 7 and 8). Therefore, we decided to perform the reactions at 0 °C as it provided the optimal conversion to product and sustained capability to control the exotherm.

Table 4. Evaluation of Conditions for the Magnesium-Bromine Exchange on Micro-scale[a]

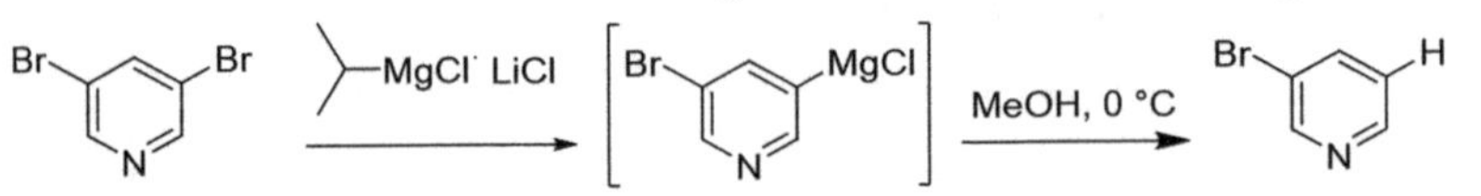

Entry	Solvents for stock solutions of **18**	Temp (°C)	Grignard (equiv)	Residence time (s)	**44** (AP)	**18** (AP)
1	THF	-20	1.2	240	80.5	14.6
2	THF	-20	1.5	240	90.1	1.7
3	THF	-20	1.5	120	89.2	<1.0
4	4:1 THF/DME	-20	1.5	120	85.5	10.2
5	4:1 THF/DME	-20	1.5	240	83.7	8.2
6	4:1 THF/DME	-20	1.5	480	91.8	3.6
7	THF	0	1.5	120	92.5	1.3
8	THF	20	1.5	120	88.4	<1.0

[a]Reproduced with permission *from ref. (52)*. Copyright 2019 American Chemical Society.

With magnesium-bromine exchange conditions optimized, the next step was to include stage 2 into the continuous process. Targeted at a multigram scale, an Uniqsis Flowsyn system was implemented and a third pump was incorporated to add SO_2Cl_2. As shown in Figure 9, two coils made of 3/8 inch stainless steel tubing were introduced as flow reactors to afford 10 mL of volume for both the metal-bromine exchange and the sulfonylation. The whole equipment train was submerged in a cold bath (0-2 °C) and the resulting reaction stream was transferred to a flask that contained *t*BuNH$_2$ (as a solution in DCM) and maintained at 0-2 °C. The residence times for the two stages were approximately 120 seconds and 90 seconds, enabling us to process 19 g input material **18** in an hour. With 1.5 equiv Grignard and 1.5 equiv SO_2Cl_2 relative to **18**, two otherwise identical experiments on a 40 and 76 g scale afforded 87% and 91% of **19**, respectively.

Compared with the previous batch process, the flow process provided a similar purity profile at the end of reaction (Figure 10). Besides the unreacted input material **18**, <1.0% (by HPLC) 3-bromopyridine **48** was typically observed, likely attributed to the protonation of Grignard intermediate by moisture. Because of its higher pKa compared to the **19**, **48** was completely removed by a treatment with 2.0 M aq HCl. Chlorosulfonamide **49** was another process-related impurity. It is likely that this impurity originated from low levels (typically 0.5-3.0%) of 3-bromo-5-chloropyridine **50** present in the starting material and was inert in the subsequent steps (*69*).

In contrast to the low levels of **48** and **49** in the process, 3-bromo-5-chloropyridine **50** was typically observed in much higher amounts. Interestingly, the amount of **50** varied from 2.2% to 12.0% with SO_2Cl_2 from different suppliers. Because SO_2Cl_2 is known to slowly decompose to SO_2 and Cl_2 upon storage, it was reasonable to postulate the reaction between Grignard intermediate **44** and Cl_2 or other electrophilic chlorine from SO_2Cl_2 led to the formation of the impurity. Though the process had excellent tolerance of the impurity by purging through crystallization and subsequent cake washes, it caused significant yield loss. Thereby sourcing high quality SO_2Cl_2 became critical to mitigate such process variability.

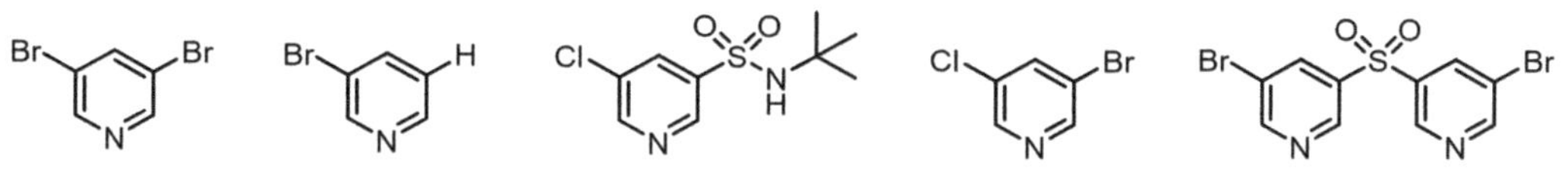

Figure 10. Major impurities in the crude reaction mixture. Reproduced with permission from ref. (*52*). *Copyright 2019 American Chemical Society.*

The only process-related impurity that caused a quality concern was sulfone **51**, because of its much lower solubility (<1.0 mg/mL) than the parent compound in all suitable solvents evaluated for the crystallization. In addition, the sulfone impurity would react in later steps of the route and the result was carried along to the API. We suspected double arylation of SO_2Cl_2 led to the formation of the impurity; however, stressed experiments employing 0.5 equiv SO_2Cl_2 instead of 1.5 equiv SO_2Cl_2 did not afford >5.0% **51** (*70*). It also appeared amounts of **51** varied from 1.5 to 6.0% with *i*-PrMgCl·LiCl purchased from different suppliers, leading us to suspect the impurity could be related to the presence of transition metal impurities in *i*-PrMgCl·LiCl (*71*)(). As the impurity impacted API quality, spiking studies were conducted, and demonstrated that the route tolerated up to 2.0% of **51**, requiring us to target below this level when crafting the control strategy.

The previous crystallization occurred during the distillation of the crude reaction mixture in THF/DCM mixtures during the exchange to *n*-heptane. Though the loss of **19** to mother liquor was minimal, crystals often grew on the reactor wall and around the agitator, resulting in up to 10% loss. In addition, the level of **51** in the isolated product was enriched because of its extremely low solubility in the crystallization solvent. Therefore, alternative crystallization conditions to purify the crude mixture were evaluated.

Based on high-throughput solubility evaluation and crystallization screenings with pure **19**, there were five solvent systems identified to afford the crystalline product while being suitable for a distillation or solvent switch of process solvents (Figure 11). Though *n*-heptane is an excellent solvent to purge in-process impurities **18** and **50**, its mixtures with other organic solvents, such as EtOAc or THF, resulted in high solubility of **19**. A targeted solubility of <5 mg/mL at the endpoint required the composition of *n*-heptane to be ≥90%, forcing us to process at high volumes to reduce loss of **19** to the mother liquor. In contrast, much lower solubility of **19** was observed in alcohol-water mixtures, achieving <2mg/mL in either 20:80 (v/v) MeOH/H_2O or *i*PrOH/H_2O.

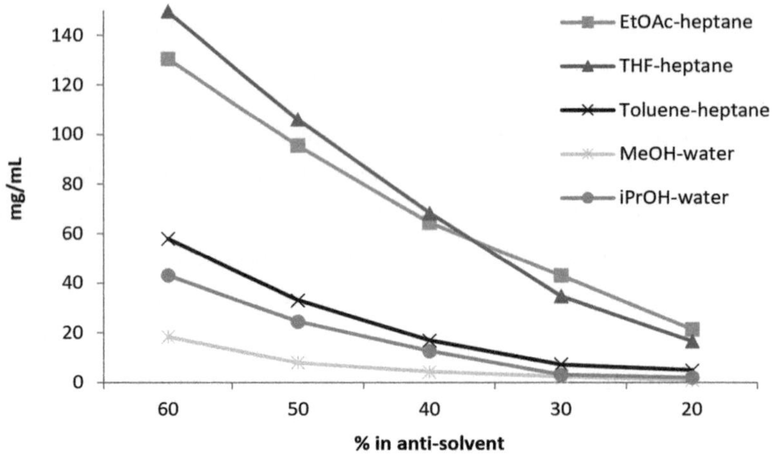

Figure 11. Solubility of arylsulfonamide 19 in suitable crystallization solvents. Reproduced with permission from ref. (52). Copyright 2019 American Chemical Society.

Table 5. Solubility of Arylsulfonamide 19 and Sulfone 51 in Different Solvents[a]

Solvent (s)	*Solubility of* ***19*** *(mg/mL)*	*Solubility of* ***47*** *(mg/mL)*
THF	>100	24.5
THF/*n*-Heptane (20:80)	16.6	2.8
EtOAc	>100	20.1
EtOAc/*n*-Heptane (20:80)	21.4	3.2
Toluene	51.4	7.8
Toluene/*n*-Heptane (20:80)	4.9	0.6
MeOH	>100	1.1
MeOH/Water (20:80)	1.0	<0.1
*i*PrOH	87.0	0.5
*i*PrOH/Water (20:80)	1.9	<0.1

[a] Reproduced with permission from ref. (52). Copyright 2019 American Chemical Society.

After collecting data with pure **19**, crystallization of crude **19** was evaluated with all five solvent systems (Table 5). It was challenging to remove **51** while minimizing product loss as the impurity was at least five times less soluble in those solvents. The best solvent system turned out to be 20% EtOAc in *n*-heptane which reduced **51** by 3-4% but led to a 25% yield loss. Realizing it was ineffective to remove **51** by crystallization, we decided to take advantage of the lower solubility of **51** by implementing filtration of the solution of the crude product prior to the recrystallization.

Accordingly, a solvent mixture of MeOH/H_2O was selected for crystallization because of the highest differences of solubility between **19** and **51** in MeOH, despite the fact that there was minimal purging of **51** in the crystallization itself. After further optimizations, the end of reaction mixture was treated with 2.0 M aq HCl to remove **48**, then washed with 1.0 M aq $NaHCO_3$. The organic mixture in THF/DCM (ca. 3:1) was distilled to minimal volume of the reactor and solvent exchanged to MeOH. The subsequent polish filtration of the mixture in 8 L/kg MeOH was effective to reduce variable levels of **48** to below 1.5%, with minimal yield loss of **19**. The filtrate was then further distilled and crystallized by addition of H_2O to achieve 98% purity of arylsulfonamide **19** while complying with the quality requirement of <2.0% sulfone **48** (1.1-1.2 % observed in isolated material).

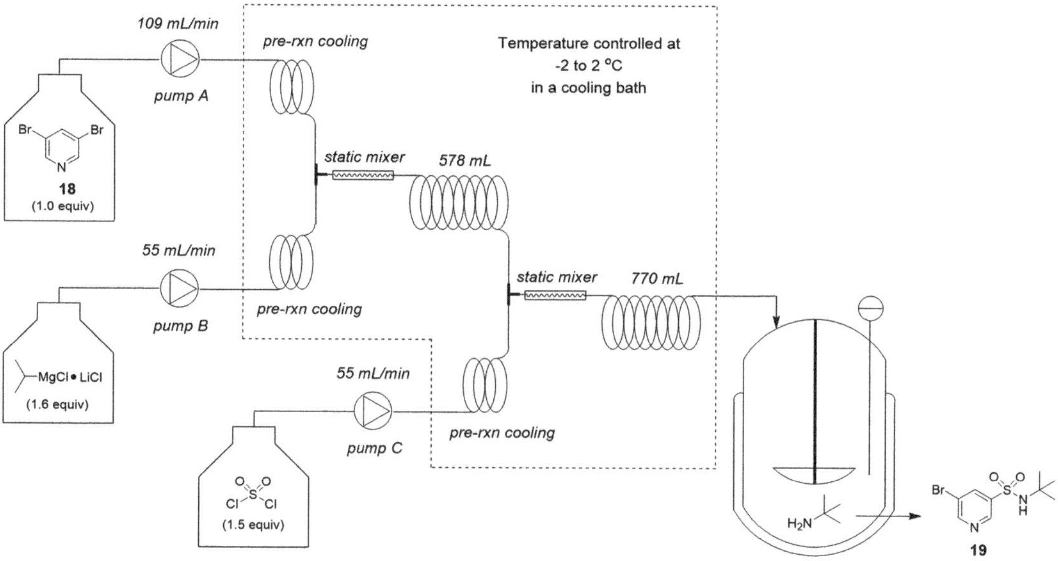

Figure 12. Design for a pilot scale continuous process to prepare arylsulfonamide 19. Reproduced with permission from ref. (52). Copyright 2019 American Chemical Society.

After optimizing the reaction conditions, work-up and purification, the process was transferred to on-scale production (Figure 12). Similar design of equipment as the set-up in the lab was implemented and perfluoroalkoxy (PFA) tubing was chosen as the continuous reactor. Further optimization indicated 3.5 min was required as the residence time for metal-halogen exchange to improve process robustness. Therefore, the process execution would require 8-10 days to transform 90 kg input material into product mixture.

During process transfer to the production facility, one major challenge was the elevated levels of impurities **50** and **51** relative to those observed during development. To meet the production timeline for API delivery, although the root cause for these deviations was not completely understood, we decided to collect reaction streams in multiple fractions to mitigate the risk of process variability. To determine if the fractions met the quality requirements and were suitable for downstream processing, analytical tests of individual fractions were implemented as IPC (in-process control). We believed the strategy to divide process streams into fractions for HPLC analysis provided information on reaction behavior and ensured meeting of quality acceptance criteria (72). During production, a jacketed 50 L reactor that contained a DCM solution of *t*-$BuNH_2$ at 0 °C was used to receive the crude mixture from the continuous reactor. After 3 h, when the volume of the output reached 45 L, the mixture was transferred from the reactor to a 50 L carboy, awaiting IPC sampling to determine quality. This was repeated for successive 50 L carboys.

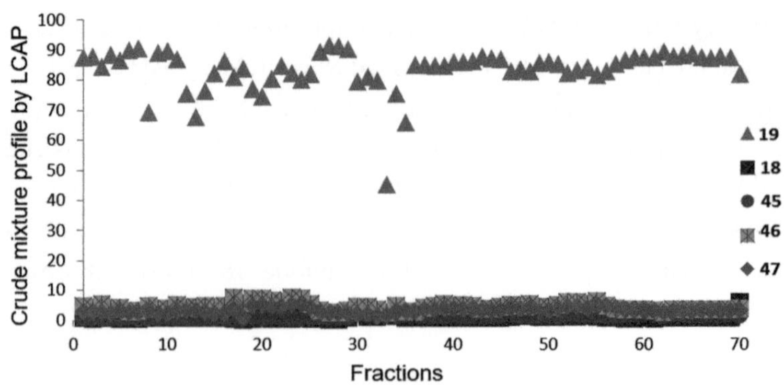

Figure 13. Purity profiles of all fractions from production. Reproduced with permission from ref. (52). Copyright 2019 American Chemical Society.

A total of 70 fractions of the crude reaction stream were collected and analyzed by HPLC (Figure 13). As expected, 94% of them contained 75-92% desired product, and only fraction no. 33 contained <65% **19**. We determined that an unplanned introduction of moisture to the continuous reactor occurred during the production of fraction no. 33. HPLC analysis indicated the formation of the elevated levels of the hydrolysis of sulfonyl chloride **45** accounted for the lower purity of the fraction. Fortunately, it was not a concern of quality as we could exclude fraction no. 33, thereby it only led to yield loss. We postulated that the formation of the impurity was because of the degradation of SO_2Cl_2 with moisture. The resulting acidic solution subsequently corroded the parts of the peristaltic pump that led to change of the flow rates. To prevent such an issue, we freshly prepared the SO_2Cl_2 solution to minimize introducing moisture during standing and frequently inspected the pump head tubing. Following this intervention, later fractions after fraction no. 35 consistently provided >79% desired product in the crude mixture.

The levels of impurities from the production run were consistent with the lab-scale experiments: impurities **18** and **49** were typically below 1.5%, and impurities **50** and **51** were at 3.2-7.6% and 1.3-5.3%, respectively. For purification of the streams collected, the fractions (except 33) were combined and split into two batches for work-up and crystallization. The crude mixture was first washed with 2.0 M aq HCl solution to remove impurity **48**, then neutralized by another wash with 1.0 M aq $NaHCO_3$ solution. The organic mixture was distilled and solvent switched to 8 L/kg MeOH. To reduce the level of fairly insoluble sulfone impurity **51** to <2.0%, the aforementioned mixture in MeOH was filtered at 25 °C and concentrated to 4 L/kg for crystallization. The nucleation occurred when H_2O was added at 45 °C, the crystals were allowed to grow for 1 h and the slurry was cooled to room temperature. After another 8 h, the suspension was filtered and the cake was washed with *n*-heptane, affording 76 kg of arylsulfonamide **19** in 63% yield, of >98.0 % yield and >97.0 wt %. For both isolated batches, levels of impurity **51** were controlled to 1.1% and 1.2%, respectively.

The next step in the route was a palladium-catalyzed cyanation (73, 74, 75) of the pyridinyl bromide. Earlier efforts on this project led to a set of reaction conditions that utilized $K_4Fe(CN)_6$ as the cyanide source (76, 77, 78, 79) with both Pd and Cu as co-catalysts to afford the desired product in 70% yield (Eq 8).

Attempts to improve the conversion with $K_4Fe(CN)_6$ to >90% were unsuccessful, and therefore, $Zn(CN)_2$ was investigated as an alternative cyanide source. A ligand screen was completed via HTE with both dioxane and DMAc as solvents (Figure 14). Dioxane was generally a poor solvent regardless of the ligand whereas DMAc provided high levels of conversion to the corresponding nitrile with several ligands. Further optimization led to the selection of XantPhos as the ligand for the cyanation.

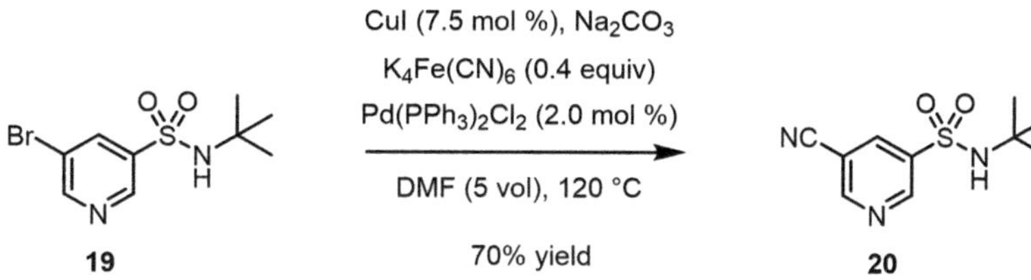

(8)

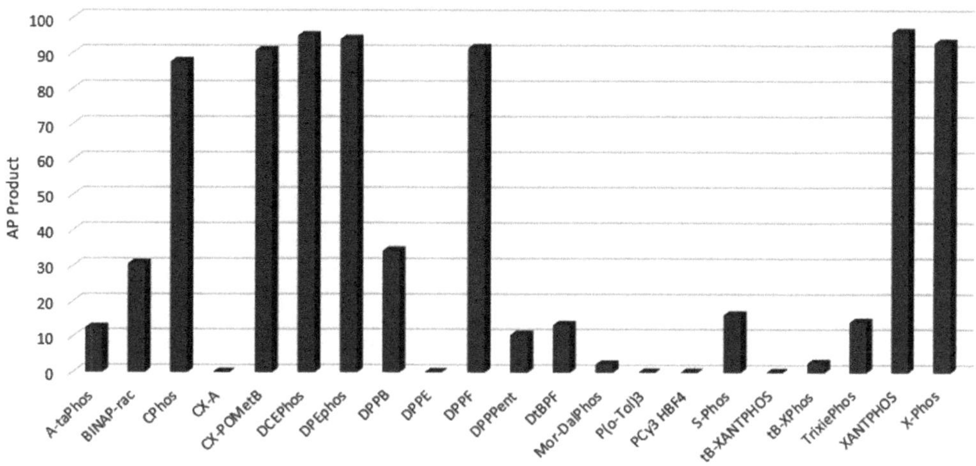

Figure 14. Conversion to nitrile vs ligand employed.

The optimized reaction conditions employed Pd_2dba_3 as the palladium source with XantPhos as the ligand. An excess of $Zn(CN)_2$ improved the robustness of the reaction, leading to an in-process yield of 98%. With optimized reaction conditions established (Eq 9), we next turned our attention to the remediation of Zn and Pd, and the isolation of **20**.

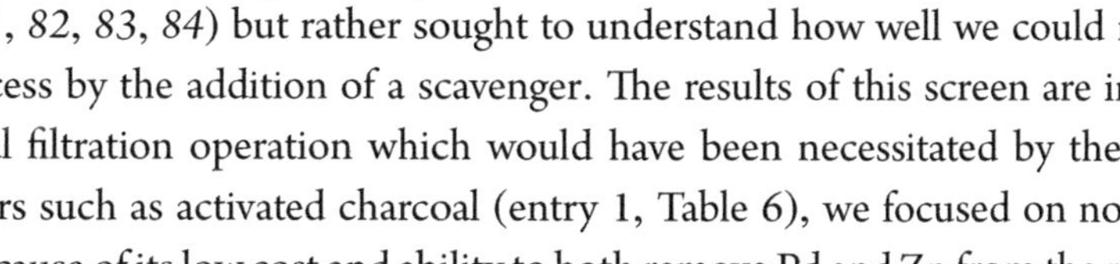

(9)

As there is another Pd-catalyzed step in the process, we did not yet seek to control Pd to its final specification (*80*, *81*, *82*, *83*, *84*) but rather sought to understand how well we could remove Zn and Pd from the process by the addition of a scavenger. The results of this screen are in Table 6. To avoid an additional filtration operation which would have been necessitated by the use of solid supported scavengers such as activated charcoal (entry 1, Table 6), we focused on non-solid supported scavengers. Because of its low cost and ability to both remove Pd and Zn from the reaction stream, ethylene diamine was chosen as the scavenger (entry 5, Table 6). When added at the end of the cyanation reaction, the slow rate of addition of ethylene diamine and water were key to prevent aggregation of the metal salts.

Table 6. Remediation of Pd and Zn from the Cyanation[a]

Entry	*Scavenger*	*Pd (ppm)*	*Zn (ppm)*
1	DARCO G60	620	<25
2	NH_4OH	140	910
3	N-Acetyl Cysteine	280	850
4	EDTA	230	1240
5	Ethylene Diamine	46	47

[a] Reproduced with permission from ref. (*15*). Copyright 2019 American Chemical Society.

We developed a process to extract the product into EtOAc, swap the solvent to IPA, and isolate by the addition of water as an anti-solvent. This process was effective in removing the Pd and Zn from the reaction but the crystallization did not purge the dba and XantPhos. The next step, however, served as a quality gatekeeper, mitigating any risk of excess dba and XantPhos carryover from this crystallization. This process was scaled to prepare 50 kg of **20**, in an average 82% yield and 96 % purity over two batches.here was only <35 ppm Pd in the isolated material.

The initial results from the condensation reaction presented several opportunities to improve the yield and process knowledge. The main issue from the outset was the stalling of the imidate reaction at 90%. Separately, two main impurities were identified, the amidine dimer **52** and the sulfone impurity **53** (Scheme 20).

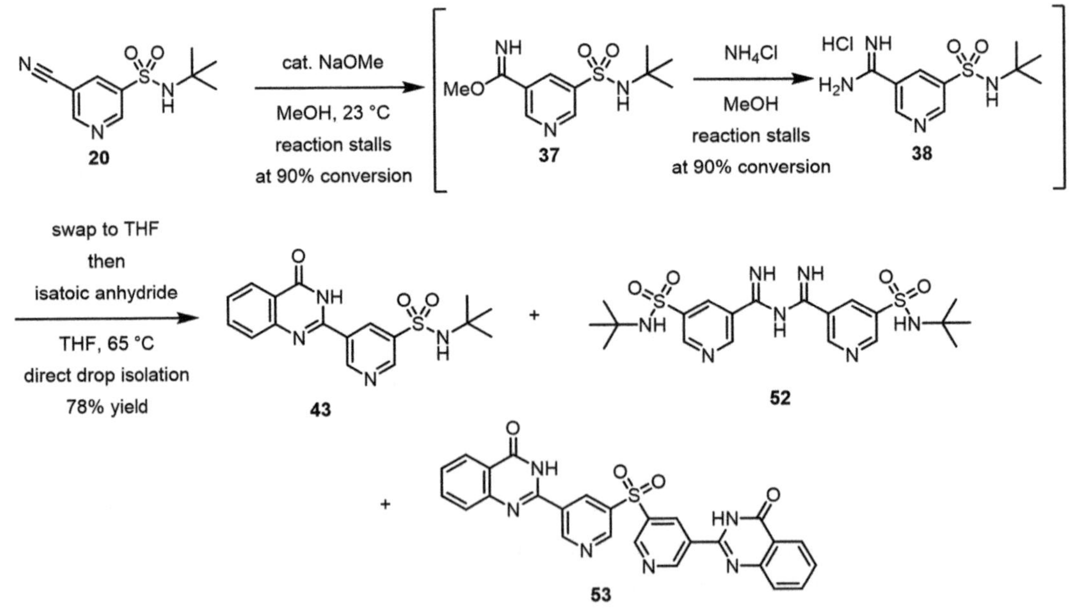

Scheme 20. Status of the condensation reaction at route selection.

The effort to understand the dynamics of the imidate formation keyed on the reaction temperature in that the reactions run at 60 °C appeared to stall around 80% while the reactions performed at room temperature appeared to give around 90% conversion. A temperature study was performed to verify these unanticipated observations. It was discovered that the conversion was directly proportional to the reaction temperature. The surprise was that colder reactions gave higher conversion! In light of this, the reaction was run initially at room temperature to achieve a rapid 95% conversion followed by cooling the reaction to 5 °C and aging for 12 h to increase the conversion to > 95% (Figure 15). It was hypothesized that the imidate formation was an entropy driven process and that controlling the equilibrium through a lowered reaction temperature to favor the imidate was important (Scheme 21). Therefore, additional factors that may contribute to an equilibrium process, such as reaction concentration and methoxide loading were investigated.

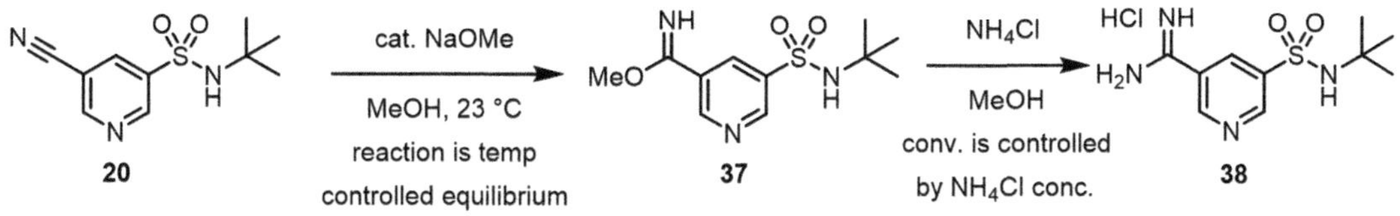

Scheme 21. Formation of amidine.

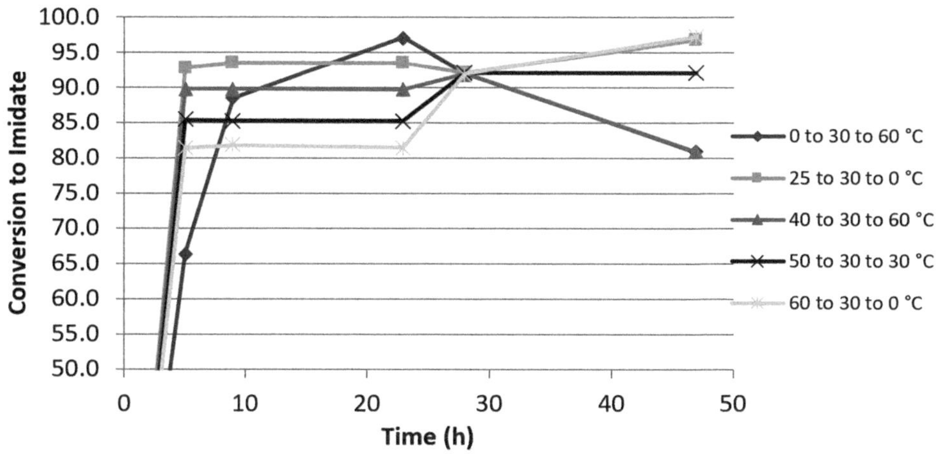

Figure 15. Impact of temperature on the equilibrium of the nitrile and imidate. Reproduced with permission from ref. (15). Copyright 2019 American Chemical Society.

A Design of Experiment (DoE) protocol was utilized to finalize the optimization of the imidate step. The factors that were chosen for evaluation were:

- final resting temperature
- mol% sodium methoxide
- reaction volumes
- exogenous water

with the response measured by remaining area percent of nitrile **20** (Figure 16). The experiment highlighted that all reactions conducted at lower temperature and low reaction volumes (more concentrated) gave the highest conversion. Utilizing 20 mol% sodium methoxide also increased the conversion; however, the effect was negligible for reactions conducted at lower temperature. It was determined that water did not significantly impact the reaction. Therefore, the final conditions moving forward were to execute the reaction with 10 mol % sodium hydroxide at seven volumes with a final age temperature of 0 °C. Although the reaction gave slightly higher conversion at five volumes there was a risk of precipitating out the nitrile under these conditions and thus the use of seven volumes mitigated this risk to the equilibrium process.

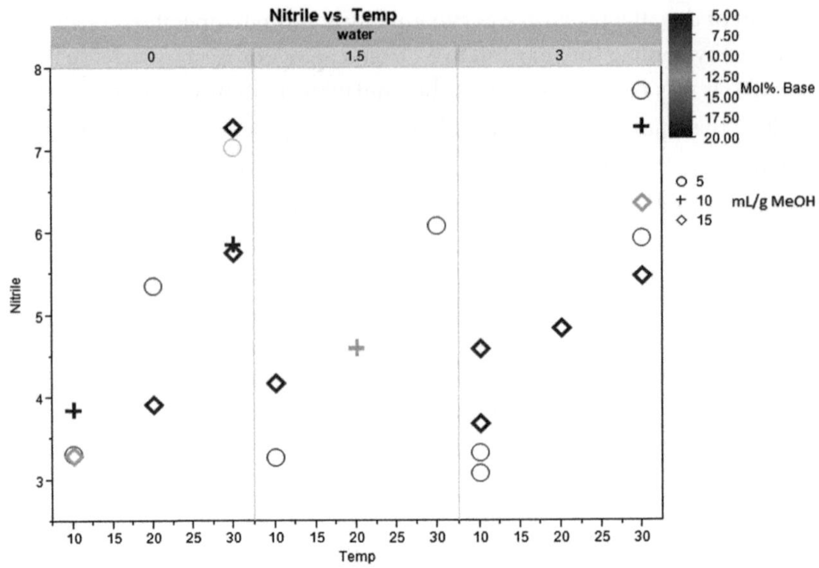

Figure 16. Optimization of the imidate step by DoE.

The next step in the process was the amidine formation using ammonium chloride. The components of focus were:

- the ammonium salt
- the quantity of ammonium salt
- temperature
- concentration

Additionally, it was anticipated that the amidine dimer impurity **52** was generated at this stage and knowledge surrounding this impurity would need to be gathered.

We sought to improve the conversion which had stalled at 90% conversion using different ammonium salts. We discovered that ammonium chloride gave the fastest conversion versus other ammonium reagents. (Figure 17). Therefore, ammonium chloride was selected as the reactant for further development.

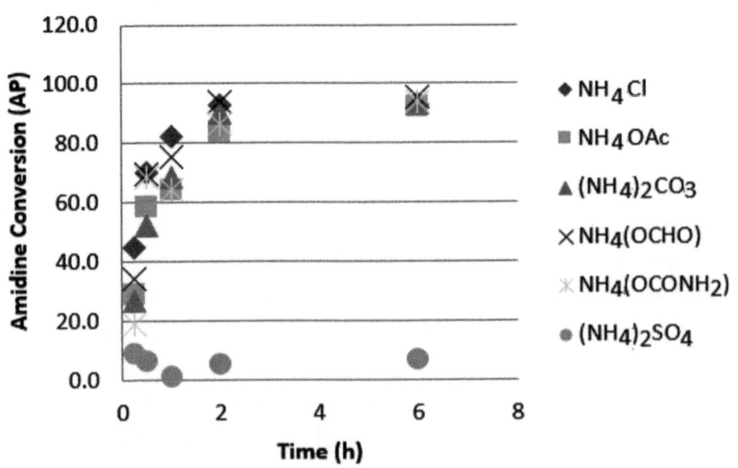

Figure 17. Amidine conversion over time.

The amidine formation with ammonium chloride was studied at different temperatures (Figure 18). The kinetics of the reaction were quite fast above 40 °C although slightly lower yields were obtained at these higher temperatures due to formation of the amidine dimer impurity **52**. Therefore, in order to minimize formation of the dimer by increasing the rate of reaction, the amidine formation was run at 50 °C. No significant benefit was realized above 1.25 equiv of ammonium chloride and a small boost in yield could be obtained by running the reaction at an increased concentration (7 volumes). This was important as the seven volumes required for this step matched the seven volumes utilized in the imidate formation step. Therefore, the imidate was not isolated. The only unit operations required to convert the imidate to the amidine were addition of the 1.25 equiv of ammonium chloride and heating to 50 °C.

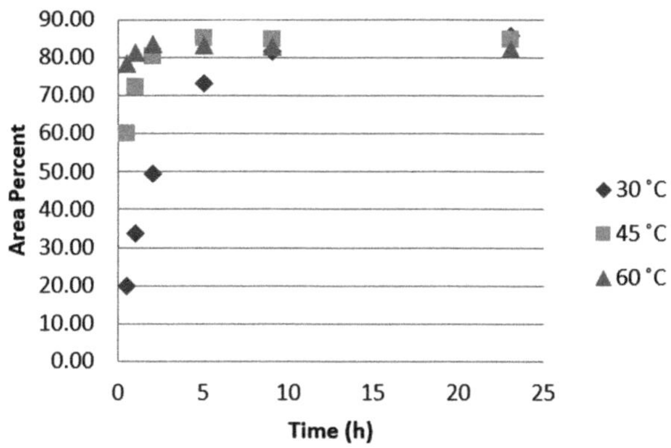

Figure 18. Amidine Conversion at different temperatures without the addition of water.

The first stage of the condensation with isatoic anhydride was to swap the solvent from methanol to tetrahydrofuran. The solvent swap was affected by adding 10 volumes of THF and distilling down to 5 volumes and repeating this process twice under atmospheric distillation conditions at 61 °C. The resulting reaction stream was confirmed to contain less than 1% methanol and studies were conducted to verify that this level of methanol did not pose a risk to the process (Figure 19). These studies determined that 20:1 MeTHF/MeOH resulted in >90% conversion to amidine.

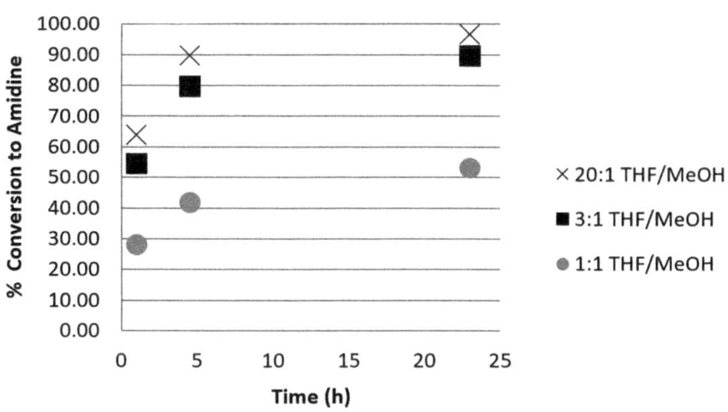

Figure 19. Methanol tolerance in the condensation.

The next item of concern was the rate of the reaction as the reaction gives off carbon dioxide gas in forming the new ring. Kinetics of the reaction displayed that the reaction reaches 70% completion during the first hour and that pressurization of reaction vessels would not be anticipated. We demonstrated as a safeguard that the isatoic anhydride could be added in portions over an extended period with no impact to the reaction profile or final isolation.

As quinazolinone **39** had poor solubility in THF, the product crystallized during the course of the reaction. The addition of 3.3 L/kg water at the 60 °C reaction temperature effectively dissolved the ammonium chloride byproduct and fortuitously the amidine dimer impurity **52**. Cooling to 23 °C and filtration followed by two 4 L/kg washes with 3:1 THF/water, and subsequent drying afforded the quinazolinone **39** in 90% yield with 99.2 % purity.

Developing a process for installation of the aminomethylpyridine fragment was the next challenge toward delivering a commercial synthesis. This strategy was featured in both of the initial routes although the S_NAr substrates were different, 2,4-dichloroquinazoline vs quinazolinone. To achieve a better perspective for a potential path forward, a high-throughput screen was conducted using the input quinazolinone **39** as it was envisioned that the quinazolinone could produce numerous reactive intermediates, including aryl chlorides (Eq 10).

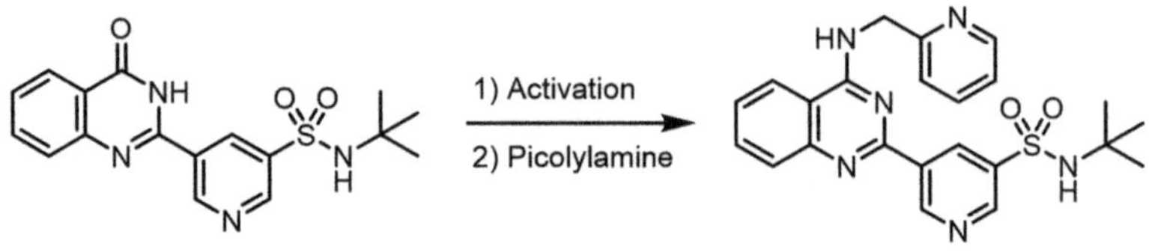

39 **27** (10)

The goal for HTE was to identify reagents that retained that activity of PyBrop (bromotripyrrolidinophosphonium hexafluorophosphate) yet afforded better PMI and greenness for the process. A logical protocol was for the array to focus primarily on chlorination reagents. The array consisted of six chlorination reagents ($POCl_3$, PCl_5, PCl_3, Vilsmeier reagent, oxalyl chloride, and the Lambert cyclopropenium reagent with oxalyl chloride), four solvents (*N*-methyl-2-pyrrolidinone, toluene, THF, and acetonitrile), and three additives (DMAP, Oxyma, DMF), in addition to a base case without any additives. The results from the study overwhelmingly demonstrated that Vilsmeier conditions in THF were best: they gave excellent reactivity and clean reaction profiles (Table 7). Having identified both the Vilsmeier reagent and the reaction solvent from the HTE array, the initial work focused on developing a process utilizing the much greener Vilsmeier reagent as a potential alternative to PyBrop.

The Vilsmeier reagent itself offered impressive reactivity without the need for any additives or bases. This is advantageous for the development of a low PMI process. However, the reagent itself carries risk as it is immensely hygroscopic and can be difficult to source on commercial scale. Therefore, development work focused on generating the reagent in situ from oxalyl chloride and DMF.

During development it was discovered that the 2-picolylamine was functioning as a base in addition to its role as a coupling partner, necessitating the use of three equivalents of this raw material to drive the reaction to completion. As such, several less-expensive bases were investigated in order to minimize the use of 2-picolylamine. We found that the potassium salt of **39** generated using KO*t*-Bu was effective in this regard although the introduction of a new *t*-butoxide impurity was observed. The use of potassium bicarbonate enabled the transformation with less 2-picolylamine without the introduction of new impurities. The potassium bicarbonate process afforded an 87% yield on initial scale up (Scheme 22). The results were encouraging for commercial process development although significant reaction engineering efforts would be required to manage handling of oxalyl chloride, the off-gassing of the reaction, and the aqueous work-up of a product of very low solubility for removing inorganic salts. Therefore, the commercial process development of the Vilsmeier option was put on hold to ensure the success of the pressing clinical delivery using the PyBrop process.

Table 7. Top 10 Results from Chlorination Reagent Screening for SnAr

Reagent	*Activator*	*Solvent*	*Product HPLC %*	*Conversion*
Vilsmeier	None	THF	99	99
Vilsmeier	DMF	THF	98	98
Oxalyl Chloride	DMF	THF	98	100
Oxalyl Chloride	DMF	MeCN	94	98
Vilsmeier	None	MeCN	89	89
Vilsmeier	None	Toluene	88	89
Oxalyl Chloride	DMF	Toluene	88	100
$POCl_3$	DMF	MeCN	83	84
Vilsmeier	DMF	Toluene	77	78
PCl_5	DMF	MeCN	72	76

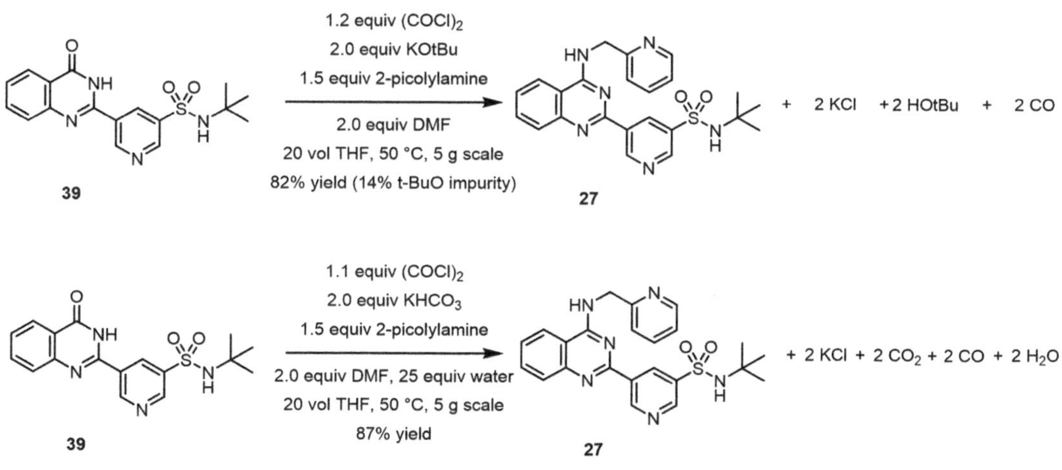

Scheme 22. S_NAr reactions of quinazolinone 36 with $(COCl)_2$.

We discovered that Py-Brop and DBU were effective reagents for activating the quinazolinone for nucleophilic substitution. The DBU was required to solubilize the otherwise remarkably insoluble quinazolinone and was also able to produce consistent kinetics and reliable reactivity. However, the challenges to this system were the impurities associated with the addition of the DBU base and the residual pyrrolidine derived from the PyBrop reagent, presumably from the reaction of these two amines with the activated quinazoline intermediate (Scheme 23).

The pyrrolidine impurity **54** could be controlled through the procurement of high quality PyBrop (>99%), which resulted in <5 % of this impurity during the process. The DBU impurity **55**, however, ensued at levels upward of 10 % after holding the activated intermediate for 3 h with 1.5 equiv of DBU (Figure 20). This was a critical issue as the length of this hold time could vary during processing. To solve this issue a DoE study was conducted that examined DBU equivalents, reaction concentration and temperature as a function of time. The study showed that by:

- lowering the charge of DBU to 1.15 equiv
- decreasing the reaction concentration from 10 vol to 20 vol
- reducing the reaction temperature from 23 to 10 °C

the desired S_NAr was favored while the DBU impurity was decreased to levels of <2%, even after holding the activated intermediate stream for 24 h (Figure 21). Under these optimized conditions, the crystallization was performed by simply adding water to afford the isolated product in 99.5% in 88% average yield when run twice at 17 kg scale.

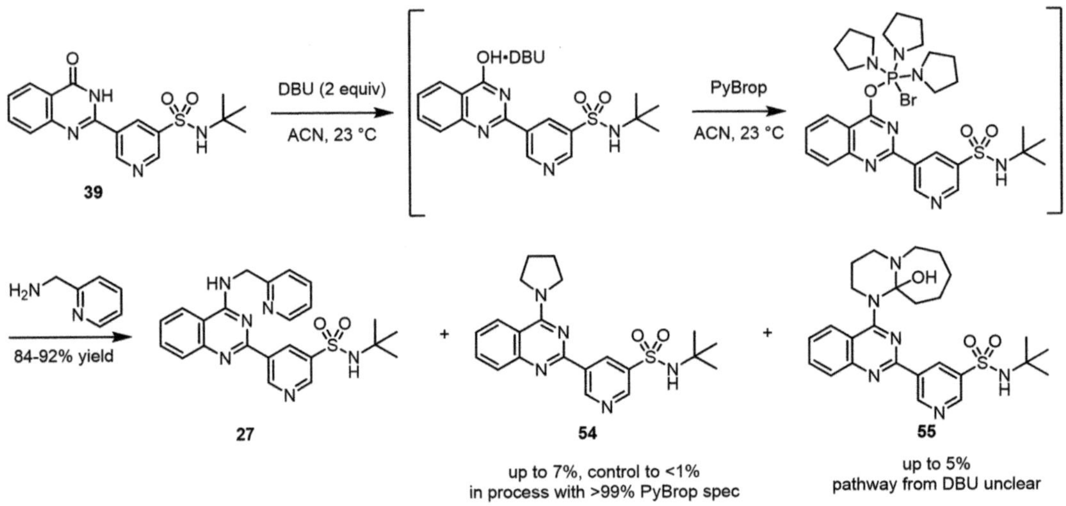

Scheme 23. Development of the S_NAr reaction of 39 with PyBrop.

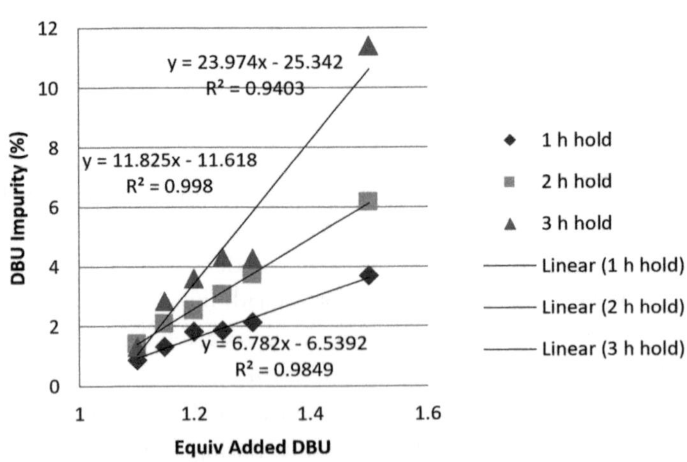

Figure 20. Amount of DBU impurity formed relative to the amount of DBU at 1-3 h hold for activation step. Reproduced with permission from ref. (15). Copyright 2019 American Chemical Society.

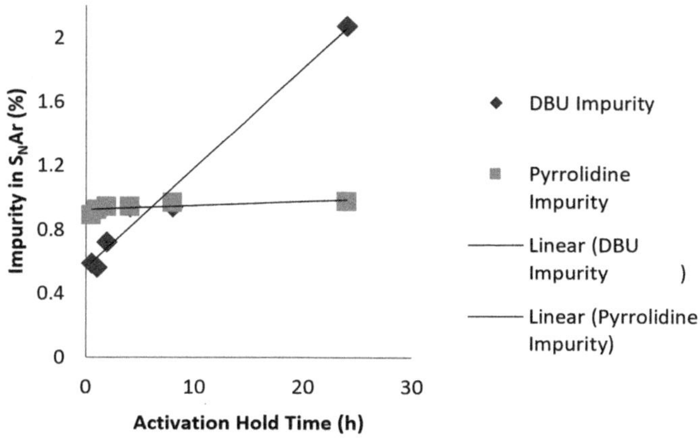

Figure 21. S_NAr activation impurity profile utilizing 1.15 equiv DBU at 10 °C.

Although proof of concept of the C-H activation has been achieved for the installation of the arene on the quinazoline core, further development was required to attain a robust, scalable process. The initial reaction optimization demonstrated that we could employ neat iodobenzene and that the reaction worked best with a carboxylate base. However, to better understand the impact of each variable in the reaction, a series of HTE screens was carried out which investigated the class-variables of solvent, base, palladium precursor, the impact of additional ligands, and arene source (bromobenzene, iodobenzene, phenyltriflate).

Several key learnings were recognized during this stage of the optimization. As the C-H activation did not proceed under ~110 °C, the options for solvent were limited, and of those, only dioxane and anisole were effective. Anisole was chosen for further development because it is a greener, less toxic solvent (*85*). Of the palladium sources, there was no significant difference in conversion as both Pd(II) and Pd(0) mediated the direct arylation. Ligands, however, were not effective for the desired transformation leading to either limited conversion (in the case of phosphines and phosphites) or slower kinetics (in the case of 1:1 L/M ratio with pyridines and diketo-compounds). Many bases were tested in the reaction, and bases that could serve as ligands for Pd such as trialkylamines did not afford high levels of conversion. Further, strong bases such as alkoxides resulted only in unreacted starting material. Carbonate bases were generally effective, with and without the addition of an acid additive (*86*, *87*, *88*, *89*, *90*), and the initial conditions utilized KOPiv as the base with anisole as the solvent (Eq 11).

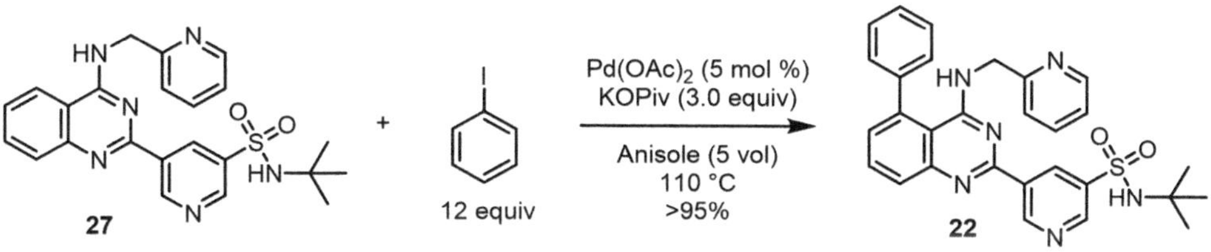

(11)

To develop a more robust process, we sought to understand the mechanism of this transformation. To that end, we performed a computational study of the catalytic cycle, calculated at the B3LYP/DefSV(P) level of theory (Figure 22). Coordination of $Pd(OAc)_2$ to the starting material followed by loss of HOAc led to the complex **56b**. C-H activation through a concerted metallation/deprotonation (CMD) process (*91*, *92*, *93*, *94*, *95*, *96*) led to a palladium complex with an empty orbital on Pd (**56d**), which was rather high in energy at 23.0 kcal. Coordination of another molecule of the starting material provided a much more stable palladium complex (**56e**). Precoordination of iodobenzene then facilitated the oxidative addition of iodobenzene in the rate-determining step to afford a Pd(IV) complex (**56h**). Reductive elimination from this Pd(IV) complex afforded the direct arylation product as a complex with Pd. The alternative pathway of first oxidative addition of the aryl iodide followed by C-H activation was significantly higher in energy.

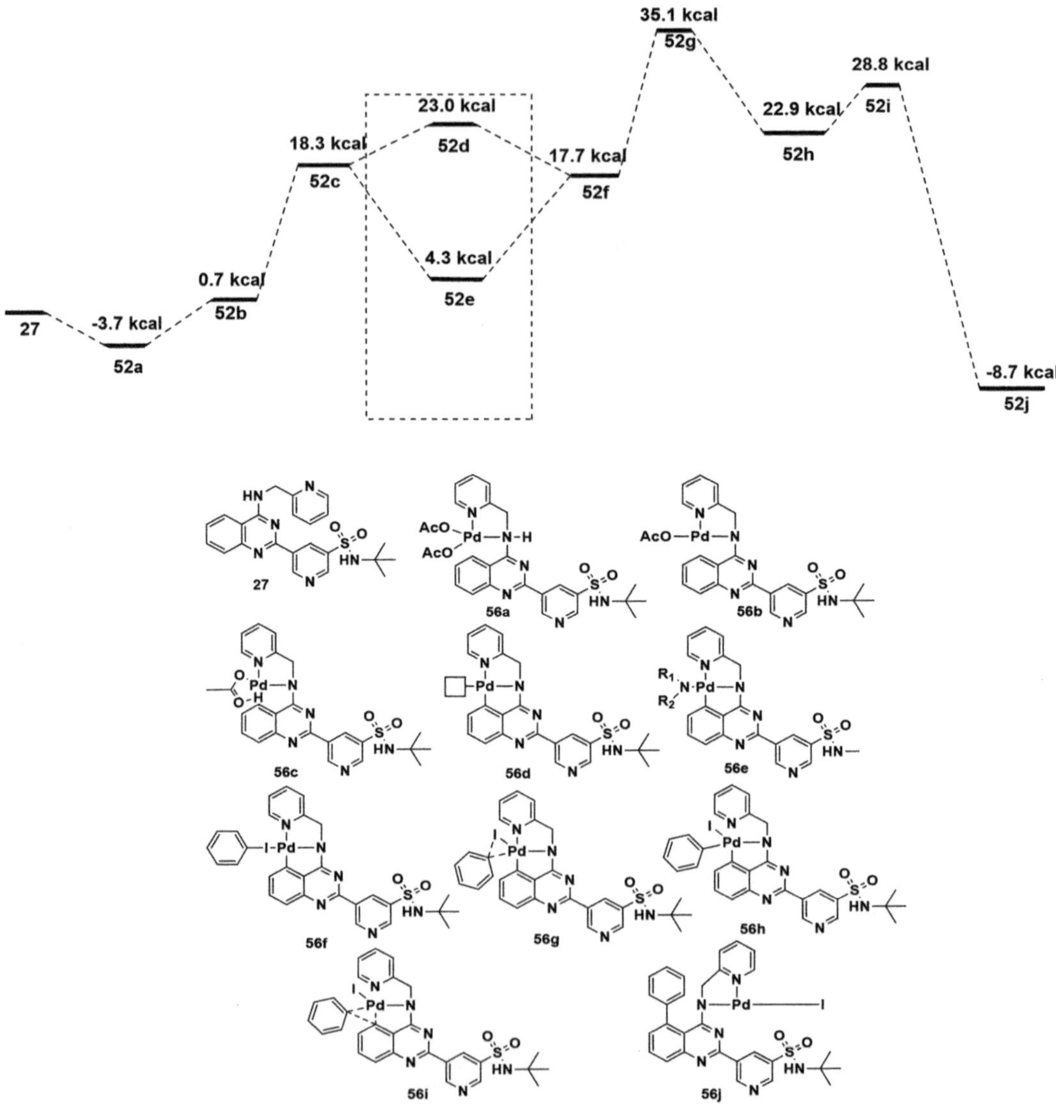

Figure 22. Computational study of the catalytic cycle (calculated at B3LYP/DefSV(P) level of theory). Reproduced with permission from ref. (16). Copyright 2019 American Chemical Society.

The presence of this empty orbital after C-H activation and formation of a more stable palladium complex explained many of the results of our optimization (**56d** and **56e**). The low energy palladium complexes that can be formed from coordination of a ligand to this metal center serves as thermodynamic sinks, rendering the catalyst ineffective and unable to undergo subsequent oxidative addition. Therefore the use of ligands, such as phosphines or pyridines, in addition of organic bases, such as triethylamine and DABCO, or coordinating solvents, such as DMAc and NMP, all inhibit the reaction and can shut down catalysis. Further, the need for the precoordination of iodobenzene to facilitate the oxidative addition helps explain why we need an excess (12 equiv) of iodobenzene, as we can push the equilibrium toward this pre-oxidative addition complex. Reducing the amount of iodobenzene had a significant impact on the rate of reaction. Lastly, the high energy of oxidative addition iodobenzene of 35.1 kcal/mol provided evidence for why stronger Ar-X bonds, such as bromobenzene and phenyl triflate were unreactive electrophiles in this transformation.

Through a collaboration with the Blackmond lab, we sought to investigate the mechanism of the direct arylation experimentally to align with our computational results. Kinetic experiments (*97, 98, 99, 100*) determined that the reaction was zero order in substrate and base and first order in iodobenzene and palladium, suggesting that both iodobenzene and palladium were involved in the turnover limiting step. Furthermore, the key palladacycle (**57**) from the C-H activation was isolated by the stoichiometric reaction of **27** with $Pd(OAc)_2$, $KHCO_3$, and pyridine, which was required to fill the empty coordination site on Pd (Figure 23). An X-ray structure of this complex confirmed that the C-H activation step occurred at a much lower temperature than the reaction. Combined with the orders of palladium and iodobenzene, this data provided experimental evidence that the oxidation addition of iodobenzene was rate limiting. Subjecting the palladacycle to the reaction conditions as the catalyst led to similar conversion and purity as with $Pd(OAc)_2$.

Having gained a better understanding of the key parts of the catalytic cycle, we turned our attention to the conversion kinetics of the reaction. As the reaction is zero order in substrate, and large excess of iodobenzene are employed, we expected linear kinetics with respect to consumption of the substrate **27**. However, we observed non-linear kinetics as the rate slowed as the reaction proceeded (Figure 24). Product inhibition was not observed from spiking studies of the product. Further, iodide poisoning of the catalyst (*101*) from the inorganic byproduct KI was also not observed. Therefore, we had to determine the cause of catalyst decomposition. We hypothesized that the catalyst decomposition was not the result of an off-cycle sink of the catalyst but rather an issue with solubility. At temperature (>100 °C), KOPiv was not completely soluble, and the product and KI crystallized out as the reaction proceeded. Therefore we proposed that the palladium was entrapped in these solids, leading to less palladium in solution and a rate decay as the reaction proceeded. To test this hypothesis, we performed hot filtrations of the direct arylation at various time courses along the reaction. We observed palladium in both the solids and in solution after 1 h of the reaction, which had required ~16 h to reach complete conversion. Therefore, to alleviate this rate decay, a re-examination of the direct arylation was required.

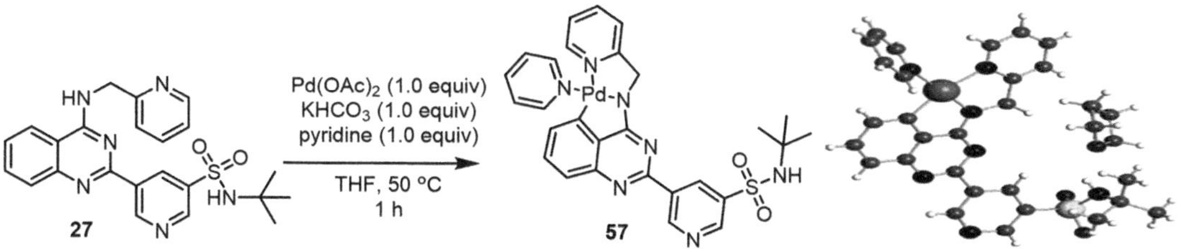

Figure 23. X-ray structure of the palladacycle. Note that one THF molecule is located in the cage of the X-ray structure. Reproduced with permission from ref. (16). Copyright 2019 American Chemical Society.

The only other effective solvent for the direct arylation was dioxane, which we chose to avoid because of its inherent toxicity relative to anisole. Therefore, we revisited our base selection. As can be seen in Figure 25, similar RAP of product was observed with 1.0 - 3.0 equiv of KOPiv. When increasing the equivalents of base with cesium as the counterion, significant levels of a new impurity were observed.

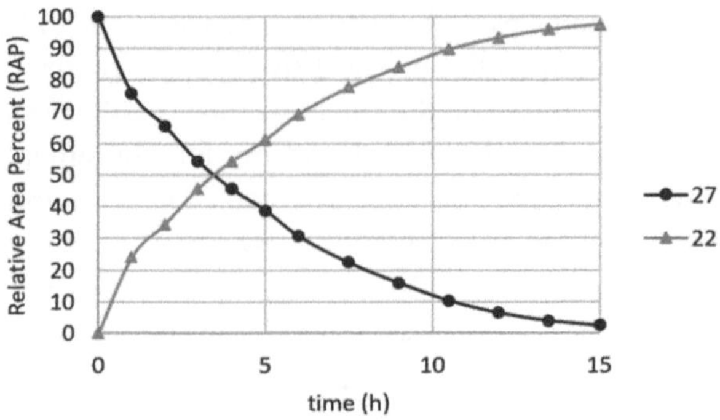

Figure 24. Kinetics of the C-H activation reaction with KOPiv.

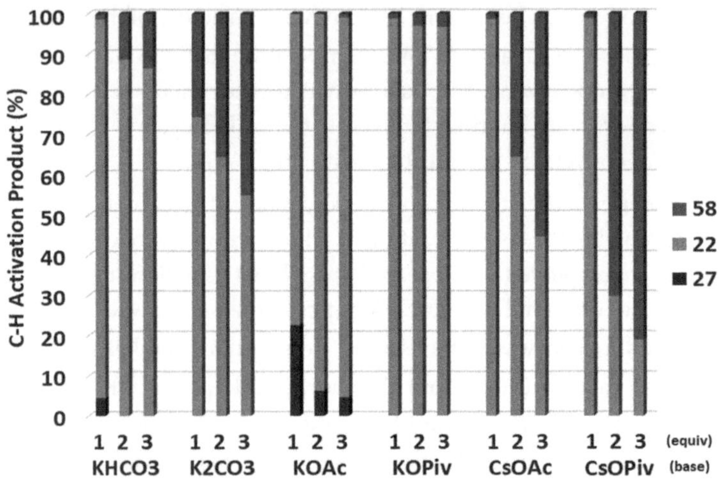

Figure 25. Revisiting base selection in the direct arylation. Reproduced with permission from ref. (16). Copyright 2019 American Chemical Society.

Elucidation of this impurity led to a tetracycle (Figure 26, **58**), the result of the product undergoing a second, C-N bond-forming C-H activation (*102, 103, 104*). Computational analysis of the product-Pd-I complex shows the C-H bond proximal to the empty orbital on Pd, thereby easily facilitating the second C-H activation.

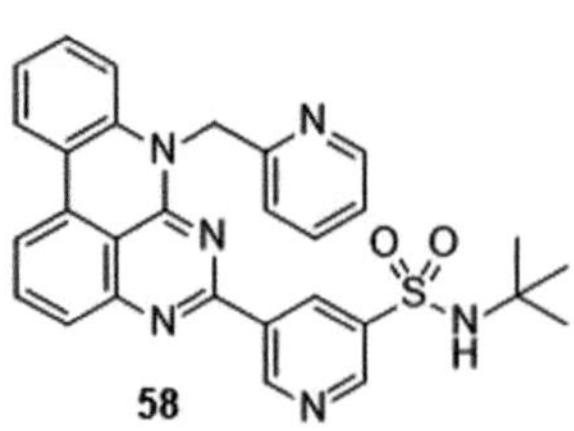

58

Figure 26. Structure of the dual C-H activation product (left) and product-Pd-I complex (right). Reproduced with permission from ref. (16). Copyright 2019 American Chemical Society.

The solubility of this tetracyclic impurity was ~10x lower than that of the desired product, and formation of this impurity must be minimized to not introduce a new impurity in the isolated material at the penultimate stage. Therefore, we sought to understand the mechanism of formation of the dual C-H activation product.

The second C-H activation does not occur without Pd as heating the product with base does not result in ring closure. Based on mechanistic studies, we proposed a catalytic cycle for the formation of the tetracycle (Figure 27). The product-Pd-I complex **59a** can undergo a second C-H activation in the presence of base. Reductive elimination of complex **59b** yields the tetracycle, presumably coordinated to a Pd(0) complex **59c**. Reoxidation of the Pd(0) occurs via a double oxidative addition/reductive elimination with iodobenzene to form biphenyl as a side product and **59c**. Ligand and base exchange will then turnover the catalyst. The reoxidation of Pd(0) has precedent in the literature as a way to form dimeric arenes (*105*, *106*). Furthermore, studies on the tetracyclic system showed that without iodobenzene, the cycle will not turn over.

In other experiments, the amount of biphenyl product forming in the reaction was tracked and corresponds to the turnover of the catalyst. Lastly, the single C-H activation pathway is also effective with Pd_2dba_3 as a catalyst, presumably through the same oxidation pathway where biphenyl is again observed as a byproduct.

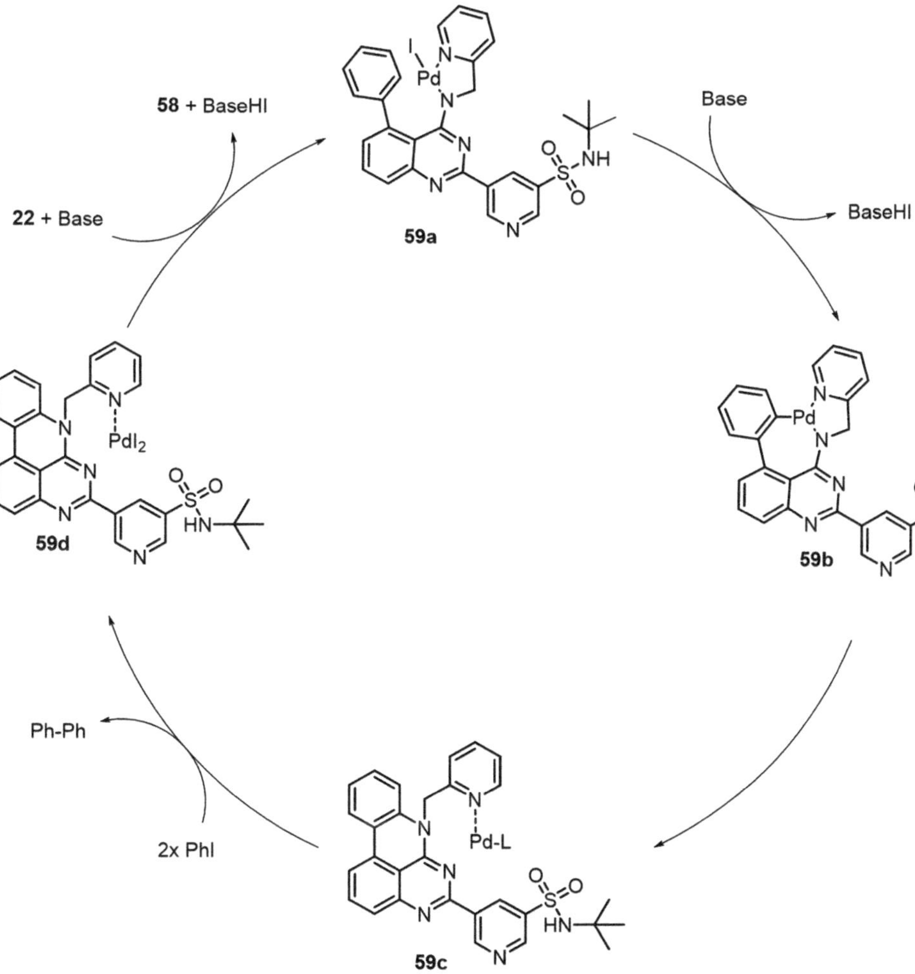

Figure 27. Proposed catalytic cycle for formation of the tetracyclic impurity **58**.

Investigating the catalytic cycle showed two ways to minimize formation of the tetracyclic impurity. Since iodobenzene was required to reoxidize Pd(0), reducing the amount of iodobenzene should then minimize the amount of biphenyl. However, we were unable to significantly reduce the amount of iodobenzene based on the precoordination required for oxidative addition. The other option was to minimize the amount of base, which was required for the second C-H activation. By utilizing slightly greater than 1 equiv of base, we were able to employ a Cs base in this transformation. Both CsOAc and CsOPiv were effective when charged at <1.05 equiv. However, CsOAc is highly hygroscopic, and therefore, CsOPiv was chosen as the base for this transformation, which was prepared in situ from Cs_2CO_3 and pivalic acid (Eq 12).

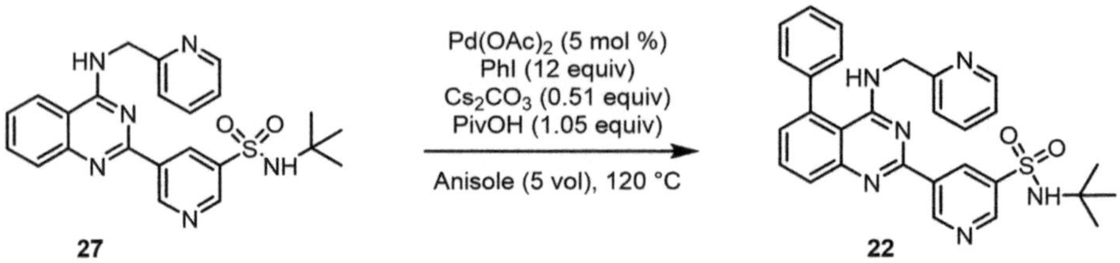

(12)

Changing the base to CsOPiv had a profound impact on the kinetics of the reaction as well. In addition to increasing the reaction temperature to 120 °C, the increased solubility of CsOPiv led to a homogeneous reaction, and the desired linear kinetics were obtained (Figure 28), providing a robust process that can be performed on kilogram scale.

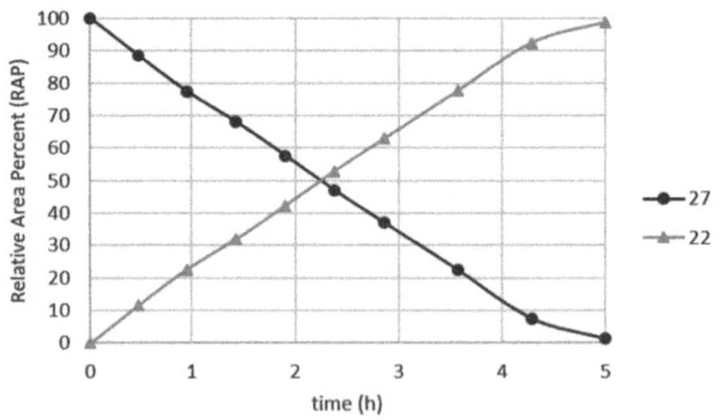

Figure 28. Kinetics of the direct arylation with CsOPiv. Reproduced with permission from ref. (16). Copyright 2019 American Chemical Society.

As we continued to develop the CsOPiv reaction conditions, we tracked the formation of four other impurities in addition to the tetracycle, all of which were product degradants. Interestingly, the product was prone to low levels of dimerization in the reaction conditions through single electron chemistry at the pseudo-benzylic position, which resulted in two diasteromeric dimers of the product. Dealkylation or oxidation to the corresponding amide of the product were also observed at <1 %, presumably from a similar single electron pathway.

Having optimized the reaction conditions for the direct arylation, development of the work-up and isolation of the C-H activation product was then completed. The product had remarkably low solubility in every solvent and combination test (<16 mg/mL), except for NMP at72 mg/mL. Because the direct arylation was a reactive crystallization, the product and resultant CsI byproduct were both out of solution at the end of the reaction. Based on these two observations, it would not be possible to perform a traditional wash to remove the inorganic salts from the reaction. Instead, we cooled the reaction stream to 0 °C and filtered the product and inorganics. Attempts to add an antisolvent to decreases losses at this stage were unsuccessful. The cake was then washed with anisole, followed by a 1:1 mixture of IPA/H_2O to remove CsI. To further improve the purity and potency, the product was recrystallized from 7:1 DMAc/H_2O, with the addition of ethylene diamine to serve as a Pd scavenger resulting in <20 ppm Pd in the isolated cake.

The direct arylation was scaled to 6.5 kg for three batches in an average of 76% yield after the recrystallization. In each batch, the key tetracyclic impurity was formed in <1.15 % and was purged to <0.15 % after the recrystallization. The dimers of the penultimate and a dealkylated product were also purged in the two crystallizations to afford high quality penultimate material.

The final step of the synthesis was the removal of *tert*-butyl protecting group on the sulfonamide to afford BMS-919373 (**1**) (Eq 13). The initial process heated penultimate intermediate **22** in 7 L/kg conc. HCl (35-37% in water) at 90 °C for two hours to achieve complete conversion. The mixture was then neutralized by the addition of a large volume of NaOH aq. solution and the API was purified by a crystallization/recrystallization process in 60% yield. The main challenge for the initial process was the HCl vapor generated under the harsh reaction conditions was incompatible with the stainless-steel equipment train. Therefore the initial development goal was to identify conditions employing a less corrosive acid. If an alternative acid was not identified, HCl would still serve as the reagent but the reaction conditions require modifications to preserve the equipment, such as decreasing the reaction temperature, diluting the concentration of HCl, or reducing the quantity of HCl.

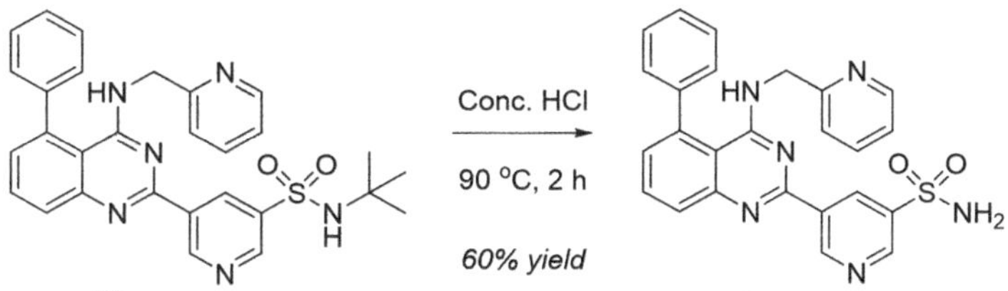

22 → 1 (13)

The first alternative acid evaluated was sulfuric acid, which offered similar reactivity but much less corrosiveness to the equipment train as it does not fume under the reaction conditions. Though the optimized reaction condition displayed slower kinetics than the previous HCl process, reaction in 10 L/kg 1.8 M H_2SO_4 proceeded to complete conversion after 20 h. Screening with different solvents and additives also suggested an enhanced reaction rate with trifluoroethanol (TFE) or acetic acid (HOAc) as the co-solvent. As we continued to develop the H_2SO_4–mediated process, however, elevated levels of product degradation were observed, including a peroxide impurity, an amide impurity and a dealkylated impurity (Figure 29). In addition, API synthesized from the H_2SO_4–mediated process also showed poorer stability by generating more peroxide impurity **60** under storage conditions. As the peroxide impurity had not been observed in the previous HCl process and the structure had alerted in-silico genotoxic evaluation, we decided to deprioritize the development of the H_2SO_4–mediated process.

Next, aiming to identify an alternative acid for deprotection, a broad range of acids was evaluated with water or TFE as the solvent (Figure 30). Based on the results of the screening, methanesulfonic acid (MSA), *p*-toluenesulfonic acid hydrate (*p*-TSA) and phosphoric acid (H_3PO_4) were identified as the leads for follow-up studies. Unfortunately, multigram scale experiments showed emergence of peroxide impurity with the reactions using MSA or *p*-TSA. Attempts to improve H_3PO_4–mediated process were also unsuccessful, as the conditions generated up to 10% *t*-Bu transfer impurity **63.**

We believed the high level of trapping of *tert*-butyl cation by the pyridine nitrogen was partially due to the higher viscosity of the reaction mixture as 20 equivalents of H_3PO_4 (85% aq solution) were required for complete conversion. Under such conditions, release of isobutylene became slower and favored equilibrium to the cation under acidic conditions. It is also possible the corresponding anion, dihydrogen phosphate, is a weaker nucleophile and a better leaving group than the chloride ion from the HCl process. Through numerous follow-up experiments, we were able to reduce the formation of the *t*-Bu transfer impurity by altering solvent composition and improving mixing, but the levels of the impurity were still variable, ranging from 2% to 8%. As we did not identify a straightforward solution to completely resolve this problem, we decided to deprioritize the H_3PO_4–mediated conditions.

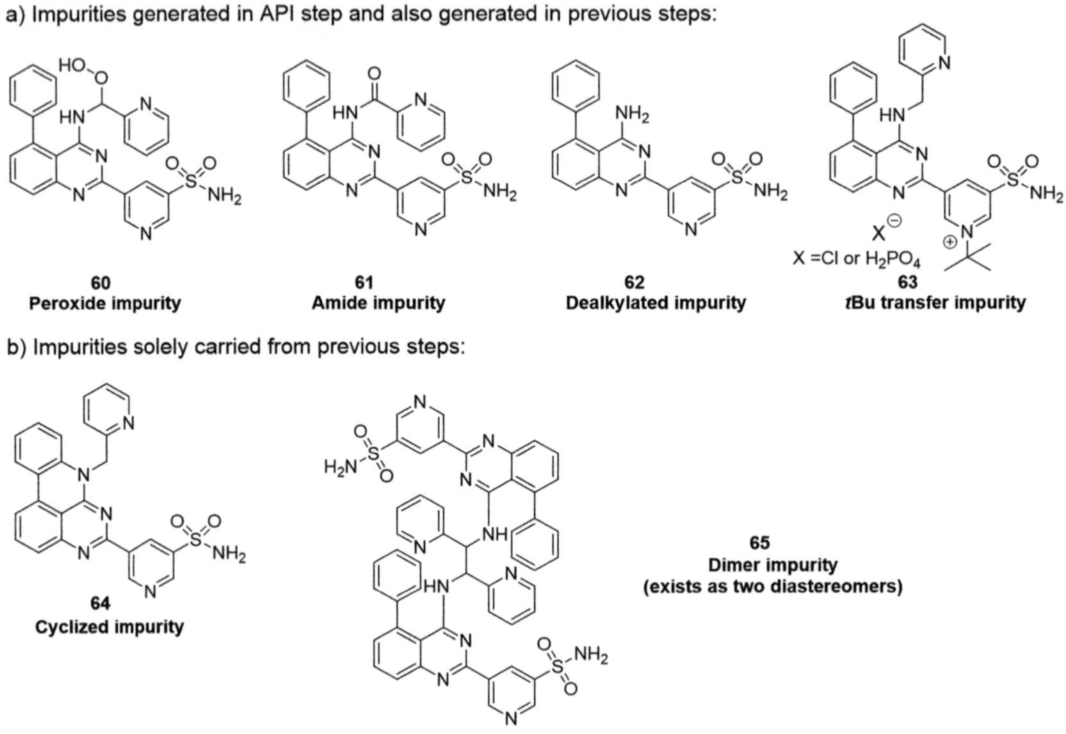

Figure 29. Impurities in the API step.

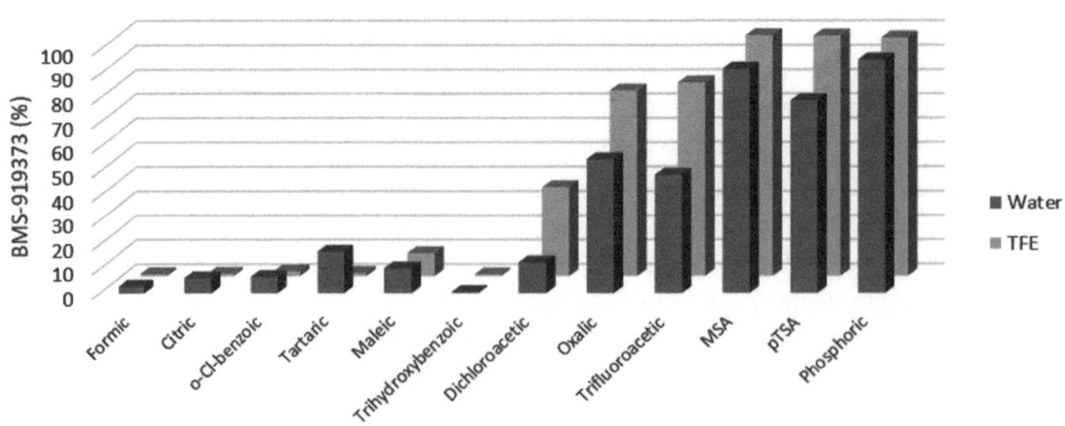

Figure 30. Deprotection reaction with alternative acids.

After screening alternative acids and solvents, we came to the conclusion that HCl was the optimal acid choice for deprotection after all. Therefore we decided to improve the previous process, targeting to resolve equipment incompatibility and reduce PMI. We initiated the development by studying the impact of several key reaction parameters on reaction performance, including temperature, equivalents of acid, concentration, and co-solvent. Our studies revealed that we can achieve a reaction performance similar to that obtained from the initial conditions of 7.0 L of conc. HCl/kg of penultimate **22** by using only 3.0 L conc. HCl/ kg of **22**. The use of less volume of conc. HCl together with the need of lower volume of base to neutralize the reaction mixture would reduce the PMI for this step significantly. Using either TFE or HOAc as a co-solvent slightly accelerated reaction rate while other protic solvents did not provide any improvement. The reaction temperature appeared critical. While the reaction proceeded to complete conversion with 3.0 L conc. HCl/kg of **22** at 40 °C in 8 h, it also formed 9% of the *t*-Bu transfer impurity **63**. Increasing the reaction temperature to 70 °C, reduced the level of the *t*-Bu transfer impurity **63** to <2%. The higher reaction temperature presumably favored the release of isobutylene from the reaction mixture, thereby offering a cleaner reaction profile. To accommodate the emergence of several new impurities, the previous crystallization condition was altered. Water and EtOH were added to the reaction mixture prior to the neutralization and the API was then crystallized during the addition of aq KOH.

The aq. EtOH crystallization effectively removed most impurities and provided API that met quality acceptance criteria, but the material isolated from the process was associated with a pale to light green color. Unfortunately, we were not able to discover any unknown signals by HPLC analysis for new impurities at various wavelengths. The presence of color was not directly relevant to the purity of API isolated from the process.

Several carbon-based scavengers were effective in purging the color under acidic conditions, but were not pursued due to concerns of equipment incompatibility with HCl. Interestingly, we noticed slight color variations of API batches from different penultimate inputs, which led us to determine that two API impurities, namely the cyclized impurity **64** and *t*-Bu transfer impurity **63**, displayed a strong green/yellow color. Recrystallization studies with first drop API delivered three conditions that afforded white crystalline product: DMF/H_2O, DMAc/H_2O, and THF/MeOH/H_2O. After further evaluations, THF/MeOH/H_2O was selected as the recrystallization conditions that enabled purity upgrade and reduction of color to pure white.

After we identified an optimized process to improve safety and PMI while meeting the acceptance criteria, the API step was scaled to 6.0 kg for three batches (Table 8). The first batch was intended to deliver material with elevated levels of impurities for toxicological studies and the subsequent two batches were aimed with optimal purity for clinical supply. To avoid any potential incompatibility of strong acidic condition with stainless steel and Hastelloy® components, two 200 L glass-lined reactors with Hicoflex® valves (for charging reagents) were used for production. The process was executed smoothly and three batches were manufactured in 82-92% yield. The new crystallization and recrystallization procedure offered excellent purging of critical impurities and provided API that met its acceptance criteria.

The route to I_{Kur} API starts from commercially available 3,5-dibromopyridine and isatoic anhydride (Scheme 24). Implementation of the direct arylation route led to a streamlined synthetic route. The overall yield of the route was increased to 28.1%, compared to 3.6% for the initial cyclization route. The overall PMI for the route was significantly improved, resulting in a reduction from >2000 in the initial cyclization route to 407.3 in the direct arylation route. The route was used to prepare ~13 kg of I_{Kur} API which met all API specifications.

Table 8. Optimized API Process for Production

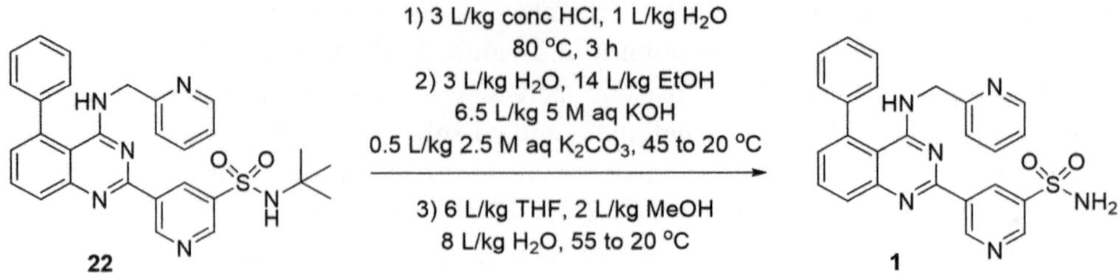

	Batch 1	Batch 2	Batch 3
Purity of API (AP)	99.21	99.88	99.93
Yield (%)	92.0	83.7	82.3
Cyclized impurity AP (**64**)	0.05	<0.05	<0.05
*t*Bu transfer impurity AP (**63**)	0.14	<0.05	<0.05
Dimer impurity 1 AP (**65**)	0.17	<0.05	<0.05
Dimer impurity 2 AP (**65**)	0.15	<0.05	<0.05

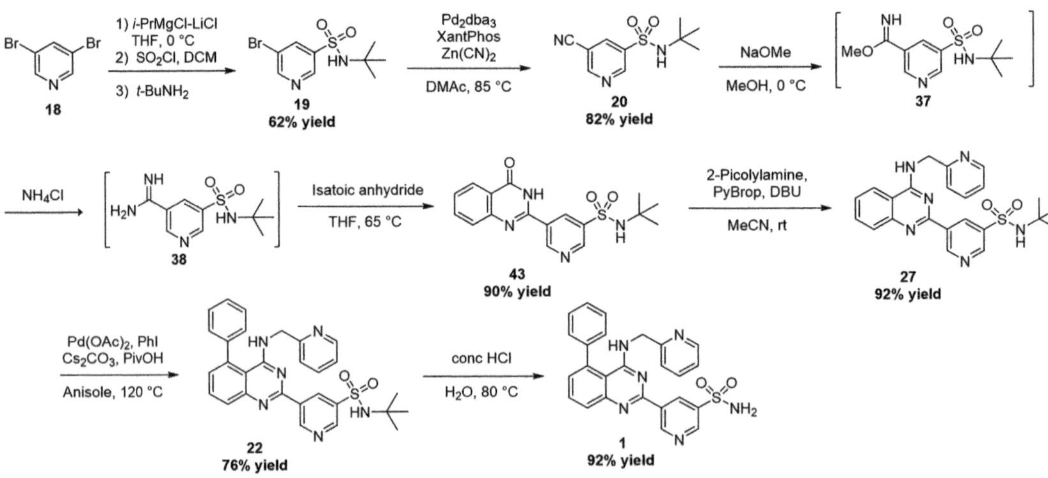

Scheme 24. Route to I_{Kur} API.

Conclusion

The I_{Kur} inhibitor program initiated in discovery chemistry with the goal of identifying a selective, and orally bioavailable potent I_{Kur} inhibitor for the potential treatment of AF. The first clinical candidate advanced to Phase I studies from this effort was from the pyrazolodihydropyrimidine series, BMS-394136. Due to the potential for formation of reactive metabolites in human, a second generation, structurally differentiated phenylquinazoline series was explored. Extensive optimization of this series for potency, selectivity, PK/PD and brain penetration led to the identification of BMS-919373. The compound had an acceptable candidate profile, completed preclinical safety studies and was advanced to Phase I clinical trials in 2011. BMS-919373 subsequently completed long term safety studies in preclinical species, single and multiple ascending dose studies in human and DDI interaction studies. With an acceptable exposure and safety profile in Phase I studies, BMS-919373 was advanced further to Phase II studies in 2014.

The program transitioned to process chemistry during this time to identify a scalable route to BMS-919373. The first route developed focused on an efficient cycloaddition reaction as part of an enabling route to afford material for the early clinical trials. Commercial route exploration then began. The key retrosynthetic disconnection for this route was a C-H activation to install the aryl group, which allowed the use of inexpensive, readily available starting materials. A condensation reaction between isatoic anhydride and amidine derived from 3,5-dibromopyridine can provide the quinazoline core in a few steps in high yield and purity. Installation of 2-picolylamine followed by direct arylation and deprotection afforded BMS-919373 that met all required specifications. Utilization of this new route led to a ~ 60x reduction in starting material requirements, 2x improvement in yield, ~3x improvement on PMI, and positioned the program for a scalable, robust route to clinical API. Unfortunately, in 2017, the Phase 2 study was terminated due to insufficient enrolment.

References

1. *Atrial Fibrillation*. https://www.cdc.gov/heartdisease/atrial_fibrillation.htm (accessed 04 Apr 2020).
2. Benjamin, E. J.; Muntner, P.; Alonso, A.; Bittencourt, M. S.; Callaway, C. W.; Carson, A. P.; Chamberlain, A. M.; Chang, A. R.; Cheng, S.; Das, S. R.; Delling, F. N.; Djousse, L.; Elkind, M. S. V.; Ferguson, J. F.; Fornage, M.; Jordan, L. C.; Khan, S. S.; Kissela, B. M.; Knutson, K. L.; Kwan, T. W.; Lackland, D. T.; Lewis, T. T.; Lichtman, J. H.; Longenecker, C. T.; Loop, M. S.; Lutsey, P. L.; Martin, S. S.; Matsushita, K.; Moran, A. E.; Mussolino, M. E.; O'Flaherty, M.; Pandey, A.; Perak, A. M.; Rosamond, W. D.; Roth, G. A.; Sampson, U. K. A.; Satou, G. M.; Schroeder, E. B.; Shah, S. H.; Spartano, N. L.; Stokes, A.; Tirschwell, D. L.; Tsao, C. W.; Turakhia, M. P.; VanWagner, L. B.; Wilkins, J. T.; Wong, S. S.; Virani, S. S. Heart Disease and Stroke Statistics—2019 Update: A Report From the American Heart Association. *Circulation* **2019**, *139*, 56–528.
3. Lippi, G.; Sanchis-Gomar, F.; Cervellin, G. Global epidemiology of atrial fibrillation: An increasing epidemic and public health challenge. *Int. J. Stroke* **2020**doi:10.1177/1747493019897870.
4. Peterson, B. E.; Al-Khatib, S. M.; Granger, C. B. Apixaban to prevent stroke in patients with atrial fibrillation: a review. *Ther. Adv. Cardiovasc. Dis.* **2017**, *11*, 91–104.
5. Lévy, S.; Maarek, M.; Coumel, P.; Guize, L.; Lekieffre, J.; Medvedowsky, J.; Sebaoun, A. Characterization of different subsets of atrial fibrillation in general practice in France: the ALFA study. The College of French Cardiologists. *Circulation* **1999**, 99, 3028–3035.
6. Nattel, S.; Carlsson, L. Innovative approaches to anti-arrhythmic drug therapy. *Nat Rev Drug Discovery* **2006**, 5, 1034–1049.
7. Yue, L.; Fengt, J.; Lit, G.; Nattel, S. Characterization of an ultrarapid delayed rectifier potassium channel involved in canine atrial repolarization. *J. Physiol.* **1996**, *496*, 647–662.
8. Laughlin, J. C.; Kowey, P. R. Dronedarone:A New Treatment for Atrial Fibrillation. *J. Cardiovasc. Electrophys.* **2008**, *19*, 1220–1226.
9. Dézsic, C. A.; Szentes, V. The Real Role of β-Blockers in Daily Cardiovascular Therapy. *Am. J. Cardiovasc. Drugs* **2017**, *17*, 361–373.
10. Gunaga, P.; Lloyd, J.; Mummadi, S.; Banerjee, A.; Dhondi, N. K.; Hennan, J.; Subray, V.; Jayaram, R.; Rajugowda, N.; Reddy, K. U.; Kumaraguru, D.; Mandal, U.; Beldona, D.;

Adisechen, A. K.; Yadav, N.; Warrier, J.; Johnson, J. J.; Sale, H.; Putlur, S. P.; Saxena, A.; Chimalakonda, A.; Mandlekar, S.; Conder, M.; Xing, D.; Gupta, A. K.; Gupta, A.; Rampulla, R.; Mathur, A.; Levesque, P.; Wexler, R. R.; Finlay, H. J. Selective I_{Kur} Inhibitors for the Potential Treatment of Atrial Fibrillation: Optimization of the Phenyl Quinazoline Series Leading to Clinical Candidate 5-[5-Phenyl-4-(pyridin-2-ylmethylamino)quinazolin-2-yl]pyridine-3-sulfonamide. *J. Med. Chem.* **2017**, *60*, 3795–3803.

11. Lloyd, J.; Finlay, H. J.; Vacarro, W.; Hyunh, T.; Kover, A.; Bhandaru, R.; Yan, L.; Atwal, K.; Conder, M.; Jenkins-West, T.; Shi, H.; Huang, C.; Li, D.; Sun, H.; Levesque, P. Pyrrolidine amides of pyrazolodihydropyrimidines as potent and selective KV1.5 blockers. *Bioorg. Med. Chem. Lett.* **2010**, *20*, 1436–1439.
12. Finlay, H. J.; Johnson, J. A.; Lloyd, J. L.; Jiang, J.; Neels, J.; Gunaga, P.; Banerjee, A.; Dhondi, N.; Chimalakonda, A.; Mandlekar, S.; Conder, M. L.; Sale, H.; Xing, D.; Levesque, P.; Wexler, R. R. Discovery of 5-Phenyl-N-(pyridin-2-ylmethyl)-2-(pyrimidin-5-yl)quinazolin-4-amine as a Potent I_{Kur} Inhibitor. *ACS Med. Chem. Lett.* **2016**, 7, 831–834.
13. Ford, J. W.; Milnes, J. T. New Drugs Targeting the Cardiac Ultra-Rapid Delayed-Rectifier Current (I_{Kur}): Rationale, Pharmacology and Evidence for Potential Therapeutic Value. *J. Cardiovasc. Pharmacol.* **2008**, *52*, 105–120.
14. Identifier: NCT02156076 and NCT02153437. ClinicalTrials.gov.
15. Wisniewski, S. R.; Steven, J. M.; Yu, M.; Fraunhoffer, K. J.; Romero, E. O.; Savage, S. A. Utilizing Native Directing Groups: Synthesis of a Selective I_{Kur} Inhibitor, BMS-919373, via a Regioselective C–H Arylation. *J. Org. Chem.* **2019**, *84*, 4704–4714.
16. Wisniewski, S. R.; Savage, S. A.; Romero, E. O.; Eastgate, M. D.; Tan, Y.; Simmons, E. M.; Plata, E.; Sowa, J. R., Jr.; Blackmond, D. G. Utilizing Native Directing Groups: Mechanistic Understanding of a Direct Arylation Leads to Formation of Tetracyclic Heterocycles via Tandem Intermolecular, Intramolecular C–H Activation. *J. Org. Chem.* **2019**, *84*, 7961–7970.
17. Unless otherwise designated, all purities are HPLC area %. In keeping with standard industrial practice for optimization, in-process yields are assessed using "area percentage" (AP, listed as %) yields. AP values are uncorrected for the difference in LC response factor for the non-solvent peaks in the reaction. These data, while approximate, give insight into general trends, and aid in determining optimal conditions for lead reactions.
18. For a recent review of direct arylation, see: Lei, A.; Zhang, H. *Transition Metal-Catalysed Direct Arylation of Unactivated Arenes with Aryl Halides In C-H and C-X Bond Functionalization: Transition Metal Mediation*; Ribas, X. , Ed.; Royal Society of Chemistry: London, 2013; pp 310−327.
19. For a recent review of direct arylation, see: Sharma, U.; Modak, A.; Maity, S.; Maji, A.; Maiti, D. *Direct Arylation via C–H Activation In New Trends in Cross-Coupling: Theory and Applications.* Colacot, R., Ed. .; Royal Society of Chemistry: London, 2015; 551-609.
20. For a recent review of direct arylation, see: Rossi, R.; Bellina, F.; Lessi, M.; Manzini, C. Cross-Coupling of Heteroarenes by C-H Functionalization: Recent Progress towards Direct Arylation and Heteroarylation Reactions Involving Heteroarenes Containing One Heteroatom. *Adv. Synth. Catal.* **2014**, *356*, 17–117.
21. Huang, L.; Li, Q.; Wang, C.; Qi, C. Palladium(II)-Catalyzed Regioselective Arylation of Naphthylamides with Aryl Iodides Utilizing a Quinolinamide Bidentate System. *J. Org. Chem.* **2013**, *78*, 3030–3038.

22. Nadres, E. T.; Santos, G. I. F.; Shabashov, D.; Daugulis, O. Scope and Limitations of Auxiliary-Assisted, Palladium-Catalyzed Arylation and Alkylation of sp^2 and sp^3 C–H Bonds. *J. Org. Chem.* **2013**, *78*, 9689–9714.
23. Li, Z.-X.; Sun, S.-Y.; Qiao, H.-J.; Yang, F.; Zhu, Y.; Kang, J.-X.; Wu, Y.-S.; Wu, Y.-J. Palladium-Catalyzed Regioselective C8–H Amination of 1-Naphthylamine Derivatives with Aliphatic Amines. *Org. Lett.* **2016**, *18*, 4594–4597.
24. Wang, L.-L.; Yang, M.-X.; Liu, X.-C.; Song, H.; Han, L.; Chu, W.-Y.; Sun, Z.-Z. Pd-catalyzed direct C-H cyanation of picolinamides via bidentate chelation assistance. *Appl. Organomet. Chem.* **2016**, *30*, 680–683.
25. Shang, R.; Ilies, L.; Nakamura, E. Iron-Catalyzed Directed C(sp2)–H and C(sp3)–H Functionalization with Trimethylaluminum. *J. Am. Chem. Soc.* **2015**, *137*, 7660–7663.
26. Iwasaki, M.; Kaneshika, W.; Tsuchiya, Y.; Nakajima, K.; Nishihara, Y. Palladium-Catalyzed peri-Selective Chalcogenation of Naphthylamines with Diaryl Disulfides and Diselenides via C–H Bond Cleavage. *J. Org. Chem.* **2014**, *79*, 11330–11338.
27. Huang, L.-H.; Sun, X.-D.; Li, Q.; Qi, C.-Z. Palladium-Catalyzed C8 Alkylation of 1-Naphthylamides with Alkyl Halides via Bidentate-Chelation Assistance. *J. Org. Chem.* **2014**, *79*, 6720–6725.
28. Odani, R.; Hirano, K.; Satoh, T.; Miura, M. Copper-Mediated Dehydrogenative Biaryl Coupling of Naphthylamines and 1,3-Azoles. *J. Org. Chem.* **2013**, *78*, 11045–11052.
29. For a Rh-catalyzed olefination on a similar naphthalene system, see: Li, X.; Gong, X.; Zhao, M.; Song, G.; Deng, J.; Li, X. Rh(III)-Catalyzed Oxidative Olefination of N-(1-Naphthyl)sulfonamides Using Activated and Unactivated Alkenes. *Org. Lett.* **2011**, *13*, 5808–5811.
30. For a Ru-catalyzed C-H activation of arenes utilizing a similar directing group, see: Aihara, Y.; Chatani, N. Ruthenium-catalyzed direct arylation of C–H bonds in aromatic amides containing a bidentate directing group: significant electronic effects on arylation. *Chem. Sci.* **2013**, *4*, 664–670.
31. For a recent C-H arylation with aryl boronic acids, see: Wei, Y.; Kan, J.; Wang, M.; Su, W.; Hong, M. Palladium-Catalyzed Direct Arylation of Electron-Deficient Polyfluoroarenes with Arylboronic Acids. *Org. Lett.* **2009**, *11*, 3346–3349.
32. For a recent C-H arylation with aryl boronic acids, see: Kirchberg, S.; Vogler, T.; Studer, A. Stereoselective Palladium-Catalyzed Carboaminoxylations of Indoles with Arylboronic Acids and TEMPO. *Angew. Chem. Int. Ed.* **2009**, *48*, 4235–4238.
33. For a recent C-H arylation with aryl boronic acids, see: Spangler, J. E.; Kobayashi, Y.; Verma, P.; Wang, D.-H.; Yu, J.-Q. α-Arylation of Saturated Azacycles and N-Methylamines via Palladium(II)-Catalyzed C(sp3)–H Coupling. *J. Am. Chem. Soc.* **2015**, *137*, 11876–11879.
34. For a recent C-H arylation with aryl boronic acids, see: Laforteza, B. N.; Chan, K. S. L.; Yu, J.-Q. Enantioselective ortho-C-H Cross-Coupling of Diarylmethylamines with Organoborons. *Angew. Chem. Int. Ed.* **2015**, *54*, 11143–11146.
35. For an example of C-H arylations with diaryliodonium salts, see: Deprez, N. R.; Sanford, M. S. Synthetic and Mechanistic Studies of Pd-Catalyzed C–H Arylation with Diaryliodonium Salts: Evidence for a Bimetallic High Oxidation State Pd Intermediate. *J. Am. Chem. Soc.* **2009**, *131*, 11234–11241.

36. For an example of C-H arylations with diaryliodonium salts, see: Storr, T. E.; Greaney, M. F. Palladium-Catalyzed Arylation of Simple Arenes with Iodonium Salts. *Org. Lett.* **2013**, *15*, 1410–1413.
37. For an example of C-H arylations with diaryliodonium salts, see: Wen, J.; Zhang, R.-Y.; Chen, S.-Y.; Zhang, J.; Yu, X.-Q. Direct Arylation of Arene and N-Heteroarenes with Diaryliodonium Salts without the Use of Transition Metal Catalyst. *J. Org. Chem.* **2012**, *77*, 766–771.
38. Smith, M. J.; Lawler, M. J.; Kopp, N.; Mcleod, D. D.; Davulcu, A. H.; Lin, D.; Katipally, K. Development of a Concise Multikilogram Synthesis of LPA-1 Antagonist BMS-986020 via a Tandem Borylation–Suzuki Procedure. *Org. Process Res. Dev.* **2017**, *21*, 1859–1863.
39. Bogolubsky, A. V.; Ryabukhin, S. V.; Plaskon, A. S.; Stetsenko, S. V.; Volochnyuk, D. M.; Tolmachev, A. A. Dry HCl in Parallel Synthesis of Fused Pyrimidin-4-ones. *J. Comb. Chem.* **2008**, *10*, 858–862.
40. Niementowksi, S. V. Selbsttätige, stetig wirkende Quecksilberluftpumpe für chemische Zwecke. *Chem. Ber.* **1894**, *27*, 1386–1394.
41. Meyer, J. F.; Wagner, E. C. The Niementowski Reaciton. The Use of Methyl Anthranilate or Isatoic Anhydride with Subsitituted Amides or Amidines in the Formation of 3-Subsittuted-4-Keto-3,4-Dihydroquinazolines. The Course of the Reaction. *J. Org. Chem.* **1943**, *8*, 239–252.
42. For a review on magnenium-halogen exchange, see: Knochel, P.; Dohle, W.; Gommermann, N.; Kneisei, F. K.; Kopp, F.; Korn, T.; Sapountzis, I.; Vu, V. A. Highly Functionalized Organomagnesium Reagents Prepared through Halogen–Metal Exchange. *Angew. Chem. Int. Ed.* **2003**, *42*, 4302–4320.
43. For an example using *i*-PrMgCl·LiCl: Krasovskiy, A.; Knochel, P. A LiCl-Mediated Br/Mg Exchange Reaction for the Preparation of Functionalized Aryl- and Heteroarylmagnesium Compounds from Organic Bromides. *Angew. Chem. Int. Ed.* **2004**, *43*, 3333–3336.
44. Several synthetic methods to sulfonamide were considered but not evaluated: Pd-catalyzed chlorosulfonylation: DeBergh, J. R.; Niljianskul, N.; Buchwald, S. L. Synthesis of Aryl Sulfonamides via Palladium-Catalyzed Chlorosulfonylation of Arylboronic Acids. *J. Am. Chem. Soc.* **2013**, *135*, 10638–10641.
45. Several synthetic methods to sulfonamide were considered but not evaluated: DABSO-mediated sulfonylation: Woolven, H.; Gonzalez-Rodriguez, C.; Marco, I.; Thompson, A. L.; Willis, M. C. DABCO-Bis(sulfur dioxide), DABSO, as a Convenient Source of Sulfur Dioxide for Organic Synthesis: Utility in Sulfonamide and Sulfamide Preparation. *Org. Lett.* **2011**, *13*, 4876–4878.
46. Several synthetic methods to sulfonamide were considered but not evaluated: Pd-catalyzed sulfination: Shavnya, A.; Coffey, S. B.; Smith, A. C.; Mascitti, V. Palladium-Catalyzed Sulfination of Aryl and Heteroaryl Halides: Direct Access to Sulfones and Sulfonamides. *Org. Lett.* **2013**, *15*, 6226–6229.
47. Pu, Y.-M.; Christesen, A.; Ku, Y.-Y. A simple and highly effective oxidative chlorination protocol for the preparation of arenesulfonyl chlorides. *Tetrahedron Lett.* **2010**, *51*, 418–421.
48. Hoffman, R. V. m-Trifluoromethylbenzenesulfonyl Chloride. *Org. Synth.* **1981**, *60*, 121–123.
49. Marron, B. E.; Fritch, P. C.; Mark-Worth, C. J.; Maynard, A. T.; Swain, N. A. *Inhibitors of Ion Channels*. Patent WO 2008/118758 A1, 2008.

50. Changi, S. M.; Wong, S.-W. Kinetics Model for Designing Grignard Reactions in Batch or Flow Operations. *Org. Process Res. Dev.* **2016**, *20*, 525–539.
51. For a discussion on the aggregation of Grignard reagents, see: Schnegelsberg, C.; Bachmann, S.; Kolter, M.; Auth, T.; John, M.; Stalke, D.; Koszinowski, K. Association and Dissociation of Grignard Reagents RMgCl and Their Turbo Variant RMgCl·LiCl. *Chem. Eur. J.* **2016**, *22*, 7752–7762and references cited therein.
52. Yu, M.; Strotman, N. A.; Savage, S. A.; Leung, S.; Ramirez, A. A Practical and Robust Multistep Continuous Process for Manufacturing 5-Bromo-N-(tert-butyl)pyridine-3-sulfonamide. *Org. Process Res. Dev.* **2019**, *23*, 2088–2095.
53. For a representative review, see: Burcham, C. L.; Florence, A. J.; Johnson, M. D. Continuous Manufacturing in Pharmaceutical Process Development and Manufacturing. *Ann. Rev. Chem. Biomol. Eng.* **2018**, *9*, 253–281.
54. For a representative review, see: Plutschack, M. B.; Pieber, B.; Gilmore, K.; Seeberger, P. H. The Hitchhiker's Guide to Flow Chemistry. *Chem. Rev.* **2017**, *117*, 11796–11893.
55. For a representative review, see: Movsisyan, M.; Delbeke, E. I. P.; Berton, J. K. E. T.; Battilocchio, C.; Ley, S. V.; Stevens, C. V. Taming hazardous chemistry by continuous flow technology. *Chem. Soc. Rev.* **2016**, *45*, 4892–4928.
56. For a representative review, see: Gutmann, B.; Cantillo, D.; Kappe, C. O. Continuous-Flow Technology—A Tool for the Safe Manufacturing of Active Pharmaceutical Ingredients. *Angew. Chem. Int. Ed.* **2015**, *54*, 6688–6728.
57. For a representative review, see: Porta, R.; Benaglia, M.; Puglisi, A. Flow Chemistry: Recent Developments in the Synthesis of Pharmaceutical Products. *Org. Process Res. Dev.* **2016**, *20*, 2–25.
58. For discussions on continuous processing in pharma, see: Baxendale, I. R.; Braatz, R. D.; Hodnett, B. K.; Jensen, K. F.; Johnson, M. D.; Sharratt, P.; Sherlock, J.-P.; Florence, A. J. Achieving Continuous Manufacturing: Technologies and Approaches for Synthesis, Workup, and Isolation of Drug Substance. May 20–21, 2014 Continuous Manufacturing Symposium. *J. Pharm. Sci.* **2015**, *104*, 781–791.
59. For discussions on continuous processing in pharma, see: Thayer, A. M. End-To-End Chemistry. *Chem. Eng. News* **2014**, *92*, 13–21.
60. For discussions on continuous processing in pharma, see: Poechlauer, P.; Colberg, J.; Fisher, E.; Jansen, M.; Johnson, M. D.; Koenig, S. G.; Lawler, M.; Laporte, T.; Manley, J.; Martin, B.; O'Kearney-McMullan, A. Pharmaceutical Roundtable Study Demonstrates the Value of Continuous Manufacturing in the Design of Greener Processes. *Org. Process Res. Dev.* **2013**, *17*, 1472–1478.
61. For discussions on continuous processing in pharma, see: Poechlauer, P.; Manley, J.; Broxterman, R.; Gregertsen, B.; Ridemark, M. *Org. Process Res. Dev.* **2012**, *16*, 1586–1590.
62. For discussions on continuous processing in pharma, see: Jiménez-González, C.; Poechlauer, P.; Broxterman, R.; Yang, B.-S.; am Ende, D. A.; Baird, J.; Bertsch, C.; Hannah, R. E.; Dell'Orco, P.; Noorman, H.; Yee, S.; Reintjens, R.; Wells, A.; Massonneau, V.; Manley, J. Key Green Engineering Research Areas for Sustainable Manufacturing: A Perspective from Pharmaceutical and Fine Chemicals Manufacturers. *Org. Process Res. Dev.* **2011**, *15*, 900–911.

63. For an example of using continuous technology to process unstable organometallic compounds that are otherwise challenging or impossible to scale up, see: Usutani, H.; Nihei, T.; Papageorgiou, C. D.; Cork, D. G. Development and Scale-up of a Flow Chemistry Lithiation–Borylation Route to a Key Boronic Acid Starting Material. *Org. Process Res. Dev.* **2017**, *21*, 669–673.
64. Evaluations of reaction calorimetry showed adiabatic temperature rises of 15.7 °C, 51.1 °C and 37 °C for steps 1, 2 and 3, respectively.
65. For an example of process safety evaluation of a magnesium-bromine exchange, see: Tang, W.; Sarvestani, M.; Wei, X.; Nummy, L. J.; Patel, N.; Narayanan, B.; Byrne, D.; Lee, H.; Yee, N. K.; Senanayake, C. H. Formation of 2-Trifluoromethylphenyl Grignard Reagent via Magnesium–Halogen Exchange: Process Safety Evaluation and Concentration Effect. *Org. Process Res. Dev.* **2009**, *13*, 1426–1430.
66. Maynard, G. D. Sulfuryl Chloride. In *Encyclopedia of Reagents for Organic Synthesis, 8 Volume Set*; John Wiley & Sons, 2001.
67. Sulfuryl Chloride. CAS No.: 7791-25-5. *OECD SIDS*; UNEP Publications, 2004.
68. For an example of using *i*-PrMgCl·LiCl for magnesium-iodine exchange under continuous mode, see: Tricotet, T.; O'Shea, D. F. Automated Generation and Reactions of 3-Hydroxymethylindoles in Continuous-Flow Microreactors. *Chem. Eur. J.* **2010**, *16*, 6678–6686.
69. Chlorosulfonamide displayed poor reactivity in the next step, which is a Pd-catalyzed cyanation.
70. Gilman, H.; Fothergill, R. E. The Reaction Between Sulfochlorides and Organomagnesium Halides. *J. Am. Chem. Soc.* **1929**, *51*, 3501–3508.
71. Fu, Y.; Zhu, W.; Zhao, X.; Hugel, H.; Wu, Z.; Su, Y.; Du, Z.; Huang, D.; Hu, Y. CuI catalyzed sulfonylation of organozinc reagents with sulfonyl halides. *Org. Biomol. Chem.* **2014**, *12*, 4295–4299.
72. In order to obtain quick response of end of reaction profile, in-line spectroscopic methods were proposed for the monitoring of production, but was not developed due to timeline constraint and equipment availability at the production site.
73. For a recent review of palladium-catalyzed cyanations, see: Anbarasan, P.; Schareina, T.; Beller, M. Recent developments and perspectives in palladium-catalyzed cyanation of aryl halides: synthesis of benzonitriles. *Chem. Soc. Rev.* **2011**, *40*, 5049–5067.
74. For a recent review of palladium-catalyzed cyanations, see: Takagi, K. Palladium-Catalyzed Cross-Coupling Involving α-Hetero-Substituted Organometals: Palladium-Catalyzed Cross-Coupling Involving Metal Cyanides. In *Handbook of Organopalladium Chemistry for Organic Synthesis*. Negishi, E.-i. , Ed.; Wiley-Interscience: New York, 2002
75. For a recent review of palladium-catalyzed cyanations, see: Sundermeier, M.; Zapf, A.; Beller, M. Palladium-Catalyzed Cyanation of Aryl Halides: Recent Developments and Perspectives. *Eur. J. Inorg. Chem.* **2003**, 3513–3525.
76. For a recent cyanation with $K_4[Fe(CN)_6]$, see: Yueng, P. Y.; So, C. M.; Lau, C. P.; Kwong, F. Y. A Mild and Efficient Palladium-Catalyzed Cyanation of Aryl Chlorides with K4[Fe(CN)6]. *Org. Lett.* **2011**, *13*, 648–651.
77. For a recent cyanation with $K_4[Fe(CN)_6]$, see: Weissman, S. A.; Zewge, D.; Chen, C. Ligand-Free Palladium-Catalyzed Cyanation of Aryl Halides. *J. Org. Chem.* **2005**, *70*, 1508–1510.

78. For a recent cyanation with $K_4[Fe(CN)_6]$, see: Schareina, T.; Zapf, A.; Beller, M. Improving palladium-catalyzed cyanation of aryl halides: development of a state-of-the-art methodology using potassium hexacyanoferrate(II) as cyanating agent. *J. Organomet. Chem.* **2004**, *689*, 4576–4583.
79. For a recent cyanation with $K_4[Fe(CN)_6]$, see: Schareina, T.; Zapf, A.; Beller, M. Potassium hexacyanoferrate(II)—a new cyanating agent for the palladium-catalyzed cyanation of aryl halides. *Chem. Commun.* **2004**, 1388–1389.
80. For a recent review of Pd removal, see: Garrett, C. H.; Prasad, K. The Art of Meeting Palladium Specifications in Active Pharmaceutical Ingredients Produced by Pd-Catalyzed Reactions. *Adv. Synth. Catal.* **2004**, *346*, 889–900.
81. Konigsberger, K.; Chen, G.-P.; Wu, R. R.; Girgis, M. J.; Prasad, K.; Repic, O.; Blacklock, T. J. A Practical Synthesis of 6-[2-(2,5-Dimethoxyphenyl)ethyl]-4-ethylquinazoline and the Art of Removing Palladium from the Products of Pd-Catalyzed Reactions. *Org. Process Res. Dev.* **2003**, *7*, 733–742.
82. Welch, C. J.; Albaneze-Walker, J.; Leonard, W. R.; Biba, M.; DaSilva, J.; Henderson, D.; Laing, B.; Mathre, D. J.; Spencer, S.; Bu, X.; Wang, T. Adsorbent Screening for Metal Impurity Removal in Pharmaceutical Process Research. *Org. Process Res. Dev.* **2005**, *9*, 198–205.
83. Busacca, C. A.; Senanyake, C. H. *Developing Palladium-Catalyzed Arylations of Carbonyl-Activated C-H Bonds In Transition Metal-Catalyzed Couplings in Process Chemistry*; Magano, J., Dunetz, J. R., Eds.; Wiley-VCH: Weinheim, Germany, 2013; pp 25-37.
84. Gallagher, W. P.; Vo, A. Dithiocarbamates: Reagents for the Removal of Transition Metals from Organic Reaction Media. *Org. Process Res. Dev.* **2015**, *19*, 1369–1373.
85. Prat, D.; Wells, A.; Hayler, J.; Sneddon, H.; McElroy, C. R.; Abou-Shehada, S.; Dunn, P. J. CHEM21 selection guide of classical- and less classical-solvents. *Green Chem.* **2016**, *18*, 288–296.
86. For an example of an inorganic base and carboxylic acid additive, see: Lafrance, M.; Fagnou, K. Palladium-Catalyzed Benzene Arylation: Incorporation of Catalytic Pivalic Acid as a Proton Shuttle and a Key Element in Catalyst Design. *J. Am. Chem. Soc.* **2006**, *128*, 16496–16497.
87. For an example of an inorganic base and carboxylic acid additive, see: Kim, J.; Hong, S. H. Ligand-Promoted Direct C–H Arylation of Simple Arenes: Evidence for a Cooperative Bimetallic Mechanism. *ACS Catal.* **2017**, *7*, 3336–3343.
88. For an example of an inorganic base and carboxylic acid additive, see: Stephens, D. E.; Lakey-Beitia, J.; Atesin, A. C.; Atesin, T. A.; Chavez, G.; Arman, H. D.; Larinov, O. V. Palladium-Catalyzed C8-Selective C–H Arylation of Quinoline N-Oxides: Insights into the Electronic, Steric, and Solvation Effects on the Site Selectivity by Mechanistic and DFT Computational Studies. *ACS Catal.* **2015**, *5*, 167–175.
89. For a selected review on carboxylate ligands in transition-metal-catalyzed C–H functionalization reactions, see: Lapointe, D.; Fagnou, K. Overview of the Mechanistic Work on the Concerted Metallation–Deprotonation Pathway. *Chem. Lett.* **2010**, *39*, 1118–1126.
90. For a selected review on carboxylate ligands in transition-metal-catalyzed C–H functionalization reactions, see: Ackermann, L. Carboxylate-Assisted Transition-Metal-Catalyzed C–H Bond Functionalizations: Mechanism and Scope. *Chem. Rev.* **2011**, *111*, 1315–1345.

91. Davies, D. L.; Donald, S. M. A.; Macgregor, S. A. Computational Study of the Mechanism of Cyclometalation by Palladium Acetate. *J. Am. Chem. Soc.* **2005**, *127*, 13754–13755.
92. García-Cuadrado, D.; Braga, A. A. C.; Maseras, F.; Echavarren, A. M. Proton Abstraction Mechanism for the Palladium-Catalyzed Intramolecular Arylation. *J. Am. Chem. Soc.* **2006**, *128*, 1066–1067.
93. Davies, D. L.; Macgregor, S. A.; McMullin, C. L. Computational Studies of Carboxylate-Assisted C–H Activation and Functionalization at Group 8–10 Transition Metal Centers. *Chem. Rev.* **2017**, *117*, 8649–8709.
94. Gorelsky, S. I.; Lapointe, D.; Fagnou, K. Analysis of the Concerted Metalation-Deprotonation Mechanism in Palladium-Catalyzed Direct Arylation Across a Broad Range of Aromatic Substrates. *J. Am. Chem. Soc.* **2008**, *130*, 10848–10849.
95. Sun, H.-Y.; Gorelsky, S. I.; Stuart, D. R.; Campeau, L.-C.; Fagnou, K. Mechanistic Analysis of Azine N-Oxide Direct Arylation: Evidence for a Critical Role of Acetate in the $Pd(OAc)_2$ Precatalyst. *J. Org. Chem.* **2010**, *75*, 8180–8189.
96. Gorelsky, S. I.; Lapointe, D.; Fagnou, K. Analysis of the Palladium-Catalyzed (Aromatic)C–H Bond Metalation–Deprotonation Mechanism Spanning the Entire Spectrum of Arenes. *J. Org. Chem.* **2012**, *77*, 658–668.
97. Blackmond, D. G. Reaction Progress Kinetic Analysis: A Powerful Methodology for Mechanistic Studies of Complex Catalytic Reactions. *Angew. Chem. Int. Ed.* **2005**, *44*, 4302–4320.
98. Mathew, J. S.; Klussmann, M.; Iwamura, H.; Valera, F.; Futran, A.; Emanuelsson, E. A. C.; Blackmond, D. G. Investigations of Pd-Catalyzed ArX Coupling Reactions Informed by Reaction Progress Kinetic Analysis. *J. Org. Chem.* **2006**, *71*, 4711–4722.
99. Blackmond, D. G. Kinetic Profiling of Catalytic Organic Reactions as a Mechanistic Tool. *J. Am. Chem. Soc.* **2015**, *137*, 10852–10866.
100. Bures, J. Variable Time Normalization Analysis: General Graphical Elucidation of Reaction Orders from Concentration Profiles. *Angew. Chem. Int. Ed.* **2016**, *55*, 16084–16087.
101. Wang, X.-C.; Hu, Y.; Bonacorsi, S.; Hong, Y.; Burrell, R.; Yu, J.-Q. Pd(II)-Catalyzed C–H Iodination Using Molecular I_2 as the Sole Oxidant. *J. Am. Chem. Soc.* **2013**, *135*, 10326–10329.
102. For an example of domino C-H activation/C-N bond formation in the synthesis of phenanthridinones, see: Wang, G.-W.; Yuan, T.-T.; Li, D.-D. One-Pot Formation of C-C and C-N Bonds through Palladium-Catalyzed Dual C-H Activation: Synthesis of Phenanthridinones. *Angew. Chem. Int. Ed.* **2011**, *50*, 1380–1383.
103. For an example of domino C-H activation/C-N bond formation in the synthesis of phenanthridinones, see: Saha, R.; Sekar, G. Stable Pd-nanoparticles catalyzed domino CH activation/CN bond formation strategy: An access to phenanthridinones. *J. Catal.* **2018**, *366*, 176–188.
104. For an example of domino C-H activation/C-N bond formation in the synthesis of phenanthridinones, see: Jaiswal, Y.; Kumar, Y.; Pal, J.; Subramanian, R.; Kumar, A. Rapid synthesis of polysubstituted phenanthridines from simple aliphatic/aromatic nitriles and iodo arenes via Pd(II) catalyzed domino C–C/C–C/C–N bond formation. *Chem. Commun.* **2018**, *54*, 7207–7210.

105. For an example of homocoupling of aryl halides, see: Hennings, D. D.; Iwama, T.; Rawal, V. H. Palladium-Catalyzed (Ullmann-Type) Homocoupling of Aryl Halides: A Convenient and General Synthesis of Symmetrical Biaryls via Inter- and Intramolecular Coupling Reactions. *Org. Lett.* **1999**, *1*, 1205–1208.
106. For an example of homocoupling of aryl halides, see: Wang, L.; Zhang, Y.; Liu, L.; Wang, Y. Palladium-Catalyzed Homocoupling and Cross-Coupling Reactions of Aryl Halides in Poly(ethylene glycol). *J. Org. Chem.* **2006**, *71*, 1284–1287.

Chapter 3

Discovery and Development of Omecamtiv Mecarbil: A Novel Cardiac Myosin Activator for the Potential Treatment of Systolic Heart Failure

Bradley P. Morgan,[1] Seb Caille,[2] and Shawn D. Walker[*,2,3]

[1]Research & Nonclinical Development, Cytokinetics, 280 East Grand Avenue, South San Francisco, California 94080, United States

[2]Drug Substance Technologies, Process Development, Amgen Inc., 1 Amgen Center Drive, Thousand Oaks, California 91320, United States

[3]Current address: Development Operations, Johnson Matthey Inc., 25 Patton Road, Devens, Massachusetts 04134, United States

[*]Email: shawn.walker@matthey.com

Omecamtiv mecarbil is a first-in-class, direct activator of cardiac myosin, the motor protein of the cardiac sarcomere that transfers chemical energy (ATP hydrolysis) into a mechanical force for contraction. By promoting the force-generating cycle of myosin, the drug acts like "more hands on the rope" to improve cardiac contractility without increasing cardiac myocyte intracellular calcium concentrations. It is currently being evaluated as a potential treatment for heart failure with the goal of establishing a new continuum of care for patients in both in-hospital and outpatient settings. The first part of this chapter will describe the discovery and candidate selection of omecamtiv mecarbil, including the initial synthetic pathway and the development of an early stage manufacturing process which allowed for rapid advancement in Phase 1/2 clinical evaluation. The second part of the chapter will describe the design of the robust synthetic process used to prepare omecamtiv mecarbil for Phase 3 clinical trials and highlight the challenges and opportunities at the drug substance/drug product interface. This includes optimization of a convergent synthesis of omecamtiv mecarbil's unsymmetrical urea core from stable precursors, as well as engineering the drug substance particle properties to aid drug product manufacturability and performance.

Introduction

Omecamtiv mecarbil (**1**) (Figure 1) is a first-in-class direct activator of cardiac myosin that is currently in Phase 3 clinical trials for the treatment of systolic heart failure (HF) otherwise called heart failure with reduced ejection fraction (HFrEF). HF is a rapidly growing cardiovascular condition with an estimated prevalence of > 38 million people globally. Among patients > 65 years old, HF is the leading cause of hospitalization and the total medical costs world-wide are estimated to exceed USD 50 billion by 2030. Systolic HF or HFrEF is thought to comprise approximately half of the overall HF population (*1*, *2*).

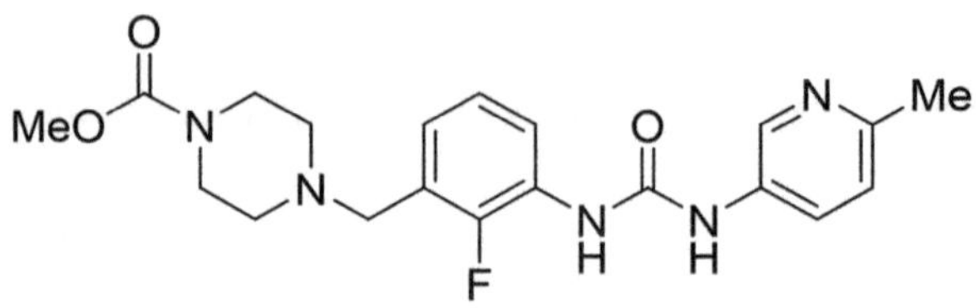

1

Figure 1. Structure of omecamtiv mecarbil.

Omecamtiv mecarbil (**1**) was designed to improve the basic problem in patients suffering from heart failure with reduced ejection fraction (HFrEF), that is, "a failure of the heart to pump blood at a rate commensurate with the requirements of the metabolizing tissue" (*3*). Historically, drugs that improve cardiac contractility do so by modulating calcium signaling in the myocardium and are classified as calcitropes (*4*). Calcitropes such as dobutamine and milrinone activate second messenger pathways that raise the levels of cardiac calcium concentrations and provide relief to heart failure symptoms. However, dobutamine and milrinone increase both oxygen consumption and heart rate, can produce cardiac arrhythmias, and are associated with higher rates of mortality in HF patients (*5–7*).

A leading hypothesis for this increase in mortality associated with dobutamine and milrinone is that the increases in intracellular calcium levels can promote arrhythmia and activate maladaptive Ca^{2+}-dependent signaling cascades making hearts less efficient (*8*). In order to avoid the adverse effects of these calcitropes, agents that increase the cardiac contractility directly at the level of the sarcomere (myotropes-(*4*)) should not perturb the intracellular calcium levels. The myotrope **1** works downstream from the second messenger pathways, instead acting directly at the level of cardiac myosin resulting in improvement of cardiac contractility without increasing cardiac myocyte intracellular calcium concentrations, myocardial oxygen consumption, or heart rate. The unique mechanism of action of **1** promotes the force-generating cycle of myosin, acts like "more hands on the rope", and provides an intriguing approach to improve cardiac contractility (*9*).

Discovery and Early Development of Omecamtiv Mercarbil (1)

Discovery of Omecamtiv Mecarbil

The discovery of omecamtiv mecarbil (**1**) has been the subject of both manuscripts (*10*) and reviews (*11*, *12*), therefore, this section will focus mainly on the unpublished story around the optimization and selection of the clinical candidate as well as the development of the initial manufacturing route to enable rapid entry into clinical studies. Additionally, we share insights into the transition from candidate selection to early development, and how aggressive timelines and continued optimization provided the foundation for the selection of the initial manufacturing route.

Screening

The primary breakthrough that allowed for the mechanism-based screening of sarcomere activators was the development of a high-throughput, reconstituted sarcomere assay (*13*). This assay relies on monitoring the adenosine triphosphatase (ATPase) activity of the system as a surrogate for muscle contractility. Furthermore, through systematic substitution of the assay components, the part of the sarcomere that is being activated, myosin or thin filaments can be established (*11*). High-throughput screening (HTS) using the aforementioned sarcomere assay yielded 4-fluoro-*N*-(3-nitro-5-(pyridin-3-yloxy)phenyl)benzamide **2** (Scheme 1) as an interesting primary hit. The benzamide **2** was deemed to be sufficiently appealing to initiate hit-to-lead optimization efforts (that is, early stage optimization to identify promising lead structures) based on the following observations:

1) there were apparent structural activity relationships (SAR) in this series from the screening data (for example, an electron withdrawing group (EWG) represented by the nitro-group in **2** was thought to be important for activity at this stage of the analysis)
2) there was excellent selectivity for the cardiac sarcomere over the fast-skeletal sarcomere and smooth muscle myosin
3) benzamide **2** increased contractility in rat cardiomyocytes at concentrations believed to be adequate to elicit a biological response in vivo in the absence of increasing calcium transients in these cardiomyocytes.

Although **2** was a promising hit based on initial biological characterization, it had some unfavorable features:

1) potentially genotoxic nitro moiety
2) sub-optimal physical properties (e.g., low solubility)
3) high intrinsic clearance
4) low free fraction
5) potential off-target liabilities.

A major focus of the optimization campaign was not only to improve these undesirable properties but to also identify a drug candidate that would be suitable for both acute IV administration and have drug metabolism and pharmacokinetics (DMPK) properties appropriate for chronic, oral administration.

Early Discovery

For the initial optimization of benzamide **2**, an abbreviated optimization sequence is shown in Scheme 1. The preliminary goal was to provide a suitable agent for IV administration for preclinical, in vivo, proof-of-concept studies. The nitro moiety of the central phenyl ring in **2** was replaced with a fluorine to maintain the electron deficient character of the central aromatic ring that, at the time, was thought to be important for the SAR. The central amide bond was supplanted with a more hydrolytically stable urea and provided a less restrictive SAR around the "right-side" of the molecule

(*10*). This enabled a swap of the 4-fluorophenyl moiety for a slightly more soluble 3-pyridyl group in urea **3**.

At this point, the current set of compounds had a combination of high clearance, low free fraction, and low solubility that prevented obtaining in vivo verification of the mechanism of action for this structural class. In order to achieve the initial proof-of-concept for this series, the focus of optimization progressed to increasing the compound's free fraction (*14*) and to further improve solubility. Both goals were achieved by saturation of the "left-side" pyridine ring of urea **3** (Scheme 1) into a piperidine ring and by acylation of the resultant piperidine to maintain a hydrogen bond acceptor. As a result, **4** was produced. Compound **4** provided a higher free fraction than previous compounds, adequate solubility at pH 4.5, and showed a robust increase in left ventricular fractional shortening in normal, anesthetized Sprague-Dawley rats after IV administration at low μM total concentrations (~ 1 μM). This gave the program the sought-after proof-of-concept for this chemical series (Figure 2).

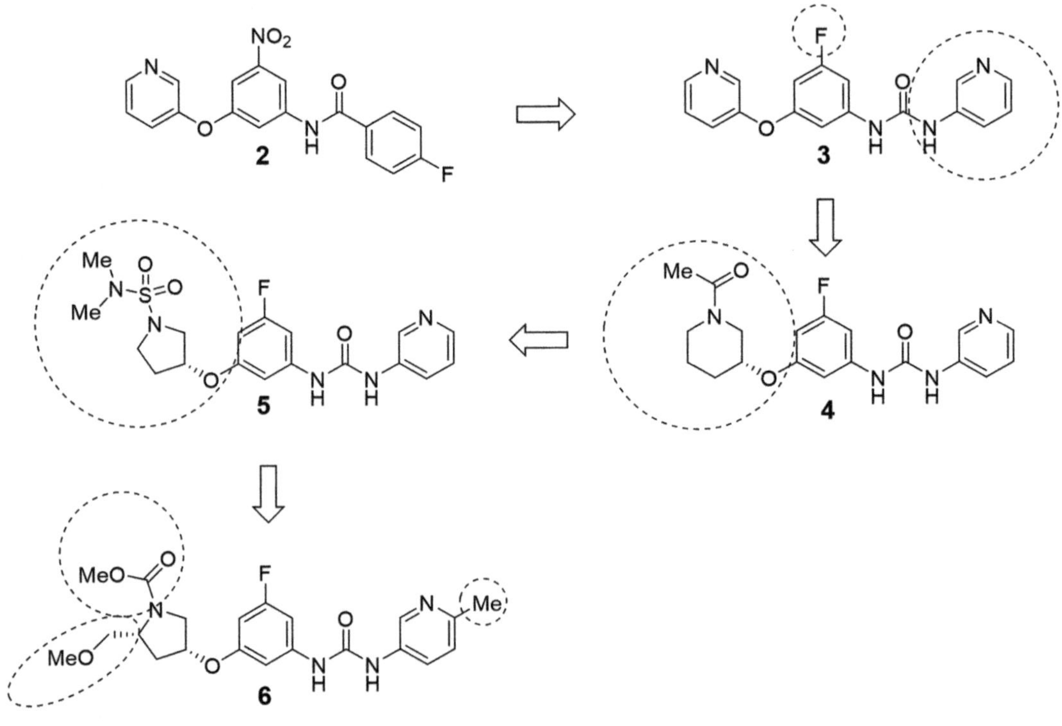

Scheme 1. Initial optimization efforts.

The in vivo characterization of **4** also revealed an off-target effect: at super-therapeutic concentrations (*i.e.*, < 10 μM), **4** directly reduced blood pressure. Through a series of careful experiments using rat aortic ring preps, the vasodilation of **4** was attributed to K_{ATP} channel opening. The K_{ATP} channel opening SAR proved to be very sensitive to manipulation as minor modifications of the "left-side" of **4** (Scheme 1) removed vasodilation in aortic rings. This "left-side" SAR exploration led to improvements in cardiomyocyte potency and a second milestone compound, **5**. Compound **5** showed a dose-response increase in left ventricular fractional shortening in normal, anesthetized Sprague-Dawley rats after IV administration and little to no effect on blood pressure or heart rate at super-therapeutic concentrations. Additionally, **5** demonstrated efficacy in a dog model of heart failure (*15*).

Thus far, the discovery efforts had optimized for several desirable properties of cardiac myosin activators:

1) specificity for cardiac muscle
2) appropriate potency in cardiomyocytes with physical properties that translated to what was believed to be clinically relevant efficacy in animal models
3) no increases in calcium transients in cardiomyocytes
4) removal of off-target activity, and
5) adequate solubility for an interim parenteral formulation.

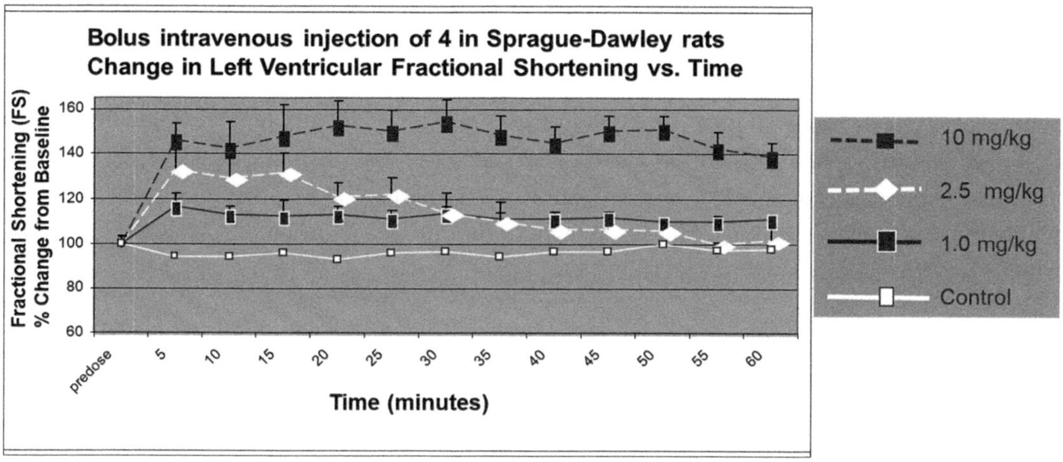

*Figure 2. Dose response increase in left ventricular fractional shortening measured by echocardiography in normal rats upon treatment with **4** via an IV-bolus injection.*

There were additional properties that were desired for further optimization. When compounds such as **5** were profiled for their potential for drug-drug interactions, they were rather potent inhibitors of cytochrome P450 1A2. Furthermore, additional improvements in solubility were desired for the IV application and enhancement in the projected human PK parameters were desired for chronic, oral administration.

Elimination of the cytochrome P450 1A2 inhibition was readily achieved by sterically blocking the "right-side" pyridine nitrogen moiety of **5** (Scheme 1) with an ortho-methyl substituent. Presumably, this methyl moiety prevents the pyridine nitrogen from chelating with the heme-bound iron of CYP 450 1A2 dramatically attenuating the inhibition (CYP 450 1A2 IC_{50}: **5** ~ 60 nM, **6** > 10 μM). Additional optimization of the "left-side" ring to increased solubility (shake flask solubility @ pH 4: **5** < 0.4 mg/mL, **6** ~ 2.3 mg/mL) and enhanced cardiomyocyte potency (cardiomyocyte fractional shortening % @ μM: **5** ~ 137 @ 0.8 μM, **6** ~ 130 @ 0.4 μM) resulted in **6**.

At this point of the optimization, there were drug candidates suitable for acute IV administration, however, the pharmacokinetic properties were not optimal for chronic, oral administration due to the high projected clearance in humans. Several compounds had moderate to good oral bioavailability in preclinical species (e.g., rat and dog) despite having clearance at or exceeding hepatic blood flow in these species. Therefore, the absorption of these structures was likely suitable for oral administration and the limiting feature to pharmacokinetic optimization was believed to be oxidative turnover mediated by CYP 450s, the probable culprit in the high clearance exhibited in preclinical species. Additionally, further improvements in solubility were desired for application for the acute IV administration as well.

Selection of Clinical Candidates

To find compounds with improvements in oxidative turnover and solubility, the discovery efforts took a step back with the hope of taking two steps forward (Scheme 2). Scouring through microsome data looking for compounds with reduced turnover revealed ether **7**, a compound with a reduced extraction ratio in human microsomal preparations (E_H ~ 0.3) compared to most analogs in this series. An additional appealing feature with **7** was that the "left-side" ring was now symmetrical thus removing the chiral center and likely decrease the ultimate cost of manufacturing.

The problem with ether **7** was that it was nearly insoluble in aqueous vehicles, thus rendering it difficult to prepare as an IV formulation and potentially resulting in solubility limiting absorption. It was hypothesized that the C-O bond in the "left-side" piperidine ring could be replaced by an N-C bond (Scheme 2, **7** to **8**) to form a "left-side" piperazine ring. The resultant "left-side" piperazine ring of **8** had improved calculated properties (*16*) compared to **7** as represented by the cLogP of 1.88 and 3.03, respectively. This was hypothesized to further reduce the oxidative clearance. Furthermore, the calculated pK_a of the resultant benzylic nitrogen in the piperazine of **8** was less than 7 (calculated pK_a ~ 6.1) thus allowing **8** to bind as a neutral compound and also allowed for improved solubility at pH suitable for an IV formulation.

The resultant synthesis of **8** brought these hypotheses to reality, that is, **8** had several improved optimization parameters over **7** such as lower human microsomal turnover (E_H **8** < 0.2, **7** ~ 0.3), greater solubility (shake flask solubility @ pH 4: **8** ~ 15 mg/mL, **6** nearly insoluble), and better cardiomyocyte potency (cardiomyocyte fractional shortening % @ μM: **8** ~ 131 @ 0.8 μM, **7** ~ 123 @ 2 μM).

The switch from an ether linkage of **7** to the benzyl amine of **8** aided the synthesis of analogs of the central benzene core of the compounds (Scheme 2) as well. To simplify this core and to probe potential SAR differences with the new core molecules containing the piperazine, the fluorine atom at position 5 of the central benzene ring of **8** was replaced with a hydrogen in compound **9**. Unlike the aforementioned SAR, the biochemical and cardiomyocyte potencies of **9** (cardiomyocyte fractional shortening % @ μM: **9** ~ 139% @ 0.8 μM) were similar to **8**. Additionally, the solubility of **9** was ample for the acute IV formulation. The pharmacokinetics projections in humans were satisfactory for a clinical candidate albeit further optimization of the projected human PK to potentially improve the half-life was desired. After extensive profiling, **9** was selected as a potential clinical candidate and moved to both preclinical assessment and GMP manufacturing.

In parallel to initiating preclinical and manufacturing activities with **9**, optimization efforts continued with the goal of further improvement in potency and projected human PK to improve the dose efficacy (Scheme 2). As mentioned above, switching the ether linkage of **7** to the benzyl amine of **8** promoted the ease of synthesis and, consequently, the SAR manipulation of the central benzene ring. Although an EWG on the central benzene ring was shown to be no longer essential in **9**, we wanted to further probe the appendant fluorine moiety to investigate its effect on potency and PK. Moving the fluorine substituent of the central benzene ring to position 2, resulted in the discovery of omecamtiv mecarbil (**1**). Compound **1** had several improvements compared to **9** including:

1) improved cardiomyocyte potency (cardiomyocyte fractional shortening % @ μM: **1** ~ 123% @ 0.2 μM, **9** ~ 139% @ 0.8 μM)
2) improved pharmacokinetic parameters (represented by dog clearance **1** ~ 7.3 mL/min/kg, **9** ~ 15 mL/min/kg).

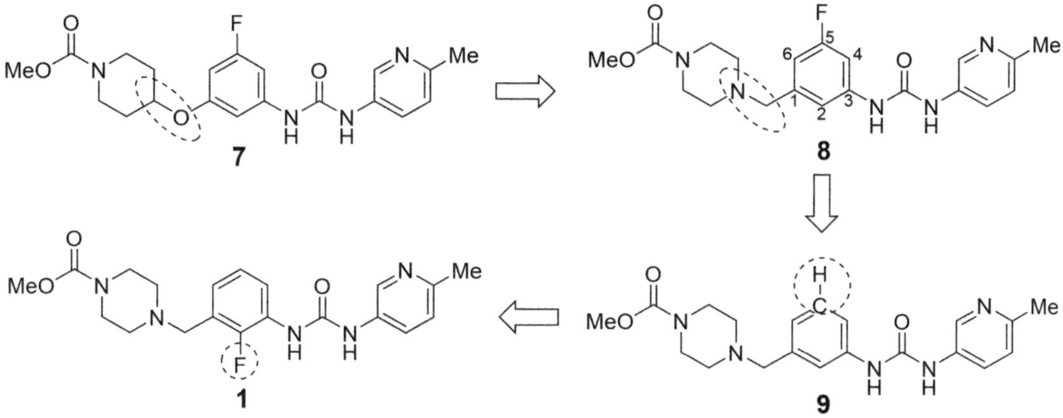

*Scheme 2. Final optimization to omecamtiv mecarbil (**1**).*

The improvements in intrinsic potency and pharmacokinetic parameters resulted in improved dose-efficacy in preclinical models of cardiac contractility (Figures 3 and 4). Compound **9** demonstrated robust dose and concentration response increases in contractility, as measured by increases in fractional shortening using echocardiography, when administered via IV infusion to Sprague-Dawley rats (Figure 3). Compound **1** also showed robust dose and concentration response increases in contractility under similar conditions (Figure 4). Additionally, **1** augmented cardiac contractility at doses and concentrations between 4-9 time lower than for **9**. The enhanced dose-efficacy in preclinical contractility models and the superior projected human pharmacokinetic parameters for **1** vs. **9** highlighted the potential superiority of **1** as a development candidate.

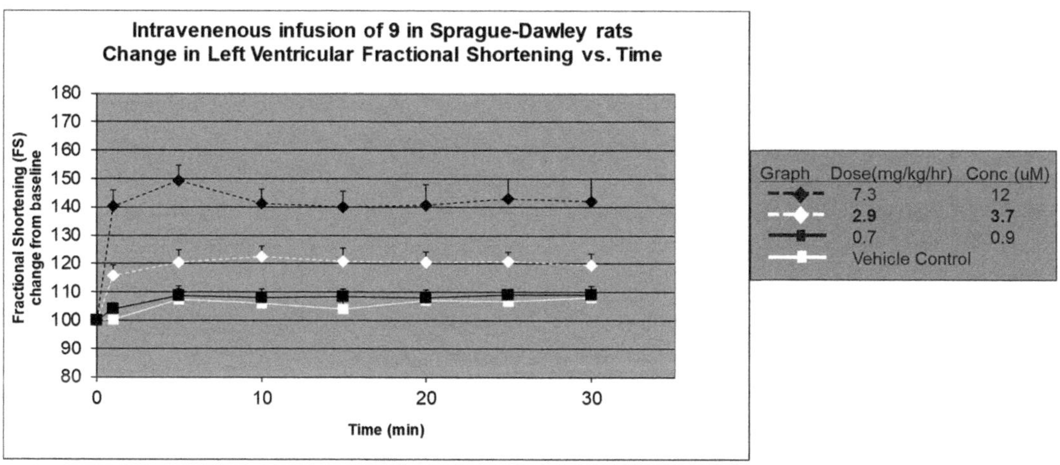

*Figure 3. Dose response for percent increase of fractional vs. time for **9** dosed by IV infusion to anesthetized rats.*

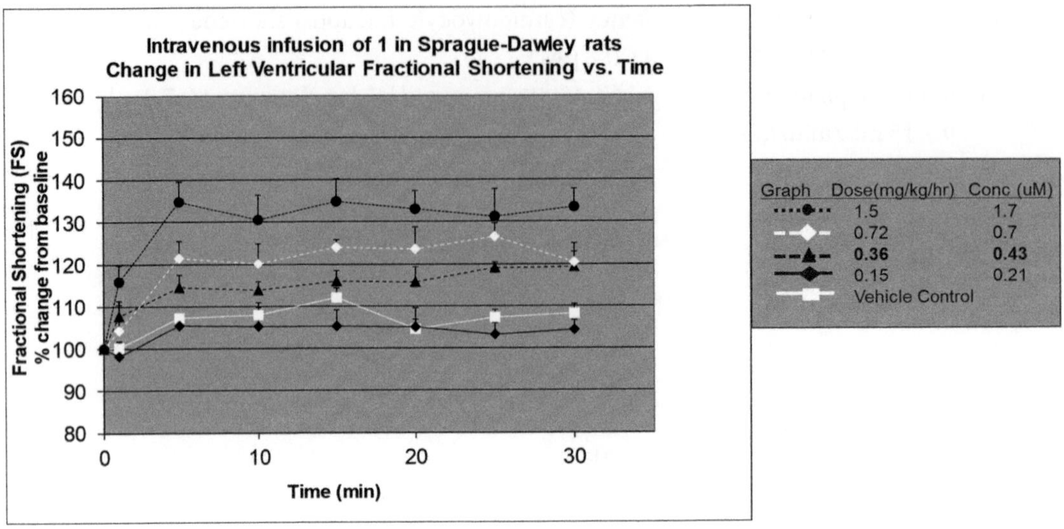

*Figure 4. Dose response for percent increase of fractional vs. time for omecamtiv mecarbil (**1**) dosed by IV infusion to anesthetized rats.*

Early Development of Omecamtiv Mecarbil (1)

At this stage of early development, we were faced with a difficult decision. Compound **9** was already successfully advanced into preclinical development, however, we believed that compound **1** could potentially be the better candidate. The question we asked ourselves was how could we accelerate **1**, so that we could "catch-up" to the already established timeline set with **9**? The remainder of this section focuses on how **1** was accelerated through preclinical and manufacturing timelines to begin development as scheduled.

Research and Initial Manufacturing Timelines

The transition from research to development is rarely seamless. Timelines for innovation are, at best, imprecise and early chemical development is usually rate-limiting for clinical development. The overall timeline for the discovery story of **1,** as discussed in the previous section, is shown in Figure 5. The screening work was run simultaneously with the research efforts towards the discovery of cardiac sarcomere activators. The HTS was initiated in Year 1. Approximately 6 months later, the first myosin activator series, creatively named "Series 1", was discovered. Series 1 provided the first preclinical, "proof-of-concept" compound as measured in isolated rat hearts and through in vivo assessment in rats (IV dosing). However, the pharmacokinetics and physical properties of Series 1 proved resistant to optimization into compounds with "drug-like" properties.

The series that led to omecamtiv mecarbil (**1**), represented by **2** in Scheme 1, was discovered approximately nine months after initiation of screening (Figure 5). Over the next year and a half, several series were uncovered through HTS and underwent optimization. The discovery of **4** (Scheme 1), consolidated the optimization efforts to this structural class of compounds. The direct vasodilation mediated by **4** through K_{ATP} channel opening was identified around seven months later. An additional three months of optimization uncovered the milestone compound **5**, that removed the K_{ATP} channel opening as discussed above. It took around eight months of additional optimization to first synthesize the initial development candidate **9** and the first manufacture of this active pharmaceutical ingredient (API) was initiated eleven months later.

The discovery of **1** came just two to three months prior to the initiation of the manufacture of **9** (Figure 5). Figure 6 shows the timeline to advance **1** from first synthesis to first dose in humans, which was approximately thirteen months. Shortly after synthesizing **1** (less than one month), the hemodynamic and pharmacokinetics assessment demonstrated the promise of **1** as a potential clinical candidate. This data provided the motivation to invest in the development of a manufacturing route and accelerated preclinical assessment of **1** leading to its selection for a clinical candidate near the year's end and the subsequent manufacturing campaign in the first quarter of the following year. The development of and intravenous (IV) formulation, IV toxicology, and IV product manufacture led to the regulatory filing of **1** approximately twelve months after the initial synthesis. The first subject was dosed one month after the initial regulatory filing.

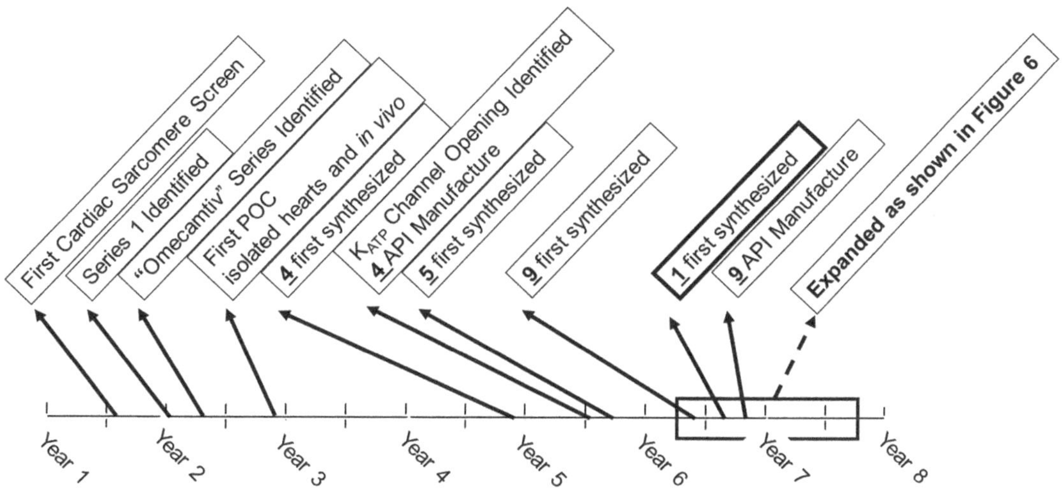

*Figure 5. Research timeline for the discovery of omecamtiv mecarbil (**1**).*

Development of a Manufacturing Route to Enter Clinical Trials

As discussed in the previous section, the development and manufacture of the API **9** was occurring simultaneously with the discovery of **1**. The synthetic route to manufacture **9** was reasonably straightforward (Scheme 3). The synthesis began with an S_N2 coupling using two commercially available fragments: 1-(bromomethyl)-3-nitrobenzene (**10**) and methyl piperazine-1-carboxylate (**11**), in dichloromethane (CH_2Cl_2) and a slight molar excess of diisopropylethyl amine (DIPEA). This resulted in the isolation of methyl 4-(3-nitrobenzyl)piperazine-1-carboxylate (**12**) as a white solid in 85-90% yield (Scheme 3, Step 1). Reduction of the nitro moiety of **12** by heterogenous hydrogenation over a palladium on carbon (Pd/C) catalyst in tetrahydrofuran (THF) buffered with solid potassium carbonate (K_2CO_3) afforded methyl 4-(3-aminobenzyl)piperazine- 1-carboxylate (**13**) as a solid in 80-85% yield (Scheme 3, Step 2). Treatment of **13** with 5-isocyanato-2-methylpyridine (**14**) in refluxing acetone resulted in **9** that crystallized directly from the reaction mixture as a white solid in 85-90% yield (Scheme 3, Step 3).

This 3-step route appeared to be nearly optimal for **9**, however, it would be preferable to replace the genotoxic intermediate isocyanate **14** with a more stable, non-genotoxic reagent for the final step of the synthetic pathway.

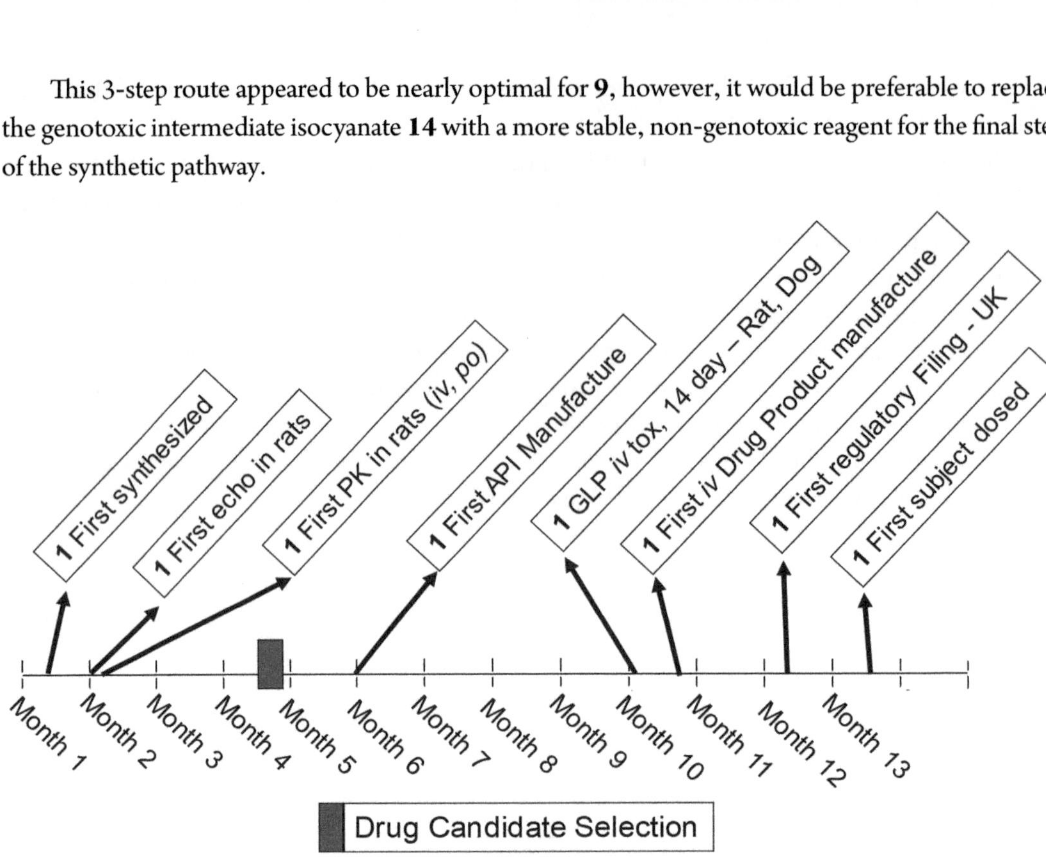

*Figure 6. Omecamtiv mecarbil (**1**): 13 m from first Medicinal Chemistry synthesis to first human subject dosed.*

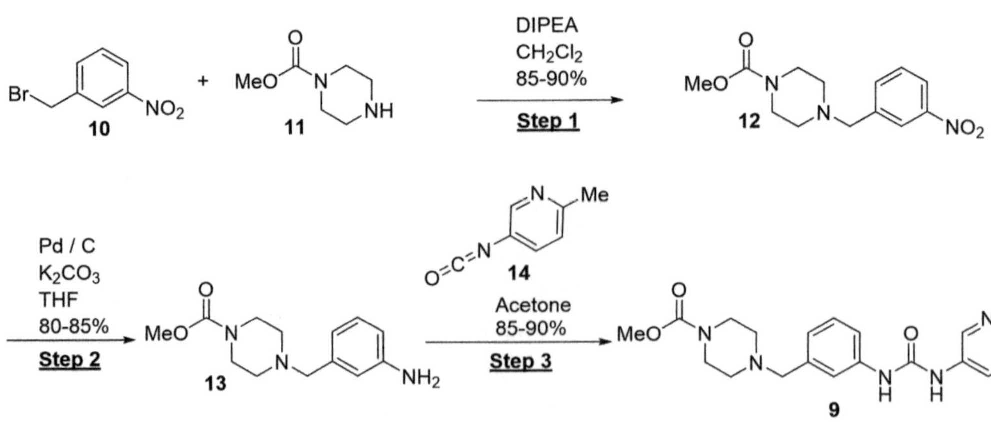

*Scheme 3. Synthetic route to manufacture **9**.*

The initial four-step medicinal chemistry synthesis of **1**, (Scheme 4), took advantage of the synthetic route developed for **9**. Step 1 in the synthesis was the free radical bromination of the commercially available 2-fluoro-1-methyl-3-nitrobenzene (**15**) using *N*-bromosuccinimide and azobisisobutyronitrile (AIBN), as a free radical initiator, in carbon tetrachloride (CCl_4). This yielded a mixture of the desired monobromide 1-(bromomethyl)-2-fluoro-3-nitrobenzene (**16**) and the dibromide 1-(dibromomethyl)-2-fluoro- 3-nitrobenzene (**17**) in a variable 2-3 : 1 mixture, respectively, dependent on slight variations in the reaction conditions. Compound **16** could be isolated in pure form in approximately 50-60 % yield after column chromatography. However, since dibromide **17** was unreactive in the subsequent S_N2 substitution, it proved most efficient to take the crude mixture forward to Step 2.

The addition of the mono/dibromide **16**/**17** mixture to carboxylate **11** under similar conditions used for the synthesis of coupled **12** (Scheme 3, Step 1), resulted in an 30-35% yield of 1-(dibromomethyl)-2-fluoro-3-nitrobenzene (**18**) after silica gel chromatography (Scheme 4, Step 2). At this stage of the synthetic development, the chromatography appeared necessary to provide high quality final material. This sequence required additional optimization to provide a robust manufacturing route.

In Step 3, the nitro moiety of **18** was reduced via heterogenous hydrogenation to produce methyl 4-(3-amino-2-fluorobenzyl)piperazine-1-carboxylate (**19**) under conditions and with yields similar to the conversion of **12** to **13** (Scheme 3, Step 2). The final step was the treatment of **19** with **14** that again mirrored the synthesis of **13** to **9**, resulting in urea **1** that crystallized directly from the reaction mixture as a white solid in 85-90% yield (Scheme 4, Step 4).

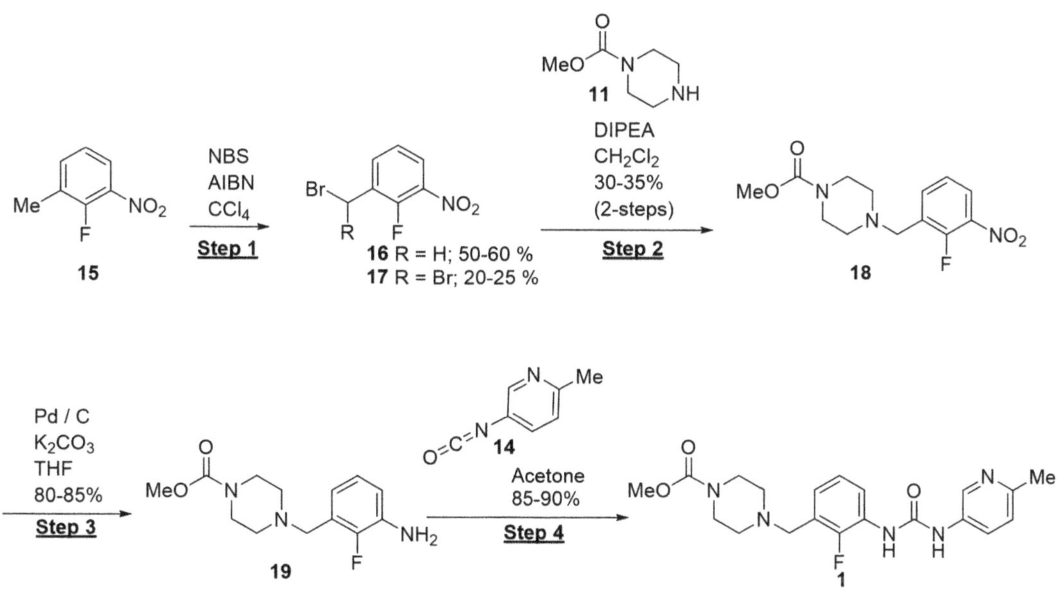

*Scheme 4. Initial medicinal chemistry synthetic route to omecamtiv mecarbil (**1**).*

Overall, the route of synthesis in Scheme 4 looked attractive for manufacturing through commercial launch after substantial development and optimization of each step (such as control of the bromination step to the desired mono bromide **15**, removal of chromatography, elimination of chlorinated solvents, control of the hydrogenation of the aryl nitro moiety, and removal of the of genotoxic intermediate isocyanate **14** as discussed previously). However, optimization of the chemistry for GMP manufacturing proved to be difficult considering the time constraints for entering first in human (FIH) clinical trials. Therefore, we opted to initiate clinical studies through a modified route (Scheme 5) that proved easier to optimize for initial manufacturing campaigns. This allowed for additional time to optimize the route in Scheme 4 for long-term and commercial manufacture of **1** as discussed in "Development of a Potential Commercial Process" section below.

The manufacturing route to produce **1** that allowed for the rapid preclinical development timeline is shown in Scheme 5. Step 1 of this reaction sequence was a nickel-mediated cyanation of commercially available 3-chloro-2-fluoroaniline (**20**) in refluxing *N*-methylpyrrolidine that produced 3-amino-2-fluorobenzonitrile (**21**) as a solid in 70-85% yield. To save time in development, this reaction was first conducted on relatively large scale (e.g., approximately 50 g scale) in fixed equipment that did not allow much light penetrance and the yield was in the range of 70%. As it turns out, it was rather a fortunate to begin at this scale where there was very little ambient light. Many months later, this reaction was run on smaller scale, e.g,, approximately 200 mg) in a light penetrant round bottom flask to facilitate a radiolabeled synthesis and the yield plummeted to less than 10%. On this 200 mg scale, the yield recovered to the 70-85% range when the reaction was run protected from direct light by simply covering the reaction mixture with foil. This empirical finding was not further investigated because the fixed equipment used during manufacturing inherently did not allow for light penetrance. We have manufactured all the API needed for Phase 1-2 clinical development using this route.

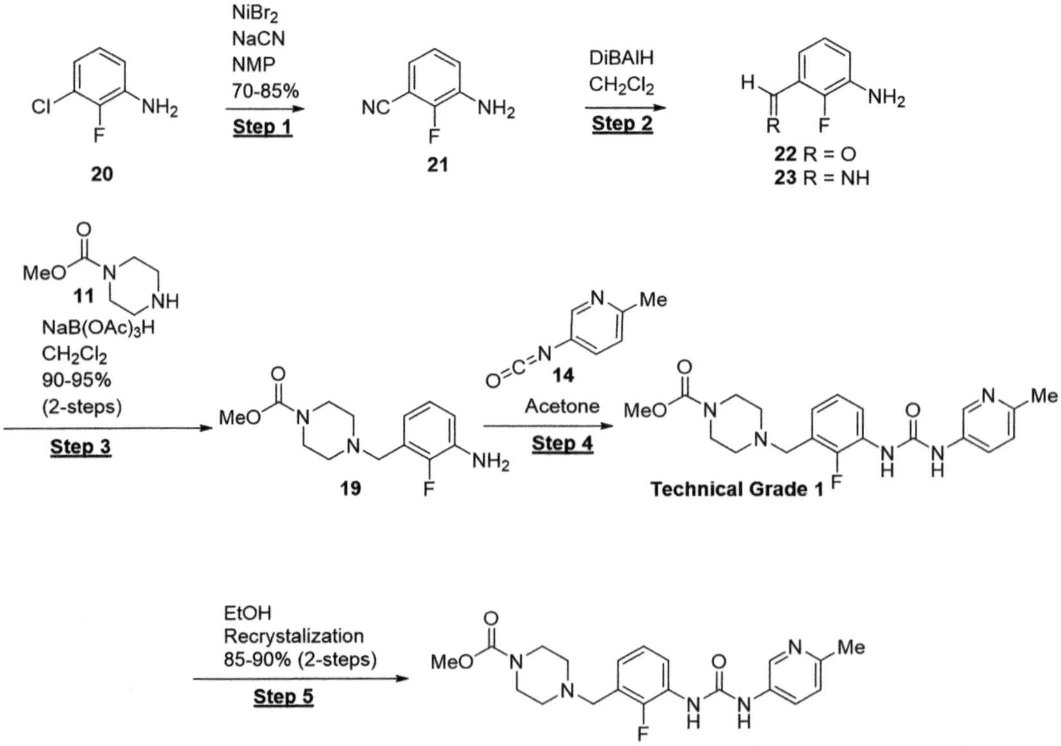

*Scheme 5. Early development synthetic route for the manufacture of omecamtiv mecarbil (**1**).*

In Step 2 (Scheme 5), the cyano moiety of **21** was reduced with di-isobutyl aluminum hydride (DiBAlH) in dichloromethane (CH_2Cl_2) between 0 and 10 °C. After quenching this reaction mixture with a solution of potassium sodium tartrate tetrahydrate (Rochelle's salt), a mixture of 3-amino-2-fluorobenzaldehyde (**22**) and 2-fluoro-3-(iminomethyl)aniline (**23**) was isolated as a solid. The aldehyde/imino mixture of **22** and **23** was taken directly into Step 3, a reductive amination (or reductive transamination of **23**) with ester **11** and sodium triacetoxyborohydride ($NaB(OAc)_3H$) as the reductant in CH_2Cl_2 to produce amine **19** in 90-95% yield over the two steps. Step 4 of this sequence was identical to that used in previous routes, that is, the coupling of **19** with **14** in refluxing acetone to provide technical grade **1** that crystallized directly from the reaction mixture as a white solid. To ensure complete removal of the genotoxic isocyanate **14** that may have been trapped in the crystal lattice of the crystallization from the reaction mixture, technical grade **1** from Step 4 was further recrystallized from ethanol (Step 5) to result in urea **1** in 85-90% yield (from **19**) as a crystalline anhydrous polymorph of the free base.

The initial clinical supply route (Scheme 5) was rapidly developed, robust, fit for purpose, and used to produce hundreds of kilograms of API to supply modified-release drug product development and proof-of-concept clinical trials through Phase 2. The continued clinical development of omecamtiv mecarbil (**1**) fueled the investment to optimize and improve the basic chemistry shown in previously in Scheme 4 into a robust synthetic process. This prepared **1** for pivotal Phase 3 clinical trials and potential commercialization. The dramatic improvements in the synthesis and manufacture of **1** in light of the many challenges and opportunities are discussed in the following section.

Development of a Potential Commercial Process

Although the fit-for-purpose process shown in Scheme 5 was suitable for supporting early clinical development, it was not considered optimal for the long-term manufacture of **1**. The following were identified as key issues for the early process that needed to be addressed in designing a large-scale factory process:

1) Use of the relatively expensive starting material **20**
2) The requirement for stoichiometric nickel reagents, long reaction time (days), high temperature (200 °C) and use of a toxic cyanide salt
3) Formation of polyalkylation byproducts in the reductive amination step that necessitated a sequence of salt formation / salt break crystallizations to eliminate the impurities
4) Instability and genotoxicity of isocyanate starting material **14**
5) Undefined drug substance crystallization control and impact on bulk properties

Although the isocyanate **14** was an excellent coupling partner for urea bond formation, it was not practical for long-term use as it was an oil that underwent self-condensation to generate a mixture of dimer and trimer impurities during storage (Figure 7) (*17–20*). The formation of these insoluble impurities also meant that reagent **14** typically needed to be filtered shortly before use. Furthermore, due to the potential mutagenicity of the isocyanate and its late stage introduction in the process, an additional drug substance recrystallization step from ethanol was added as a control measure to consume or remove any residual isocyanate.

In practice, this involved dissolution of **1** in methanol, polish filtration, solvent switch to ethanol and cooling to complete the crystallization. Although this afforded **1** in excellent purity (>99.5 LC area %), the relatively rapid cooling crystallization (necessary due to instability of the urea in hot ethanol) provided a mixture of small needles and agglomerated bundles (Figure 8) that led to long filtration and drying times (*21*, *22*). For example, a 42 kg batch of drug substance produced via this process required three days for filtration and cake washing and over one week of drying time. Thus, for the next generation of the process, it was necessary to identify a stable isocyanate equivalent and to improve the drug substance crystallization (and particle properties) for large scale manufacture.

Scheme 6 depicts the key changes anticipated in moving from the early clinical route to a potential commercial process. The key aniline intermediate **19** would continue to be leveraged along with a more stable 'masked isocyanate' (e.g. carbamate or alternative), and the reductive amination sequence was replaced with direct alkylation of methyl 1-piperazine carboxylate **11**. The later step would be enabled by selective benzylic bromination of the relatively inexpensive 2-fluoro-3-nitrotoluene (**15**).

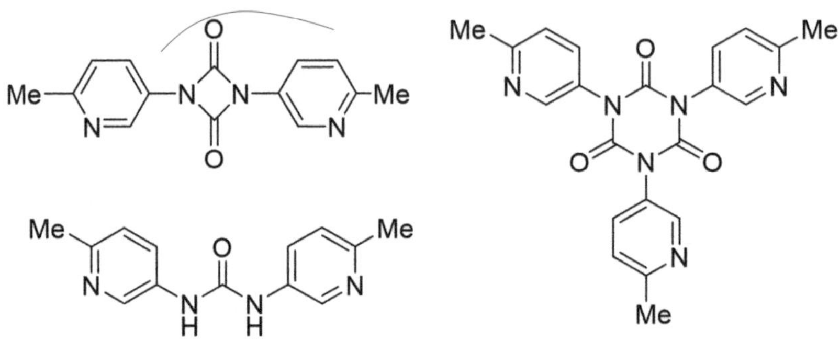

Figure 7. Degradation impurities from isocyanate ***14****.*

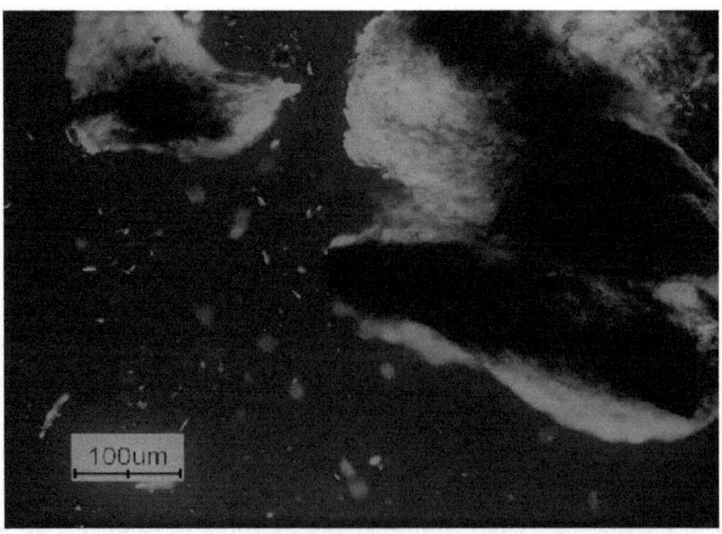

Figure 8. Small needle-like particles and agglomerated bundles of the anhydrous free base of ***1****.*

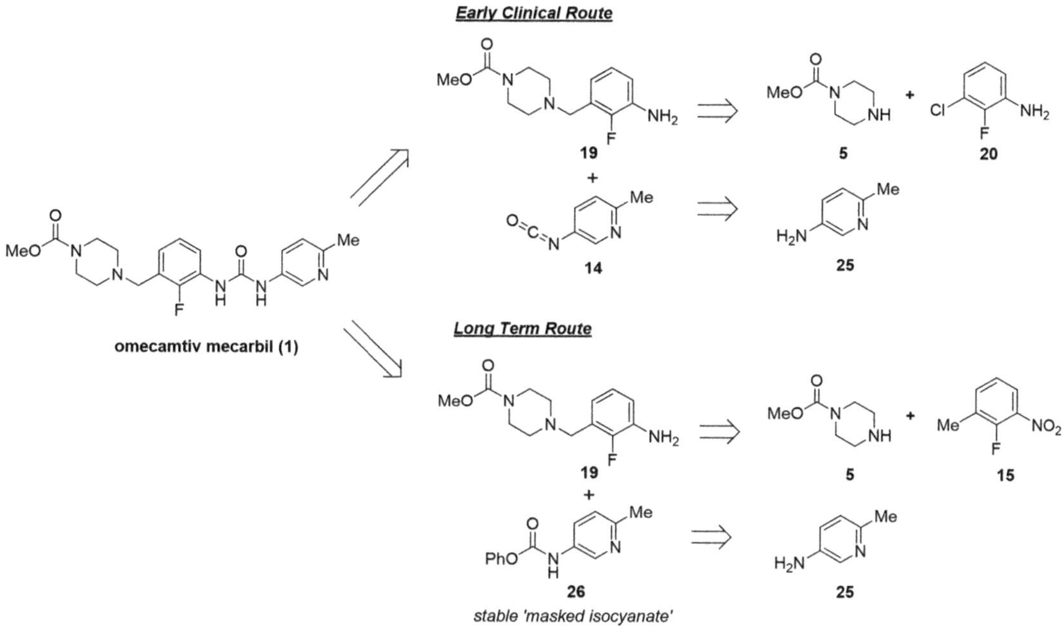

*Scheme 6. Key intermediates in the manufacture of omecamtiv mecarbil (**1**).*

Development of a Plant Process for Aniline 19

A more robust and cost-effective synthesis of aniline **19** was devised which started from 2-fluoro-3-nitrotoluene (**15**). Using the same bond disconnections shown in Scheme 4, we proposed that radical bromination of nitrobenzene **15** followed by alkylation with methyl piperazine-1-carboxylate (**5**) and subsequent nitro reduction would represent an attractive route for large scale production of **19** (Scheme 6). The benzylic bromination was evaluated using a variety of brominating reagents (bromine, *N*-bromosuccinimide, dibromo-dimethylhydantoin), solvents (carbon tetrachloride, cyclohexane, acetic acid, acetonitrile, isopropyl acetate) and initiators (benzoyl peroxide, azobisisobutyronitrile). The conditions shown in Scheme 7 were determined to be optimal. Thus, treatment of **15** with *N*-bromosuccinimide in acetic acid, in the presence of catalytic (3%) benzoyl peroxide at 83 °C for twelve hours afforded a mixture of monobrominated and dibrominated products **16** and **17** in a ~ 2:1 ratio.

A robust procedure to conduct this transformation was developed involving constant rate addition of a slurry of *N*-bromosuccinimide in acetic acid to a solution of **15** and benzoyl peroxide, thus enabling control of the reaction exotherm and limiting the formation of polymerization side-products derived from nitrobenzene **15**. Importantly, due to the extremely corrosive nature of the bromination mixture, it was critical to ensure that there were no exposed metal parts in the reactor. Since even Hastelloy was extensively corroded under the bromination conditions, glass-lined vessels with no wetted metal parts were required. Reaction vessels for the process were also vented through a circulating scrubber system containing 5 N aqueous sodium hydroxide to ensure exiting headspace gases (containing e.g. bromine or HBr) were effectively neutralized.

At the end of the bromination reaction, phosphorus acid was charged to quench residual peroxide, followed by addition of water, toluene and separation of the phases. The toluene phase was washed with aqueous sodium hydroxide to remove acetic acid and provide a toluene solution of the benzylic bromides. This mixture of bromides could be employed directly into the alkylation step as only the monobromide **16** reacted with methylpiperazine-1-carboxylate (**11**) to afford nitro piperazine **18**. The crude product was converted into the hydrochloride salt **18-HCl** by treatment with HCl in a mixture of toluene and 2-propanol. The product precipitated from the reaction medium as a crystalline solid. The crystallization process was effective in eliminating high levels of process impurities (including dibromide **17**) to provide **18-HCl** in 63% overall yield and >99% purity on a 20 kg scale (Scheme 7).

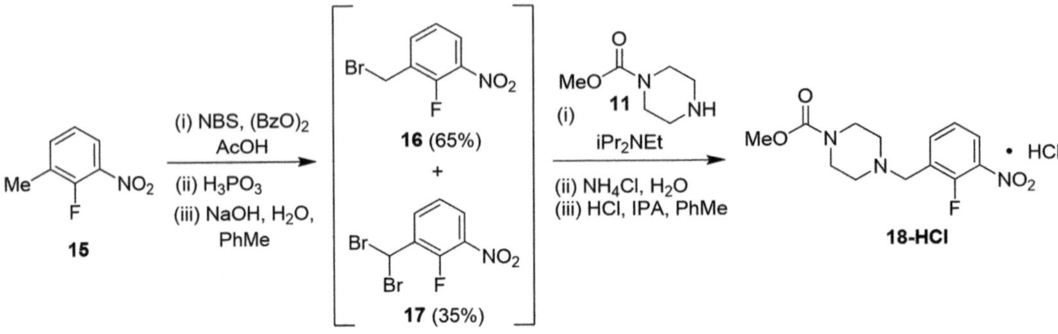

*Scheme 7. Kilo-lab process for **18-HCl**.*

Although further reaction optimization was unable to reduce levels of the gem-dibromide **17**, selective mono-debromination of this material could be achieved via treatment of the mixture of bromides **16** and **17** with diethyl phosphite (*23, 24*). Thus, the crude polybrominated mixture in toluene was treated with diethyl phosphite and diisopropylethylamine in methanol for 3 h at 40 °C to afford >95% assay yield of the desired monobromide **16**. Telescoping this solution into the alkylation step, followed by work-up, treatment with HCl and crystallization, provided **18-HCl** in 88% overall yield and >99% purity from **15** on 77 kg scale.

To achieve a clean alkylation reaction, the piperazine content in starting material **11** had to be strictly controlled to low levels (< 0.5%). Refrigerated storage of **11** was also necessary to suppress degradation that generated additional impurities, including piperazine. Unfortunately, small levels of residual piperazine were converted downstream into the dialkylation product **24** which could not be readily rejected via crystallization (Figure 9).

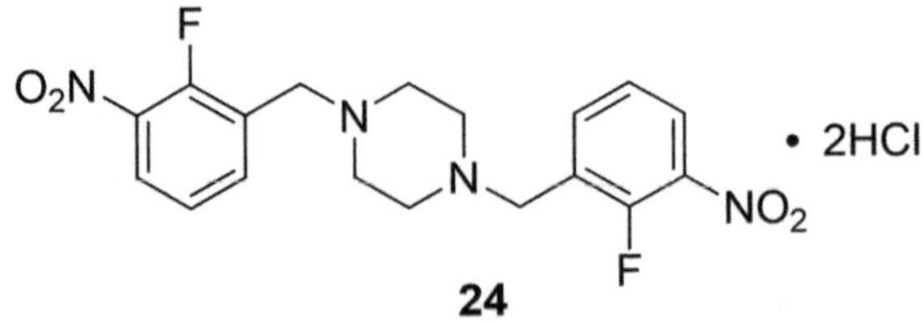

24

*Figure 9. Dialkylation impurity **24** formed from residual piperazine in **11**.*

Based on this problem with the piperazine, a salt screen was conducted and the oil **11** was ultimately replaced with an improved crystalline phosphate salt hydrate (**11-phosphate**) for large-scale manufacturing (Scheme 8). This phosphate salt was stable at room temperature, generated in high purity containing only <0.5% piperazine, and could be used directly in the alkylation process in the presence of diisopropylethylamine base to neutralize the acid. Using **11-phosphate**, nitro **18-HCl** was successfully prepared in 81% yield on 400 kg scale (Scheme 9) from **15**.

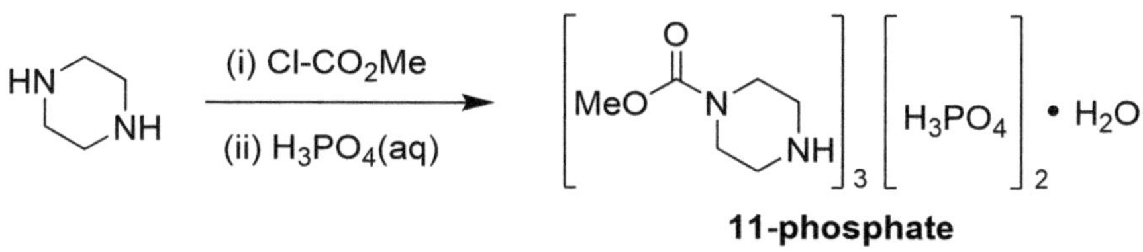

*Scheme 8. Stable **11-phosphate** salt hydrate for use in factory process.*

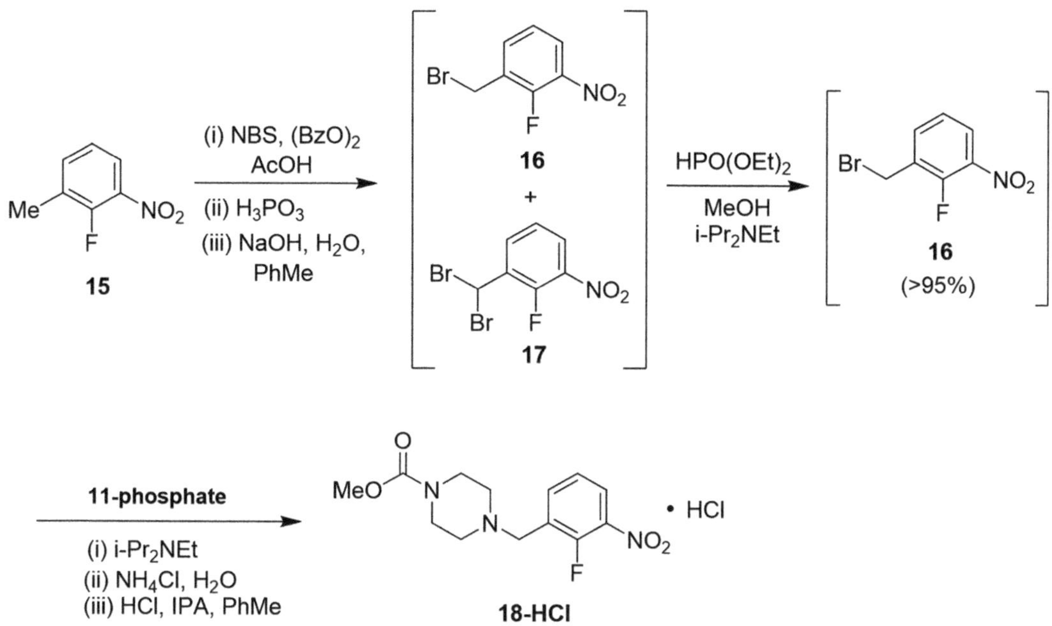

*Scheme 9. Optimized plant process for nitro **18-HCl**.*

Next, efforts focused on cleanly converting nitro **18-HCl** to the target aniline **19** (*25–29*). Hydrogenation of nitro groups are known to proceed via a stepwise process (via nitroso and hydroxylamine intermediates). In addition to these intermediates, formed along the primary pathway (Path A in Scheme 10), several disproportionation byproducts (e.g. azoxy, azo, hydrazo compounds) can also be generated (Path B in Scheme 10) before final reduction to the aniline. As these intermediates are considered potentially mutagenic impurities (PMIs), the reaction design and isolation process must assure that they are controlled to very low levels in the final drug substance (*30*).

Screening of hydrogenation catalysts included the evaluation of several Pt, Pd and Ni catalysts and a Pd(Fe)/C catalyst was initially selected due to its successful performance in reducing intermediate hydroxylamine levels in a previous development program (*31*). Nitro salt **18-HCl** could be conveniently used directly in the hydrogenation process, employing 1% Pd(Fe)/C at 60 psi hydrogen in a 1:1 mixture of methanol/water. Under these conditions, good conversion was obtained at 20-30 °C over several hours but the batch required aging overnight to consume the various intermediates and reach completion. The process ran as a slurry-to-solution transformation since **18-HCl** had poor solubility in the solvent system, but the corresponding aniline salt was fully soluble under the reaction conditions. Therefore, at the end of the reaction, the catalyst was removed via filtration, and the pH of the resulting solution was neutralized to crystallize the freebase **19**.

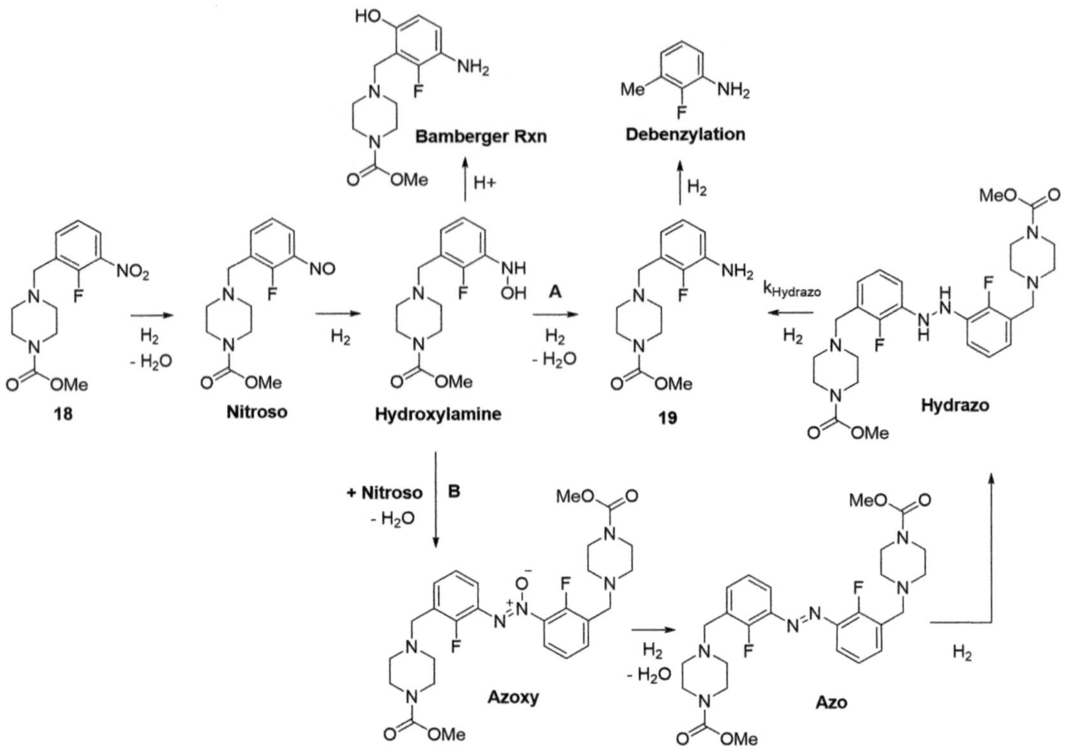

*Scheme 10. Intermediates and byproducts formed during hydrogenation of nitro **18**.*

After scaling up this process (Scheme 11), aniline **19** was isolated in 87% yield (containing < 1 ppm residual palladium) on 9.1 kg scale.

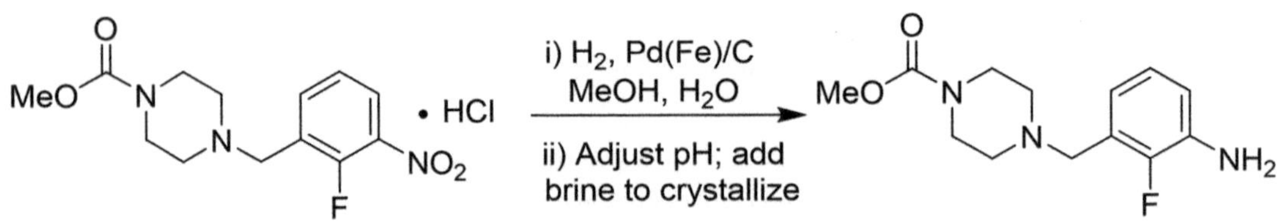

18-HCl **19**

*Scheme 11. Kilo-lab Process for aniline **19**.*

Although the process was successful at this intermediate scale, several issues needed to be addressed to improve robustness before moving to larger scale runs:

1) Iron was found to leach from the Pd(Fe)/C catalyst possibly leading to dark product solutions. Because of this dark color, after catalyst filtration and crystallization of the product, careful cake washing was needed to prevent discoloration of the isolated product
2) The iron-modified palladium catalyst was not effective at preventing hydroxylamine accumulation (>50% by LC when nitro **18** had been completely consumed) before final conversion to the amine, and an alternative system was desired
3) The crystallization from methanol-water was challenging to perform (prone to oiling out) and concentrated aqueous brine (antisolvent) was needed for good product recovery; a controlled crystallization from an organic solvent system was preferred for larger scale

4) Use of the hydrochloric acid salt, **18-HCl**, in the reduction process promoted formation of an aminophenol impurity, presumably via Bamberger rearrangement of the intermediate hydroxylamine (*32, 33*). Although this was a minor impurity, it carried forward through the urea formation step into the final drug substance at ~0.1 LC area % (Scheme 12). Moving forward, we wanted to identify neutral nitro reduction conditions that did not promote the rearrangement
5) Poor solubility of **18-HCl** under the reaction conditions complicated mass-transfer considerations and a simplified system with soluble starting material, intermediates and product was preferred for long-term development

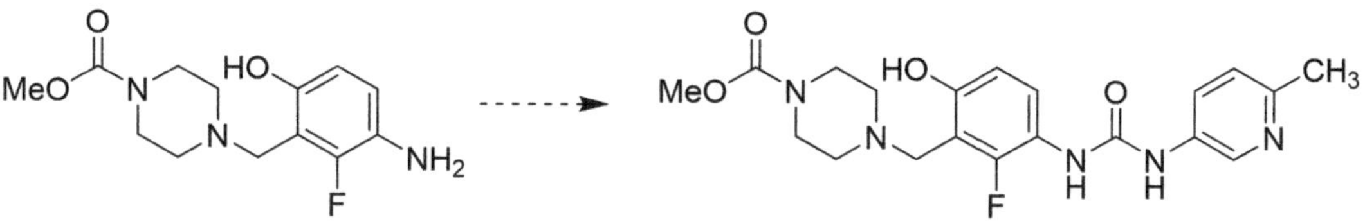

Bamberger rearragement impurity in **19**

Impurity in **1**

*Scheme 12. Drug substance impurity resulting from aminophenol byproduct in aniline **19**.*

Further experimentation revealed that a Pd/C catalyst could be used instead of the modified Pd(Fe)/C catalyst. And by switching to the freebase **18** for the reduction step, no Bamberger rearrangement byproduct was observed. As an improvement to the original crystallization of **19** from an aqueous mixture, an orthogonal crystallization process from toluene/*n*-heptane was developed that was robust and provided good rejection of impurities.

To streamline operations, toluene solvent was also used in hydrogenation step along with ethanol cosolvent to increase the solubility of **18** and to prevent phase separation of water generated during the reaction. Thermal hazards evaluation calorimetry confirmed that the reaction was exothermic with a heat of reaction of -566 kJ/mol and adiabatic temperature rise (ATR) of 78 °C. Catalyst loading and operating temperature were the most important parameters affecting the rate of hydrogenation under the optimized conditions. To suppress the rate of competing hydrogenolysis to 2-fluoro-3-methylaniline, we preferred to maintain the batch temperature below 25 °C. As the debenzylation byproduct (Scheme 10) was an oil, it was well rejected in the crystallization and did not impact product quality.

To limit PMIs resulting from incomplete hydrogenation, it was important to drive the reaction to completion (*34*). Kinetic studies, spiking experiments and batch history were used to complete the process design. As shown in Figure 10, reaction progress in a small-scale run was monitored by in situ Fourier Transform Infrared Spectroscopy (ReactIR) and it was demonstrated that **18** was consumed after ~5 h to afford ~35% hydroxylamine and ~65% **19**, with the additional reaction time serving to complete the transformation to **19**.

For the plant process (Scheme 13), **18-HCl** in toluene was treated with aqueous sodium hydroxide to liberate the freebase **18**. Hydrogenation was conducted in six volumes of toluene/ethanol at 15-20 °C for twelve hours using 0.7% Pd/C catalyst and 60 psi hydrogen. The batch was filtered to remove the catalyst and the collected filtrate was distilled to remove water and ethanol. The batch was then cooled, seeded, and *n*-heptane antisolvent was added to complete the crystallization.

The aniline **19** was isolated in 90% yield (99 LC area % purity, containing <1 ppm palladium) on 249 kg scale.

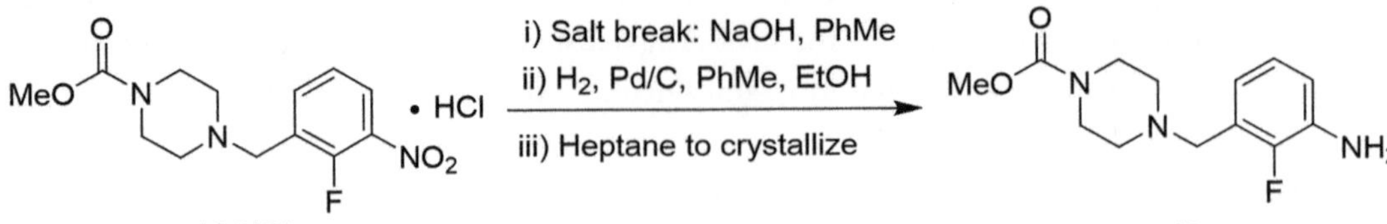

18-HCl **19**

Scheme 13. Plant process for aniline ***19****.*

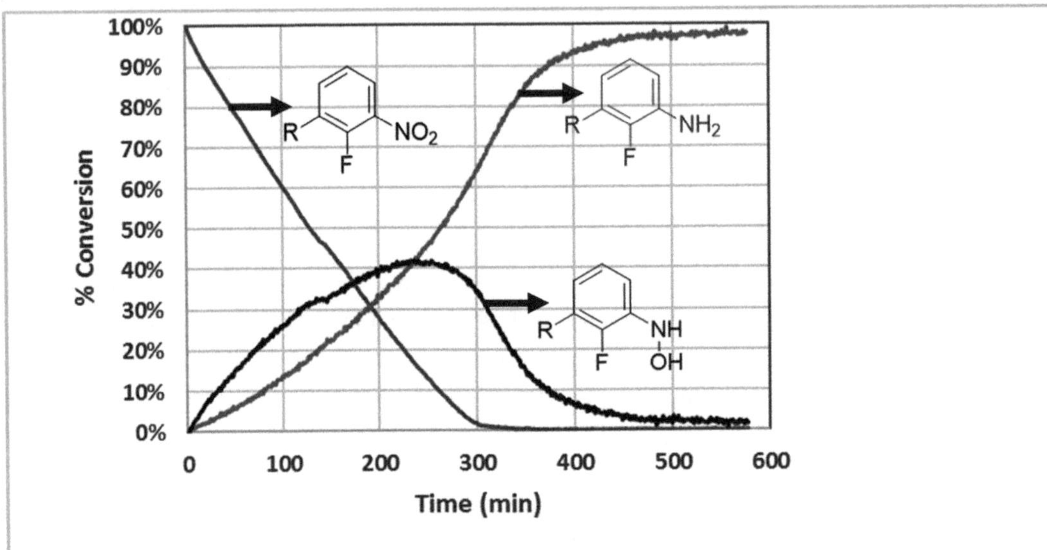

Figure 10. Reaction profile for reduction of nitro ***18*** *with 0.7% Pd/C at 15 °C, 60 psi* H_2.

Development of a 'Masked Isocyanate' and Urea Coupling

In searching for a stable replacement for isocyanate **14**, a variety of carbamate alternatives were assessed including substituted phenyl-, isopropyl-, *N*-hydroxysuccinimidoyl- and *tert*-butoxycarbamates (Figure 11) (*35–40*).

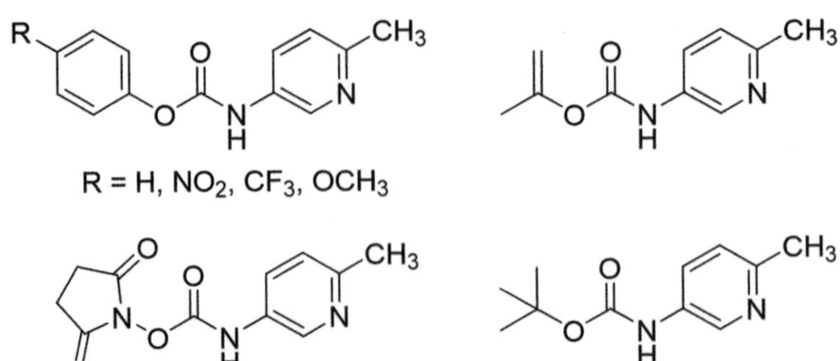

Figure 11. Carbamates evaluated to replace isocyanate ***14****.*

The phenyl carbamate was ultimately selected as it offered the best balance of stability, ease of preparation, cost, and reactivity in the coupling with **19** to afford urea **1**. Importantly, it was also discovered that a crystalline hydrochloric acid salt of carbamate **26** was more stable than the corresponding freebase. Carbamate **26-HCl** was stable at 40 °C/75% relative humidity for at least six months and could be stored at room temperature for >2 years. Use of the hydrochloride salt also simplified manufacturing as **26-HCl** could be prepared by a direct-drop isolation process from aminopyridine **25** and phenyl chloroformate (Scheme 14) (*41*). In practice, phenyl chloroformate was slowly added to a solution of **25** in 15 volumes of acetonitrile at 20 °Cover ~3 h to maintain the batch temperature < 30 °C (*42, 43*). Overall, the reactive crystallization of carbamate **26-HCl** was well controlled by the feed rate and the product was isolated in high yield and purity after filtration/drying. Employing this one-pot process, **26-HCl** was produced in 95% yield and >99% purity on 290 kg scale (*44*).

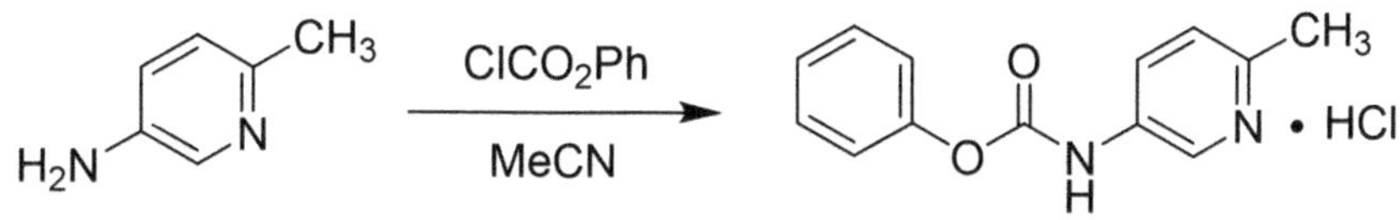

25 **26-HCl**

*Scheme 14. Plant process for phenyl carbamate **26-HCl**.*

Aminolysis of carbamate **26-HCl** with aniline **19** to afford omecamtiv mecarbil (**1**) was studied with a variety of base and solvent combinations. The coupling proceeded as a heterogeneous (slurry-to-slurry) transformation under many conditions, with the product **1** often precipitating during the reaction. This heavy precipitation of salts and product in many solvent systems, generated thick mixtures that made agitation difficult. However, eventually it was determined that heating a mixture of **19** (1.0 equiv), carbamate **26-HCl** (1.1 equiv) and diisopropylethylamine (1.2 equiv) in 5 volumes of 1:1 THF/acetonitrile at 55 °C for 20 h, provided a clean solution of **1** in >95% assay yield. Excess diisopropylethylamine was necessary for high conversion and purity, serving to both neutralize the HCl introduced with the carbamate and presumably promoting the in situ release of isocyanate via an E1cB mechanism (*45*).

The crystallization of **1** was completed by charging five volumes of water, seeding, and then slowly adding another ten volumes of water over three hours (Scheme 15). Agglomeration of the product needles produced large oval particles during the addition of the water (Figure 12). Formation of these large particles (~200-2000 μm length) led to fast filtration times, providing the crystalline free base of **1** in >99.5% purity and 82% yield on 287 kg scale.

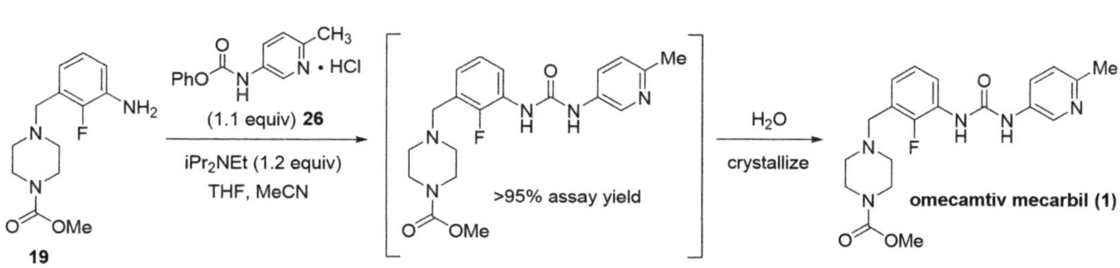

Scheme 15. Urea formation and crystallization of omecamtiv mecarbil.

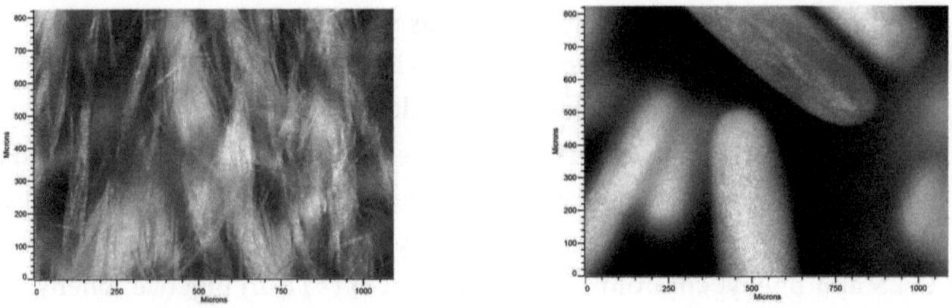

(a) (b)

*Figure 12. Agglomeration of omecamtiv mercabil (**1**) free base needles during crystallization: (a) PVM microscopy image after 1 h of water addition; (b) PVM microscopy image after 3 h of water addition.*

Process Design with the End in Mind – Drug Substance Final Form, Crystallization, and Bulk Properties

During the drug development program, five different oral modified release (MR) formulations, an oral immediate release (IR) formulation and an IV formulation were evaluated. Given the risk of excessive pharmacological effects at high plasma concentrations of omecamtiv mecarbil, the MR formulations were developed with the goal of preserving oral bioavailability while lowering C_{max} and maintaining the peak-to-trough fluctuation at steady state (*46*). These spanned a range of drug delivery technologies including osmotic pump / swellable core tablets, matrix tablets, and multiparticulate capsules (*47–49*).

Given the challenges in contemporaneously developing these multiple drug products, our objective was to enable formulation performance and minimize variability introduced from the drug substance powder properties. Therefore, the drug substance needed to be prepared with consistent bulk properties including particle size distribution, low aspect ratio, adequate powder flow, and good stability during tablet/capsule manufacture (*50*). Unfortunately, matrix tablet formulation experiments revealed that the free base form of omecamtiv mecarbil (**1**) reacted with acidic excipients (e.g. citric acid, fumaric acid) in the formulations to generate high levels of amorphous citrate/ fumarate salts (*51*). To avoid undesired chemistry and form conversions during drug product processing, salt screening was conducted to identify an alternative drug substance form that would be stable during tablet manufacture.

As a result, the omecamtiv mecarbil dihydrochloride salt hydrate was identified, characterized and ultimately selected for long-term development. The salt crystallized in high yield from 2-propanol / water mixtures as long rods. Polymorph screening on the dihydrochloride salt identified a single form (Figure 13) containing ~3.4 wt% water that was non-hygroscopic and existed as a channel hydrate per single crystal X-ray analysis. Importantly, this form did not convert to other solid forms/salts during matrix tablet manufacturing. A scalable salt process was developed that involved treatment of a solution of omecamtiv mecarbil in 2-propanol with aqueous hydrochloric acid (2.2 equiv), followed by cooling and addition of more 2-propanol to complete the crystallization of the dihydrochloride hydrate salt (Scheme 16). In-line wet milling with a high-shear rotor-stator mixer was employed to reduce the aspect ratio of the crystals, improve bulk properties and aid downstream drug product processing (*52, 53*).

During development runs, in-line Focused Beam Reflectance Measurement (FBRM) was used to monitor progress of the milling and time to reach steady state. Slurry samples were collected periodically to profile the milling by off-line PSD analysis via Malvern laser diffraction. Using a wet mill in recycle mode, the milling was typically complete in three batch turnovers (Figure 14). Rotor tip speed was maintained constant during scale-up to achieve a consistent terminal particle size and the milling end point was confirmed via in-process PSD testing. Post wet milling, the slurry was filtered, washed and dried. Using this optimized process, a 248 kg batch of omecamtiv mecarbil dihydrochloride hydrate was isolated in 93% yield and >99.9% purity.

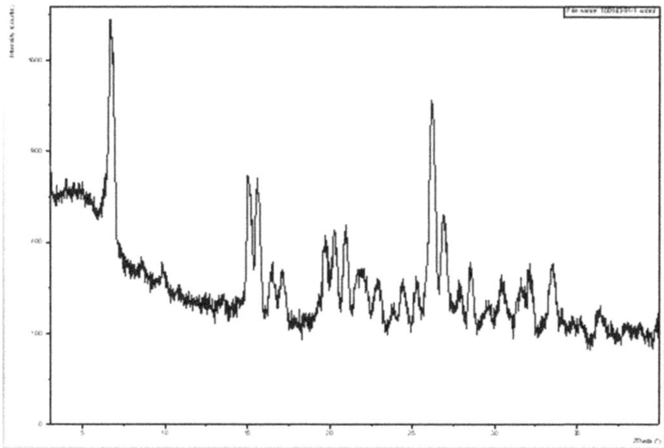

Figure 13. XRPD of omecamtiv mecarbil di-HCl hydrate.

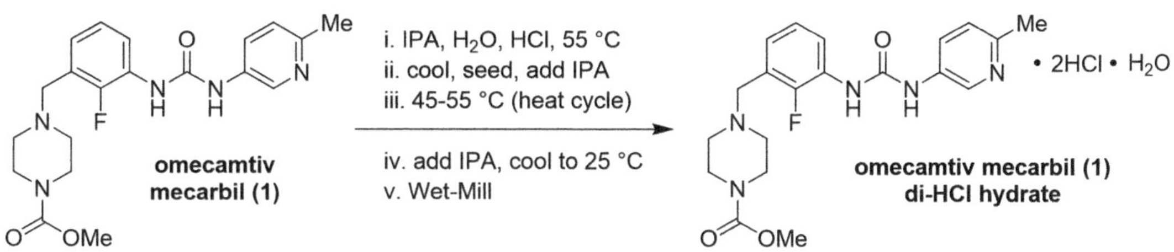

Scheme 16. Omecamtiv mecarbil Di-HCl hydrate salt formation and isolation.

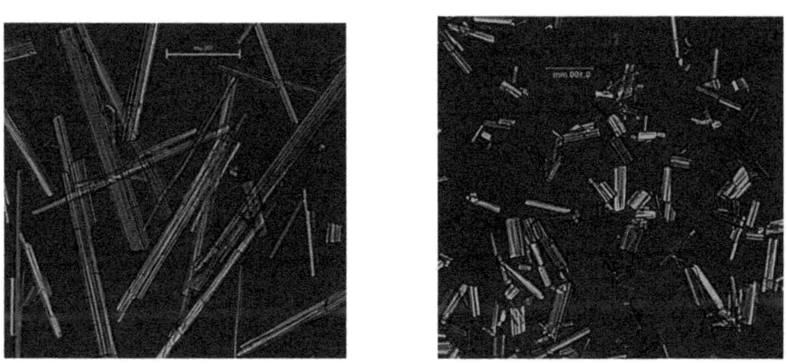

(a) **(b)**

Figure 14. Crystallization of omecamtiv mecarbil di-HCl hydrate (a) before wet milling; (b) post wet milling.

In summary, an efficient and robust process to manufacture the novel cardiac myosin activator, omecamtiv mecarbil, was developed (Scheme 17) (*54*). The longest linear sequence consisted of six steps, including a final salt formation step to afford the drug substance in ~55% overall yield. Relative to the initial route used to supply preclinical and early clinical studies, the E-factor for the optimized process was reduced from 308 to 72 (>75% reduction). Additionally, due to amorphous conversion of the free base form in the presence of excipients, a more stable dihydrochloride hydrate form of omecamtiv mecarbil was developed. Overall, the process has been successfully implemented on large scale to afford >1 metric ton of drug substance to support registrational studies (*55, 56*).

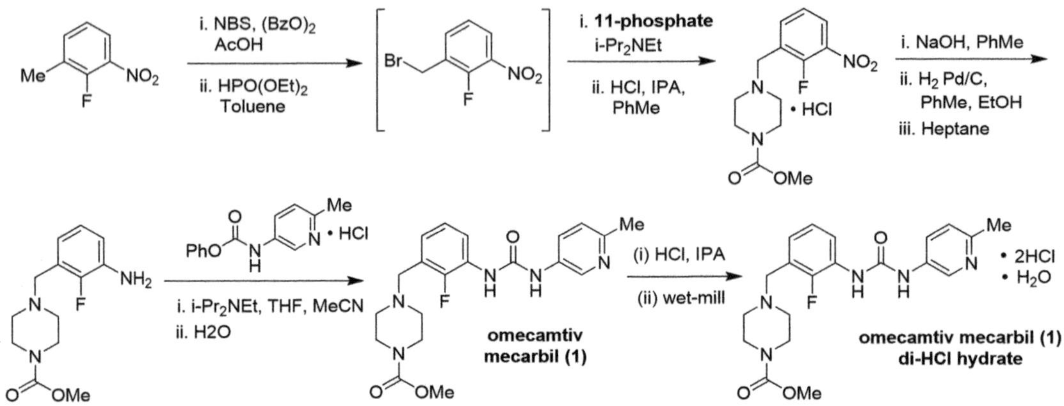

Scheme 17. Summary of omecamtiv mecarbil di-HCl hydrate final process.

Clinical Results and Summary

The discovery and development efforts around omecamtiv mecarbil (**1**) were accomplished with the goal of increasing cardiac contractility in patients suffering from reduced cardiac performance in HFrEF and translating this mechanism of action into improving patient outcomes and exercise capacity. The clinical research began with early trials with **1** designed to evaluate both the intravenous and oral formulations in a diversity of volunteers for tolerability, safety, and pharmacodynamic and pharmacokinetic profiles. This led to the key Phase 2 clinical trial with MR oral formulations, COSMIC-HF (*57*) (Chronic Oral Study of Myosin Activation to Increase Contractility in Heart Failure). COSMIC-HF demonstrated **1** to have statistically significant improvements in pre-specified secondary measures of cardiac function and dimensions (*58–60*) in HFrEF patients compared to placebo. Additionally, patients treated with **1** compared to placebo showed a statistically significant improvement in levels of NT-pro-BNP, a biomarker associated with the severity of heart failure (*61, 62*). Adverse events were comparable between **1** and placebo. The results from COSMIC-HF may translate into improved clinical outcomes for patients with HFrEF which supported advancement of **1** into Phase 3 clinical trials.

Will directly improving cardiac contractility at the level of the cardiac sarcomere through selective activation of cardiac myosin lead to improved patient outcomes? The ultimate answer to the effectiveness of this therapeutic approach will be answered with the ongoing Phase 3 clinical trials, GALACTIC-HF (Global Approach to Lowering Adverse Cardiac Outcomes Through Improving Contractility in Heart Failure) and METEORIC-HF (Multicenter Exercise Tolerance Evaluation of Omecamtiv mecarbil Related to Increased Contractility in Heart Failure). If successful, omecamtiv mecarbil may represent to a new continuum of care for patients in both in-hospital and out-patient settings. Accordingly, this chapter was intended to provide the reader with the molecule's journey from idea to candidate selection, from fit-for-purpose early clinical supply route to scalable factory process designed with performance of the final product in mind.

References

1. Ziaeian, B.; Fonaroe, G. C. *Nat. Rev. Cardiol.* **2016**, *13*, 368–378.
2. Heidenreich, P. A.; Albert, N. M.; Allen, L. A.; Bluemke, D. A.; Butler, J.; Fonarow, G. C.; Ikonomidis, J. S.; Khavjou, O.; Konstam, M. A.; Maddox, T. M.; Nichol, G.; Pham, M.; Piña, I. L.; Trogdon, J. G. *Circ.: Heart Failure* **2013**, *6*, 606–619.
3. Braunwald, E., Grossman, W. Clinical Aspects of Heart Failure. In *Heart Disease: A Textbook of Cardiovascular Medicine*, 4th ed.; Braunwald, E., Ed.; Philadelphia: Saunders, 1992; pp 444–463.
4. Psotka, M. A.; Gottlieb, S. S.; Francis, G. S.; Allen, L. A.; Teerlink, J. R.; Adams, K. F.; Rosano, G. M. C.; Lancellotti, P. *J. Am. Coll. Cardiol.* **2019**, *73*, 2345–2353.
5. Kass, D. A.; Solaro, R. J. *Circulation* **2006**, *113*, 305–315.
6. Cohn, J. N.; Goldstein, S. O.; Greenberg, B. H.; Lorell, B. H.; Bourge, R. C.; Jaski, B. E.; Gottlieb, S. O.; McGrew, F.; DeMets, D. L.; White, B. G. *New Engl. J. Med.* **1998**, *339*, 1810–1816.
7. Packer, M.; Carver, J. R.; Rodeheffer, R. J.; Ivanhoe, R. J.; DiBianco, R.; Zeldis, S. M.; Hendrix, G. H.; Bommer, W. J.; Elkayam, U.; Kukin, M. L. *New Engl. J. Med.* **1991**, *325*, 1468–1475.
8. Yano, M.; Kohno, M.; Ohkusa, T.; Mochizuki, M.; Yamada, J.; Kohno, M.; Hisaoka, T.; Ono, K.; Tanigawa, T.; Kobayashi, S.; Matsuzaki, M. *Am. J. Physiol. Heart Circ. Physiol.* **2000**, *279*, H1898–H1905.
9. Malik, F. I.; Hartman, J. J.; Elias, K. A.; Morgan, B. P.; Rodriguez, H.; Brejc, K.; Anderson, R. L.; Sueoka, S. H.; Lee, K. H.; Finer, J. T.; Sakowicz, R.; Baliga, R.; Cox, D. R.; Garard, M.; Godinez, G.; Kawas, R.; Kraynack, E.; Lenzi, D.; Lu, P.; Muci, A.; Niu, C.; Qian, X.; Pierce, D. W.; Pokrovskii, M.; Suehiro, I.; Sylvester, S.; Tochimoto, T.; Valdez, C.; Wang, W.; Katori, T.; Kass, D. A.; Shen, Y.; Vatner, S. F.; Morgans, D. J. *Science* **2011**, *331*, 1439–1443.
10. Morgan, B. P.; Muci, A.; Lu, P.; Qian, X.; Tochimoto, T.; Smith, W. W.; Garard, M.; Kraynack, E.; Collibee, S.; Suehiro, I.; Tomasi, A.; Valdez, S. C.; Wang, W.; Jiang, H.; Hartman, J.; Rodriguez, H. M.; Kawas, R.; Sylvester, S.; Elias, K. A.; Godinez, G.; Lee, K.; Anderson, R.; Sueoka, S.; Xu, D.; Wang, Z.; Djordjevic, N.; Malik, F. I.; Morgans, D. J. *ACS Med. Chem. Lett.* **2010**, *1*, 472–477.
11. Romero, A.; Morgan, B. P. Cardiac Sarcomere Activation and its Therapeutic Application in Heart Failure. In *2017 Medicinal Chemistry Reviews*; Editor-in-Chief: Bronson, J. J.; Section Editor: Dow, R. L. ; Volume 52, Chapter 5, pp 109–123.
12. Malik, F. I.; Morgan, B. P. *J. Mol. Cell Cardiol.* **2011**, *4*, 454–461.
13. Hartman, J. J., Malik, F., Sakowicz, R., Finer, J. T. U.S. Patent Application 2009/539,164.
14. Benet, L. Z.; Hoener, B. *Clin. Pharm. Therapeutics* **2002**, *71*, 115–121.
15. Shen, Y. T.; Malik, F. I.; Zhao, X.; Depre, C.; Dhar, S. K.; Abarzúa, P.; Morgans, D. J.; Vatner, S. F. *Circ.: Heart Failure* **2010**, *3* (4), 522–527.

16. Property calculation primarily used software from Advanced Chemistry Development, Inc., (ACD/Labs, https://www.acdlabs.com/) and from the Chemical Computing Group (MOE: Molecular Operating Environment (https://www.chemcomp.com/).
17. Isocyanates are known to undergo self-condensation reactions to form cyclic dimers (uretidinediones) and trimers (isocyanurates), see references 16-19: Buckles, R. E.; McGrew, L. A. *J. Am. Chem. Soc.* **1966**, *88*, 3582–3586.
18. Malwitz, N.; Kresta, J. E. *J. Cell. Plas.* **1988**, 495–523.
19. Nawata, T.; Kresta, J. E.; Frisch, K. C. *J. Cell. Plast* **1975**, 267–278.
20. Wirpsza, Z. *Polyurethanes: Chemistry, Technology and Application*; Ellis Horwood: London, U.K., 1993.
21. For a detailed summary of the solid-state characteristics of the anhydrous free base form of omecamtiv mecarbil and comparison to a hemihydrate form see: Morrison, H.; Quan, B. P.; Walker, S. D.; Hansen, K. B.; Nagapudi, K.; Cui, S. *Org. Process Res. Dev.* **2015**, *19*, 1842–1848.
22. Prolonged refluxing of omecamtiv mecarbil in EtOH during the recrystallization can lead to degradation which lowers yield. Characterization studies demonstrated that heating omecamtiv mecarbil in ethanol at reflux for 6 h resulted in an erosion of solution chemical purity from >99% to 86% by LC analysis.
23. Hirao, T.; Masunaga, T.; Ohshiro, Y.; Agawa, T. *J. Org. Chem.* **1981**, *46*, 3745–3747.
24. Liu, P.; Chen, Y.; Deng, J.; Tu, Y. *Synthesis* **2001**, 2078–2080.
25. Lead references and reviews on nitroarene reductions, see the refs 24−28: Shen, M.; Liu, H.; Yu, C.; Yin, Z.; Muzzio, M.; Li, J.; Xi, Z.; Yu, Y.; Sun, S. *J. Am. Chem. Soc.* **2018**, *140*, 16460–16463.
26. Song, J. J.; Huang, Z. F.; Pan, L.; Li, K.; Zhang, X. W.; Wang, L.; Zou, J. J. *Appl. Catal., B* **2018**, *227*, 386–408.
27. Blaser, H. U.; Steiner, H.; Studer, M. *ChemCatChem* **2009**, *1*, 210–221.
28. Tafesh, A. M.; Weiguny, J. *Chem. Rev.* **1996**, *96*, 2035–2052.
29. Kasparian, A. J.; Savarin, C.; Allgeier, A. M.; Walker, S. D. *J. Org. Chem.* **2011**, *76*, 9841–9844.
30. The ICH M7 TTC-based acceptable intake of a mutagenic impurity present in pharmaceuticals for long-term (>10 years) treatment is 1.5 μg/day.
31. Metal additives (including iron, vanadium, copper, cobalt) have been reported to reduce the build-up of intermediates during nitro group reductions; In the screening phase for this project, an Fe-doped catalyst was used instead of adding Fe powders or salts to existing catalyst systems. For a lead reference on modifiers in catalytic metal supported hydrogenations see: Baumeister, P.; Blaser, H.-U.; Studer, M. *Catalysis Lett.* **1997**, *49*, 219–222.
32. Bamberger, E. *Chem. Ber.* **1894**, *27*, 1548–1557.
33. Robertson, G. R.; Evans, R. A. *J. Org. Chem.* **1940**, *5*, 142–145.
34. Bacterial reverse mutation test (Ames test) for mutagenicity: Nitro & Hydroxylamine - positive; Hydrazo - negative for mutagenicity.
35. For the synthesis of ureas from phenyl carbamates, see refs 34 and 35: Thavonekham, B. A *Synthesis* **1997**, 1189–1194.
36. Adamiak, R. W.; Stawinski, J. *Tetrahedron Lett.* **1977**, *22*, 1935–1936.

37. For the use of isopropenyl carbamates see: Gallou, I.; Eriksson, M.; Zeng, X.; Senanayake, C.; and Farina, V. *J. Org. Chem.* **2005**, *70*, 6960–6963.
38. For the use of N-hydroxysuccinimide carbamates see: Morpurgo, M.; Bayer, E. A.; Wilchek, M. *J. Biochem. Biophys. Methods* **1999**, *38*, 17–28.
39. For the use of t-butoxycarbamates see: Lamothe, M.; Perez, M.; Colovray-Gotteland, V.; Halazy, S. *Synlett* **1996**, 507–508.
40. For a discussion on the relative reactivity of various carbamates in producing unsymmetrical ureas as well as the electrochemical preparation of trifluoroethylcarbamates see: Matsumura, Y.; Satoh, Y.; Onomura, O.; Maki, T. *J. Org. Chem.* **2000**, *65*, 1549–1551.
41. For a paper highlighting examples of the bottom-up approach to process development, including several direct isolation processes, see: Chen, C.-K.; Singh, A. K. *Org. Process Res. Dev.* **2001**, *5*, 508–513. See also: Anderson, N. G. *Practical Process Research & Development*, 2nd ed.; Academic Press: Oxford, U.K., 2012.
42. Process Hazard Evaluation Testing: Adiabatic Temperature Rise (ATR) = 54 °C.
43. Aminopyridine **25-HCl** typically precipitated during addition of the first ~20% of phenylchloroformate before dissolving.

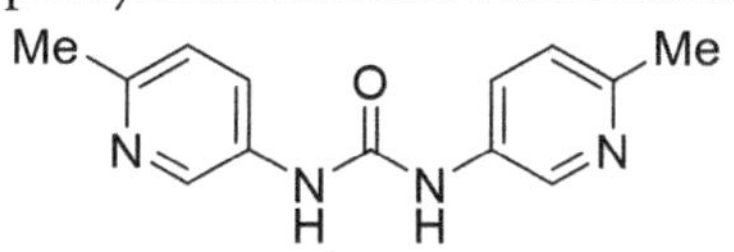

44. **R-Urea** Typical process impurities are residual aminopyridine **6** (entrained in the product as the corresponding **5-HCl** salt) and R-Urea (shown to the left).
45. Ilieva, S.; Nalbantova, D.; Hadjieva, B.; Galabov, B. *J. Org. Chem.* **2013**, *78*, 6440–6449.
46. Palaparthy, R.; Banfield, C.; Alvarez, A.; Yan, L.; Smith, B.; Johnson, J.; Monsalvo, M. L.; Malik, F. *Int. J. Clin. Pharmacol. Ther.* **2016**, *54*, 217–227.
47. Herrlich, S.; Spieth, S.; Messner, S.; Zengerle, R. *Adv. Drug Delivery Rev.* **2012**, *64*, 1617–1627.
48. Kumar, S.; Jeet, K.; Ashish, B. *Asian J. Pharm.* **2015**, *9*, S13–S25.
49. Dash, T. R.; Verma, P. *Int. J. Pharma Res. Rev.* **2013**, *2*, 12–24.
50. For a study correlating drug substance particle size distribution with other bulk properties (such as surface area and powder flow) to enable a drug product formulation process see: Morrison, H. G.; Tao, W.; Trieu, W.; Walker, S. D.; Cui, S.; Huggins, S.; Nagapudi, K. *Org. Process Res. Dev.* **2015**, *19*, 1076–1081.
51. Crystalline and amorphous content in the tablet formulations were quantified by ^{19}F solid state NMR.
52. For some lead references on the application of wet-milling to pharmaceuticals see references 52-53: Harter, A.; Schenck, L.; Lee, I.; Cote, A. *Org. Process Res. Dev.* **2013**, *17*, 1335–1344.
53. Acevedo, D.; Kamaraju, V. K.; Glennon, B.; Nagy, Z. K. *Org. Process Res. Dev.* **2017**, *21*, 1069–1079.
54. Allgeier, A. M.; Bernard, C.; Correll, T. L.; Cosbie, A.; Crockett, R. D.; Cui, S.; Faul, M. M.; Hansen, K. B.; Huggins, S.; Langille, N.; Mennen, S. M.; Morgan, B. P.; Morrison, H.; Muci, A.; Nagapudi, K.; Quasdorf, K.; Ranganathan, K.; Roosen, P.; Shi, X.; Thiel, O. R.; Wang, F.; Tvetan, J. T.; Woo, J. C. S.; Wu, S.; Walker, S. W. *Org. Process. Res. Dev.* **2019**, *23*, 1558–1567.

55. For an alternative Pd-catalyzed synthesis of unsymmetrical ureas, including omecamtiv mecarbil, see references 55-56: Breitler, S.; Oldenhuis, N. J.; Fors, B. P.; Buchwald, S. L. *Org. Lett.* **2011**, *13*, 3262–3265.
56. Vinogradova, E. V.; Fors, B. P.; Buchwald, S. L. *J. Am. Chem. Soc.* **2012**, *134*, 11132–11135.
57. Teerlink, J. R.; Felker, G. M.; McMurray, J. J. V.; Solomon, S. D.; Adams, K. F.; Cleland, J. G. F.; Ezekowitz, J. A.; Goudev, A.; Macdonald, P.; Metra, M.; Mitrovic, V.; Ponikowski, P.; Serpytis, P.; Spinar, J.; Tomcsányi, J.; Vandekerckhove, H. J.; Voors, A. A.; Monsalvo, M. L.; Johnston, J.; Malik, F. I.; Honarpour, N. *Lancet* **2016**, *388*, 2895–2903.
58. Biering-Sørensen, T.; Roca, G. Q.; Hegde, S.; Shah, A.; Clagett, B.; Mosley, T. H.; Butler, K. R.; Solomon, S. D. *J. Card. Failure* **2015**, *21*, S84.
59. Weissler, A. M.; Harris, W. S.; Schoenfeld, C. D. *Circulation* **1968**, *37*, 149–159.
60. Kramer, D. G.; Trikalinos, T. A.; Kent, D. M.; Antonopoulos, G. V.; Konstam, M. A.; Udelson, J. E. *J. Am. Col. Cardiol.* **2010**, *56*, 392–406.
61. Rahimi, K.; Bennett, D.; Conrad, N.; Williams, T. M.; Basu, J.; Dwight, J.; Woodward, M.; Patel, A.; McMurray, J.; MacMahon, S. *JACC: Heart Failure* **2014**, *2*, 440–446.
62. Maisel, A.; Mueller, C.; Adams, K.; Anker, S. D.; Aspromonte, N.; Cleland, J. G. F.; Cohen-Solal, A.; Dahlstrom, U.; DeMaria, A.; Di Somma, S.; Filippatos, G. S.; Fonarow, G. C.; Jourdain, P.; Komajda, M.; Liu, P. P.; McDonagh, T.; McDonald, K.; Mebazaa, A.; Nieminen, M. S.; Peacock, W. F.; Tubaro, M.; Valle, R.; Vanderhyden, M.; Yancy, C. W.; Zannad, F.; Braunwald, E. *Eur. J. Heart Failure* **2008**, *10*, 824–839.

Chapter 4

Identification and Development of AMG 986: A Potent and Efficacious APJ Agonist for the Treatment of Heart Failure

Matthew R. Kaller,[1] Paul Dransfield,[2] Robert P. Farrell,[*,3] and Ted C. Judd[3]

[1]Department of Medicinal Chemistry, Amgen Inc., One Amgen Centre Drive, Thousand Oaks, California 91320, United States

[2]H3 Biomedicine Inc., 300 Technology Square, Cambridge, Massachusetts 02139, United States

[3]Department of Process Development, Amgen Inc., One Amgen Centre Drive, Thousand Oaks, California 91320, United States

[*]Email: rfarrell@amgen.com

This chapter describes the discovery and development of AMG 986. The rationale for targeting an agonist of the apelin receptor is discussed as well as the structure-activity relationship and Medicinal Chemistry efforts pursued that led to AMG 986. To alleviate a severe synthetic bottleneck, early collaboration between Medicinal Chemistry and Process Chemistry on a novel route to a key chiral sulfonamide fragment was initiated prior to final clinical candidate selection. The discovery, optimization, and scale-up of the improved stereospecific route along with the results from multi-kilogram scale-up of the improved synthetic route to AMG 986 will be discussed.

Introduction

Heart failure (HF) is a degenerative and terminal disease that affects 26 million people globally, with one million new cases diagnosed every year and 330,000 deaths per year in the US alone (*1–4*). In heart failure, the heart in not able to supply enough blood and oxygen to meet the metabolic demands of the body. However, it is adaptable and able to meet demand in the short term by remodeling itself to increase the volume of blood pumped. If this increased demand on the heart is not short term, continued ventricular remodeling will eventually lead to reduced contractility and stroke volume. Eventually, failure of the heart will result. The current standard of care targets circulating hormones and their resulting implications. This alleviates symptoms but does not slow down the progression of the disease (*5–7*). Taken together, the prevalence of heart failure and the current treatment options clearly demonstrate a severe unmet medical need in the treatment of heart failure.

Apelin is the endogenous ligand of the APJ receptor (APJ), a Gi coupled G-protein receptor closely related to the angiotensin II type 1 receptor (AT1R) (*8*, *9*). APJ expression is induced by hypoxia and biological derangements associated with heart failure. It is expressed in endothelial and smooth muscle cells of the coronary and pulmonary vasculature as well as in the myocardium. In the coronary system, apelin induces vasodilation, lowers blood pressure, and has a positive ionotropic effect (*10*). Administration of pyr-apelin-13 (a pyroglutamate form of apelin-13), the most potent endogenous ligand, improves cardiac function acutely in healthy volunteers as well as in heart failure patients (*11*, *12*), suggesting that stimulation of the apelin/APJ system could be beneficial for the treatment of heart failure.

Apelin is a 77-amino acid prepropeptide that is further processed into 36–, 17–, 13–, and 12–amino acid active isoforms. The active isoforms of apelin all have a short half-life, limiting the efficacy of these peptides in a chronic disease. We sought to discover a long-acting and specific APJ agonist that could have substantial benefit in the treatment of chronic human heart failure culminating in the advancement of AMG 986 as our clinical candidate (Figure 1).

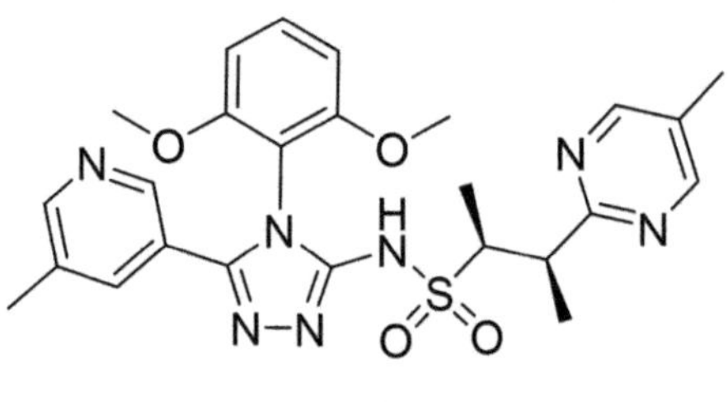

AMG 986

Figure 1. AMG 986.

To enable identification of novel chemical matter, a high-throughput screening (HTS) campaign was undertaken. This resulted in the discovery of triazole **1** (Table 1) exhibiting sub-micromolar potency in a cyclic adenosine monophosphate (cAMP) assay (measure of inhibition of forskolin-induced cAMP production). However, **1** displayed high apparent intrinsic clearance in both rat and human liver microsomes. For comparison, the endogenous ligand pyr-apelin-13 displayed a potency of 0.0016 μM (E_{max} fold reduction of 96%) in the cAMP assay and was used as a benchmark in all described assays.

Table 1. Structure of HTS Hit 1

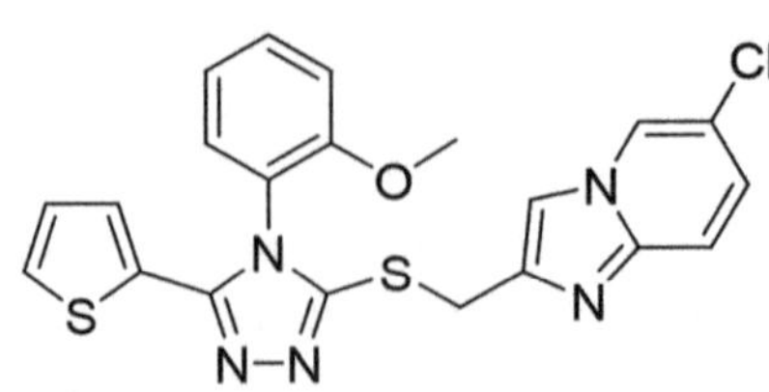

Compound	*cAMP EC_{50} (μM)[a]*	*cAMP E_{max} (%)*	*RLM / HLM $Cl_{int,app}$ (μL/(min*mg))*
1	0.89	99	>399 / >399

[a]Mean of at least two runs.

Discovery of AMG 986

The trisubstituted triazole core of **1** lent itself to a systematic exploration of each substituent to address both potency and metabolic stability. The starting point of our structure-activity relationship (SAR) efforts was to replace the 6-chloroimidazo[1,2-*a*]pyridyl moiety with a synthetically simpler substituent. Introduction of a 4-chlorophenyl moiety (compound **2** in Table 2) was detrimental to potency by 30-fold and had no effect on microsomal stability. To investigate the role of the sulfide in the metabolism of the molecule, replacements were evaluated. Introduction of an amide **3** brought about no improvements in either potency or microsomal clearance. However, substitution of the sulfide with the sulfonamide **4** not only resulted in a large increase in potency, but more importantly provided a significant reduction in microsomal clearance for both the rat and human isoforms. Further optimization of the linker length found that addition of an extra methylene unit to produce **5** provided a further 50-fold improvement in potency while maintaining moderate microsomal clearance values.

Table 2. Sulfide Replacement Effects on Potency and Microsomal Stability

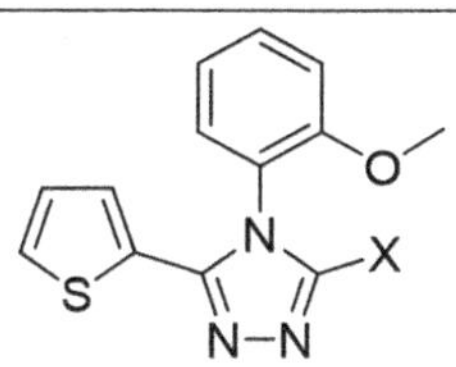

Compound	*X*	*cAMP EC_{50} (μM)[a]*	*cAMP E_{max} (%)*	*RLM / HLM $Cl_{int,app}$ ($\mu L/(min*mg)$)*
2	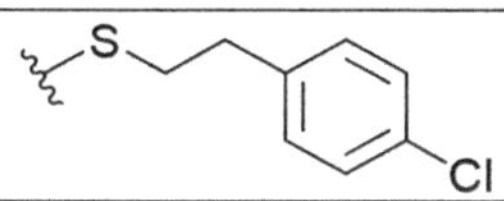	30	80	>399 / >399
3	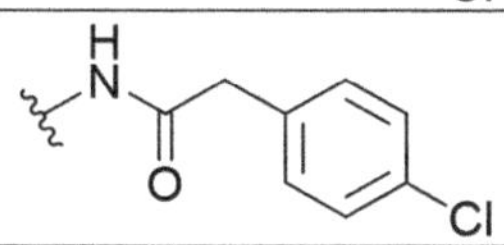	>30	60	>399 / >399
4	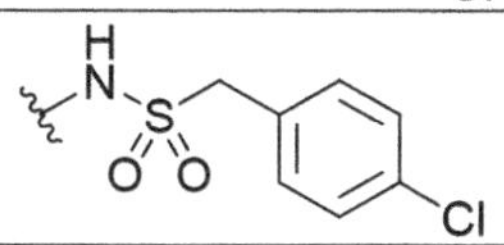	0.099	99	82 / 37
5	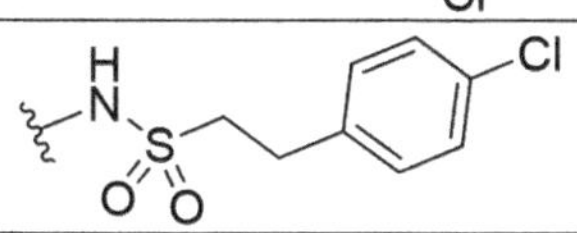	0.002	102	75 / 58

[a]Mean of at least two runs.

At this point in the program, we switched our primary assay to a human guanosine 5'-*O*-[γ-thio]triphosphate (hGTPγS) assay (a direct measure of GPCR activation which is independent of any downstream events). This created a larger window for the triage of compounds for SAR purposes (Table 3). Sulfonamide **5** exhibited a potency of 0.120 μM in the primary hGTPγS assay and recognizing the potential for atropisomers, we sought to investigate its effects on potency.

Changing the 2-methoxy phenyl ring to a 2-methoxy-6-methyl phenyl ring resulted in a mixture of atropisomers that were separable although not assignable. The separated isomers **6** and **7** displayed a 10-fold difference in potency, though the addition of the 6-methyl substituent only realized a 2-fold overall improvement in potency over **5**. The gains in potency were outweighed by the additional step required for separation of atropisomers. Thus, we sought to remove the atropoisomeric property through modification of the 6-position to render a symmetrical ring. This resulted in the symmetrical 2,6 dimethoxyphenyl **8** which to our delight exhibited a 30-fold improvement in potency over **5**, while maintaining moderate rat and human intrinsic clearance in microsomes.

Table 3. Aryl Ring Substituents Effect on Potency and Atropisomerism

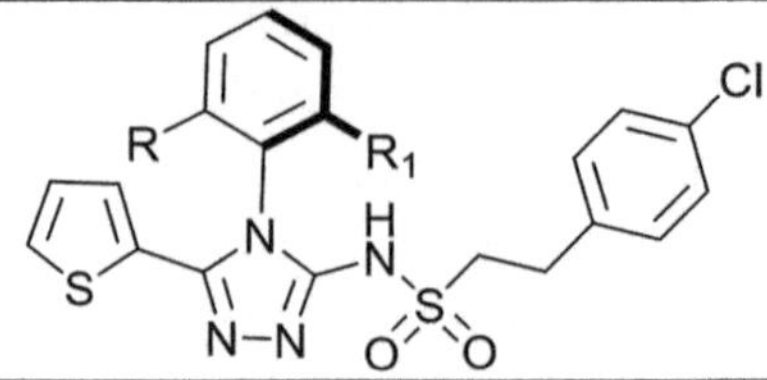

Compound	*R*	R_1	*hGTPγS* EC_{50} *(μM)*[a]	*hGTPγS* E_{max} *(%)*	*RLM / HLM* $Cl_{int,app}$ *(μL/(min*mg))*
5	H	OMe	0.12	167	75 / 58
6[b]	Me	OMe	0.064	226	N/A
7[b]	OMe	Me	0.74	237	N/A
8	OMe	OMe	0.004	142	103 / 74

[a]Mean of at least two runs. [b]The absolute stereochemistry was not determined for these two atropisomers and the depicted stereochemistry was assigned arbitrarily. The compounds could be separated by chiral chromatography and had different optical rotations.

In our experience, thiophene rings have the potential to be a site of metabolism and furthermore, the overall lipophilicity of the molecule was higher than desired (cLogP = 5.44). For these reasons we sought to replace the thiophene ring with amore polar variant (Table 4). Replacement of the thiophene with a thiazole to yield **9** served to dramatically improve microsomal clearance, albeit with a nearly ten-fold loss in potency (Table 4). The thiazole regioisomer **10** exhibited no additional benefits for either potency or microsomal clearance. Installation of an oxazole **11** only served to further erode the potency another two-fold relative to the thiazole **9** congener. The furan **12** retained all the potency of the thiophene, but also importantly significantly lowered the microsomal turnover in comparison to **8**. Introduction of a phenyl ring **13** as a classic bio-isostere of thiophene and furan, resulted in a modest loss of potency and microsomal stability.

Noticing a trend where substituents that lowered the overall cLogP of the molecule led to improved microsomal turnover, we explored adding a nitrogen to the phenyl ring. The 2-pyridyl **14** resulted in a further two-fold loss in potency over **13**, but a marked improvement in microsomal clearance in both rat and human liver microsomes. We noted that introduction of the 2-pyridyl moiety had lowered the cLogP from 4.92 for the furan **12** to 4.35 for 2-pyridyl **14**. Walking the nitrogen around the ring further to the 3-pyridyl **15**, served to further reduce the cLogP to 4.14. This maintained the microsomal stability displayed with **14** but did rescue some of the lost potency.

Furan **12** was potent in our in vitro assay and displayed acceptable pharmacokinetic properties allowing us to start exploring the SAR around the chlorophenyl substituent. As noted previously, the overall lipophilicity of the molecule was not only a concern for microsomal stability, but also for CYP3A4 induction. Further efforts to lower the lipophilicity involved removing the chlorine atom and adding polar functionality (Table 5). Simple replacement of the chlorine atom with a fluorine atom to produce **16** had a minimal effect on potency and microsomal clearance. In combination with a more polar functionality at the 2-position, it served to reduce the cLogP and improved potency two to three -fold. Addition of a nitrile **17** (cLog P = 3.93) or a methyl sulfone **18**, (cLog P = 2.84) improved the microsomal stability. Replacement of the 2-substituted phenyl ring with a 2-pyridyl ring **19** maintained the cLogP (2.5) but resulted in a six-fold reduction in potency. Replacement of the phenyl **16** with a pyridine **19** further reduced microsomal clearance on the rat isoform. Introduction of an additional nitrogen in the ring, namely formation of a pyrimidine ring **20**, maintained excellent microsomal clearance on both rat and human isoforms. It also recovered the potency loss observed with **19** resulting in a potency of 3 nM on the hGTPγS assay.

Table 4. Thiophene Analogs

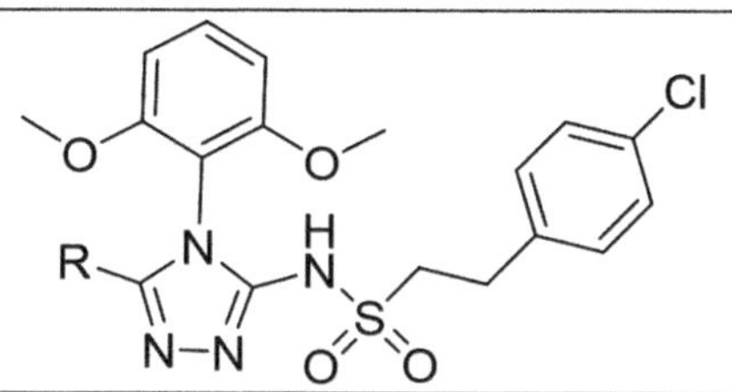

Compound	*R*	*hGTPγS EC_{50} (μM)*[a]	*hGTPγS E_{max} (%)*	*RLM / HLM $Cl_{int,app}$ (μL/(min*mg))*
8	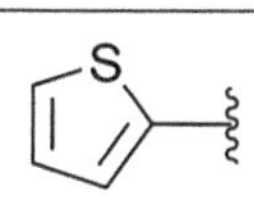	0.004	142	103 / 74
9	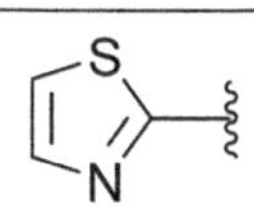	0.035	171	<14 / <14
10	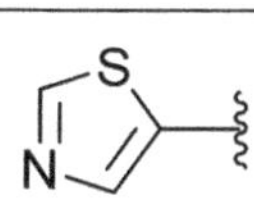	0.030	159	42 / 28
11	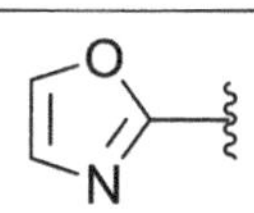	0.071	219	<14 / <14
12	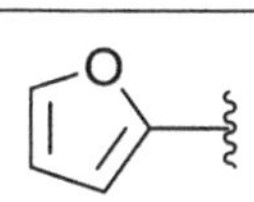	0.007	134	35 / 22
13	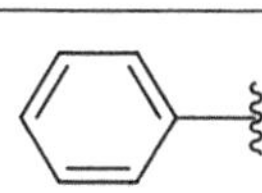	0.026	157	79 / 32
14	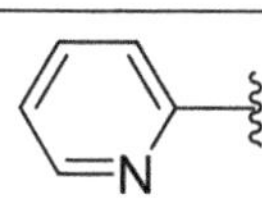	0.059	157	23 / <14
15	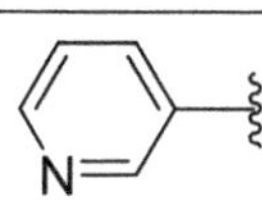	0.016	141	29 / 23

[a]Mean of at least two runs.

The furan moiety was a reasonable replacement for thiophene; however furans also have the potential for oxidative metabolism. To minimize any potential oxidation, a methyl group was installed at the C-5 of the furan ring to block this site of metabolism. The methyl furan **21** (Table 6) resulted in an unexpected ~ten-fold improvement in potency while maintaining good microsomal clearance in both species.

Table 5. Replacements for 4-Chlorophenyl Substituent

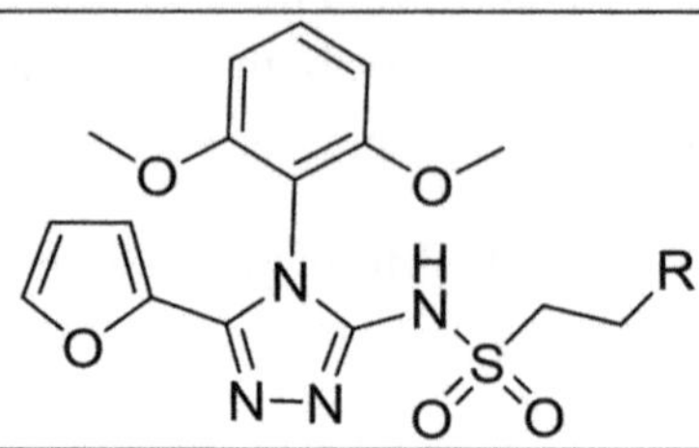

Compound	*R*	*hGTPγS EC_{50} (μM)*[a]	*hGTPγS E_{max} (%)*	*RLM / HLM $Cl_{int,app}$ (μL/(min*mg))*
12	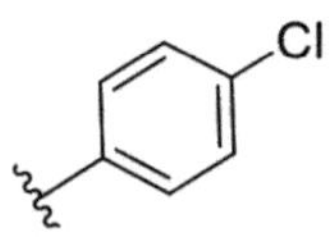	0.007	134	35 / 22
16	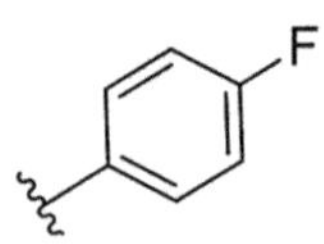	0.016	156	23 / 28
17	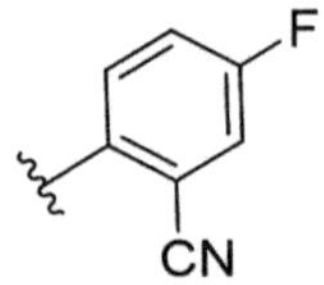	0.003	140	34 / 21
18	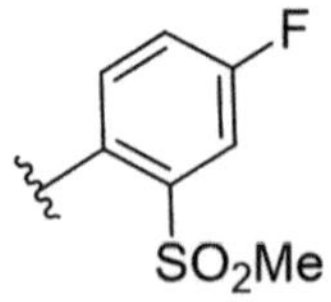	0.002	132	34 / <14
19	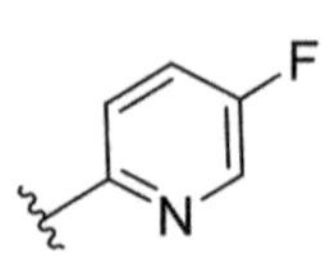	0.012	146	<14 / <14
20	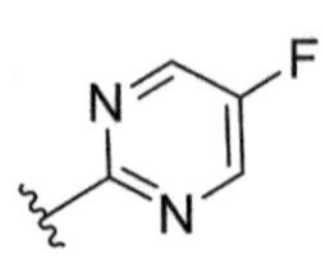	0.003	130	<14 / <14

[a]Mean of at least two runs.

The last area for SAR optimization that we explored was the ethylene linker. It was clear that the pyrimidine substituent exhibited a favorable interaction with the binding pocket. However, we believed that due to the flexibility of the ethylene linker, this motif would sit in several conformational states of which some would be high in energy. Thus, we hypothesized that if we could reduce the number of high energy conformations and populate more low energy conformations through restricted rotation, we could increase the percentage of bioactive conformations and improve the potency of the compounds.

Addition of a methyl group α to the sulfonamide yielded a pair of enantiomers and upon separation, we observed a significant difference in potency favoring the (*2S*) enantiomer **23**, over the (*2R*) enantiomer **22** (Table 6). Further, addition of a methyl group beta to the sulfonamide yielded a diastereomeric mixture and again separation of isomers revealed a further preference for one of the diastereomers; namely **25** exhibiting (*2S, 3R*) stereochemistry and a further boost in potency over diastereomer **24**. The additional methyl groups of **25** were a definitive gain for potency (80 pM in the hGTPγS assay), but microsomal stability for the rat isoform was eroded, while stability in the human isoform was maintained.

Table 6. Introduction of Restricted Rotation

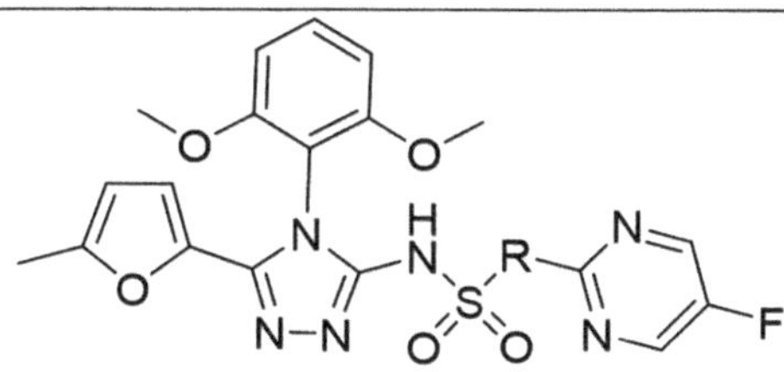

Compound	*R*	*hGTPγS EC$_{50}$ (μM)^{a}*	*hGTPγS E$_{max}$ (%)*	*RLM / HLM Cl$_{int,app}$ (μL/(min*mg))*
21	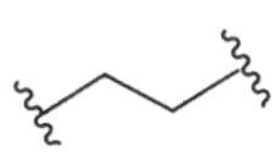	0.0003	134	<14 / <14
22	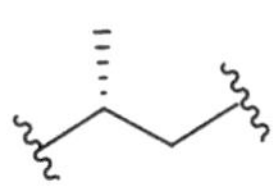	0.002	140	41 / 15
23	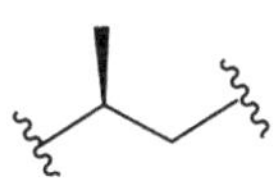	0.0002	132	34 / <14
24	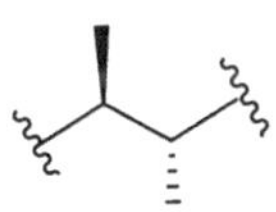	0.001	146	117 / 15
25	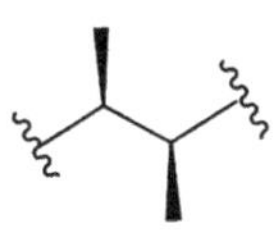	0.00008	149	129 / <14

[a]Mean of at least two runs.

The improved potency and microsomal stability of **25** identified it as a potential lead compound for proof-of-concept studies. Further evaluation of the pharmacokinetic properties of the compound in both rat and dog confirmed it was suitable for in vivo studies (*13*). Isolated perfused rat hearts (Langendorff preparation) were infused with **25** and its effects on dP/dt, a measure of the rate of rise of pressure in the left ventricle and a surrogate for myocardial contractility, were measured and found to be comparable to pyr-apelin 13, but at a five-fold lower EC_{50}. Results of a study undertaken to evaluate the in vivo effects of **25** on cardiac function in an ischemic cardiomyopathic rat model (*14*) showed a greater than 20% increase in ejection fraction following IV infusion with no observed change in heart rate.

Encouraged by these positive results, the molecule was advanced to a canine model of heart failure that is induced by coronary microembolization and is conducted under anesthesia. Infusion of escalating doses of **25** resulted in a 17% increase in ejection fraction (*15*) when compared to vehicle at a dose of 0.1 mg/kg and a free drug concentration of 16 nM (*16*). Measured cardiac output was increased, with importantly no increases in heart rate or blood pressure following the infusion with **25**.

With these positive results for **25**, we sought to address some of the outstanding issues of this lead compound, specifically regarding the induction of CYP3A4 mRNA (84 percent of control (POC) at a concentration of 10 μM). SAR observations indicated that this issue was apparent in a number of furan-containing analogues. Thus, we looked to move away from the furan ring and replace it with a polar functionality to reduce the propensity for induction. Our aim was to reduce the induction signal to less than 30 POC.

Reviewing the existing SAR, it was noted that 3-pyridyl analog **15** (Table 4), exhibited good microsomal turnover and only a modest loss in potency compared to its furan analogue. Introduction of a 3-pyridyl moiety into the more optimized scaffold based on **25** led to a 7.5-fold loss in potency but achieved a 2-fold reduction in CYP3A4 induction. Attempts to further refine the properties of the molecule, balancing potency and lipophilicity, led to 2*S*-methyl sulfonamide derivative **27**. This showed a further reduction in CYP3A4 induction, but also a two-fold loss in potency relative to **26** (Table 7). Noting the past improvement in potency achieved when adding a methyl substituent to the furan, we were able to improve the potency of **26** by adding a methyl group to the 5-position of the pyridyl ring **28**. This displayed an improved potency and lowered the CYP3A4 induction to less than 30%. Further reduction of the CYP3A4 induction could be gained by modifying the fluoropyrimidine to a methylpyrimidine (AMG 986), resulting in a compound with excellent potency, microsomal stability, and with acceptable CYP3A4 induction of 19%.

Striving to evaluate the improvements of AMG 986, we evaluated the pharmacokinetic properties of the molecule in both rat and dog (*17*). This analysis predicted the molecule to be suitable for once daily dosing in humans (*18*). After evaluation of AMG 986 in an acute rat myocardial infarction (MI) model (*15*), the compound progressed to the canine model of heart failure to evaluate it in larger animals. At a dose of 0.01 mg/kg, we observed a 13% increase in ejection fraction at a free drug concentration of 0.77 nM. Again, no changes in heart rate or blood pressure were observed. To achieve an efficacious unbound plasma concentration of 1.0 nM in our canine model of heart failure, coupled with our predicted human PK that was derived through allometric scaling, we projected a clinical efficacious dose of 17 mg QD. Based on the in vivo efficacy and safety assessment, AMG 986 was selected for clinical development.

Table 7. Evaluation of Furan Substitutions upon CYP3A4 Induction

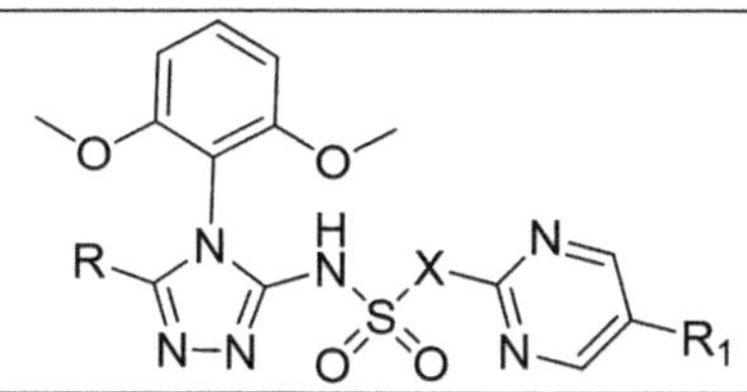

Compound	*R*	*X*	*R1*	*hGTPγS EC50 (μM)[a]*	*hCYP3A4 mRNA induction (10 μM, POC)*
25	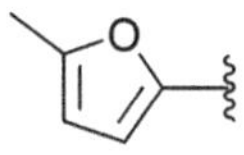	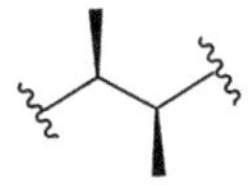	F	0.00008	84%
26	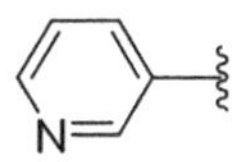	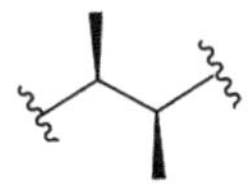	F	0.0006	45%
27	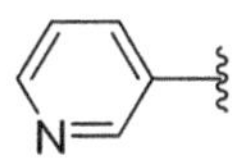	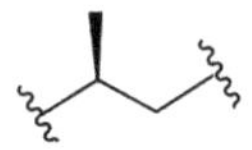	F	0.0013	22%
28	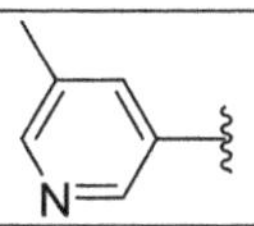	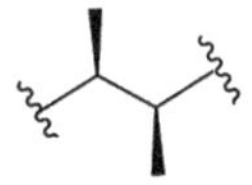	F	0.0003	28%
AMG 986	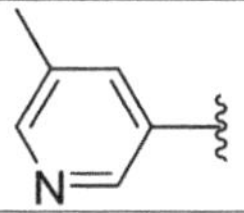	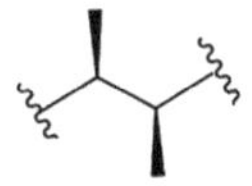	Me	0.0003	19%

[a]Mean of at least two runs; [b] % E_{max} comparable to **25** and pyr apelin-13.

From the triazole HTS hit, a systematic examination of each substituent around the core provided improvements to the compound. Finding atropisomers and then designing them out of the molecule led to a 30-fold improvement in potency. Efforts to remove the problematic thiophene led to a methyl furan that improved the potency a further ten-fold. Deliberate efforts to reduce the number of conformations the molecule populated resulted in a specific chiral sulfonamide fragment and further improved the potency.

Throughout the optimization, the lipophilicity of the molecule was lowered through the introduction of polarity throughout the scaffold. These optimization efforts allowed us to leverage the large gains in potency to help fine tune the metabolic stability and CYP3A4 induction providing AMG 986.

Medicinal Chemistry Route to Synthesis of AMG 986

The initial Medicinal Chemistry route evolved out of the desire to incorporate specific structural motifs into the substituents around the triazole core of the HTS hit to improve potency and reduce metabolic stability. The route allowed for installation of the sulfonamide fragment to occur in one step at the end, but changes to other substituents unfortunately required installation early in the synthetic sequence.

The route to the sulfonamide (Scheme 1) started with 2-chloro-5-fluoropyrimidine (**29**). A Suzuki-Miyaura cross-coupling with the Molander salt of 2-butene gave the trisubstituted alkene

30 (Scheme 1). The alkene was converted to the sulfonic acid **31** in a moderate yield, but the overall reaction time was long and required a tedious purification. The sulfonic acid was generated in an unfavorable 3.5 to 1 ratio of the anti to syn isomers. The dimethoxybenzyl (DMB) protected sulfonamide **32** was generated via the sulfonyl chloride, which was generated with oxalyl chloride. The sulfonamide could be deprotected under acidic conditions to give a diastereomeric mixture of sulfonamides **33-36** which could be separated under chiral SFC conditions to provide the individual stereoisomers.

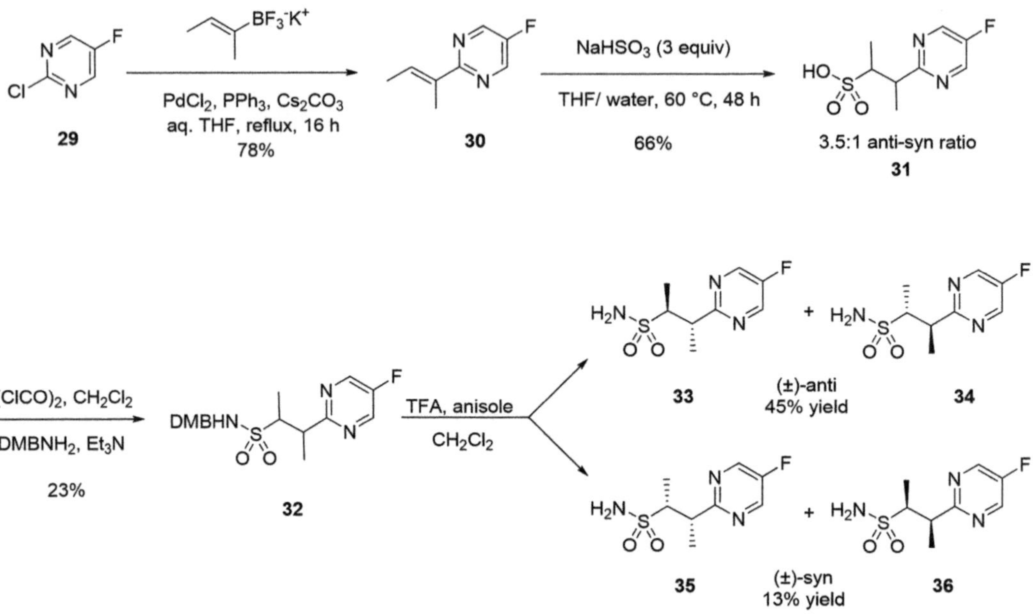

Scheme 1. Synthesis of diastereomer mixture of sulfonamide ***33-36****.*

The Medicinal Chemistry route to the triazole core was initiated with the acylation of 2,6-dimethoxyaniline (**37**) with furan-2-carbonyl chloride (**38**) to provide amide **39** in high yield (Scheme 2). The amide was converted to the corresponding thioamide via Lawesson's reagent, which was subsequently treated with hydrazine hydrate to yield **40** in 98% yield. The core triazole could be completed by treating hydrazonamide **40** with cyanogen bromide. The furan was then brominated to produce bromide **41**. Further elaboration of the bromo furan to the methyl furan was accomplished under Suzuki-Miyaura cross-coupling conditions with trimethylboroxine. Conversion of the amino triazole to the bromo triazole **42** was accomplished in a moderate yield via a Sandmeyer type reaction under nonaqueous conditions. The final step required the sulfonamide to be installed under copper-mediated conditions to produce triazole **25** in 90% yield. To facilitate the throughput of analogues, the racemic sulfonamide was separated into individual stereoisomers on gram scale using chiral SFC prior to the coupling. Until a stereospecific route to the sulfonamide could be developed, this separation allowed the analogues to be rapidly processed through the project flow scheme.

An efficient strategy for assembling these molecules came via a three-component coupling strategy (Scheme 3). The isothiocyanate **43**, available in one step from the aniline precursor, could be treated with the single enantiomer of sulfonamide **44** followed by 5-methylnicotinohydrazide (**45**) and silver nitrate to generate the intermediate **46** in one pot. The triazole core could be cyclized by heating with methanesulfonic acid (MSA) in acetonitrile to produce AMG 986 in 55% yield over the two steps.

While this chemistry was amenable for early discovery, particularly around different portions of the molecule and determining the desired stereoisomer of the sulfonamide fragments, the scale up of the sulfonamide intermediates left us looking for an improved route. As it became apparent during lead optimization that AMG 986 might achieve internal milestones requiring more material for pre-clinical testing, and consistent with our standard practices, we engaged our colleagues in Process Development and in the Catalysis lab to begin process research to identify a scalable route.

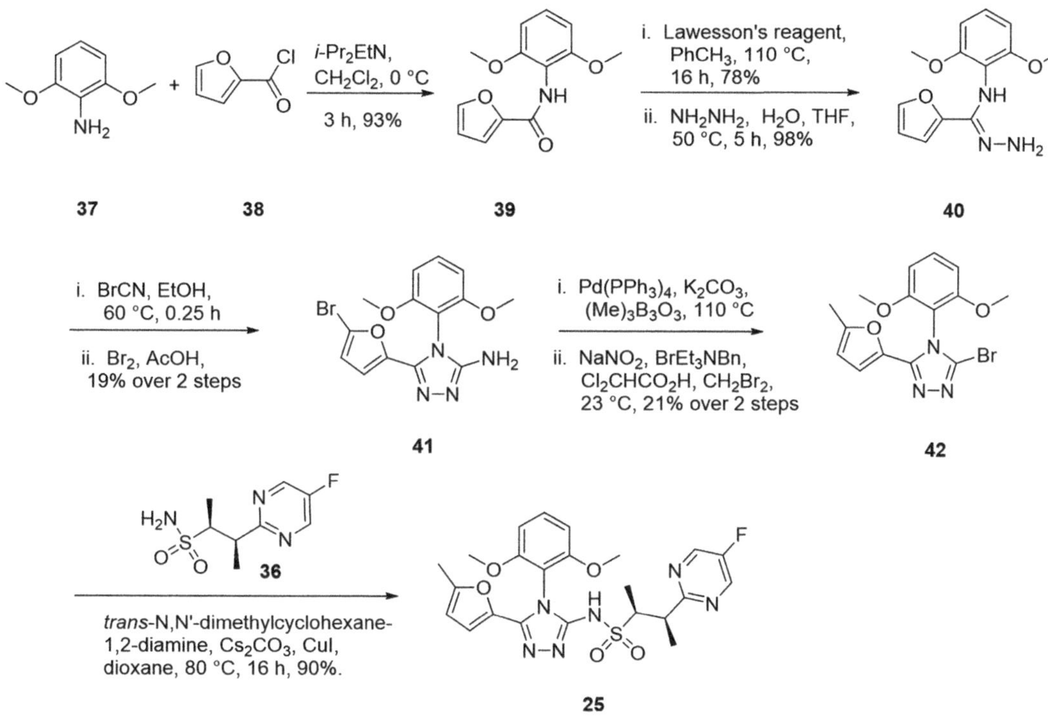

Scheme 2. Early Medicinal Chemistry route.

Process Development of AMG 986

Prior to the final selection of AMG 986 as the clinical candidate for the APJ program, the Process Development (PD) and Medicinal Chemistry teams partnered to explore an alternative synthesis of a late stage intermediate. Specifically, an improved route to chiral, non-racemic α,β-cis-dimethylsulfonamides (Figure 2) was sought to alleviate a synthesis bottleneck impacting SAR and multi-gram scale-up for in vivo studies. The goal of that collaboration was to accelerate final molecule selection while at the same time, position PD to rapidly supply multi-kilogram quantities of the drug substance for FIH clinical trials.

The chiral α,β-dimethyl moiety was key to improving the potency of the lead molecules and with that space optimized, final SAR efforts would focus on the peripheral ring substituents. The original Discovery route (Scheme 4) generated a mixture of stereoisomers that were separated by chiral column chromatography for relative biological activity assessment of the individual isomers.

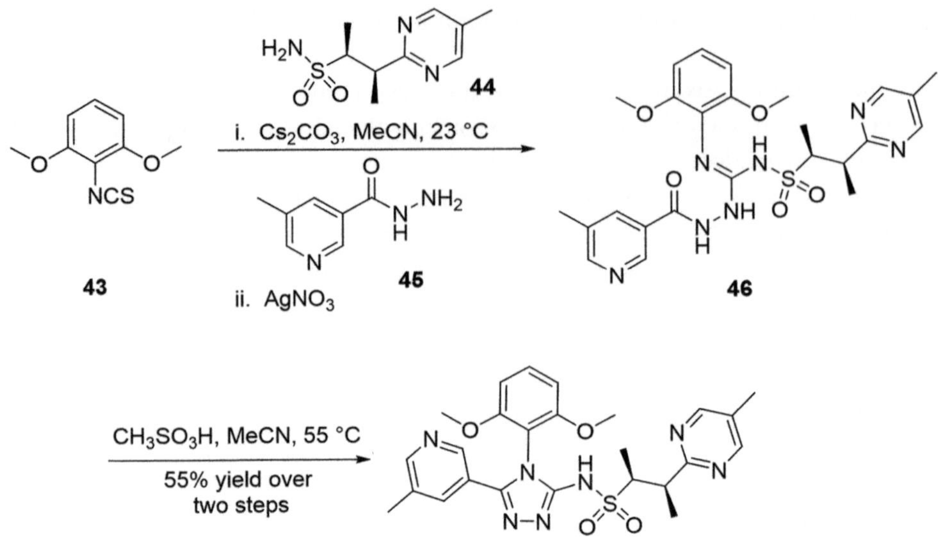

AMG 986

Scheme 3. Three-component coupling synthesis of AMG 986.

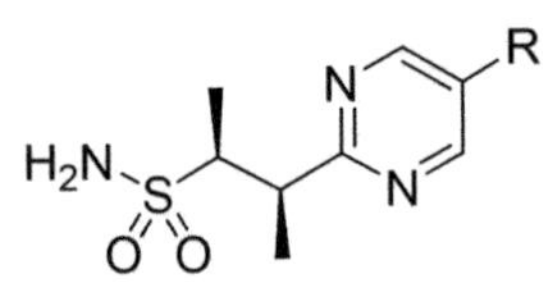

Figure 2. Target late stage intermediate chiral α,β-cis-dimethylsulfonamide.

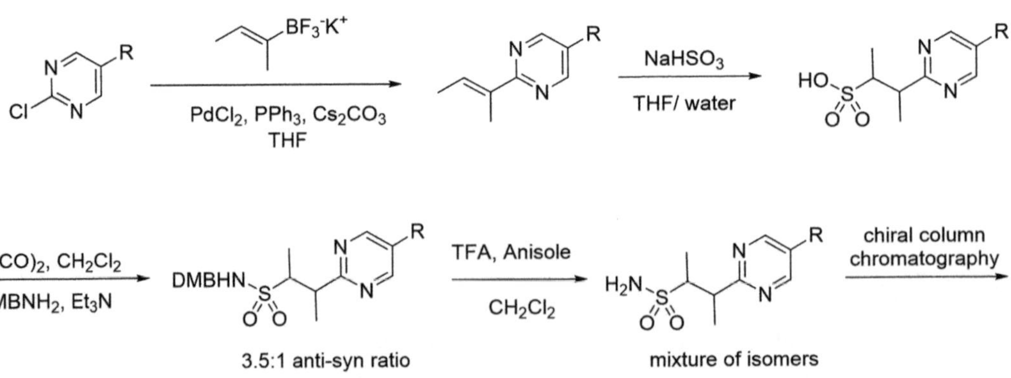

Scheme 4. Original route to α,β-dimethyl sulfonamides.

Multi-kilogram chiral column chromatographic separation of racemic mixtures is an option to advance single-enantiomer drug candidates to clinical studies. However, that approach is increasingly wasteful if the separation step occurs towards the later stage of the route and the unwanted isomer cannot be recycled. It also reduces flexibility in the manufacturing space as specialized equipment is required for large scale manufacturing. Modern pharmaceutical industry initiatives of 'operational sustainability' and 'green chemistry' provide a sound rationale for investment into more efficient, stereospecific approaches earlier in the process development cycle despite the risk that programs are often terminated in Phase 1 clinic trials.

With a well-planned high-throughput experimentation (HTE) approach, success in catalytic stereoselective approaches can be realized in much shorter timeframes than traditional univariable approaches. When considering a stereoselective approach to the target sulfonamide fragments, a suitable pro-chiral substrate was required for which the Discovery group proposed a compelling option. Having recently devised and published (*19*) a novel approach to primary sulfonamides, the team was able to apply that approach towards unsaturated α,β-dimethyl sulfonamides for SAR evaluation for the APJ program. These olefins were now viewed as substrates for a potential stereoselective hydrogenation to the corresponding enantiomerically pure sulfonamides (Scheme 5).

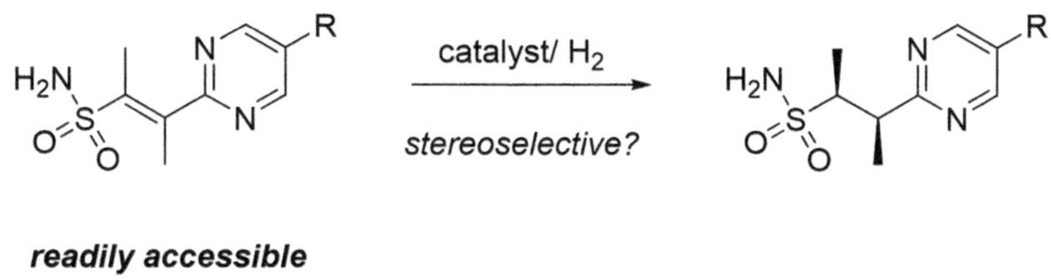

Scheme 5. Proposed stereoselective transformation to enable direct access to α,β-dimethyl sulfonamides.

Notable examples of asymmetric hydrogenations of substituted alkenyl sulfones to enantiomerically pure alkyl sulfones had been described in the literature utilizing ruthenium (*20*) and iridium (*21*) catalyst systems to generate a single chiral center. However, to our knowledge, examples of asymmetric hydrogenations of substituted alkenyl sulfonamides towards enantiomerically pure alkyl sulfonamides has not been published and generation of two adjacent stereocenters from a tetra substituted alkenyl sulfonamide was without precedent.

A proof of concept project was initiated by our Catalysis group for a first-tier multivariate screen. Metal catalysts were limited to ruthenium, iridium, and rhodium combined with variety of chiral phosphine ligands in either CH_2Cl_2 or MeOH as solvent. From this limited screen, a catalyst system consisting of 1:1 $Rh(COD)BF_4$ / (*R*)-Phanphos gave a modest yet encouraging ee of desired sulfonamide **36** in 45% in high dr (Scheme 6).

Using these conditions, the conversion stalled at 43% and extended reaction time at higher temperature and/or pressure was unproductive. Indeed, in some instances, slightly lower ee at the expense of further conversion was observed. Doubling the catalyst loading to 4 mol% resulted in >90% reaction conversion. However, due to the negative impacts on efficiency, cost of goods, and work-up, such high loadings of catalyst were undesirable.

This prompted the exploration of additives which might stabilize or otherwise improve catalyst turnover. $Zn(OTf)_2$, HBF_4, and Et_3N were added to the next screening attempt. To our delight, incorporation of 0.2 equiv of $Zn(OTf)_2$ or HBF_4 (at 2 mol% catalyst loading), resulted in conversions of 99% and 94%, respectively. Additionally, product ee and dr's were in line with those achieved from the prior screening in the absence of additives. Of note, the same loading of Et_3N shut down the reaction almost entirely (3% conversion).

Moving forward, zinc triflate was chosen over tetrafluoroboric acid for its slightly better performance, but more importantly for its superior safety profile and ease of handling. HBF_4 is a category 1 corrosive and health hazard. Minor optimization of the existing reaction was evaluated to arrive at 2.0 mol% $Rh(COD)_2BF_4$, 2.5 mol% (*R*)-Phanephos, 0.2 equiv $Zn(OTf)_2$, in MeOH (0.4

M) at 30 bar H_2 and a temperature of 23 °C. The hydrogenation was carried out for 16 h, producing >99% conversion in 53% ee.

Having this process in place five weeks from project initiation served to speed up final SAR efforts which in turn soon led to final candidate selection. For our lead candidate, the fluorine substitution on the pyrimidine substrate used in our hydrogenation screening was replaced with a methyl group to establish AMG 986. That change did not impact the enantioselectivity or reaction conversion.

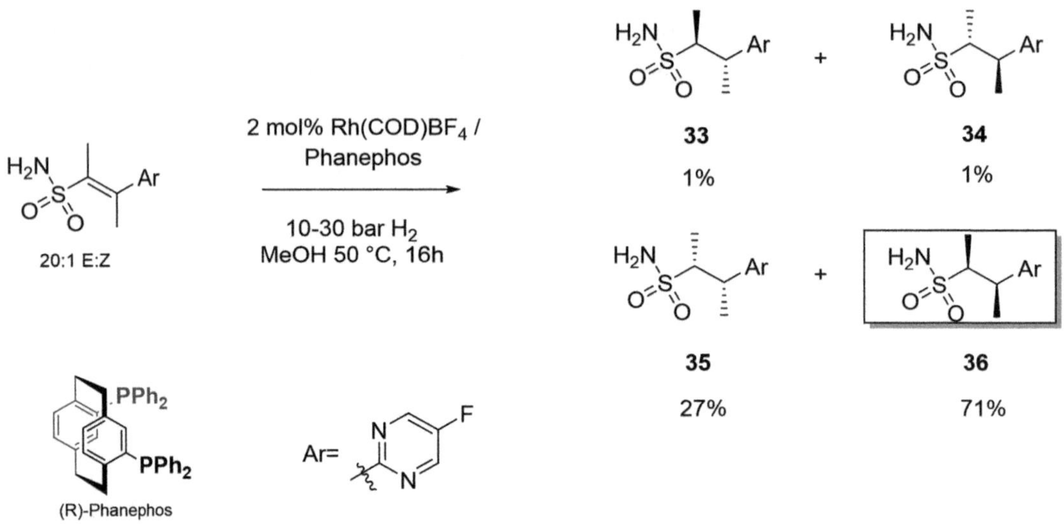

Scheme 6. Initial screening hit for chiral hydrogenation of dimethyl trans-olefin.

While the selectivity was not high enough to establish a scalable process at this point, recrystallization(s) of the crude material from ethanol could be used to upgrade enantiomeric purity to high levels to supply the early multi-gram scale lots needed for exploratory toxicology studies. Additionally, multi-gram quantities could now be readily synthesized to begin development of the downstream steps toward the target molecule while efforts to further improve the enantioselectivity of the hydrogenation in process development ramped up. An effort to quickly refine conditions to the pro-chiral vinyl sulfonamide **51** (Scheme 7) was initiated.

The approach was deemed suitable for multi-kilogram scale-up as each step had been demonstrated on multi-gram scale in good to moderate yields using unoptimized conditions. As such, our intent was to generate a steady supply of olefin **51** to support the continued optimization of the hydrogenation step while simultaneously developing scalable workups and isolations for **51**. The identification of any critical reaction parameters impacting successful scale-up was also key to ensure success in a single multi-kilogram lot to support our planned GMP campaign under tight timelines.

Starting with a Pd-catalyzed Suzuki-Miyaura coupling of in situ derived vinyl boronic acid with chloropyrimidine **47**, the vinyl pyrimidine **48** was obtained in 84% yield. Subsequently, pyrimidine-2-thiol was converted to the sulfenyl chloride using sulfuryl chloride and used directly to convert vinyl pyrimidine **48** to the racemic sulfide **49**. After initial success of this step on small scale, both purity and yields dropped off significantly when we began to scale up. This issue was traced back to the water content of pyrimidine-2-thiol that was sourced from an alternative vendor. To remediate this situation and allow for selection of additional vendors, an azeotropic distillation from toluene of the raw material was implemented to ensure adequate supply for the campaign. When dry thiol was utilized, a yield of 78% was achieved after crystallization from *n*-heptane.

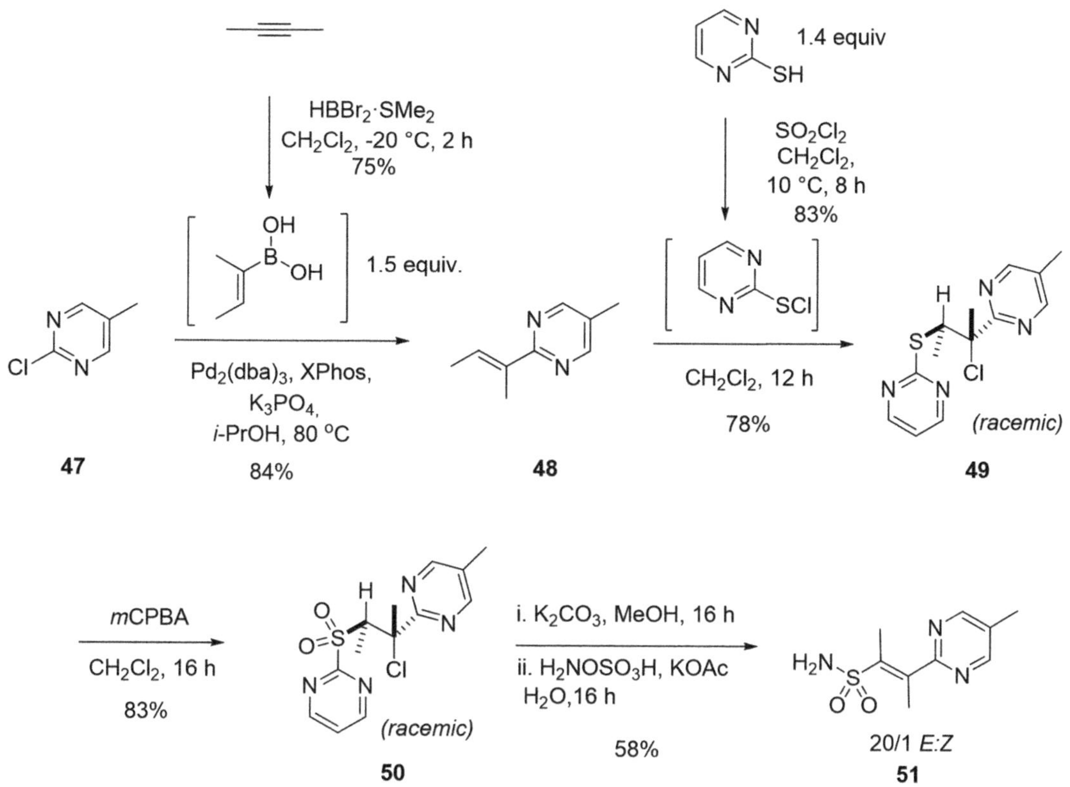

*Scheme 7. Five linear steps scale up route to vinyl sulfonamide **51** from commercially available raw materials.*

Next, racemic **49** was smoothly oxidized to the corresponding sulfone **50** using *m*-CPBA in CH_2Cl_2 at 25 °C and the product crystallized from MTBE following aqueous workup. Lastly, treatment of **50** with potassium carbonate in MeOH at 30 °C for 16 h effected elimination to generate an intermediate sulfinate which upon treatment with hydroxyamine-*O*-sulfonic acid, gave the sulfonamide **51** as a crystalline solid from MeOH/ pet ether in 58% yield over two steps.

In order to meet material requirements for GLP toxicology and FIH clinical studies, a 9.5 kg batch of the olefin was prepared in a single run starting from 9 kg of 2-chloro-5-methylpyrimidine in 99.6% HPLC area purity.

In parallel with the development and scale up of the vinyl sulfonamide **51**, a second-tier ligand screen for the asymmetric hydrogenation was initiated to improve the stereoselectivity. This screen was focused on the identification of an improved ligand while holding the pressure constant for a head to head comparison. An additional 70 ligands were evaluated from which thirteen ligands (Table 8) exceeded the 58% ee baseline obtained with (*R*)-Phanephos.

The data in Table 8 indicated that Josiphos SL-J216-2 (Table 8, entry 13) was the clear ligand choice; its use resulted in >99% conversion and delivered product **44** in the highest ee of 91%. Additionally, this ligand is commercially available for large scale applications. A final optimization study was carried out to further lower raw material costs while also reducing the amount of spent catalyst to remove in the workup and isolation (Table 9).

Table 8. Second Round Ligand Screen Employing $Rh(COD)_2$

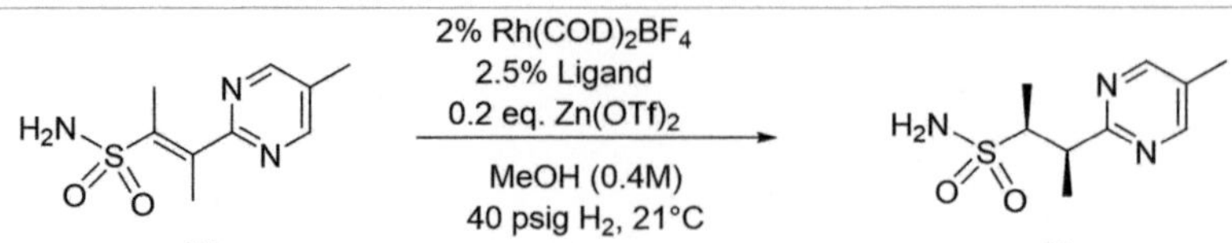

51 → **44**

Entry	*Ligand*	*Conversion to Product 44*	*Entry*	*Ligand*	*Conversion to Product 44*
1	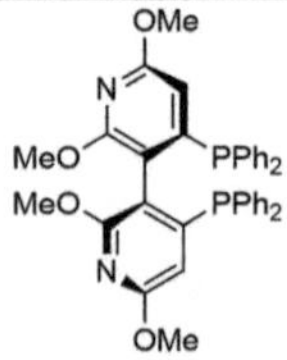 CTH-(*R*)-PPhos	-64% ee 45% conv.	**2**	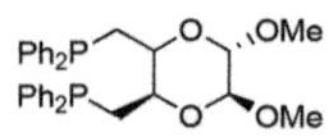 *R,R,R,R*-TPhos	-65% ee 36% conv.
3	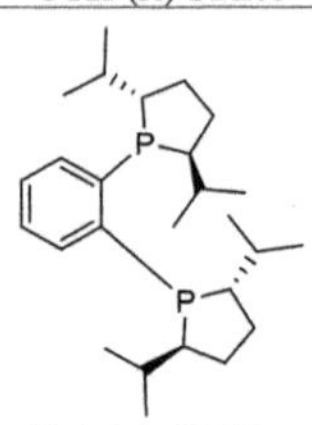 (*S,S*)-*i*-Pr-DuPhos	-68% ee >99% conv.	**4**	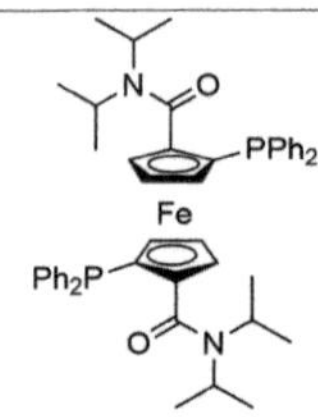 (*R*)-Jafaphos	-85% ee 48% conv
5	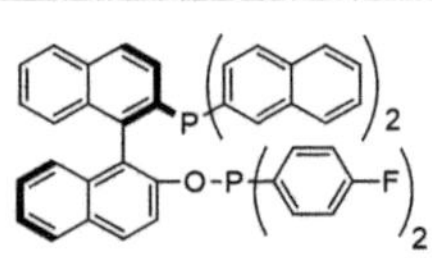 (*R*)-2-Nap-BIPNITE-p-F	-97% ee 81% conv.	**6**	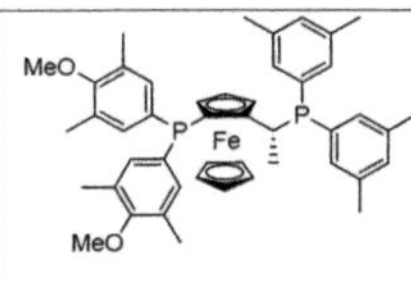 Josiphos J418-1	-56% ee >99% conv.
7	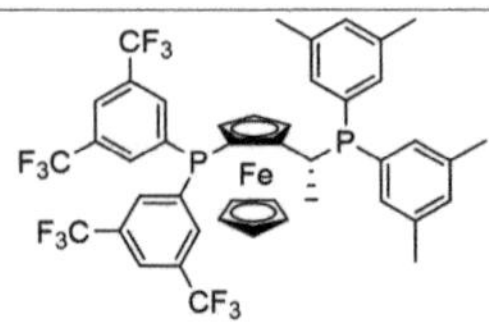 Josiphos J008-1	-60% ee >99% conv.	**8**	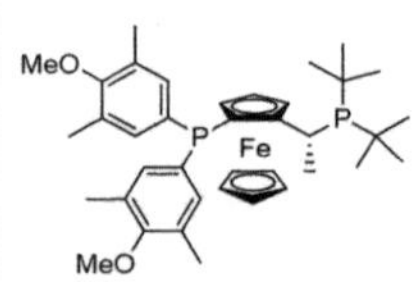 Josiphos J013-1	-70% ee >99% conv.
9	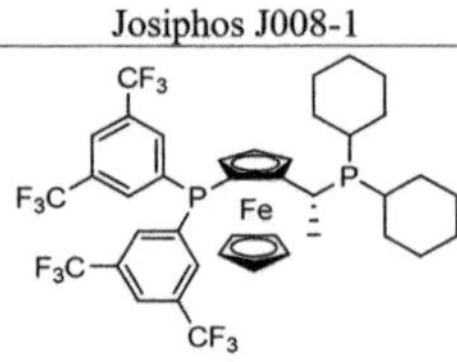 Josiphos J006-1	71% ee >99% conv.	**10**	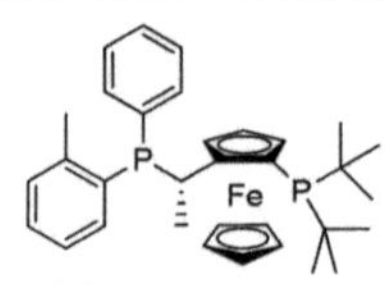 Josiphos J505-2	-71.0% ee >99% conv.
11	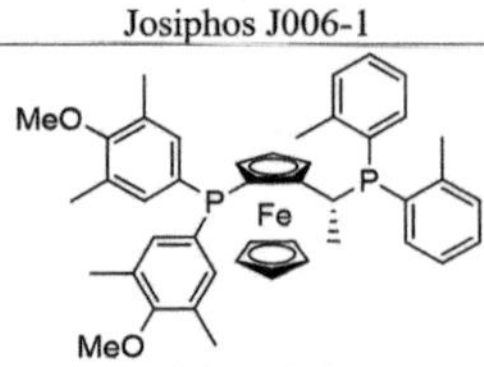 Josiphos J425-1	-76% ee >99% conv.	**12**	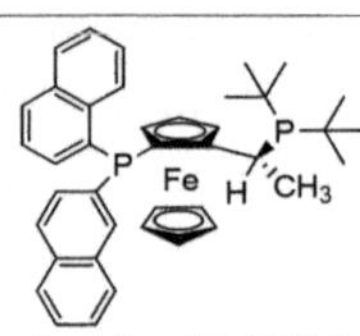 Josiphos SL-J216-1	-91% ee >99% conv.
13	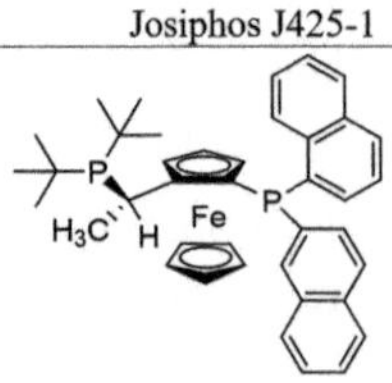 Josiphos SL-J216-2	91% ee >99% conv.	**Note:** Negative ee values highlighted indicate the excess of the undesired enantiomer.		

Table 9. Hydrogenation Optimization of Catalyst Loading, Pressure and Temperature

Entry	*Rh(COD)$_2$BF$_4$ (mol%)*	*SL-J216-2 (mol%)*	*Temp °C*	*Pressure (psig)*	*% Conversion*	*ee %*
1	0.5	0.625	50	40	86	89.7
2	1	1.25	21	40	79	92.3
3	1	1.25	50	40	100	90.1
4	1	1.25	50	150	98	78.8
5	2	2.5	21	40	100	91.0

At 50 °C, catalyst loading could be lowered to 1 mol% (on a metal basis) while maintaining 90% ee (Table 9, entry 1). Under those same conditions, an increase in pressure led to erosion of ee (Table 9, entry 4). With limited time for further optimization, and in order to reduce the risk of incomplete conversion, the original loading conditions of 2 mol % rhodium (21 °C; 40 psig) were implemented (Table 9, entry 5) to support initial clinical development lots.

The isolation strategy for sulfonamide **44** was to first scavenge the bulk of the metal catalyst and then isolate semi-pure material. This would be purified by a subsequent reslurry or recrystallization to upgrade the enantiomeric purity while minimizing losses. Downstream metal remediation required < 20 ppm of rhodium in the final API as a parenteral formulation was planned in addition to an oral tablet for FIH trials. On a 50 g development run, treatment of the crude hydrogenation product stream with 20% (w/w) activated carbon at 45 °C for 1 h followed by cooling to 5 °C and a polish filtration provided a clear solution. The product was crystallized by the addition of water affording **44** in 84% yield and 90% ee. A final reslurry in three volumes of IPA was implemented, upgrading both enantiomeric purity (to 94% ee) and chemical purity (to 99.91 HPLC area %) with only a 2% loss of product to the mother liquor and a rhodium content of 70 ppm (determined by ICP-MS). Downstream isolations were now in place to further purge residual rhodium but a target of <100ppm at this stage was adequate so as not to unnecessarily impact yield. The hydrogenation process was scaled up in a 300 L hydrogenation vessel starting with 9.7 kg of olefin **51** to give **44** in 86% ee. A reslurry with IPA yielded 7.48 kg (77%) in 91.2% ee, 99.9% HPLC area %, and 61 ppm Rh by ICP-MS.

While development and scale up of the improved synthesis of sulfonamide **44** required a significant investment of resources and time, securing intermediates **43** and **45** (Figure 3) for the three-component convergent synthesis of AMG 986 was relatively simple. Both isothiocyanate **43** and hydrazide **45** where obtained in one step from commercially available raw materials.

Compound **43** was synthesized on 5.5 kg scale and isolated as crystalline solid in 99.2% purity following aqueous workup and solvent exchange from CH_2Cl_2 into MeOH. Compound **45** was delivered on 5.9 kg scale after direct crystallization of the MeOH reaction mixture upon cooling followed by an ethanol wash to give a white solid in 99.0% purity.

Development of the downstream GMP steps toward AMG 986 leveraged the Medicinal Chemistry second-generation synthesis of AMG 986 (Scheme 8). The highly telescoped approach was carried out with a single crystalline isolation.

Addition of sulfonamide **44** to isothiocyanate **43** using cesium carbonate (Cs_2CO_3) in acetonitrile generated the corresponding thiourea **52** which was then treated without isolation with hydrazide **45** in the presence of 2.0 equiv of $AgNO_3$. This subsequently required multiple passes of the product through silica gel to remove excess $AgNO_3$ and silver sulfide. The crude penultimate intermediate **46** was carried forward without isolation.

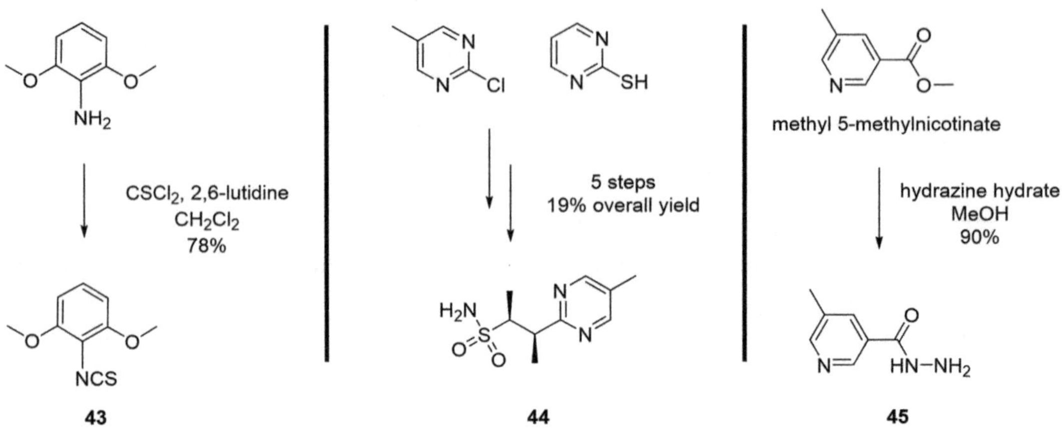

Figure 3. API starting materials for the synthesis of AMG 986 from commercial raw materials.

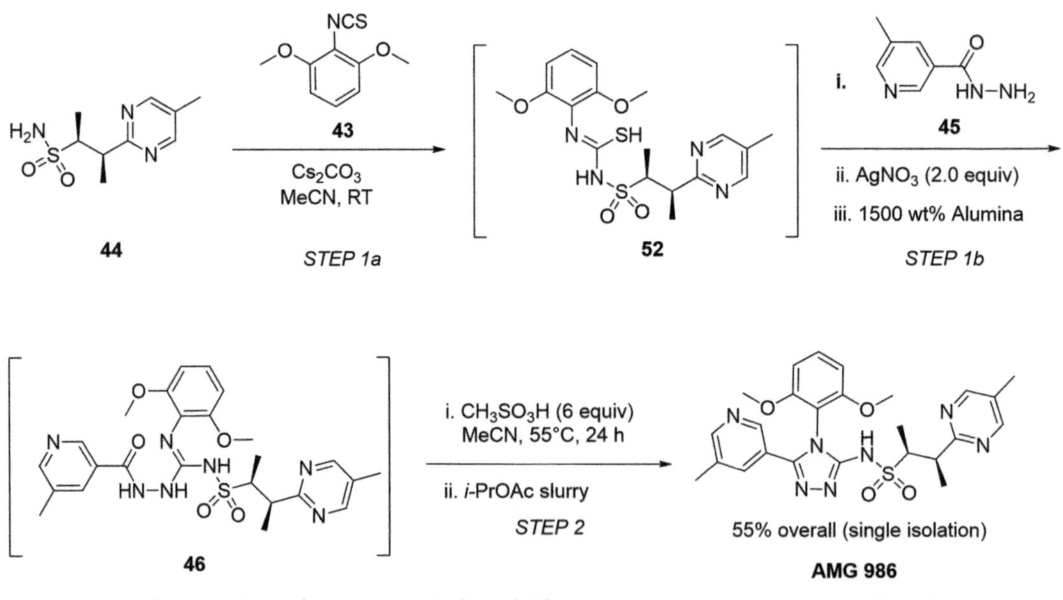

Scheme 8. Second generation Medicinal Chemistry convergent route to AMG 986.

The cyclization of crude penultimate **46** to the triazole was achieved using 6.0 equiv $MeSO_3H$ in MeCN. Following work up and chromatography, the isolated crude was slurried in *i*-PrOAc to further purify the crystalline product. This expedient route was effective for the delivery of material for final SAR activities as well as for in vivo studies. However, several issues needed to be addressed to enable scale-up for clinical supply. Reaction stalling became more pronounced at gram scale for both steps 1a and 2, requiring kicker charges of base and acid, respectively, to achieve high conversions. Removal of Ag salts from the step 1b reaction mixture presented a challenge for scale-up as the most effective method involved adsorption onto alumina (neutral, Brockmann I) charged directly to the reactor with overnight stirring. Glass reactors were prone to staining with silver deposits which would require additional cleaning measures with fixed glass reactors.

Additionally, crude penultimate **46** was processed forward into the step 2 cyclization without any purification. This carried over impurities into the final step and risked contaminating the final product. The cyclization step was also sluggish and required a large excess of $MeSO_3H$. Attempts to enhance the reaction rate using a higher reaction temperature led to poor purity crude while also potentially generating genotoxic acetamide under these reaction conditions. A final *i*-PrOAc slurry was utilized to assist with impurity purging of the crystalline product. This was repeated twice to meet the final purity target for small scale deliveries.

Most critical for PD was to replace column chromatography operations with crystallizations to improve process robustness and enable consistent delivery of product meeting our purity specifications. An added isolation point at Step 1b along with a final product polish filtration and crystallization would serve to meet this objective. Once isolation conditions were finalized, we expected a reproducible crystalline form and particle size distribution.

Beginning development with the Step 1a thiourea intermediate formation, a limited screen of alternative base/solvent combinations was carried out to address conversion issues. For initial development, equimolar amounts of sulfonamide and isothiocyanate were used. Replacement of Cs_2CO_3 with the cheaper K_2CO_3 gave unsatisfactory results in either 2-MeTHF, acetonitrile, or NMP. Reaction conversions ranged from 40-60% at 55 °C after 16 h while implementation of higher temperatures led to increased impurity content. Potassium bis(trimethylsilyl)amide (KHMDS) was ineffective as a base in this case, giving less than 1% desired product while KO*t*Bu, DBU, and K_3PO_4 also performed sluggishly. NaOH in anhydrous powdered form using NMP as solvent encouragingly gave full conversion to product at 60 °C. However, the crude purity of the thiourea intermediate was significantly lower than that of the existing conditions.

With limited time for additional screening, Cs_2CO_3 was chosen as the base. NMP was selected as a better solvent over MeCN as it provided an improved purity profile and eliminated stalling issues, presumably due to the higher solubility of Cs_2CO_3 in NMP than in MeCN. The reaction performed well in as little as six volumes of NMP with 1.3 equiv of Cs_2CO_3; however, higher concentrations led to inefficient stirring issues on scale. The thickness of the slurry increased during the reaction to the point where effective mixing was lost. This situation was first observed in a 2 L reactor in four volumes of NMP wherein poor mixing led to low conversions. The reaction was rescued by the addition of another two volumes of NMP which improved mixing and resulted in a complete reaction.

Direct crystallization of the thiourea from NMP upon water addition afforded a poorly filtered solid, with a poor purity profile. Alternatively, a water quench followed by neutralization and extraction with an appropriate solvent followed by crystallization would add significant processing steps and waste streams. In this case, a telescoped process in which the thiourea was taken directly into the next step (step 1b) was deemed the best approach.

Moving into step 1b, the top concern was to find a suitable replacement for $AgNO_3$ given the difficulty of its removal from the process stream highlighted earlier and the cost of a silver reagent. We soon discovered that carbodiimides such as EDCI can facilitate the reaction under equally mild conditions. Carbodiimides have been reported as effective thiophiles in the synthesis of cyanoguanidines from thioureas (*22*). Addition of hydrazide **45** (1.0 equiv) to the step 1a crude reaction mixture followed by the addition of solid EDCI (1.2 equiv) yielded the penultimate **46** in nearly quantitative conversion after 1 h. It is noteworthy that this order of addition resulted in significantly higher crude purity of penultimate **46** than in experiments where EDCI was added first followed shortly thereafter with hydrazide **45**. The EDCI addition is exothermic and generates CO_2

due to the freebasing of the EDCI •HCl with excess Cs_2CO_3 from the first step. RC-1 calorimetry data was collected revealing a heat of reaction upon EDCI addition of -58.21 kJ/mol with an associated adiabatic temperature rise of only 12 °C. This addition was therefore not a safety concern as it could also be also carried out at 15 °C in a dose-controlled manner so as to control the evolution of CO_2.

Stoichiometry considerations of the starting fragments **44**, **43**, and **45** were initially aimed at limiting the use of sulfonamide **44**. First, hydrazide **45** was flagged as a potential genotoxic impurity in an in silico model and the use of excess reagent presented additional risk of impurity in the API. Purging to ppm level in the final product would be mandated. Second, use of excess isothiocyanate **43** resulted in the intermediate thiourea product reacting with a second isothiocyanate molecule to generate a bis adduct impurity **53** (Scheme 9).

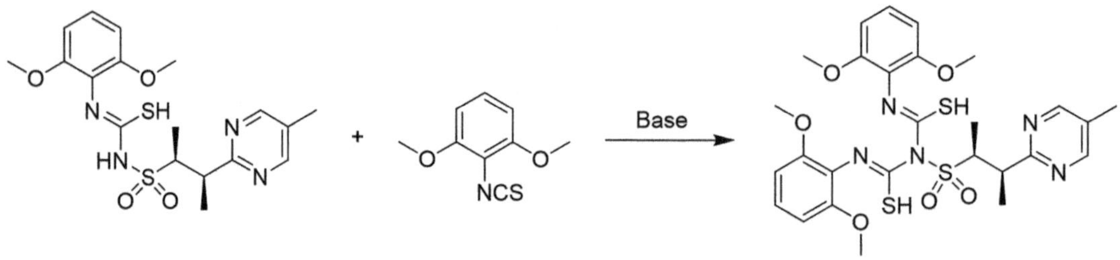

52 **43** **53**

*Scheme 9. Observed Impurity **53** arising from excess isothiocyanate **43** in step 1a.*

Therefore, despite the desire to fully convert all of sulfonamide **44** due to its relatively high cost of manufacture, a 1:1:1 ratio of all three starting materials was implemented. Each of the three starting materials was well purged in the isolations, and an analytical testing method was developed and implemented for GMP scale-up to be able to detect ppm levels of hydrazide in the final API. Analysis of development lots using this method showed levels of hydrazide controlled under 13 ppm.

Isolation of penultimate **46** was best achieved through development of a direct crystallization from the reaction mixture in NMP. This eliminated the need for extra operations, such as a second solvent to extract the product and then to identify a suitable antisolvent for crystallization or a solvent swap into a new solvent for crystallization. This approach also had the advantage of increasing throughput by reducing the maximum volume of an extractive workup.

Initially, dilute aqueous phosphoric acid was charged to neutralize the unreacted Cs_2CO_3 in the mixture thus creating a metastable solution that upon seeding with compound **46** induced crystallization. Further controlled addition of water lowered the solubility of **46** down to < 3 mg/mL. The EDC thiourea byproduct, inorganic salts, as well as product-related organic impurities were all efficiently purged using this crystallization. Optimal product recovery was achieved by targeting a pH of 5 (3.0 equiv H_3PO_4) during the quench phase and following the water addition with a half saturated (12.5%) brine solution. The product slurry filtered well on a standard Nutsche type filter and the filter cake was washed with water to clear residual salts, EDCI thiourea, and NMP. The process was successfully scaled for the first time using 3.1 kg sulfonamide **44** in a 250 L glass-lined reactor. Analytical results from this first scale-up are highlighted in Table 10.

Step 2 of the GMP campaign included the synthesis of AMG 986 through an intramolecular cyclization of the penultimate Step 1 intermediate **46** to the substituted triazole ring portion of the molecule (Scheme 10). The initial procedure developed in Medicinal Chemistry studies included a telescoped procedure whereby crude **46** was cyclized under acidic conditions using methanesulfonic acid in acetonitrile solvent at 55 °C for 24 h.

Table 10. Initial GMP Scaleup of the AMG 986 Steps 1a/b (3.1 kg Input)

Test Parameter	*Result*
Step 1 in-process test for conversion	2.8% area isothiocyanate **43** (pass)
Step 1 crude rxn product	91.25% area penultimate **46**
Mother liquor product losses	2.7 mg/g; (3.6% product loss)
Isolated product purity	99.0% area; 96.8 wt%; 95.4% ee
Isolated yield	84.2%

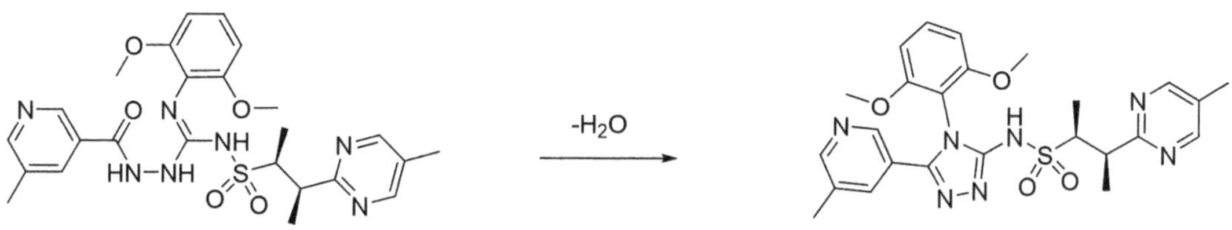

*Scheme 10. Dehydrative ring closure of penultimate **46** to AMG 986.*

These conditions provided unacceptable impurity profiles and required careful control of reaction temperature for satisfactory conversion as discussed previously. Typically, HPLC analysis of reactions on a >100 g scale revealed modest product purity of approximately 77 HPLC area % after heating for 24 h at 55 °C. We realized isolated yields of 60-70% following neutralization with aqueous base and purification by slurrying in *i*-PrOAc. Additionally, the formation of genotoxic impurities such as alkyl mesylates and acetamide were of concern for the production of clinical material.

Given these results, initial efforts for process development for step 2 of the GMP campaign began with the screening of alternative acid and solvent combinations for the acid-promoted intramolecular cyclization. Aqueous solutions of HCl, H_2SO_4, or AcOH in conjunction with organic solvents were evaluated at varying temperatures necessary for reaction progression. Under these conditions, however, varying amounts of degradation through elimination to the 2-butenyl pyrimidine fragment of the starting material **46** prior to cyclization occurred (as seen by LC/MS) generating impurity **54** (Figure 4).

Since we were unable to identify suitable acid-mediated cyclization conditions to convert **46** to AMG 986, we investigated alternative reaction processes. The preparation of triazoles via this dehydrative cyclization procedure was not well precedented in the literature. However, a detailed report by Dolzhenko et al. described an analogous intramolecular cyclization reaction to form triazoles in water at elevated temperatures using structurally simple substrates derived from substituted benzohydrazides and free guanidine (23–25). The prospect of using water as a reaction solvent, aside from the obvious advantage of non-toxicity, was appealing for the potential of controlling the solubility of the final API through pH adjustment in the final crystallization and isolation. Solubility studies determined that AMG 986 had a low solubility in water at neutral pH. However, given the sulfonamide hydrogen contained in the structure has a pKa of 6.29, both the final product and precursor **46** were highly soluble in basic aqueous solutions where they were fully deprotonated.

With this in mind, we sought reaction conditions using an aqueous base with an appropriate co-solvent. Test reactions with aqueous NaOH solutions (1.1 equiv base) as a 1:1 v/v mixture with alcoholic solvents including *n*-PrOH, EtOH and *i*-PrOH demonstrated a clean conversion to the desired product upon heating over 18 h at 80 °C with no detected degradation byproducts. With successful reaction conditions identified, solubility studies of the final API were carried out in each of the alcohol solvents used to determine the ideal solvent for the final crystallization. *i*-PrOH was the best suited of the solvents studied as AMG 986 demonstrated the lowest solubility at 1.26 mg/mL at ambient temperature, (Table 11), and was chosen for the final reaction and crystallization development.

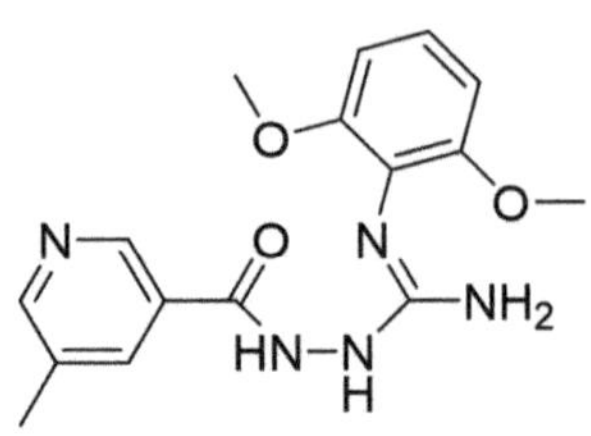

54

*Figure 4. Major impurity **54** observed in step 2 under acid conditions at elevated temperatures.*

Table 11. Solubility of AMG 986 in Selected Solvents

Solvent	*Solubility (mg/mL) at Ambient Temperature*
***n*-PrOH**	2.7
***i*-PrOH**	1.26
EtOH with 5% *i*-PrOH	6.5
MeCN	9.04
MeOH	19.3

Final reaction conditions involved adding 1.2 equivalents of a 5 N aqueous solution of NaOH to a final 1:1 v/v mixture of IPA/water (10 volumes total). Full conversion of **46** to AMG 986 was achieved upon heating for 20 h at 80 °C. By comparison with the prior acid-mediated process that afforded reaction profiles of only 77% purity on scale, reactions on 100 g scale using these new basic aqueous conditions allowed reaction purities of 97 HPLC area %.

With a scalable process identified for complete conversion to the desired API, efforts next focused on developing the pH-driven crystallization for final API isolation. Solubility studies were performed on varying volume ratios of *i*-PrOH and water at neutral pH to identify the optimum solvent ratio for the final stage of crystallization (Table 12). These studies determined a 1:3 ratio of *i*-PrOH/water provided the targeted range of desired supernatant concentration of <3 mg/mL.

With the desired final solvent ratio identified, studies were carried out to identify an appropriate aqueous acid for neutralization of the crude reaction that contained basic *i*-PrOH/water. As the final API AMG 986 contained both the acidic sulfonamide functionality as well as a pyridyl ring with a conjugate acid pKa of 3, a final pH of 5.5 was targeted for isolating AMG 986 as this pH value insured both the sulfonamide functionality remained protonated and the pyridyl ring remained as the free base.

Due to the low solubility of phosphate salts in *i*-PrOH/ water mixtures (as observed following phosphoric acid charges in development studies), a 2 M aqueous solution of citric acid was chosen for both the buffering capacity as well as solubility across the range of *i*-PrOH/water mixtures with this acid. The stepwise pH-driven crystallization was designed to ensure maximum impurity rejection and minimize any entrapment of sodium salts of AMG 986 before complete neutralization.

Table 12. Solubility of AMG 986 in *i*-PrOH Water Mixtures

i-PrOH/Water Ratio (V/V)	*Solubility of AMG 986 (mg/mL) at Ambient Temperature*
1:1	8.87
1:2	4.38
1:3	1.92

Thus, following reaction completion at 80 °C, the batch was cooled to 40 °C whereby the initial charge (0.35 equiv) of 2 M aqueous citric acid was added over a short period resulting in a neutral pH reaction solution. At this point, the API in solution was in the metastable region and a charge of seed of the final API at ~ 2 wt% was added as a water slurry, thus preventing any self-seeding and uncontrolled crystallization that may occur on standing. Following initial crystallization, the seed bed was aged at pH 7 for >2.5 h to propagate further crystallization of the final API. A second charge of 2 M citric acid (0.35 equiv) was added over 1 h while maintaining 40 °C to afford the final target pH of 5.5. Final water addition (6.5 volumes) was carried out over 1.5 h to reach the optimal *i*-PrOH-water ratio as discussed previously. Finally, the slurry was cooled slowly over >2 h to achieve the targeted <3.0 mg/mL supernatant concentration. The final product was filtered and washed with water to provide the final API in desired purity and form. As in step 1, the process was successfully scaled using a 250 L reactor at Amgen's GMP pilot plant using 6.1 kg penultimate **46**. Table 13 highlights the analytical results from this first scaleup.

Table 13. GMP Scaleup Batch 1, Step 2, Final Cyclization and Isolation of AMG 986 (6.1 kg Input)

Test Parameter	*Result*
Step 2 in-process test for conversion	0.83% area penultimate **46** (Target < 2%)
Mother liquor product losses	2.2 mg/mL; (3.3% product loss)
Isolated product purity	99.7% HPLC area%; 100 wt%; 96.0% ee
Isolated yield	89.1%

The overall sequence starting from 3.3 kg of sulfonamide **44**, delivered 5.4 kg of AMG 986 in 99.7% HPLC area % purity and 77% overall yield (Scheme 11) which was utilized for both GLP toxicology studies and FIH clinical trials.

Process Development optimization resulted in a 22% yield improvement from the original Discovery conditions while also improving impurity rejection and control through the newly implemented crystallizations of the penultimate **46** and final API. With a successfully demonstrated process in hand, future batches could be made by synthesizing the three regulatory starting materials

43, **44**, and **45** in parallel followed by assembly to AMG 986 in two weeks running a single shift in the plant. In preparation of those deliveries, further refinement of the GMP steps continued for reload campaigns supporting both clinical and drug product formulation development needs.

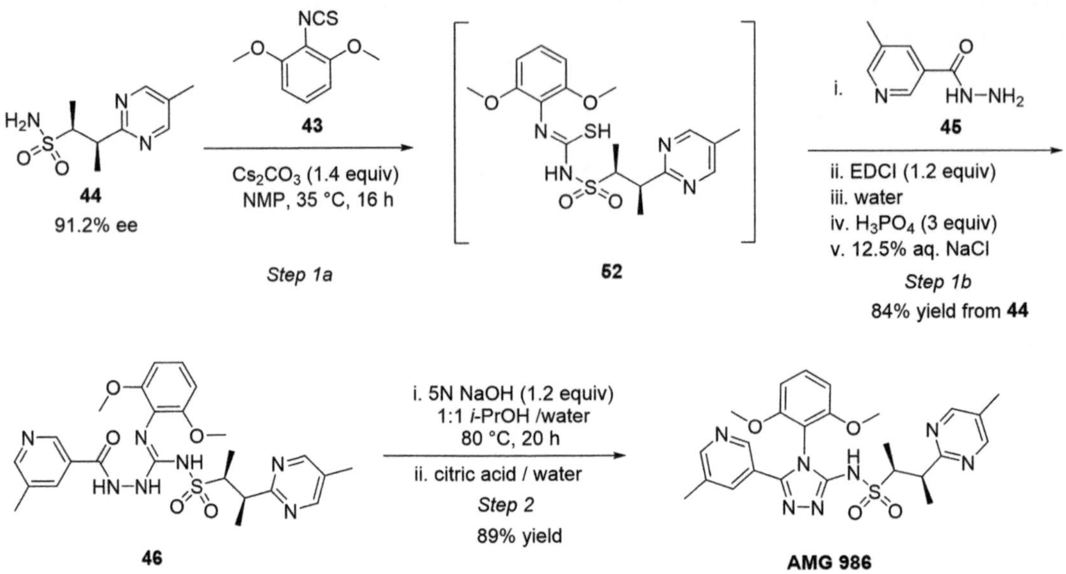

Scheme 11. The synthesis of 5.4 kg of AMG 986 for FIH clinical trials.

The primary focus was to further control the enantiomeric purity and the level of residual rhodium in the final drug substance as well as to continue to characterize the process through failure mode analysis. Tightening the specification of enantiomeric impurity **55** (Figure 5) to below 1% in the drug substance was planned now that the impurity was qualified in GLP tox. For the FIH scale-up, partial rejection of the undesired enantiomer introduced as the minor enantiomeric impurity of sulfonamide **44** (91.2% ee) was achieved through the downstream crystallizations resulting in an enantiopurity of 96.0% ee for the API.

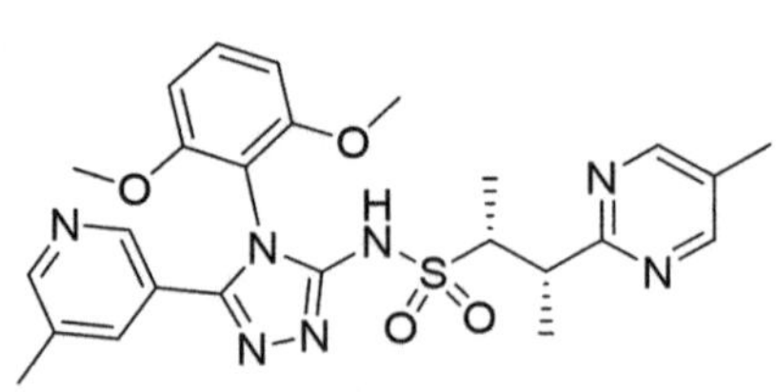

55

*Figure 5. Undesired enantiomer **55** of AMG 986.*

Fate and purge analysis utilizing sulfonamide **44** of lower enantiomeric purity (88.6% ee) helped gauge the necessary purity target going into the GMP sequence. A significant enantiomeric upgrade was observed at the penultimate **46** isolation with this material (94.2% ee) however, the step 2 isolation did not significantly reject the undesired enantiomer **55** further. Whereas sulfonamide of 96% ee was utilized in the initial scale up, additional work was required to improve the purity of sulfonamide **44** through improvements to the crystallization process of that material.

Stress studies on step 1a/b, focused on the effect of trace water on the reaction performance. Spiking with 1.0 equiv water was found to have a negative effect on the impurity profile and required controls were established for water content of starting materials, reagents and solvents. We envisioned that unintentional cyclization of penultimate **46** during step 1b reaction and workup was possible at longer hold times under basic conditions and lead to yield loss during crystallization. However, aging the crude step 1b product mixture for 16 h before the quench did not result in any detectable AMG 986. Likewise, slowing down the quench addition did not have negative impacts due to the slow rate of cyclization. Only when HPLC samples of the unquenched reaction mixture (prepared in 1:1 MeCN/ water) were re-analyzed after 24 h did small amounts of AMG 986 (1-2% area) form.

As highlighted earlier, a tight limit of residual metals was desired as a backup parenteral formulation was in consideration. A carbon treatment was therefore implemented at the final step to serve also as the polish filtration. This change also served to remove off-white color from AMG 986 which was derived from slightly colored precursor isothiocyanate **43** (tan to light brown appearance). A screen of commercially available activated charcoals identified Ecosorb C-947 (Graver Technologies) as the most efficient at removing residual rhodium. Darco-100 mesh was also effective and was implemented in its place as it was already on hand in needed quantity.

The minor changes to the GMP steps were implemented and scaled up in a 1500 L reactor starting with 16.0 kg of sulfonamide **44** to generate the largest single batch of AMG 986 manufactured: 29.7 kg in 81.6% overall yield and 99.6 HPLC area% purity as a white crystalline solid (Table 14). The enantiomeric impurity was reduced to only 0.28% owing to the optimization of the sulfonamide **44** recrystallization.

In total, over 50 kg of AMG 986 was manufactured using the refined process (Scheme 12).

Table 14. Final Analytical Results for 30 kg Scaleup

Step 1 Test Parameter	*Result*
In-process test for conversion	< 0.5% isothiocyanate **43** (pass)
Step 1 crude rxn product	87.3% penultimate **46** (HPLC)
Isolated product purity	99.4% area (HPLC); 99.2% enantiomeric purity; 100 wt%
Isolated yield	91.7%
Step 2 Test Parameter	*Result*
In-process test for conversion	0.7% area penultimate **46** (Target < 2%)
Mother liquor product losses	0.9 mg/mL; (% product loss)
Isolated product purity	99.6% area; 100 wt%; 98.4% ee
Isolated yield	89.4%

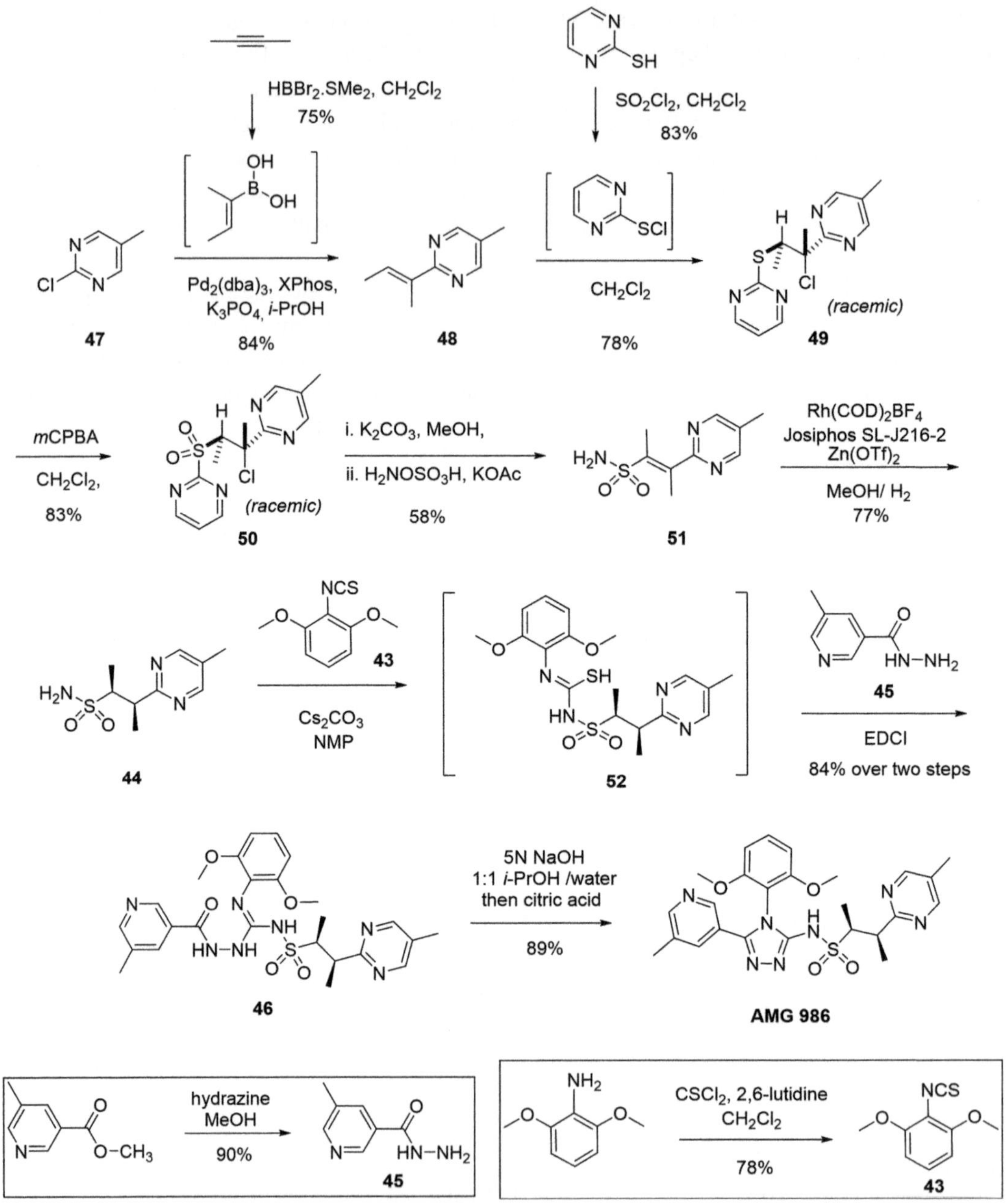

Scheme 12. Manufacturing process to AMG 986 from commercial raw materials.

Conclusions

The identification of AMG 986, a potent and efficacious agonist of APJ is described. From the initial HTS hit, the poor microsomal stability and unacceptable CYP3A4 induction were overcome by addressing the high lipophilicity and functionality prone to metabolism through modification of substituents surrounding the core triazole. The early Medicinal Chemistry route was replaced with a second-generation convergent approach well suited to Process Chemistry. The development of a scalable process towards AMG 986 was achieved under tight timelines owing to the collaborative effort by Discovery and Process Development to explore novel synthetic pathways prior to molecule selection. This investment resulted in a superior stereospecific approach for a key building block which allowed rapid scale-up to support downstream development and API supply.

Major accomplishments achieved from the development work include:

- Discovery and implementation of a novel asymmetric hydrogenation of a tetrasubstituted olefin achieved in 77% yield and 91% ee on multi-kilogram scale
- Replacement of $AgNO_3$ with EDCI to streamline isolation of the penultimate intermediate
- Removal of the use of methanesulfonic acid in MeCN to eliminate potential genotoxic impurity concerns in the final step
- Identification of a mild base-catalyzed ring closure to convert the penultimate intermediate **46** to AMG 986
- Development of two crystallization procedures to isolate and deliver AMG 986 in >99.5% purity
- 22% yield improvement in the last two steps of the synthesis

Having built a sound process for the initial multi-kilogram scale-up, the resupply campaigns were at a far less risk of manufacturing issues and costly delays. Ultimately, the AMG 986 program was discontinued in Phase 1, but the chemistry improvements accelerated the decision point as the timeline from discovery to clinical evaluation shortened.

Additionally, the project has served our organization as an example on how to rapidly advance new molecules into the clinic and be best positioned to deliver on ever accelerating timelines while simultaneously making positive strides towards efficiency and sustainability in API manufacture.

Acknowledgments

The authors wish to thank Simon Hedley, Richard Crockett and Athi Arunachalampillai for their many contributions to support the PD work. Amgen Inc. provided all research and development funding for the work described. All authors listed are either current or past employees of Amgen Inc.

References

1. Ponikowski, P.; AlHabib, K. F.; Cowie, M. R.; Force, T. L.; Hu, S.; Jaarsma, T.; Krum, H.; Rastogi, V.; Rohde, L. E.; Samal, U. C.; Shimokawa, H.; Siswanto, B. B.; Sliwa, K.; Filippatos, G. Heart failure: preventing disease and death worldwide. *ESC Heart Failure* **2014**, *1*, 4–25.
2. Benjamin, E. J.; Muntner, P.; Alonso, A.; Bittencourt, M. S.; Callaway, C. W.; Carson, A. P.; Chamberlain, A. M.; Chang, A. R.; Cheng, S.; Das, S. R.; Delling, F. N.; Djousse, L.; Elkind, M. S. V.; Ferguson, J. F.; Fornage, M.; Jordan, L. C.; Khan, S. S.; Kissela, B. M.; Knutson, K. L.; Kwan, T. W.; Lackland, D. T.; Lewis, T. T.; Lichtman, J. H.; Longenecker, C. T.; Loop, M. S.; Lutsey, P. L.; Martin, S. S.; Matsushita, K.; Moran, A. E.; Mussolino, M. E.; O'Flaherty, M.; Pandey, A.; Perak, A. M.; Rosamond, W. D.; Roth, G. A.; Sampson, U. K. A.; Satou, G. M.; Schroeder, E. B.; Shah, S. H.; Spartano, N. L.; Stokes, A.; Tirschwell, D. L.; Tsao, C. W.; Turakhia, M. P.; VanWagner, L. B.; Wilkins, J. T.; Wong, S. S.; Virani, S. S. Heart Disease and Stroke Statistics-2019 Update: A Report From the American Heart Association. *Circulation* **2019**, *139*, e56–e528.

3. Gomez-Soto, F. M.; Andrey, J. L.; Garcia-Egido, A. A.; Escobar, M. A.; Romero, S. P.; Garcia-Arjona, R.; Gutierrez, J.; Gomez, F. Incidence and mortality of heart failure: a community-based study. *Int. J. Cardiol.* **2011**, *151*, 40–45.
4. Tanai, E. Pathophysiology of Heart Failure. *Compr. Physiol.* **2016**, *6*, 187–214.
5. Metra, M.; Teerlink, J. R. Heart failure. *Lancet* **2017**, *390*, 1981–1995.
6. Ponikowski, P.; Voors, A. A.; Anker, S. D.; Bueno, H.; Cleland, J. G.; Coats, A. J.; Falk, V.; Gonzalez-Juanatey, J. R.; Harjola, V. P.; Jankowska, E. A.; Jessup, M.; Linde, C.; Nihoyannopoulos, P.; Parissis, J. T.; Pieske, B.; Riley, J. P.; Rosano, G. M.; Ruilope, L. M.; Ruschitzka, F.; Rutten, F. H.; van der Meer, P.; Authors/Task Force, M.; Document, R. 2016 ESC Guidelines for the diagnosis and treatment of acute and chronic heart failure: The Task Force for the diagnosis and treatment of acute and chronic heart failure of the European Society of Cardiology (ESC). Developed with the special contribution of the Heart Failure Association (HFA) of the ESC. *Eur. J. Heart Fail.* **2016**, *18*, 891–975.
7. Rogers, C.; Bush, N. Heart Failure: Pathophysiology, Diagnosis, Medical Treatment Guidelines, and Nursing Management. *Nurs. Clin. North Am.* **2015**, *50*, 787–799.
8. Barnes, G. Translational Promise of the pelin APJ system. *Heart* **2010**, *96*, 1011–1016.
9. Wysocka, M. B.; Pietraszek-Gremplewicz, K.; Nowak, D. The Role of Apelin in Cardiovascular Diseases, Obesity and Cancer. *Front. Physiol.* **2018**, *9*, 557–571.
10. Lu, L.; Wu, D.; Li, L.; Chen, L. Apelin/APJ system: A bifunctional target for cardiac hypertrophy. *Int. J. Cardiol.* **2017**, *230*, 164–170.
11. Japp, A. G.; Cruden, N. L.; Barnes, G.; van Gemeren, N.; Mathews, J.; Adamson, J.; Johnston, N. R.; Denvir, M. A.; Megson, I. L.; Flapan, A. D.; Newby, D. E. Acute cardiovascular effects of apelin in humans: potential role in patients with chronic heart failure. *Circulation* **2010**, *121*, 1818–1827.
12. Barnes, G. D.; Alam, S.; Carter, G.; Pedersen, C. M.; Lee, K. M.; Hubbard, T. J.; Veitch, S.; Jeong, H.; White, A.; Cruden, N. L.; Huson, L.; Japp, A. G.; Newby, D. E. Sustained cardiovascular actions of APJ agonism during renin-angiotensin system activation and in patients with heart failure. *Circ. Heart Fail.* **2013**, *6*, 482–491.
13. rat = Fu: 0.027; oral dose: 2 mg/ kg; oral AUC: 1.4 μM·h; Cl: 0.86 L/h/kg; Vdss: 0.27 μM; iv t1/2; 1.03 h; %F: 30, dog = Fu: 0.024; oral dose: 1 mg/ kg; oral AUC: 13.2 μM·h; Cl: 0.27 L/h/kg; Vdss: 0.49 μM; iv t1/2; 4.1 h; %F: 100. Administered at a dose of 0.5 mg/kg in 100% DMSO, iv for both rat and dog. 1.0 mg/kg in 100% DMSO, po for dog. 2.0 mg/kg in 0.5% methyl cellulose, 1% tween80, 98.5% water, iv for rat. Data are expressed as mean values (n = 3).
14. Berry, M. F.; Pirolli, T. J.; Jayasankar, V.; Burdick, J.; Morine, K. J.; Gardner, T. J.; Woo, Y. J. Apelin has in vivo inotropic effects on normal and failing hearts. *Circulation* **2004**, *110*, II187–II193.
15. Measured from echocardiography.
16. A maximal effect at a dose of 0.1 mg/kg was observed.
17. Ason, B.; Chen, Y.; Guo, Q.; Hoagland, K. M.; Chui, R. W.; Fielden, M.; Sutherland, W.; Chen, R.; Zhang, Y.; Mihardja, S.; Ma, X.; Li, X.; Sun, Y.; Liu, D.; Nguyen, K.; Wang, J.; Li, N.; Rajamani, S.; Qu, Y.; Gao, B.; Boden, A.; Chintalgattu, V.; Turk, J. R.; Chan, J.; Hu, -M.; Swaminath, G.; Khakoo, A. Y. Cardiovascular response to small-molecule APJ activation. *JCI Insight* **2020**, *5*, e132898.

18. predicted human PK of CL = 0.03 L/hr/kg, Vss: 0.49 L/kg,(16) T1/2: 11 hr, K01: 5.5 hr-1, F: 85%. Ranges: Cl = 0.015 - 0.06 L/h/kg, Vss = 0.34 - 0.64 L/kg.
19. Johnson, M. G.; Gribble, M. W.; Houze, J. B.; Paras, N. A. *Org. Lett.* **2014**, *16*, 6248–6251.
20. Zhou, T.; Peters, B.; Maldonado, M. F.; Govender, T.; Andersson, P. G. *J. Am. Chem. Soc.* **2012**, *134*, 13592–13595.
21. Yuasa, Y.; Yuasa, Y.; Tsuruta, H. *Can. J. Chem.* **1998**, *76*, 1304–1307.
22. Atwal, K. S.; Ahmed, S. Z.; O'Reilly, B. C. *Tetrahedron Lett.* **1989**, *30*, 7313–7316.
23. Dolzhenko, A. V.; Pastorin, G.; Dolzhenko, A. V.; Chui, W. K. *Tetrahedron Lett.* **2009**, *50*, 2124–2128.
24. A.V. Dolzhenko, A. V.; Pastorin, G.; Dolzhenko, A. V.; Chui, W. K *Tetrahedron Lett.* **2008**, *49*, 7180–7183.
25. Dolzhenko, A. V.; Dolzhenko, A. V.; Chui, W. K. *Tetrahedron* **2007**, *63*, 12888–12895.

Chapter 5

Melflufen: A Journey from Discovery to Multi-Kilogram Production

Fredrik Lehmann[1] and Johan Wennerberg[*,2]

[1]Oncopeptides AB, Västra Trädgårdsgatan 15, SE-111 53 Stockholm, Sweden
[2]Dupont Chemoswed, Celsiusgatan 35, SE-212 14 Malmö, Sweden
[*]Email: johan.wennerberg@redglead.com

Ethyl (2*S*)-2-[(2*S*)-2-amino-3-[*bis*-(2-chloroethyl)amino]phenyl]propaneamido]-3-(4-fluorophenyl)propanoate hydrochloride, (melphalan flufenamide or Melflufen), is an alkylating agent intended for the treatment of multiple myeloma. Initially only milligram quantities were synthesized, following a route starting from pharmaceutical-grade melphalan. Along with the pharmaceutical development, adjustments were made to the original medicinal chemistry route. This resulted in material for early clinical trials, but it became obvious that further development was necessary. Development resulted in a route in which two phenyl alanine derivatives were coupled to give a dipeptide. This intermediate was further manipulated to give an aniline which could be converted into the desired compound melflufen. The aniline derivative was converted to the corresponding *N,N*-*bis*-chloroethylaniline using chloroacetic acid and borane. Deprotection and conversion to the hydrochloride gave melflufen in good yield and excellent purity. Production was performed without chromatography at multi-kilogram scale to supply the API for Phase III studies and commercial validation batches.

Introduction

Treatment of cancer with chemotherapy can be traced back to the use of chemical warfare agents during World War I. One of the devastating agents used was mustard gas. Despite the ban of chemical warfare by the 1925 Geneva Protocol, a possible re-introduction in World War II was feared. Because of this, research both for developing protective measures and discovery of even more efficient agents was undertaken. During this period, it was discovered that nitrogen mustards, which are more stable analogues than sulfur mustards, ceased the division of some types of normally fast-growing somatic cells. This property suggested that such substances might be able to suppress certain types of cancer. Indeed, it was revealed that several nitrogen mustards could exert cytotoxic activity on different tissues, and this could be a potential means for controlling cancers.

In 1942, Gilman and co-workers at Yale discovered that *bis*(2-chloromethyl) methylamine, or chlormethine, (Figure 1) had antitumor effects in animal models (*1*). The same year clinical trials were initiated with promising results and some cancers demonstrated regression after treatment. Considerable excitement was generated when extended clinical studies on patients with terminal malignant diseases produced successful outcomes. All findings were classified because of the war and were not revealed for almost two decades when they were reviewed by one of the discoverers (*2*). Interesting to note is that similar clinical trials with nitrogen mustards were conducted in the UK during wartime but for similar reasons, the results were not published until after the war (*3*).

Eventually chlormethine became the first substance to be used for chemotherapy. This was followed by a continuous stream of cancer-therapy candidate alkylating agents (Figure 1) based on nitrogen mustards, many of which are still in clinical use today. Noteworthy of comment, there is a minimal change in the clinical spectrum, (which cancer forms that can be treated), compared to the original drug chloromethine, however the toxicity is different (*4*).

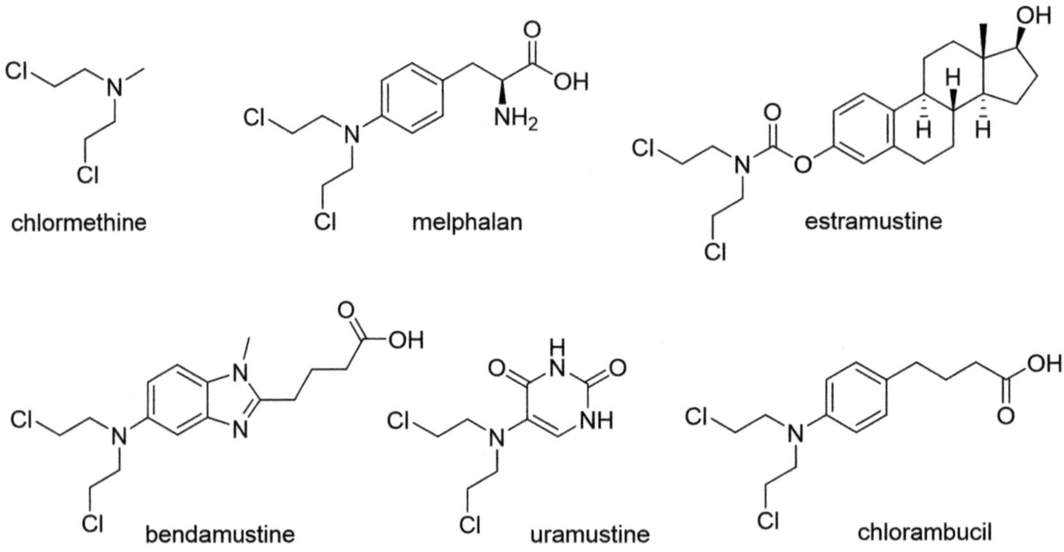

Figure 1. Clinically useful nitrogen mustard derivatives.

Nitrogen alkylating agents are strong electrophiles which react with nucleophilic nitrogen atoms in DNA bases, such as guanine, and form covalent bonds (*5*). In excess, this produces a toxic reaction. Since nitrogen mustards have two alkylating groups, they can react with guanine on each strand and cross-link the DNA helix (Figure 2). As a result, the crosslinking will cause a disruption of replication and transcription of DNA.

It is also possible that two guanine molecules on the same strand are cross-linked. This will mask that portion of the DNA chain and prevent enzymes necessary for DNA function from obtaining access. Unfortunately, nitrogen mustards may react with other nucleophilic nitrogen atoms such as those in proteins. Despite the lack of selectivity, this class of drugs is useful for the treatment of cancer. Many cancer cells divide much faster than normal cells and are more sensitive to disruption of DNA function. It should be noted that nitrogen mustards themselves may be harmful and give rise to secondary malignancies because healthy cells are also affected.

The parent drug chlormethine is too reactive for oral use because it reacts with water in the stomach and gut and must therefore be administered intravenously. The high reactivity of these agents can be reduced by modifications of the molecule. In several nitrogen mustard derivatives, an aromatic ring is placed adjacent to the nitrogen atom (Figure 1). The lone-pair of electrons on the nitrogen will interact with the aromatic ring π-electrons and will thus be less prone to displace chloride ion to form a transient aziridinium ring The intermediate aziridinium ion is more difficult to form and only strong nucleophiles like guanine will be able to react with the nitrogen mustard derivative (Figure 2). Such modifications render the drug more stable and consequently some of such drugs can be administrated orally.

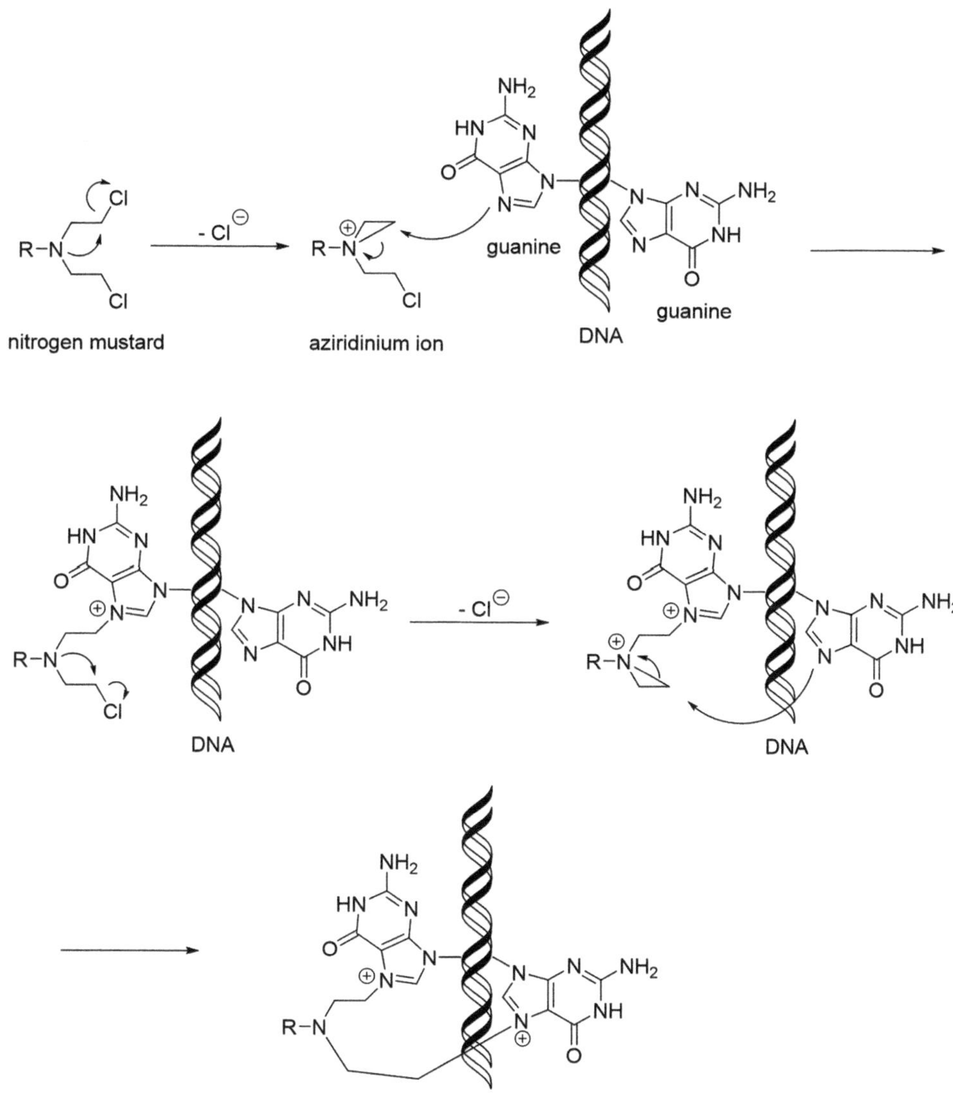

Figure 2. Nitrogen-based alkylating agents are highly reactive molecules that form aziridinium ion intermediate when one of the chlorides is intramolecularly displaced by the adjacent nitrogen. The intermediate aziridinium ion then alkylates e. g. the N-7-nitrogen of a guanine base of a DNA-strand. A similar reaction with the N-7-nitrogen in another guanine on the second DNA-strand will create a covalent cross-linking between the two DNA chains.

Despite of the possible side effects of the nitrogen based alkylating agents, they are still widely used for the treatment of a wide range of malignancies. This is what encouraged us to develop novel drugs based on alkylating compounds. A recent example is the development of melphalan flufenamide or melflufen (**1a**) by the Swedish pharma company Oncopeptides. Melflufen has several advantages compared to other nitrogen-based cytotoxic drugs (*6*). In this chapter we present the discovery and development of this anti-cancer agent. Melflufen has a modified reactivity due to the aromatic ring, but also a unique structure that results in a much higher selectivity towards cancer cells.

Discovery of Melflufen

In the late 1960s in Italy, hundreds of peptides based on sarcolysin [racemic form of melphalan (Figure 1) were evaluated for their anti-tumor activities. Out of these peptides, six analogues with anticancer activity were combined into a cocktail called peptichemio (*7*). Peptichemio was in essence a mixture of six different nitrogen-alkylating di- and tri-peptides. This mixture was shown to have good effect in the treatment of various malignant diseases, such as neuroblastoma (*8*). Despite being highly efficacious, the development of peptichemio was halted in 1979 due to refocusing of the Istituto Medico Milanese, and the mixture eventually ceased to be available. Around this point in time, other alkylating agents such as melphalan and bendamustin started to be used in its place (*9, 10*).

In the late 1990s, researchers in Sweden decided to revisit the original peptichemio mixture and undertook a systematic investigation of the peptides. They initiated their research by making small amounts of almost fifty well-defined alkylating di- and tri-peptides and tested them for their anti-tumor activities *in vitro* and *ex vivo* (see Figure 3 and Table 1 for selected examples).

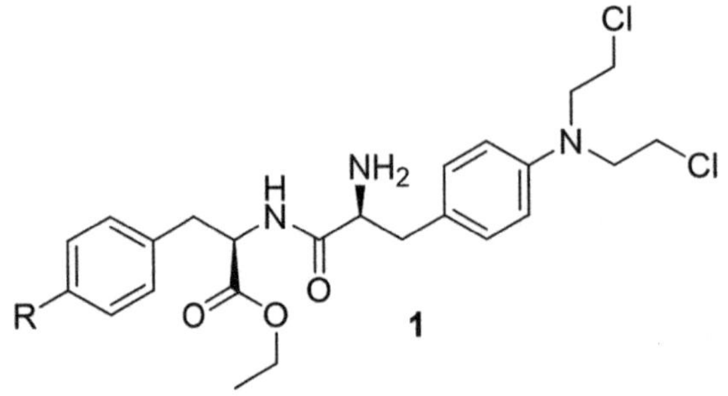

1

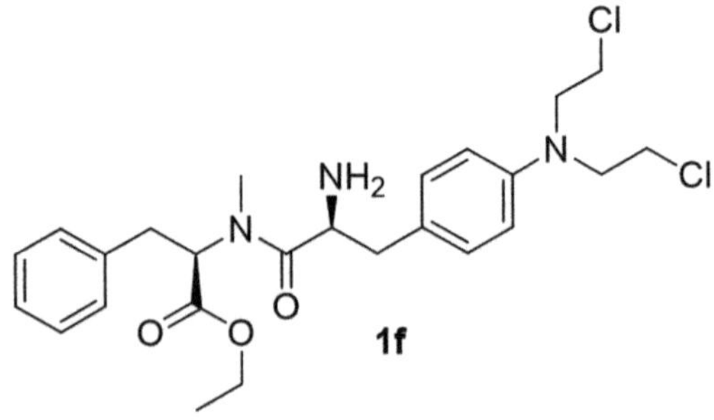

1f

*Figure 3. Melflufen analogues and N-methylmelflufen **1f**.*

Somewhat surprisingly, the activity did not differ significantly between analogues within the phenylalanine substitutions (Table 1, entry 2-6). This is explained in that hydrolytic activity of peptidases and esterases determined the activity of analogues, rather than the substitution pattern of the aromatic ring. Although the pharmacological activity was very similar for the analogues, melflufen (R = F) was chosen as the candidate drug. Since it contained a 4-F substituent, it was thought to have a slower and more benign metabolic profile as it blocks the potential para-CYP oxidation.This is common to phenyl alanine derivatives compared to the other substituents that were tried.

Table 1. Data for Selected Melflufen Analogues

Entry	*Compound*	*R*	*Cell lines (log IC_{50} uM)*		
			RPMI 8226	*CCRF-CEM*	*NCI-H69*
1	Melphalan	NA	10.0	0.92	8.9
2	Melflufen (**1a**)	F	1.0	0.050	0.064
3	**1b**	OH	1.1	0.048	0.078
4	**1c**	OCH_3	0.65	0.035	0.035
5	**1d**	NH_2	0.90	0.060	0.090
6	**1e**	NO_2	1.1	0.052	0.058
7	*N*-methylmelflufen **1f**		25	0.85	30

(NA: not applicable).

Among the findings reported was that the activity of the peptides could be correlated to lipophilicity, with an optimum point around logP 4-5. Due to its high lipophilicity, melflufen rapidly enters cells and because of the peptide and ester bonds within the molecule that are recognized by metabolic enzymes, (peptidases and esterases), it forms hydrophilic metabolites that are much less able to penetrate the cell membrane. They are thereby trapped within the cells (*11*). This results in a potency increase (melflufen versus other alkylators) ranging from 10- to several 100-fold in different patient derived in vitro models where the IC50 clearly shows the benefit of more lipophilic di-peptides compared to melphalan. (Table 1, entry 1-6).

The importance of metabolic enzymes was further validated by incubating the cells with either the peptidase inhibitor bestatin or the esterase inhibitor ebelactone before exposing the cells to melflufen (*12*). In this case, the potency of melflufen decreased by a factor 10-50 (depending on the cell-line) further validating the importance of the metabolic enzymes. It is important to note that melflufen did not become inactive when the metabolic routes were blocked, it just became less active.

This enzyme dependence was even further validated by synthesizing the *N*-methyl amide analogue **1f**, which cannot undergo enzymatic hydrolysis, because the *N*-methyl group blocks enzymatic hydrolysis (Figure 3). Once again, this molecule was indeed active, with IC50:s comparable to melphalan but much less potent than melflufen itself.

It was found that melflufen triggered rapid and irreversible DNA damage which may account for its ability to overcome alkylator-resistance in various malignant cells (*13*, *14*). The selectivity of melfufen stems in part from the fact that peptidases are overexpressed in certain types of cancer (*15*). The process is depicted in Figure 4 for peptidases, but is also relevant for esterases (*16–18*)

Melflufen has been evaluated in several clinical trials (*19*), and in 2017 the first study in multiple myeloma was reported with very good results compared to other drugs in development. Most striking was the long overall survival seen for the melflufen patients (*20*). The sample size was the smallest amongst comparable trials, but overall response rate (ORR), progression free survival (PFS) and overall survival (OS) were superior for melflufen (ORR 31%, PFS: 5.7 months, OS: 20.7 months). This was compared to corresponding studies in similar patient populations for the thalidomide derivative pomalidomide (ORR: 23%, PFS: 3.6 months, OS: 12.4 months) (*21*), proteasome inhibitor carfilzomib (ORR: 24%, PFS: 3.7 months, OS: 15.6 months) (*22*) and CD38 monoclonal antibody daraturamab (ORR: 29%, PFS: 3.7 months, OS: 17.5 months) (*23*). This encouraging data prompted further development for melflufen and it is currently being evaluated in a handful of clinical Phase II and Phase III trials. Currently (summer 2020), melflufen is under review by the FDA for accelerated approval due to favorable Phase II data. It is awaiting complete Phase III data to support a final approval.

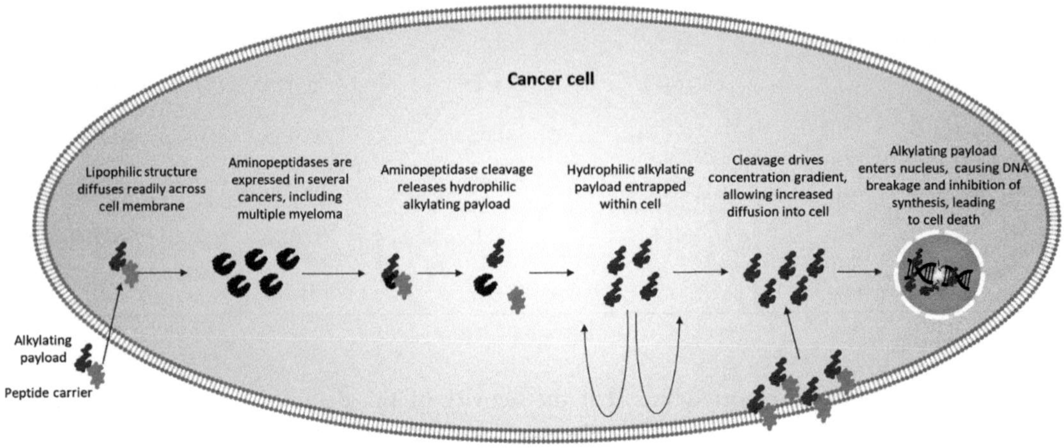

Figure 4. Mechanism of action for melflufen.

Synthetic Development

The medicinal chemistry route for the synthesis of melflufen (**1a**) is depicted in Scheme 1 (*11*). The α-amino group of pharmaceutical-grade L-melphalan was protected with *tert*-butyl dicarbonate to produce carbamate **2** in 48% yield. EDC-mediated coupling with 4-fluoro-L-phenylalanine ethyl ester generated Boc-protected melflufen **3** in 43% yield after chromatography. Deprotection with HCl in ethyl acetate and salt formation yielded melflufen in 70% yield. This route was used as a general method for the preparation of melflufen and its analogues (Table 1). It was expedient for this purpose since it was a straightforward procedure and allowed the synthesis of analogues for SAR studies.

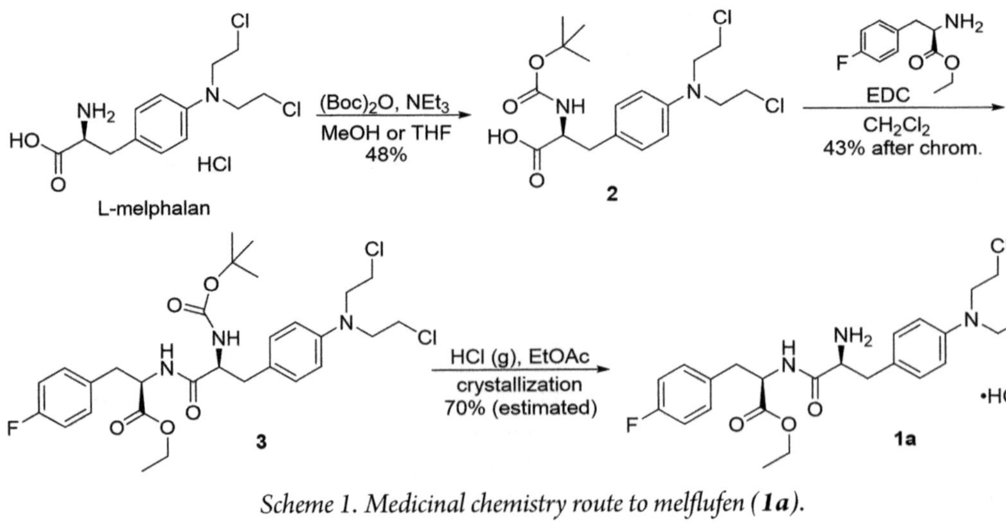

*Scheme 1. Medicinal chemistry route to melflufen (**1a**).*

Minor Improvements

The medicinal chemistry route summarized above was only used for synthesis of melflufen at milligram scale. To supply preclinical development and early clinical trials, a temporary solution was to make minor modifications to the present route. Among them was replacing the coupling reagent and omitting chromatography for purification after Boc-protection.

By these modifications, it was possible to produce a few kilograms for continued work, but the protocol was still far from optimal for large-scale synthesis. In addition, it turned out to be difficult to source the starting material melphalan of acceptable purity. We discovered that pharma grade melphalan assay is typically $\leq$ 93%. Additional research was required to derive a multi-kilogram preparation.

Route Scouting

The Swedish company DuPont Chemoswed, (now Magle Chemoswed) was given the task to produce melflufen for continued clinical development and it was soon clear that a different synthetic strategy was required for scale-up of the process. The existing route to synthesize melflufen from melphalan was an obvious choice. The route was short and gave the desired material in an acceptable purity and yield for the purpose: preclinical studies. It also allowed the preparation of analogues with different substitution patterns on the mono-substituted benzene ring. Unfortunately, the existing route to melflufen suffered from a number of practical constraints. Purification of melphalan prior to use was required via two chromatographies which created toxic waste streams.

Even after some modifications, the existing route was still not acceptable for larger scale demands. The starting material, melphalan, the product melflufen, and intermediates **2** and **3,** indeed all compounds of the route, carry the nitrogen mustard pharmacophore (Scheme 1) that rendered sophisticated personal protective equipment (PPE) mandatory. These substances are highly potent, with a threshold level value (TLV) of 1.0 $\mu g/m^3$ for melphalan and 1.6 $\mu g/m^3$ for melflufen. The consequence of this was that all steps would need to be performed under containment and that large volumes of hazardous waste would be created.

Furthermore, cleaning the facilities and procurement of specialized equipment would be challenging. The use of chromatographic purifications and a low overall yield have added to the difficult situation. Finally, there was the fact that it was difficult to source pharma grade melphalan of acceptable purity. This led us to a new approach involving a late introduction of the *N,N*-bis-chloroethylamine moiety into the core structure.

One argument for a novel route was to avoid containment during all steps of the synthetic route. However, containment was necessary in the final two steps since it is impossible to get below TLV 1.6 $\mu g/m^3$ for melflufen in air with open handling. The manufacturing company, DuPont Chemoswed,had access to a combined stainless-steel reactor/filter dryer of Nutrex type from Rosenmund. This reactor can be used for chemical reactions, crystallization, filtration and drying. The product is withdrawn from the reactor into special containers via a butterfly valve and samples are taken via a second valve. The reactor is furthermore connected to two other reactors which enables the possibility to perform multi-step reactions under containment (Figure 5).

Anilines can be converted to the corresponding *bis*-chloroethyl derivatives by various methods, and so our approach was to first construct the core structure with an aniline functionality. Melflufen carries two phenylalanine derivatives and by coupling L-Boc-4-nitrophenylalanine (**4**) with L-4-fluorophenylalanine ester (**5**) using EDC/HOBt, the core nitrodipeptide structure **6** was formed

(Scheme 2). An advantage of using EDC in combination with HOBt (1-hydroxybenzotriazole) is that epimerization of amino acids are usually suppressed.

2-Methyltetrahydrofuran (2-MeTHF) and cyclopropyl methyl ether were evaluated in the coupling. When triethylamine was used as base, the former gave an 80-90% yield and the latter produced less than 30% yield together with unconsumed starting material. While the the 80-90% yield was encouraging, disappointing was the formation of a sticky gum in all cases. However, when we changed to DMF, we produced a clean reaction in high yield. The removal of DMF in the work-up was a problem despite extensive experimental examination. Finally, chromatographic purification produced pure (>98 area%, all subsequent purities are reported as area% unless otherwise specified) material in 83% yield.

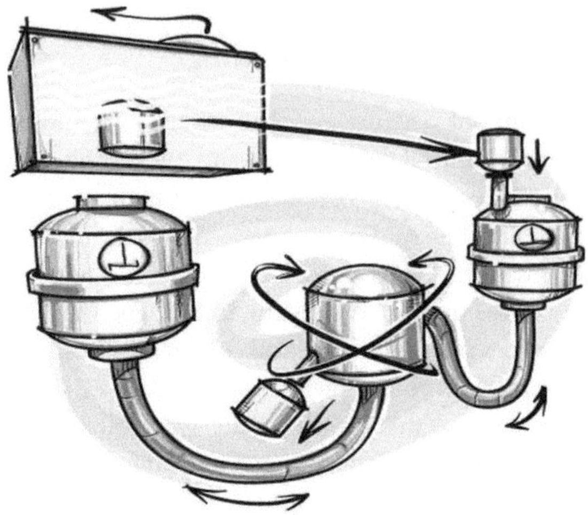

Figure 5. Pictorial presentation of a system for contained production of potent substances. In the middle, the central reactor which contains a filter and can be turned upside down for filtration. A container for loading and emptying is connected via a butterfly valve. This container is later handled in a controlled area. With this system, multistep reactions can be performed without exposure and sampling can be made via a sampling valve.

As 2-MeTHF, DMF and cyclopropyl methyl ether had produced less than optimal results, we turned to dichloromethane as solvent. Poor results were obtained when dichloromethane was initially used. Three significant TLC impurities appeared and the yield was just 50% after chromatography. But upon replacing the base triethylamine with *N*-methylmorpholine, and reducing the amount of HOBt from 1.0 to 0.1 equiv, an 80-82% yield of pure product was obtained. These overall conditions turned out to be the optimal in spite of a higher yield when using 2-MeTHF.

Reduction of the nitro group was performed with hydrogen and wet palladium on charcoal (10% Pd and 50% moist) to give aminodipeptide 7 (Scheme 2). When 5% (by weight) palladium was used, the reaction delivered pure (according to TLC and ^{1}H NMR spectroscopy) 7 in 74% yield within 10 min. Reducing the palladium charge to 2.5 mol% led to a 80% yield and 1% furnished 96-97 % yields of pure (according to TLC and ^{1}H NMR spectroscopy) 7 at 3 barand 40 °C.

With aminodipeptide 7 in hand, it was now possible to investigate the key transformation to form the reactive tail of the molecule. The conversion of an aromatic primary amine to the corresponding *N,N-bis*-chloroethyl derivative is not a single operation. A common technique is to add ethylene oxide or a haloethanol to the aromatic amine, followed by chlorination of the intermediate *bis*-hydroxyethyl compound.

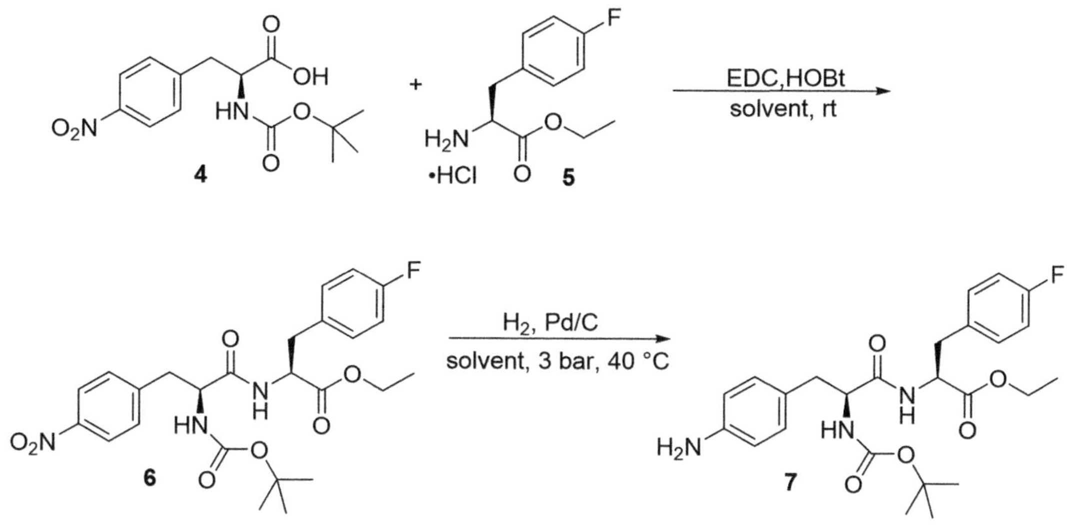

Scheme 2. Formation of the core structure aminodipeptide 7.

We commenced the work with ethylene oxide, which may be used for such a reaction under various reaction conditions (*24*). For practical reasons, we used a 2.5 M solution in THF. Aminodipeptide 7 was heated with two equivalents of ethylene oxide in THF at 55 °C for 24 h without any conversion. However, when methanol was added and the temperature was increased to 60 °C, the reaction occurred readily. Initially two products appeared corresponding to *mono*- and *bis*-alkylation, but after 24 h *bis*-alkylated diol **8** was obtained as the sole product (Scheme 3) (*25, 26*).

The transformation may also be performed under acidic conditions. When 7 and ethylene oxide were stirred in a 50:50 mixture of acetic acid and water, all the starting material was consumed and the major product was diol **8** together with a small amount of mono-alkylated amine (Scheme 3). The acid sensitive Boc-group was stable under these weakly acidic conditions. There was some variability in that some experiments led to the appearance of two additional impurities, one less polar and one more polar than the product by TLC. Therefore, the method using acidic conditions is more unpredictable than the corresponding reaction under basic conditions. Thus basic conditions were chosen for future work.

When aminodipeptide 7 was heated with potassium carbonate and chloroethanol (*27*) in excess for 10 h, *bis*-hydroxyethyl compound **8** was formed in quantitative yield (Scheme 3). After work-up and chromatographic purification, the product was obtained as a colorless oil which crystallized and yielded off-white crystals. A concern is if the ethyl ester of either 7 or **8** might undergo transesterification with chloroethanol.

Chlorination of *bis*-hydroxyethyl **8** was initially performed with thionyl chloride (*28, 29*) in dichloromethane over 30 min (Scheme 3). One of the side products from the thionyl chloride treatment, hydrogen chloride, may cause cleavage of the Boc-protecting group to give the free amine which in turn would be converted to hydrochloride salt with the excess HCl present. The reagent phosphorous oxychloride (*30, 31*) also chlorinated well but the reaction conditions were harsher as they required higher temperature and longer reaction times (Scheme 3). Boc group cleavage from the HCl present could be problematic as it might be difficult to control the quality of the product by telescoping all these steps. Handling $POCl_3$ on large scale may also be a safety concern (*32*).

In an effort to examine if there was a chlorination method which did not cleave the Boc-protecting group, diol **8** was treatedwith methanesulfonyl chloride and triethylamine followed by treatment with sodium chloride in DMF at 120 °C (Scheme 3) (*33*, *34*). This did in fact produce the dichloride Boc-protected melflufen **3**. Unfortunately, the crude material also contained excessive impurities and this approach was not further investigated. However, the exercise demonstrated that an intermediate with an intact Boc group could be obtained.

A double reductive amination/chlorination from an aminogroup to a *N,N*-bis-chloroethyl moiety in one pot may be possible via double alkylation using chloroacetaldehyde followed by a reducing agent such as sodium cyanoborohydride. This appeared to be an attractive approach to melflufen even though yields were reported to be modest (*35–37*). A further perceived disadvantage was that chloroacetaldehyde is only available as a water solution due to its instability. The presence of water under the acidic reaction conditions will cleave the Boc-protecting group which might save a step but introduced complexity in its current form.

Instead, we were inspired by a novel method using chloroacetic acid in combination with borane as the reductant to convert an amine directly to a *bis*-chloroethyl compound (*38*). We applied this method by treating aminodipeptide **7** with chloroacetic acid (20 equiv) and 1 M borane-THF solution (7 equiv) (Scheme 3). By this method it was possible to produce melflufen in 53% yield as the Boc-deprotected chloroacetate salt. As the amino group resulting from deprotection was now also accessible for reductive alkylation, a new impurity was theoretically possible. Due to this complex impurity profile situation, purification turned out to be difficult. However, this was solved as follows.

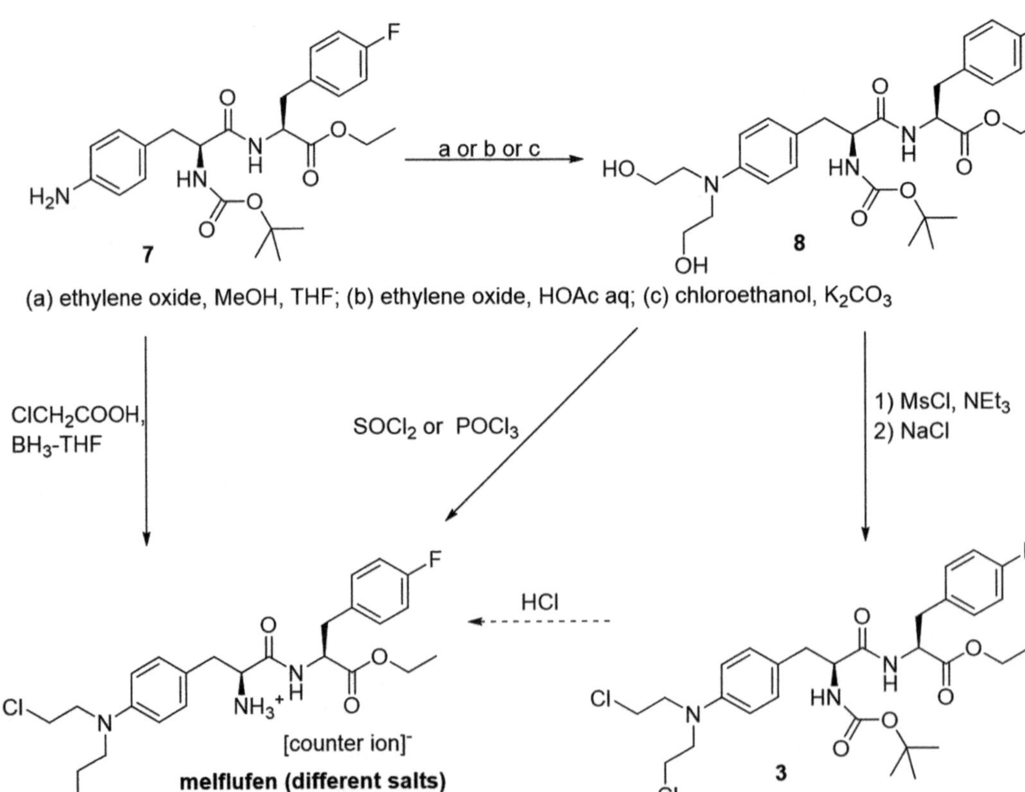

Scheme 3. Alternative pathways to ***1a*** *from aminodipeptide* **7**.

Conversion from the chloroacetate salt after Boc cleavage to the hydrochloride (melflufen) would require an extra step. A controlled stepwise process seemed more attractive and would require less acidic conditions. Screening bases such as sodium hydroxide, *N*-methylmorpholine and sodium chloroacetate as acid-scavengers failed to yield viable results. However, when sodium chloroacetate was used in large excess (15 equiv), suppression of the Boc deprotection reaction resulted. With this, a purer product was obtained and we decided that this approach should be the subject for process development (*26*).

Process Development

A plethora of coupling reagents are available for creation of the amide bond for joining the two halves of the melflufen core (*39*). Since we have had extensive positive experiences with EDC, we decided to continue on this track but would search for a replacement for the class 2 solvent, dichloromethane. DMF and ethyl acetate both worked well but the latter was chosen because it would allow using the same solvent in the subsequent hydrogenation. It was our intention to telescope the peptide coupling with the hydrogenation and avoid isolation of the intermediate at the end of the coupling step.

Thus, coupling of L-4-fluorophenylalanine ethyl ester hydrochloride with L-Boc-4-nitrophenylalanine using EDC (1.1 equiv.), HOBt (0.1 equiv.) and *N*-methylmorpholine (3 equiv) in ethyl acetate gave nitrodipeptide **6** in >99% conversion and >98% purity. Both L-4-fluorophenylalanine ethyl ester hydrochloride and nitrodipeptide **6** possess low solubility in ethyl acetate and even at 20 volumes, the product was partitioned between the solution and a sticky gum. This was workable at laboratory scale but of course a concern at scale. When 2-MeTHF was tried in place of ethyl acetate, with either *N*-methylmorpholine or Hünigs base, the same sticky residue appeared and conversion was reduced.

A significant reaction improvement was accomplished by a solvent switch to acetone. The high solubility of the reactants in acetone resulted in a high conversion to coupled product without the formation of a sticky residue. Indeed, it was possible under these conditions to subsequently perform the hydrogenation using hydrogen and Pd/C to form aminodipeptide **7** in good yield without isolation of the intermediate nitrodipeptide **6**. Unfortunately, we discovered that very high substrate purity was needed for a hydrogenation for full conversion of the nitrodipeptide **6**. Otherwise, even small amounts of residual nitrodipeptide **6** in aminodipeptide **7** would be difficult to purge downstream. Thus, nitrodipeptide **6** required recrystallization prior to use for a successful reduction and the concept of telescoping was consequently shelved for now.

Optimizing the coupling reaction of **4** and **5** became a priority to achieve full conversion to intermediate **6**. Thus, when equimolar amounts of the two building blocks were used, about 3-4% of unreacted L-Boc-4-nitrophenylalanine (**4**) remained in the product. Using 10% molar excess of L-4-fluorophenylalanine ethyl ester hydrochloride (**5**) resulted in the full conversion of L-Boc-4-nitrophenylalanine (**4**). However, this came with an added cost because L-4-fluorophenylalanine ethyl ester hydrochloride (**5**) is rather expensive. Fortunately, we discovered that the reaction performs better when using only 1.02 equiv of **5** in combination with 10% excess of EDC, a reasonably priced reagent, to provide intermediate **6** with satisfying results.

Since acetone is water-soluble, we needed a second solvent for extractive work-up, in this case 2-MeTHF. To further increase the efficiency of the extractive work-up, acetone was removed by vacuum distillation prior to work-up. The slight excess of L-4-fluorophenylalanine ethyl ester was removed by washing with 2 M sulfuric acid solution, whereas HOBt was removed by washing with sodium bicarbonate solution. These washings also effectively removed the urea byproduct originating from EDC. By adding *n*-heptane to the solution of nitrodipeptide in 2-MeTHF, the intermediate **6** was precipitated and isolated in high yield (89%) and high purity (99.9%) (Scheme 4).

With the discovery of a viable and process-friendly coupling reaction, we began exploring conditions for the reduction of the nitro group. This was surveyed in THF, ethanol and ethyl acetate, using 1-4% palladium on charcoal and hydrogen pressures between 1 and 3 bar at 40 °C. With the revived intention to perform a telescope of steps 1 and 2, ethyl acetate and ethanol were considered as solvents. The use of ethyl acetate, however, gave rise to impurities, which were not observed when ethanol was used. The disadvantage for ethanol was the poor solubility of both aminodipeptide **7** and nitrodipeptide **6** in it, requiring large volumes for acceptable solubility.

The answer was obtained when a 70/30 mixture of ethanol and ethyl acetate was used in place of each solvent alone. The best of both worlds resulted in that the solubility of the aminodipeptide **7** was much higher and no impurities were detected. When nitrodipeptide **6** was hydrogenated in this solvent mixture using 1.5% Pd/C (10% Pd, 50% moist) at 7 bar and 40 °C, complete conversion was achieved within 1 h with a purity of ~96%. The purity was enhanced by precipitation with *n*-heptane after filtration of the reaction mixture. This procedure gave beautiful white crystals of **7** in 75% yield and 99.2% purity.

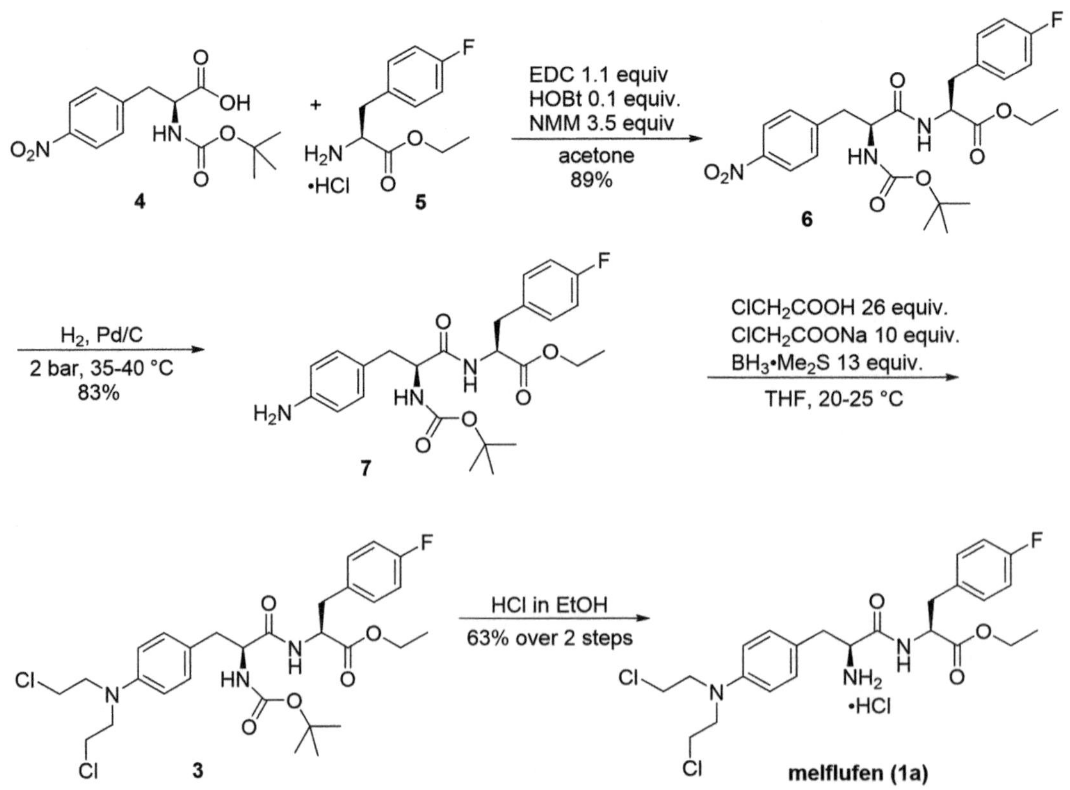

Scheme 4. Commercial route to melflufen ***1a****.*

Later, the solvent mixture of ethanol/ethyl acetate was replaced with 2-MeTHF alone which benefitted the reaction with even a higher solubility. With this modification the reaction became faster and full conversion was achieved within 30-45 min. Indeed, it was too rapid. Because the reduction of a nitro group to an amino group is one of the more exothermic reactions known, (heat formation of 560 kJ/mol), this rate was too high to be considered safe at large-scale. This reduction, which occurs in multiple steps, is dependent on mass transfer. The reaction rate is dependent on the amount of catalyst, hydrogen pressure and stirring speed (*40*). For these reasons, a Design of Experiments (DoE) was set up in order to discover reaction conditions that would reduce the reaction rate to a safe level.

The DoE study was performed with a center point of 2.25% palladium catalyst (5% Pd, 50% moist), 35 °C and 3 bar hydrogen pressure. With the aid of experimental design, we discovered conditions that gave a longer reaction time and was satisfactory from a safety point of view. Conditions used at scale were 5% palladium (50% moist), 35 °C, and 2 bar of hydrogen gas which resulted in an acceptable reaction time of 4 h. Somewhat surprisingly, the hydrogen pressure had the least impact on the reaction rate. Temperature was more influential, but the amount of charged catalyst had the greatest effect. Reaction conversion and the purity of the crude material were in general high with an exception for reaction conditions when temperature and catalyst loading were low. In these cases, conversion was incomplete as starting material remained along with several percent of the intermediate hydroxylamine.

Table 2. Removal of Palladium in Aminodipeptide 7

Entry	*Treatment*[a]	*Additive (w/w)%*	*Palladium (ppm)*
1	No treatment	-	21
2	Extraction with water	-	21
3	Activated charcoal	15	15
4	SEM26[b] and SPM32[c]	1.5 each	6
5	Si-trisamine scavenger[d]	3	2
6	Si-thiol scavenger[e]	3	7
7	MP-TMT scavenger[f]	3	5
8	Si-TMT scavenger[g]	3	<1
9	Si-TMT scavenger[g]	0.6	6
10	Si-TMT scavenger[g], 40 °C	3	<1
11	Si-TMT scavenger[g], 2 h	3	<1

[a] 2-Me-THF (16L/kg), 12 h, 21 °C. [b] 0.8 mmol/g. [c] 0.8 mmol/g. [d] 2.7 mmol N/g. [e] 1.4 mmol S/g. [f] 0.81 mmol N/g. [g] 0.45 mmol TMT/g.

The maximum amount of palladium allowed in the final product, melflufen hydrochloride, was set to 10 ppm. In order to avoid or minimize palladium removal during latter steps, the same Pd limit was assigned to aminodipeptide 7. When there was no purification or water treatment, the palladium levels were too high and therefore different options for removal of palladium were evaluated (Table 2, entries 1-2). Activated charcoal did reduce the palladium level, but to an insufficient degree (Table 2, entry 3). A selection of commercially-available metal-scavengers was examined (Table 2, entries 4-

11). The best result was obtained with the silica-based trimercaptotriazine, (Si-TMT) which reduced the palladium level significantly after overnight stirring. When 3% Si-TMT was used, the palladium level dropped from 21 ppm to ≤ 1 ppm. This could be done both at room temperature (23 °C) and elevated temperature (40 °C). For a better fit regarding manufacturing cycle time, the residence time of the scavenger contact was reduced to 2 h with the same satisfying result. After hydrogenation in 2-MeTHF, the reaction mixture was treated with 3% Si-TMT for 2 h at 35 °C. By the addition of *n*-heptane to the filtered reaction mixture, aminodipeptide 7 crystallized from the worked-up solution and resulted in 83% yield and 99.7% purity (Scheme 4).

Reductive Alkylation

During route scouting for the next two steps, we decided to use a stepwise approach for reductive alkylation and deprotection rather than a telescoped reaction. A practical problem was the large excess of reagents (15 equiv for the borane) necessary for the reductive alkylation Attempts to reduce the excess amounts gave significantly lower yields. As borane-tetrahydrofuran complex was only available at a maximum molarity of 1 M, large volumes of reaction mixtures resulted. Instead, borane-dimethyl sulfide (2 M in THF), and neat borane-dimethyl sulfide complex (10 mol/L) were evaluated. When tested, neat borane-dimethyl sulfide complex gave similar or slightly better results in terms of both conversion and purity, compared to the dilute solutions. Thus, neat borane-dimethyl sulfide complex was the hydride source of choice.

The odorous dimethyl sulfide vapor was eliminated by oxidation with 10% aqueous sodium hypochlorite solution in a scrubber. The order of reagent addition had a strong influence on the outcome of the reductive alkylation. When a solution of aminodipeptide 7 was added to the other reagents (borane-dimethyl sulfide, chloroacetic acid and sodium chloroacetate) 20% of Boc-cleaved chloroacetamide byproduct **9** was formed due to incomplete reaction (Figure 6).

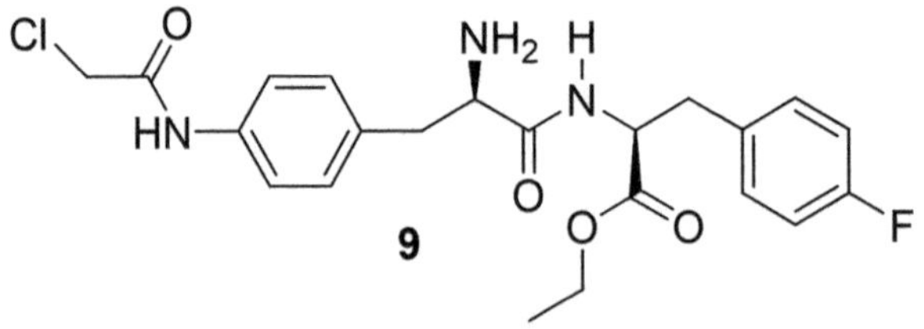

*Figure 6. Chloroacetamide byproduct **9**.*

This was circumvented by a simple alteration of the order of addition. To a solution of aminodipeptide 7 in THF was first added chloroacetic acid for protonation, followed by sodium chloroacetate. Finally, neat borane-dimethyl sulfide complex was added over an extended time of 3 h. Acidic conditions are a requirement for the reaction as a higher pH results in an incomplete reaction and decreased purity. By balancing the three reagents chloroacetic acid, sodium chloroacetate, and dimethyl sulfide complex using 30, 10 and 15 equiv, respectively, it was possible to suppress premature cleavage of the Boc-group. The slow addition of borane dimethyl sulfide and a higher temperature during addition resulted in a fast and very clean reaction.

It was possible to reduce the charge of reagents to 26, 10 and 13 equiv of chloroacetic acid, sodium chloroacetate, and dimethyl sulfide complex, respectively, but such lower levels led to an impure product dominated by the impurity chloroacetamide **9**. During development in lab scale the addition of borane-dimethyl sulfide complex was performed at 20-35 °C and finally held at 40 °C. When run at a 4.5 kg scale, borane-dimethyl sulfide complex was added at 5-10 °C and kept at 25 °C to better control the exothermic nature of the reaction. Excess borane-dimethyl sulfide complex was quenched by the addition of ethanol, and precipitation of the product was effected by the addition of aqueous potassium carbonate to produce Boc-melflufen **3** in 94% yield and 98.4% purity. Purity was increased to 99.6% by crystallization from 2-MeTHF and *n*-heptane (Scheme 4).

Deprotection and Salt Formation

With a new route established, it was important to secure a similar solid-state and impurity profile compared to material used in initial clinical trials. With this stepwise approach for reductive alkylation and deprotection established, we needed to address the Boc deprotection next. While we could make use of the protocol used in early development, however, gaseous hydrogen chloride was not our choice at this larger scale. Instead a 2.7 M hydrogen chloride solution in ethanol was used. Experiments revealed that a large excess of hydrogen chloride was necessary for a fast and clean reaction. The drawback with this large HCl excess was that the subsequent crystallization was negatively impacted.

This problem was solved by treating Boc-melflufen **3** with seven equivalents of the ethanolic hydrogen chloride at ambient temperature overnight. This was followed by repeated addition of ethanol and evaporation to remove excess HCl. After the second evaporation, melflufen HCl salt **1a** crystallized out of solution. In order to increase the yield, the ethanol was further diluted with methyl-*tert*-butyl ether. The product was isolated on a filter and dried, which gave the final product in 63% yield over the reductive alkylation/salt formation steps in 99.6% purity (Scheme 4).

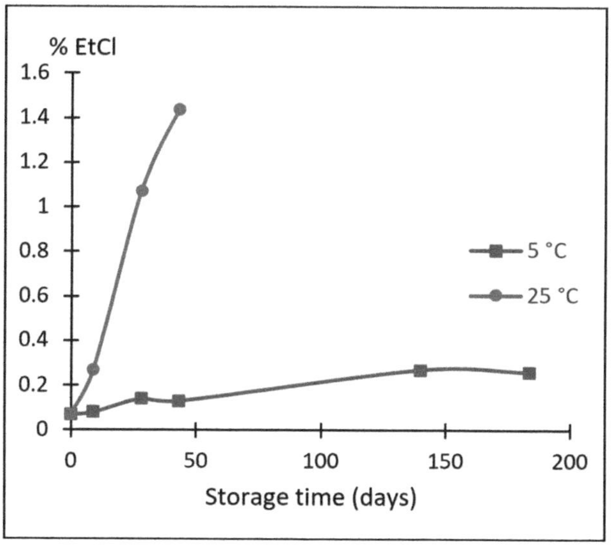

Figure 7. Temperature impact on formation of ethyl chloride in ethanolic hydrogen chloride.

While working with the final step, we experienced a quality problem with the commercial hydrogen chloride solution. While the hydrogen chloride in ethanol was of USP quality and should have been of high purity, we found substantial amounts of ethyl chloride in the product. It is known that hydrogen chloride reacts to some extent with ethanol to form ethyl chloride, and consequently our reagent solution had become degraded to contain a compound which is considered to be genotoxic and must be controlled (*41–44*).

We were able to demonstrate that up to 2% ethyl chloride was efficiently purged during the crystallization and drying of melflufen **1a**. Degradation of the ethanolic HCl was faster than anticipated but strongly temperature dependent. At 25 °C, 1% ethyl chloride was formed within a month and the degradation continued in a linear fashion. On the other hand, at 5 °C the level of ethyl chloride only increased from 0.1% to 0.2% over a period of six months (Figure 7) enabling a practical storage protocol for the solution.

Starting Materials

The two building blocks, Boc-4-nitro-L-phenylalanine (**4**) and 4-fluoro-L-phenylalanine ethyl ester (**5**) are commercially available and were sourced for the initial pilot scale batches. Unfortunately, each compound contained a small amount (around 0.5%) of the corresponding 2-substituted isomer as impurities (impurities **10** and **11** respectively, Figure 8). These impurities can react in the same manner as the building blocks **4** and **5** giving rise to impurity analogues to melflufen which are difficult to remove from the intermediates and/or target molecule. If the level of either one of these impurities exceeds the set specification 0.15 area% in a purchased batch, the batch containing the impurity could not be used. Therefore, to mitigate this risk, we decided to manufacture the building blocks ourselves with higher purity in multi-kilogram scale.

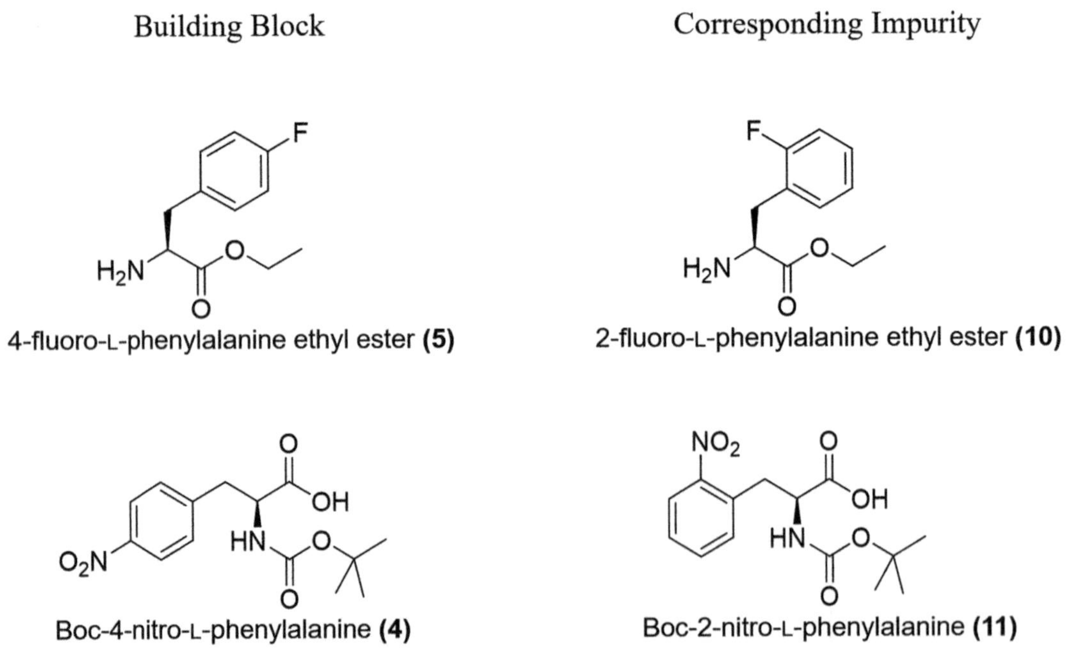

*Figure 8. Starting materials 4-fluoro-L-phenylalanine ethyl ester (**5**) and Boc-4-nitro-L-phenylalanine (**4**) and their corresponding impurities **10** and **11**.*

The synthesis of 4-fluoro-L-phenylalanine ethyl ester commenced from 4-fluorobenzyl chloride. This was treated with the enolate of diethyl acetamido malonate (**12**) to give diethyl 2-acetamido-2-[(4-fluorophenyl)methyl]propanedioate (**13**) (Scheme 5). The product from this reaction was telescoped into the next step. The hydrolysis and decarboxylation were accomplished by slow addition of potassium hydroxide and simultaneous distillation of the ethanol. After all the ethanol was distilled off, the pH was adjusted to 2 by adding 6 M HCl at a rate where evolution of carbon dioxide was controlled. The pure product was isolated by filtration to give racemic **14** in good yield (87%).

To get access to the enantiomerically pure material, a resolution method using Amano acylase was employed. Racemic **14** was dissolved in aqueous sodium hydroxide, the pH was adjusted to 8.5, and Amano acylase was added. After the enzymatic resolution was complete, the pH was adjusted to 0.7-0.8 which resulted in precipitation of the major part of the unreacted undesired *R*-isomer **(*R*)-14** which was separated by filtration. Extractive work-up was performed by adding 2-MeTHF to the filtrate containing **(*S*)-15**. After phase separation, the pH was adjusted with NaOH to 5.5 whereupon **(*S*)-15** precipitated. After isolation, the crude product was slurried in a small volume of ethanol and filtered. This operation eliminated the ~5% **(*R*)-14** remaining in the product as well as residual water and color. The yield was 35% (theoretically 50%) and no remaining **(*R*)-14** could be detected.

We were also able to demonstrate the racemization of the unwanted isomer by treatment with acetic anhydride and sodium carbonate via formation of an oxazolone-intermediate (*45*). This was not used for the present delivery since there was insufficient time to develop the method into a process, but could have been further resolved. This will be considered for future campaigns.

Amino acid **(S)-15** was slurried into a mixture of toluene/ethanol and heated to 55 °C after which thionyl chloride in toluene was added. The temperature was raised to 60-65 °C and full esterification was obtained within 2-3 h. Excess thionyl chloride and ethanol were removed by vacuum distillation followed by the addition of methyl-*tert*-butyl ether to the thick slurry. After cooling to 0 °C and stirring, the crystallized product was isolated by filtration. Drying at 30 °C under reduced pressure gave 4-fluoro-L-phenylalanine ethyl ester (**5**) in 98% yield in 99.7% purity (*45*).

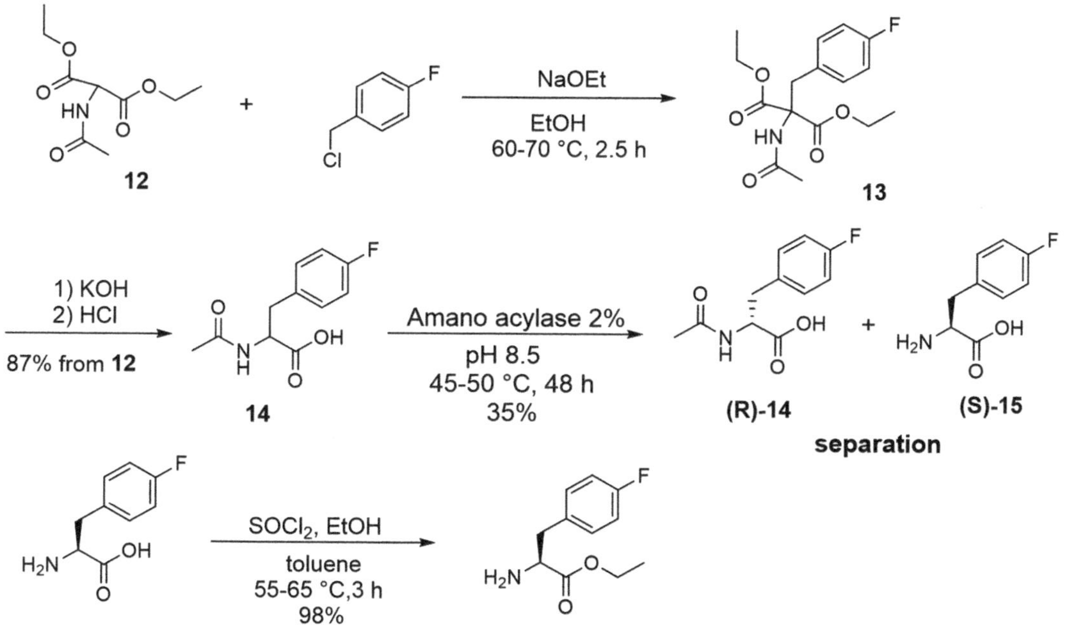

Scheme 5. Synthesis of 4-fluoro-L-phenylalanine ethyl ester.

The purification of the nitrophenylalanine was considerably simpler. Purity was enhanced by pH-controlled crystallization in that the starting material was dissolved into aqueous sodium hydroxide and thereafter precipitated by the addition of hydrochloric acid. The product was isolated by filtration and dried. The product had a purity of 98% and contained < 0.5% of the 2-nitrophenylalanine. The pure 4-nitro-L-phenylalanine was then Boc-protected using di-*tert*-butyl dicarbonate in aqueous sodium carbonate. Extractive work-up and pH-adjustment followed by crystallization gave Boc-4-nitro-L-phenylalanine (**4**) of 99.7% purity and with less than 0.15% **11** (Scheme 6) (45).

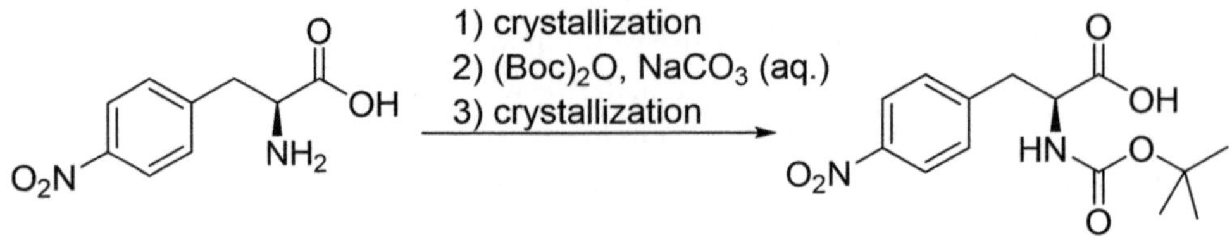

*Scheme 6. Synthesis of Boc-4-nitro-L-phenylalanine (**4**).*

Conclusion

The origin of melflufen can be traced back to the 1960s but significant development did not start until around 2000 when development with preclinical and clinical trials as well as process development was initiated. The research revealed a unique mechanism of action that resulted in a higher selectivity than other drugs in this class. Along with successful clinical development, a method for commercial manufacturing at multi-kilogram scale was developed. This work includes the late introduction of the chloroalkyl moiety as well as a novel method for bis-chloroalkylation onto acid-sensitive substrates. To mitigate risks in manufacturing, in-house methods for supply of strategic starting materials were developed.

Acknowledgments

We thank Dr Carl-Magnus Andersson for valuable comments on this manuscript.

References

1. Gilman, A.; Philips, F. S. The biological actions and therapeutic applications of the -chloroethylamines and sulfides. *Science* **1946**, *103*, 409–415.
2. Gilman, A. The initial clinical trial of nitrogen mustard. *Am. J. Surg.* **1961**, *105*, 574–578.
3. Wilkinson, J. F.; Fletcher, R. Effect of β-chloroethylamine hydrochlorides in leukemia, Hodgkin's disease and polycythaemia vera: report on eighteen cases. *Lancet* **1947**, *2*, 540–545.
4. Papac, R. J. Origins of cancer therapy. *Yale J. Biol. Med.* **2001**, *74*, 391–398.
5. Valu, K. K.; Gourdie, T. A.; Boritzki, T. J.; Gravatt, G. L.; Baguley, B. C.; Wilson, W. R.; Wakelin, L. P. G.; Woodgate, P. D.; Denny, W. A. DNA-directed alkylating agents. 3. Structure-activity relationships for acridine-linked aniline mustards: consequences of varying the length of the linker chain. *J. Med. Chem.* **1990**, *33*, 3014–3019.
6. Wickström, M.; Nygren, P.; Larsson, R.; Harmenberg, J.; Lindberg, J.; Sjöberg, P.; Jerling, M.; Lehmann, F.; Richardsson, P.; Anderson, K.; Chauhan, D.; Gullbo, J. Melflufen – a peptidase-potentiated alkylating agent in clinical trials. *Oncotarget* **2017**, *8*, 66641–66655.
7. De Barbieri, A.: Vittorio, P.; Golferini, A.; Maugeri, M.; Peronne, F.; Tassi, G. C.; Temelcou, O.; Zapelli, P. Investigations on the effect of antitumor peptides and mixtures of the same; in Leukemien und maligne Lymphome. In *Pathophysiologie Klinik, Chemo- und Immunotherapie*; Urban & Schwarzenberg: Wien, 1972; pp 20—25.

8. De Bernardi, B.; Pastore, G.; Carli, M.; Ceci, A.; Paolucci, G.; Madon, E.; Mancini, A.; Tamaro, P.; Rebuffi, L.; Colella, R.; Cozzutto, C. Effect of peptichemio in nonlocalized neuroblastoma. *Cancer* **1982**, *50*, 10–17.
9. Bayraktar, U. D.; Bashir, Q.; Qazilbash, M.; Champlin, R. E.; Ciurea, S. O. Fifty years of melphalan use in hematopoietic stem cell transplantation. *Biol Blood Marrow Transplant* **2013**, *19*, 344–356and references cited therein.
10. Tageja, N.; Nagi, J. Bendamustine: something old, something new. *Cancer Chemother. Pharmacol.* **2010**, *66*, 413–423and references cited therein.
11. Gullbo, J.; Tullberg, M.; Våbenø, J.; Ehrsson, H.; Lewensohn, R.; Nygren, P.; Larsson, R.; Luthman, K. Structure-activity relationship for alkylating dipeptide nitrogen mustard derivatives. *Oncol. Res.* **2003**, *14*, 113–132.
12. Strese, S.; Wickström, M.; Fuchs Fredlund, P.; Fryknäs, M.; Gerwins, P.; Dale, T.; Larsson, R.; Gullbo, J. The novel alkylating prodrug melflufen (J1) inhibits angiogenesis *in vitro* and *in vivo*. *J. Biochem. Pharmacol.* **2013**, *86*, 888–895.
13. Gullbo, J.; Wickström, M.; Tullberg, M.; Ehrsson, H.; Lewensohn, R.; Nygren, P.; Luthman, K.; Larsson, R. Activity of hydrolytic enzymes in tumour cells is a determinant for anti-tumour efficacy of the melphalan containing prodrug J1. *J. Drug. Target* **2003**, *11*, 355–363.
14. Gullbo, J.; Lindhagen, E.; Bashir-Hassan, S.; Tullberg, M.; Ehrsson, H.; Lewensohn, R.; Nygren, P.; de la Torre, M.; Luthman, K.; Larsson, R. Antitumour efficacy and acute toxicity of the novel dipeptide melphalanyl -p-L-fluorophenylalanine ethyl ester (J1) *in vivo*. *Invest. New Drugs* **2004**, *22*, 411–420.
15. Hitzerd, S. M.; Verbrugge, S. E.; Ossenkoppele, G.; Jansen, G.; Peters, G. J. Positioning of aminopeptidase inhibitors in next generation cancer therapy. *Amino Acids* **2014**, *46*, 793–808.
16. Ray, A.; Ravillah, D.; Das, D. S.; Song, Y.; Nordström, E.; Gullbo, J.; Richardson, P. G.; Chauhan, D.; Anderson, K. C. A novel alkylating agent Melflufen induces irreversible DNA damage and cytotoxicity in multiple myeloma cells. *Br. J. Haematol.* **2016**, *174*, 397–409.
17. Chauhan, D.; Ray, A.; Viktorsson, K.; Spira, J.; Paba-Prada, C.; Munshi, N.; Richardson, P.; Lewensohn, R.; Anderson, K. C. *In vitro* and *in vivo* antitumor activity of a novel alkylating agent, melphalan-flufenamide, against multiple myeloma cells. *Clin. Cancer Res.* **2013**, *19*, 3019–3031.
18. Wickström, M.; Haglund, C.; Lindman, H.; Nygren, P.; Larsson, R.; Gullbo, J. The novel alkylating prodrug J1: diagnosis directed activity profile ex vivo and combination analyses *in vitro*. *Invest. New Drugs* **2008**, *26*, 195–204.
19. Berglund, Å.; Ullén, A.; Lisyanskaya, A.; Orlov, S.; Hagberg, H.; Tholander, B.; Lewensohn, R.; Nygren, P.; Spira, J.; Harmenberg, J.; Jerling, M.; Alvfors, C.; Ringbom, M.; Nordström, E.; Söderlind, K.; Gullbo, J. First-in-human, phase I/IIa clinical study of the peptidase potentiated alkylator melflufen administered every three weeks to patients with advanced solid tumor malignancies. *Invest. New Drugs* **2015**, *33*, 1232–1241.
20. Richardson, P. G.; Bringhen, S.; Voorhees, P.; Plesner, T.; Mellqvist, U. H.; Reeves, B.; Paba-Prada, C.; Zubair, H.; Byrne, C.; Chauhan, D.; Anderson, K.; Nordström, E.; Harmenberg, J.; Palumbo, A.; Sonneveld, P. Melflufen plus dexamethasone in relapsed and refractory multiple myeloma (O-12-M1): a multicentre, international, open-label, phase 1-2 study. *The Lancet. Haematology* **2020**, *7*, 395–407.

21. Miguel, J. S.; Weisel, K.; Moreau, P.; Lacy, M.; Song, K.; Delforge, M.; Karlin, L.; Goldschmidt, H.; Banos, A.; Oriol, A.; Alegre, A.; Chen, C.; Cavo, M.; Garderet, L.; Ivanova, V.; Martinez-Lopez, J.; Belch, A.; Palumbo, A.; Schey, S.; Sonneveld, P.; Dimopoulos, M. Pomalidomide plus low-dose dexamethasone versus high-dose dexamethasone alone for patients with relapsed and refractory multiple myeloma (MM-003): a randomised, open-label, phase 3 trial. *The Lancet. Oncology* **2013**, *14*, 1055–1066.
22. Siegel, D. S.; Martin, T.; Wang, M.; Vij, R.; Jakubowiak, A. J.; Lonial, S.; Trudel, S.; Kukreti, V.; Bahlis, N.; Alsina, M.; Chanan-Khan, A.; Buadi, F.; Reu, F. J.; Somlo, G.; Zonder, J.; Song, K.; Stewart, A. K.; Stadtmauer, E.; Kunkel, L.; Wear, S.; Wong, A. F.; Orlowski, Z.; Jagannath, S. A phase 2 study of single-agent carfilzomib (PX-171-003-A1) in patients with relapsed and refractory multiple myeloma. *Blood* **2012**, *120*, 2817–2825.
23. Usmani, S. Z.; Weiss, B. M.; Plesner, T.; Bahlis, N. J.; Belch, A.; Lonial, S.; Lokhorst, H. M.; Voorhees, P. M.; Richardson, P. G.; Chari, A.; Sasser, A. K.; Axel, A.; Feng, H.; Uhlar, C. M.; Wang, J.; Khan, I.; Ahmadi, T.; Nahi, H. Clinical efficacy of daratumumab monotherapy in patients with heavily pretreated relapsed or refractory multiple myeloma. *Blood* **2016**, *128*, 37–44.
24. Li, Q.; Rukavishnikov, A. V.; Petukhov, P. A.; Zaikova, T. O.; Keana, J. F. W. Nanoscale 1,3,5,7-tetrasubstituted adamantanes and p-substituted tetraphenyl-methanes for AFM applications. *Org. Lett.* **2002**, *4*, 3631–3634.
25. Wahlström, N.; Wennerberg, J. *Process for preparation of nitrogen mustards derivatives.* Patent WO2016/180740A1, May 8, 2015.
26. Cotton, C.; Bäckström, B.; Fritzon, I.; Lehmann, F.; Monemi, T.; Oltner, V.; Sölver, E.; Wahlström, N.; Wennerberg, J. Development of a safe process for manufacturing of the potent anticancer agent melflufen hydrochloride. *Org. Process. Res. Dev.* **2019**, *23*, 1191–1196.
27. Huang, J.-H.; Wen; Sun, Y.-Y.; Chou, P.-T.; Fang, J.-M. Two-stage sensing property via a conjugated donor-acceptor-donor constitution: application to the visual detection of mercuric ion. *J. Org. Chem.* **2005**, *70*, 5827–5832.
28. Cammack, T.; Reeves, P. C. Synthesis of ketobemidone precursors via phase-transfer catalysis. *J. Heterocycl. Chem.* **1986**, *23*, 73–75.
29. Fang, K.; Dong, G.; Wang, H.; He, S.; Wu, S.; Wang, W.; Sheng, C. Improving the potency of cancer immunotherapy by dual targeting of IDO1 and DNA. *ChemMedChem* **2018**, *13*, 30–36.
30. Elderfield, R. C.; Covey, I. S.; Geiduschek, J. B.; Meyer, W. L.; Ross, A. B.; Ross, J. H. Synthesis of potential anticancer agents. i. nitrogen mustards derived from p-[n,n-bis(2-chloroethyl)amino]- benzaldehyde. *J. Org. Chem.* **1958**, *23*, 1749–1753.
31. Goren Keskin, S.; Stanley, J. M.; Cowley, A. H. Synthesis, characterization and theoretical investigations of molybdenum carbonyl complexes with phosphorus/nitrogen/phosphorus ligand as bidentate and tridentate modes. *Polyhedron* **2017**, *138*, 206–217.
32. Achmatowicz, M. M.; Thiel, O. R.; Colver, J. T.; Hu, J.; Silva Elipe, M. V.; Tomaskevitch, J.; Tedrow, J. S.; Larsen, R. D. Hydrolysis of phosphoryl trichloride ($POCl_3$); characterization, in situ detection and safe quenching of energetic metastable intermediates. *Org. Process Res. Dev.* **2010**, *14*, 1490–1500.
33. Gourdie, T. A.; Valu, K. K.; Gravatt, G. L.; Boritzki, T. J.; Baguley, B. C.; Wakelin, L. P. G.; Wilson, W. R.; Woodgate, P. D.; Denny, W. A. DNA-directed alkylating agents. 1.

structure–activity relationships for acridine-linked aniline mustards: consequences of varying the reactivity of the mustard. *J. Med. Chem.* **1990**, *33*, 1177–1186.

34. Kamal, A.; Kumar, B. A.; Suresh, P.; Agrawal, S. K.; Chashoo, G.; Singh, S. K.; Saxena, A. K. Synthesis of 4β-*N*-polyaromatic substituted podophyllotoxins: DNA topoisomerase inhibition, anticancer and apoptosis-inducing activities. *Bioorg. Med. Chem.* **2010**, *18*, 8493–8500.
35. Delfornue, E.; Kiss, R.; Le Corre, L.; Dujols, F.; Bastide, J.; Collignon, F.; Lesur, B.; Frydman, A.; Darro, F. Synthesis and *in vitro* antitumor activity of ring C and D-substituted phenanthroline-7-one derivatives, analogues of the marine pyridoacridine alkaloids ascidemin and meridine. *Bioorg. Med. Chem.* **2004**, *12*, 3987–3984.
36. Delfornue, E.; Darro, F.; Portefaix, P.; Galaup, C.; Bayssade, S.; Bouteillé, A.; Le Corre, L.; Bastide, J.; Collignon, F.; Lesur, B.; Frydman, A.; Kiss, R. Synthesis and *in vitro* antitumor activity of novel ring D analogues of the marine pyridoacridine ascididemin: structure-activity relationship. *J. Med. Chem.* **2002**, *45*, 3765–3771.
37. Palani, A.; Shapiro, S.; McBriar, M. D.; Clader, J. W.; Greenlee, W. J.; Spar, B.; Kowalski, T. J.; Farley, C.; Cook, J.; Heek, M.; Weig, B.; O'Neill, K.; Graziano, M.; Hawes, B. Biaryl ureas as potent and orally efficacious melanin concentrating hormone receptor 1 antagonists for the treatment of obesity. *J. Med. Chem.* **2005**, *48*, 4746–4749.
38. Chen, J.; Przyuski, K.; Roemmele, R.; Bakale, R. P. Discovery of a novel, efficient, and scalable route to bendamustine hydrochloride: the api in treanda. *Org. Process Res. Dev.* **2011**, *15*, 1063–1072.
39. Valeur, E.; Bradley, M. Amide bond formation: beyond the myth of coupling reagents. *Chem. Soc. Rev.* **2009**, *38*, 606–631.
40. Stoessel, F. Experimental study of thermal hazards during the hydrogenation of aromatic nitro compounds. *J. Loss. Prev. Process Ind.* **1993**, *6*, 79–85.
41. *Guideline on the Limits of Genotoxic Impurities*; EMEA, 2006.
42. *Guidance for Industry – Genotoxic and Carcinogenic Impurities in Drug Substances and Products: Recommended Approaches*; FDA, Draft, 2008.
43. *ICH S9 Nonclinical Evaluation for Anticancer Pharmaceuticals*; European Medicines Agency, 2009.
44. *ICH M7 Assessment and Control of DNA Reactive (Mutagentic) Impurities in Pharmaceuticals to Limit Potential Carcinogenic Risk*; European Medicines Agency, 2014.
45. Unpublished results.

Chapter 6

Discovery and Chemical Development of S 44563, a Dual Bcl-2/Bcl-xL Inhibitor

Christophe Hardouin*,[1] and Jérôme-Benoît Starck[2]

[1]Industrial Research Centre, Oril Industrie, 13 rue Desgenétais, 76310 Bolbec, France

[2]Centre of Excellence Chemistry, Servier Research Institute, 125 Chemin de Ronde, 78290 Croissy-Sur-Seine, France

*Email: hardouin.christophe@yahoo.com

This chapter describes the discovery and chemical development of S 44563, a potent inhibitor of anti-apoptotic proteins involved in programmed cell death. Medicinal chemistry efforts resulted in the design of a conformation restriction strategy and in the synthesis of the first grams of this promising drug candidate. In order to supply pre-clinical and potential early Phase 1 studies, a first generation synthesis was developed at multikilogram scale. Limitations were identified and investigations pursued to address and solve them. Alternative approaches were envisioned to improve the access to the drug substance and several key steps were successfully developed.

Introduction

Apoptosis or programmed cell death is a cell suicide mechanism by which multicellular organisms remove damaged or unwanted cells in order to maintain normal life development and homeostasis (*1*). The failure of apoptosis systems plays a causative role in carcinogenesis as well as the chemoresistance of tumor cells (*2–4*). B-Cell lymphoma 2 (Bcl-2) family proteins are crucial regulators of apoptosis in the mitochondria-mediated pathway. This family's proteins comprise two classes: pro-survival or anti-apoptotic and pro-apoptotic members with diametrically opposed functions. Their interactions modulates the balance between cell survival and cell death (*5*). The anti-apoptotic proteins such as Bcl-xL, Bcl-2, Bcl-w, and Mcl-1 suppress apoptosis by interacting with pro-apoptotic proteins Bak and Bax. Those interactions are antagonized by BH_3-only proteins such as Bad, Bim, Bid, and Noxa (*6–8*).

As inhibition of protein–protein interactions (PPIs) between the anti- and pro-apoptotic proteins restores the apoptotic process, this strategy have been investigated as a therapeutic option. In 2005, a major preclinical breakthrough in the discovery of PPI inhibitors for the Bcl-2 family was the disclosure of ABT-737 **1**, a dual Bcl-2/Bcl-xL inhibitor. Later, this acylsulfonamide-based series have been exemplified by ABT-263 **2** (Figure 1) (*9*). However, the clinical dose-limiting toxicity of ABT-263 was peripheral thrombocytopenia, caused by Bcl-xL inhibition (*10*, *11*). Subsequently, further optimization efforts led to the discovery, development and approval of the Bcl-2 selective ABT-199 **3** (venetoclax) (*12*).

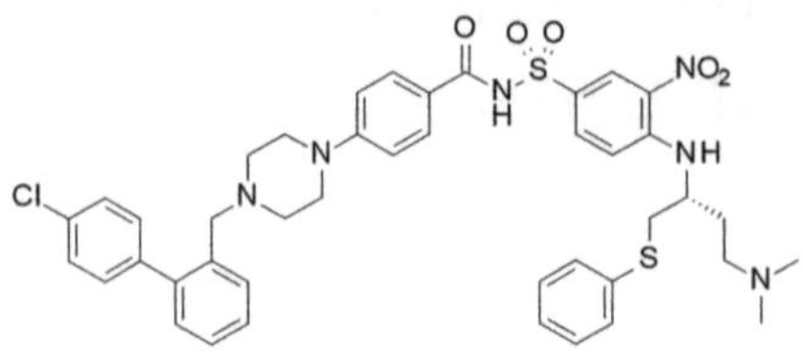

1, ABT-737

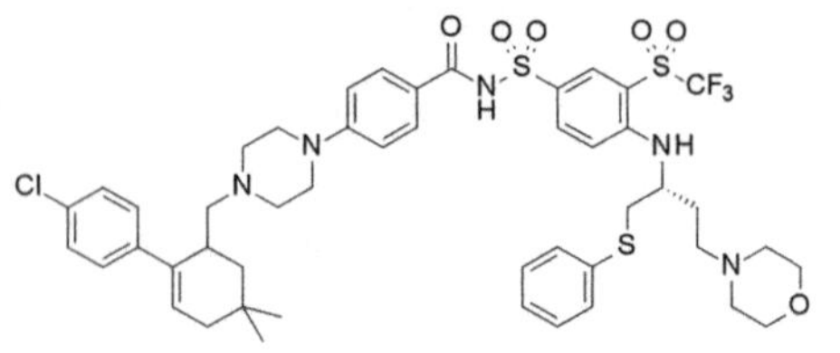

2, ABT-263

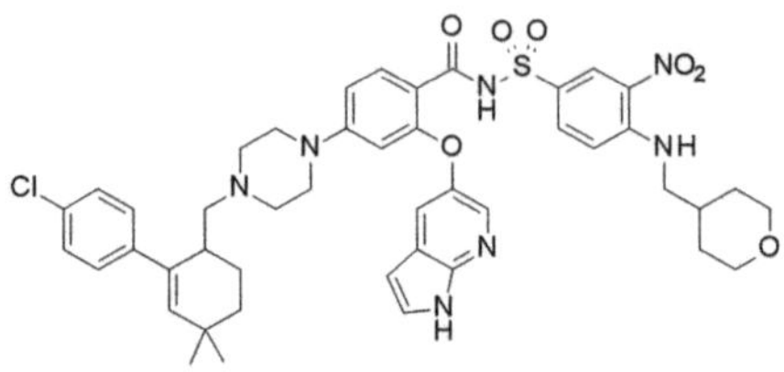

3, ABT-199

Figure 1. Chemical structures of ABT-737, ABT-263, and ABT-199.

Discovery of S 44563

As part of our efforts in the discovery of anti-apoptotic therapeutic agents, we embarked on high-throughput screening (HTS) campaign of a natural product library to identify novel chemotypes which inhibit Bcl-xL and/or Bcl-2. An original starting point which attracted our attention was abietic acid **4**, a tricyclic abietane diterpenoid (*1*, *13*). We further investigated this chemotype postulating that its binding mode may be comparable to other published carboxylic acid-containing fragments. We hypothesized that the carboxylic acid moiety may make an electrostatic interaction with Arg139 to assist in binding (*14*).

In order to test this hypothesis, we synthesized the hybrid acylsulfonamide compound **5**, coupling abietic acid **4** with the sulfonamide side-chain from ABT-737 **1** (Table 1). The selectivity profile within the Bcl-2 family, coupled with its relatively low potency was not sufficiently encouraging to pursue the abietic acid scaffold for further work to design a dual Bcl-2/Bcl-xL inhibitor.

Inspired by the concept tricyclic fragment binding to Bcl-xL, we subsequently decided to rigidify the published aryl-piperazine scaffold of **1**: we hoped that introducing a conformation restriction by cyclizing between the aryl and piperazine moieties could induce the ligand to adopt a more favorable conformation for binding, potentially leading to enhanced potency and a structurally new chemotype (*15*, *16*). Moreover, analysis of the crystal structure of the Bcl-xL complex with **1** indicated that some room may be available around the arylpiperazine moiety, in the vicinity of residue 101 (Figure 2). As a consequence, the incorporation of a cyclized moiety in this region would not impact the affinity to Bcl-xL. Furthermore, some published examples describing the efficient use of tricyclic conformational restriction in the design of CB2 agonists (*17*), GRP40 full agonists (*18*) or Second Mitochondria-Derived Activator of Caspases (Smac) mimetic (*19*) motivated us to consider this approach.

Table 1. Abietic Acid and the Corresponding Sulfonamide's Activity

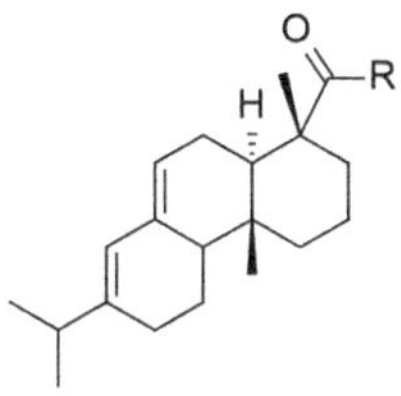

#	R=	FP (Bcl-2) IC_{50} *	FP (Bcl-xL) IC_{50} [a]
4	OH	-	171 μM
5	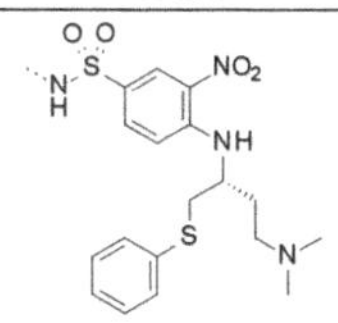	>100 μM	0.8 μM

[a]Fluorescence Polarization (FP) assay using recombinant Bcl-2 or Bcl-xL and a fluorescent PUMA BH3 peptide.

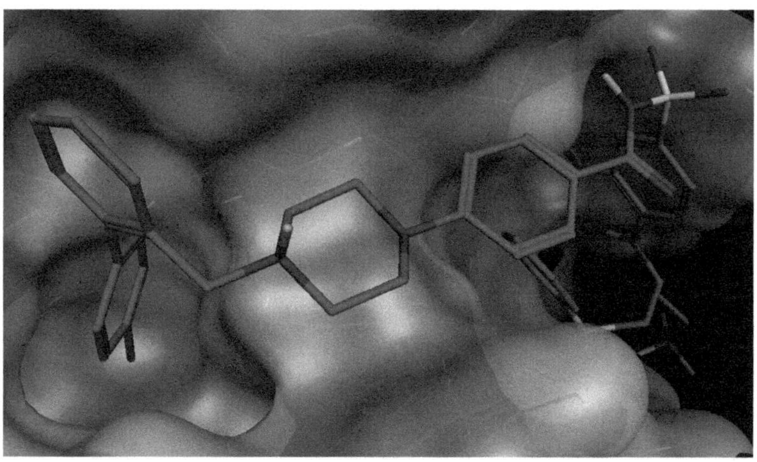

Figure 2. Crystal structure of ABT-737 in complex with Bcl-xL (PDB code 2YXJ; square highlights room for additional cyclization).

Regarding the optimal length of connecting bridge, the SAR rapidly focused on adding two carbon atoms, leading to a tricyclic scaffold containing three adjacent six-membered rings, initially exemplified by prototype molecule **6** (Figure 3). This molecule was initially synthesized as a mixture of diastereomers (a mixture of two rather than four diastereomers since the absolute configuration of the sulfonamide side-chain is *R*) and displayed a potent and balanced affinity profile for Bcl-2 and Bcl-xL. Thereafter, the separation of the two diastereomers was perfomed (see below), leading to molecules 7 (S 44563) and **8**, both potent and dual binders of Bcl-2 and Bcl-xL. Molecule S 44563 was preferred, being more potent on a set of Bcl-xL and/or Bcl2 dependent cell lines (*20*).

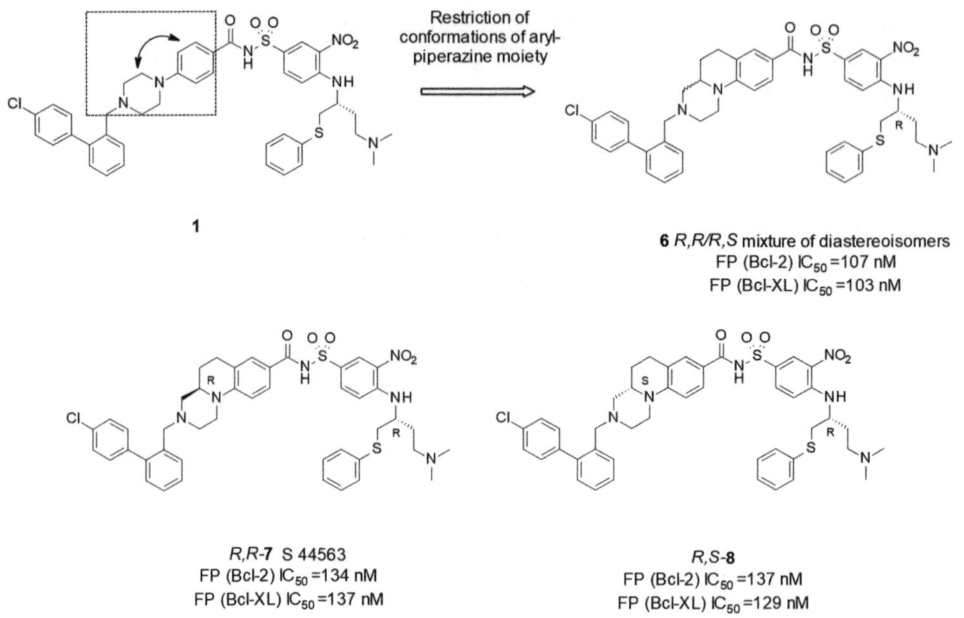

*Figure 3. Discovery of S 44563. Molecule **6** has been assayed as a mixture of 2 diastereoisomers R,R and R,S.*

Initial Research Synthetic Route to Molecule 6

The strategy for the construction of molecule **6** is outlined in the retrosynthesis shown in Scheme 1. The acylsulfonamide motif could be prepared by coupling acid **9** and sulfonamide **10** which may result from *N*-alkylation of tricyclic core **11** with an activated equivalent of biphenyl **12**. We followed the previously described synthetic methods to prepare building blocks **10** and **12** (*21*). We initially envisioned that the tricyclic moiety **11** might be accessible from commercially available 6-methoxy-2-methylquinoline (**13**).

The construction of the tricyclic structure was originally performed in a five-step sequence involving oxidation of the benzylic position of quinazoline **13** mediated by selenium dioxide, condensation of the resulting aldehyde with 1,2-ethanolamine and subsequent reduction of the imine with $NaBH_4$ to afford secondary amine **14**. This was followed by reduction of the pyridine ring and intramolecular cyclization, providing the tricyclic compound **15** (Scheme 2). Secondary amine protection was carried out to obtain *N*-benzyl **16.** Three more steps were necessary to transform methyl ether **16** into the methyl ester **18** using a palladium-mediated carbonylation sequence (*22, 23*).

After debenzylation of racemic **18** using ammonium formate in the presence of palladium on carbon, *N*-alkylation with dichlorobiphenyl compound **12** followed by ester hydrolysis under acidic conditions provided carboxylic acid **9.** The final peptide-type coupling afforded the desired **6** as a mixture of two diastereoisomers (*R,R* and *R,S*). The overall yield was 4% over 12 steps.

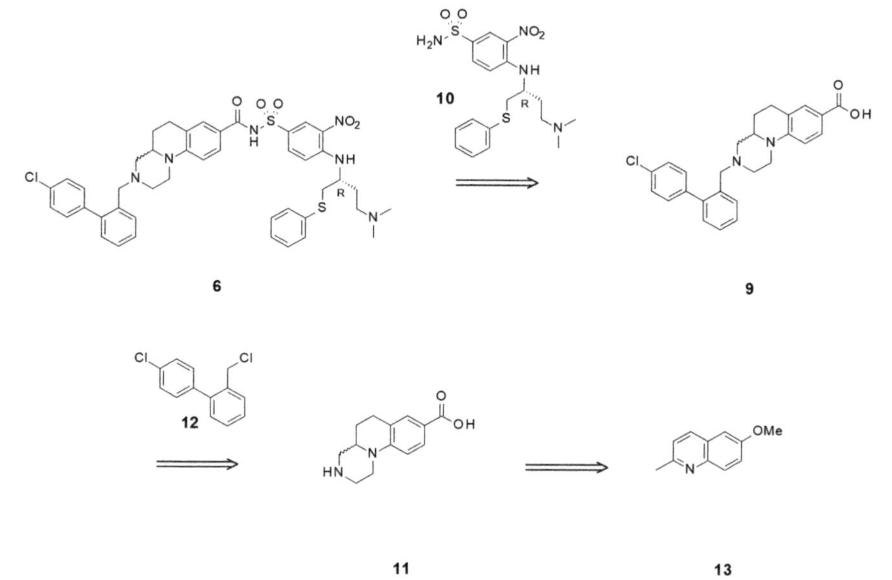

Scheme 1. Initial retrosynthesis of molecule ***6****.*

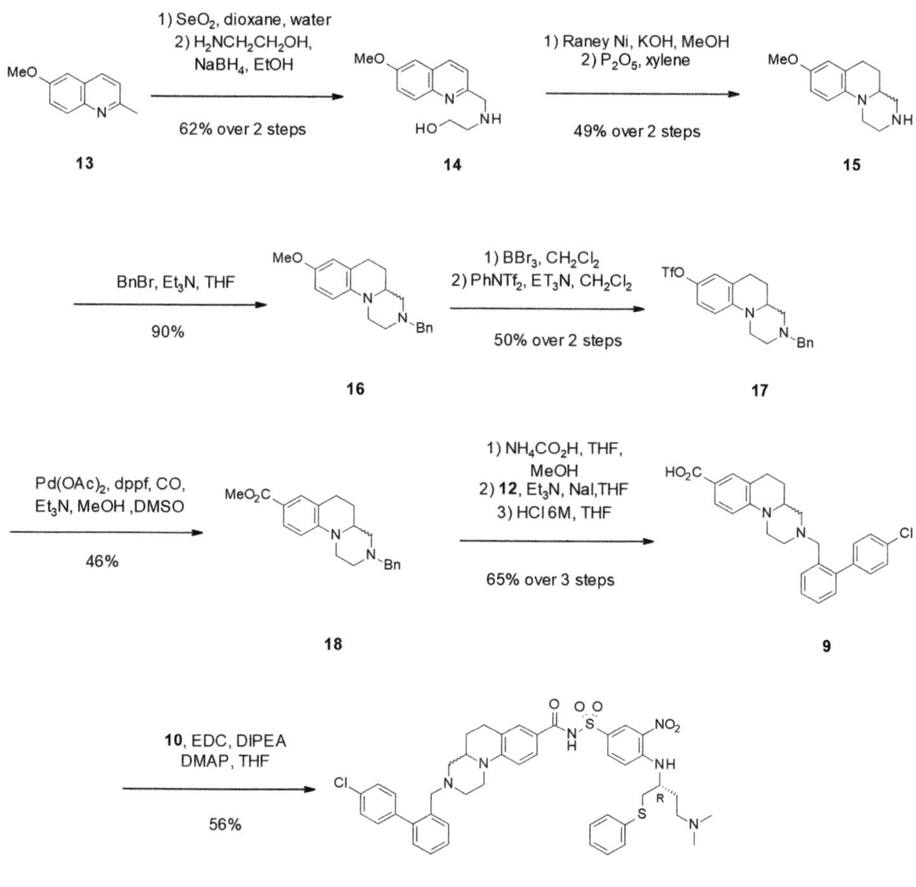

R,R/R,S **6**

Scheme 2. Laboratory route to R,R/R,S ***6****.*

Medicinal Chemistry Synthetic Route to S 44563

The desired compound, S 44563 (7) was initially accessed from racemic key intermediate **15** (readily available at that time at ten gram scale in the laboratory) via a chiral derivatization approach using (*S*)-methoxyphenylacetic acid leading to the corresponding amides *S/S*-**19** and *R/S*-**19** (Scheme 3) (*24*). These diastereoisomers were separated by chromatography and treated with potassium tert-butoxide in THF to obtain enantiomerically pure *R*-**15** and *S*-**15**. Both of these intermediates were then submitted to the similar sequence described in Scheme 2 affording respectively S 44563 and **8**.

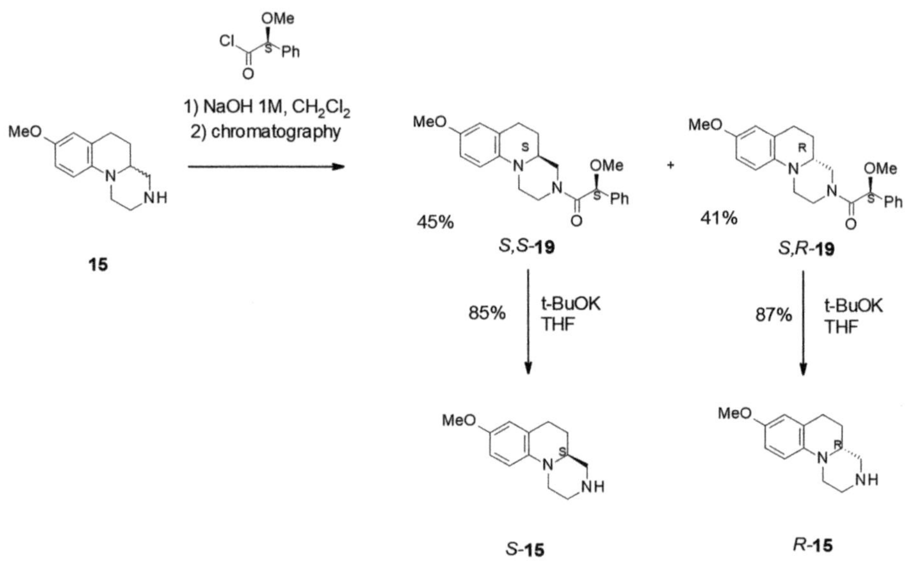

*Scheme 3. Chiral separation of racemic amine **15** leading to S 44563.*

S 44563, is a novel and potent dual inhibitor of Bcl-2 and Bcl-xL, inducing apoptosis in Bcl-2 dependent Small Cell Lung Cancer (SCLC) cell lines. Then, we demonstrated that S 44563 induces the mitochondrial caspase activation pathway in H146 cells. Further evaluation of this molecule displayed significant antitumor activity when administrated intraperitoneally once a day for 21 days in a H146 xenograft model. The tumor regression lasted for more than 50 days without weight loss, suggesting anticancer effects with acceptable level of toxicity. Furthermore, it was proven that its in vivo response in small-cell-lung cancer models was enhanced by radiotherapy (*20*).

Based on its in vitro profile and its in vivo efficacy, S44563 was rapidly accelerated into preclinical development (*25*) and was well-tolerated in preclinical toxicology assessments (*20*).

Because of promising preliminary results, larger quantities of S 44563 were required and a first-generation process chemistry route to make batches between 1 and 3 kg of S 44563 was implemented to supply the early studies (*26, 27*).

First Generation Process Chemistry Route

Manufacture of the Biphenyl Acid Moiety

The first generation process chemistry route to tricyclic amine **27** is depicted in Scheme 4. A Pd/C-mediated hydrogenation of quinaldic acid under basic conditions provided the racemic 1,2,3,4-tetrahydroquinoline derivative **21** in water without any residue of over-reduction. A better conversion rate was observed when the reaction mixture was heated at 40 °C. Sodium carboxylate **21** was obtained in CH_2Cl_2 and the pH was adjusted to 4 ± 0.2 by adding 10 N HCl. A solvent swap to MeOH followed by the addition of 1.2 equiv of $SOCl_2$ provided ester **22** that was easily isolated as its hydrochloride salt. The corresponding free base was obtained in EtOAc after treatment with aqueous NaOH followed by concentration to dryness to produce **23** as a brown oil. Water content was critical and had to be kept at <0.1% to avoid any degradation of **23** on storage.

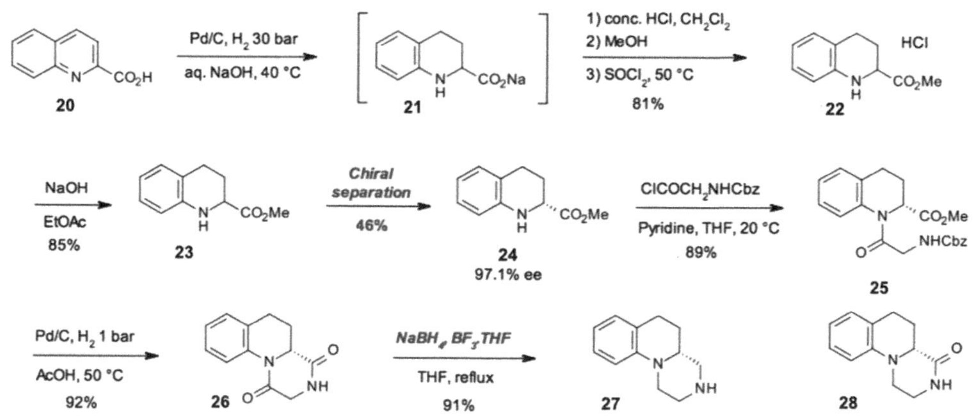

*Scheme 4. Initial process chemistry route starting from quinaldic acid **20** to manufacture the first kilogram of **27**.*

Chiral separation on a simulated moving bed (SMB) apparatus was performed on 2.5 kg scale batches in a mixture of acetonitrile and MeOH to provide desired enantiomer *R*-**24** with 97.1% ee (28). The cyclic amide ring was subsequently installed following a two-step sequence by treating Cbz-glycine acid chloride (obtained by the reaction of Cbz-glycine with $SOCl_2$ in a mixture of THF, pyridine and catalytic DMF) with **24** followed by Cbz removal mediated with Pd/C and hydrogen.

However, batch-to-batch variations in terms of conversion were noticed after the deprotection step and were attributed to residual sulfur poisoning the metal. Replacement of $SOCl_2$ with oxalyl chloride was beneficial in terms of both process robustness and chemical purity of **25** (increase from 60 to 96.9% after acidic workup and evaporation to dryness). The Cbz group was smoothly removed at 16 kg scale under one bar of hydrogen at 50 °C in a mixture of MeOH and glacial acetic acid.

Condensation between the resulting free amine and the acid chloride occurred and the product **25** cyclized spontaneously to provide diketopiperazine **26.** This was isolated in 98.8% purity and 98.0% ee after a solvent swap to EtOAc . Conversion to the key tricyclo intermediate **27** was initially carried out using a mixture of 4 equiv of $NaBH_4$ and three equiv of BF_3•THF to generate borane in situ. Under these conditions, reduction of the tertiary amide proceeded well whereas the secondary amide was more reluctant to reduce and led to the formation of up to 10% of impurity **28** in the crude product. After boron decomplexation with refluxing aqueous HCl, **27** was obtained in a disappointing 80% purity on 4.5 kg scale.

Moreover, a safety assessment carried out in an RC1 apparatus revealed that both BF_3 addition and hydrolysis were highly exothermic (estimated ΔT_{ad} of 84 and 101 °C, respectively, see Figure 4 and Figure 5). The latter was accompanied by a vigorous gas evolution culminating at 4.2 L/h/g of starting material. Although this process was implemented on a maximum scale of 4.5 kg to manufacture the first preclinical batches, we decided to not to proceed further until we investigated alternative reducing conditions for the synthesis of **27**.

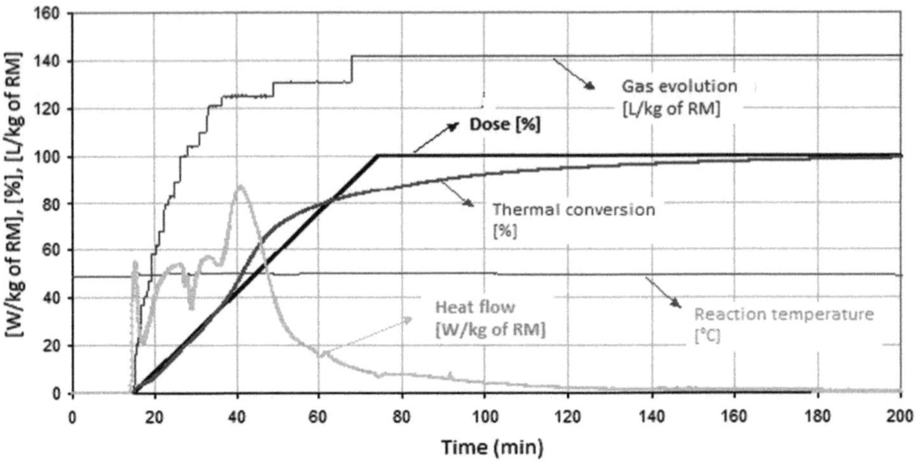

*Figure 4. RC1 trace for addition of BF_3•THF complex to **27**.*

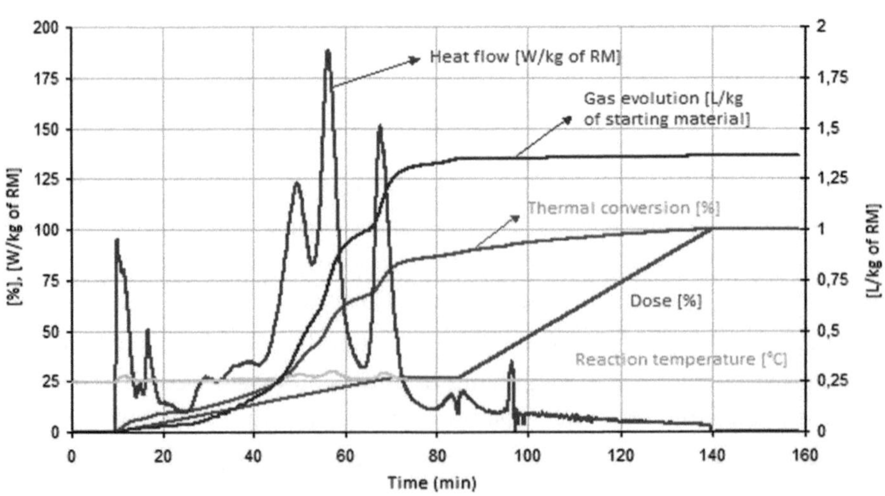

*Figure 5. RC1 trace of quench of reduction step to prepare **27** using $NaBH_4/BF_3$•THF complex.*

The conversion of amides into corresponding amines is well documented in the literature (*29–34*). Several reducing agents were evaluated for the transformation of **26** to **27** (Table 2). Except for entries 5 and 6, consumption of the diamide **26** was observed in most cases, along with the formation of partially reduced amide **28** ranging from 1 to 33%. Encouraging results were obtained with $LiAlH_4$ (entry 8) but we noticed a loss of enantiomeric purity during the workup (up to 10% with $LiAlH_4$ upon heating at 40 °C). In its place, Red-Al was selected (entry 9) as no racemization of the stereocenter was observed. At this point, we focused on optimization of the reaction conditions and workup. Exchanging THF for toluene was beneficial for the phase separation step after hydrolysis of the hydride in the reaction mixture. The amount of Red-Al was surveyed (4 to 8 equiv) leading to the use of 5 equiv as providing the best impurity profile.

As seen before, HPLC monitoring confirmed that the tertiary amide was easy to reduce (complete conversion at 20 °C) whereas the secondary amide was more reluctant (the reduction went to completion only upon warming the reaction to 50 °C) (35). Nevertheless, this new procedure was satisfactory on scale and despite a moderate 87% HPLC area chemical purity, it provided the desired tricyclic amine **27** which was used in the next step without further purification.

Table 2. Survey of Reducing Agents for Reducing Diamide 26

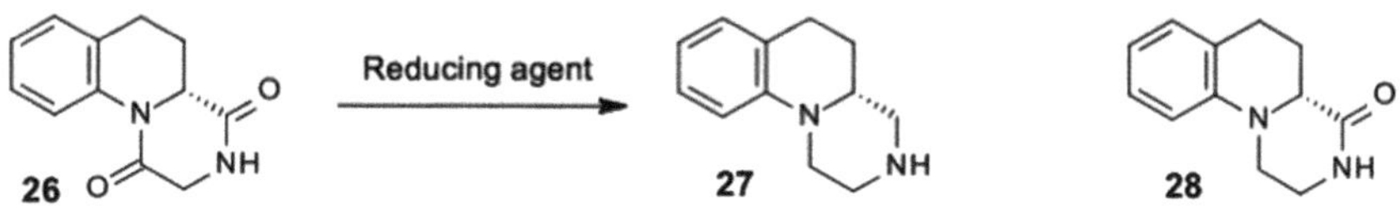

Entry	Reaction conditions	**26** (%)[a]	**27** (%)[a]	**28** (%)[a]
1	$NaBH_4$ (4 equiv), BF_3.THF (3 equiv), THF, 2 h at 50 °C	0	85	5
2	$NaBH_4$ (8.7 equiv), $ZrCl_4$ (2.2 equiv), THF, 3 h at 50 °C	0	47	15
3	9-BBN (4.7 equiv), THF, 16 h at 50 °C	0	12	33
4	Et_3SiH (8 equiv), $Ru_3(CO)_{12}$ (0.1 equiv), toluene, 18 h reflux	0	50	32
5	Ph_2SiH_2 (5 equiv), $RhH(CO)(PPh_3)_3$ (0.01 equiv), THF, 6 h at 50 °C	97	2	0
6	PMHS (7.5 equiv), H_2PtCl_6 (0.01 equiv), THF, 18 h reflux	98	0	0
7	Dibal-H (4 equiv), THF, 4 h at 50 °C	0	20	32
8[b]	$LiAlH_4$ (4 equiv), THF, 4 h at 20 °C	0	67	2
9[c]	Red-Al 65% (4.7 equiv), THF, 1 h at 50 °C	2	80	1

[a]Determined as HPLC area %. [b]On 10 g scale. [c]On 1 g scale.

Conversion of the tricyclic amine **27** to the ester **32** was then achieved following a three-step sequence (Scheme 5). The secondary amine was Boc-protected under classical conditions (Boc_2O and Et_3N in CH_2Cl_2) and subsequently brominated with NBS. Purification by flash column chromatography was required and provided **30** with a satisfactory purity (36).

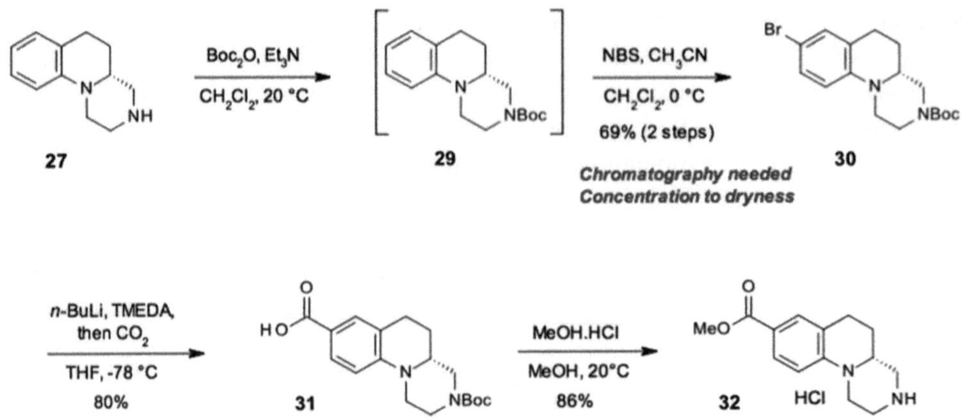

Scheme 5. Synthesis of ester **32**.

Carboxylation was originally performed with CO and catalytic $PdCl_2$(dppf) in refluxing MeOH. Since we could not use CO on large scale in our facilities, a lithium-halogen exchange reaction was developed to generate an activated species that could be trapped by CO_2 in situ. When bromide **30** was subjected to *n*-BuLi/TMEDA/THF at -78 °C followed by subsequent addition of CO_2, carboxylic acid **31** was obtained after acidic workup and filtration with a satisfactory 99% HPLC area purity. However, batch-to-batch variations combined with a debromination rate that could rise to 20% convinced us that this procedure could be problematic for the subsequent campaigns. When **31** was treated with anhydrous hydrogen chloride in methanol, we observed concomitant Boc removal and esterification providing methyl ester **32**, isolated as a salt.

The biaryl core was synthesized via a two-step sequence involving a Suzuki coupling between bromide **33** and boronic acid **34**, both commercially available, followed by alcohol chlorination with $SOCl_2$ (Scheme 6). The product **12** was obtained in low purity of 60-70 HPLC area % and required chromatographic purification on silica gel. This was a burden as in-house Ames testing produced positive results. *N*-Alkylation of **32** with purified **12** was achieved in refluxing MeCN using catalytic KI (to enhance the reaction rate) and an excess of K_2CO_3 to provide **36** (37).

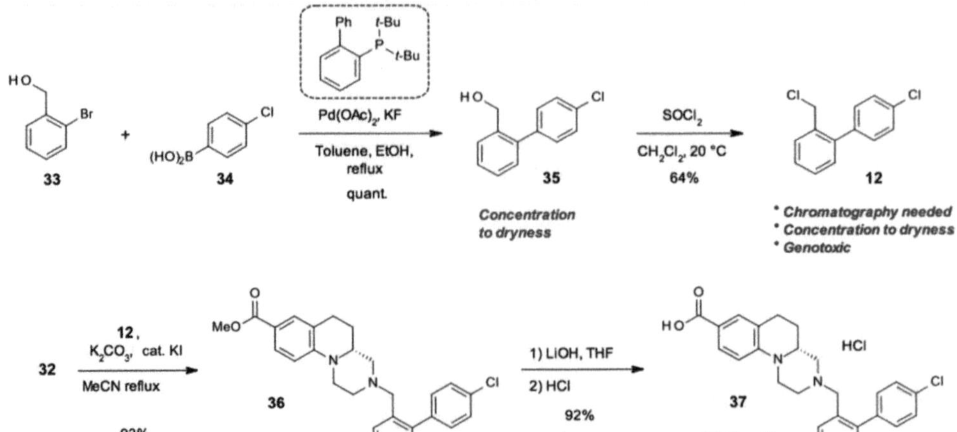

Scheme 6. Synthesis of biphenyl acid **37**.

Treatment of **36** with aqueous HCl in dioxane was originally applied to convert ester **36** into acid **37** but even after an extended reaction time, up to 4% of starting material remained in the crude. This prompted us to investigate alternative hydrolysis conditions (Table 3). After disappointing results with NaOH (entries 1 and 2), switching to the more nucleophilic LiOH provided a better conversion rate especially in a mixture of alcohol and water (entries 5 to 9). We noticed a better solubilization of the starting material in THF (entries 6 and 8) and the best results were obtained with four equiv of LiOH (entry 8, incomplete conversion if less base was charged). Isolation of the pentacycle **37** as the hydrogen chloride salt in EtOH was preferred because it provided an easier filtration combined with a better purity. The process was demonstrated on 5.6 kg scale and provided **37** in 99.2 HPLC area % chemical purity and 98.7% ee.

Table 3. Saponification of Ester 36[a]

Entry	*Nucleophile*	*Solvents*	*Conversion (%)*[b]
1	NaOH	*i*-PrOH, water	29
2	NaOH	*i*-PrOH, THF, water	42
3	LiOH	2-MeTHF, water	0
4	LiOH	Acetone, water	33
5	LiOH	*i*-PrOH, water	92
6	LiOH	*i*-PrOH, THF, water	98
7	LiOH	EtOH, water	92
8	LiOH	EtOH, THF, water	99
9[c]	LiOH	EtOH, THF, water	73

[a] Reactions run at 50 °C on 0.3 g scale with 4 equiv of nucleophile. [b] Determined as HPLC area % of species. [c] With 2.5 equiv of LiOH.

Manufacture of Sulfonamide 10 Bearing the Enantiopure 1,3-Diamine

The development of a procedure for the synthesis of the enantiopure 1,3-diamine moiety was one of the tougher challenges of this project. Enantiopure 1,3-diamines are important building blocks in the synthesis of ligands and bioactive compounds (*38, 39*). In contrast to vicinal diamines, there are very few documented methods describing the direct synthesis of 1,3-diamines (*40*). Precedents in the literature described the access of sulfonamide **10** and are illustrated in Scheme 7. For instance, Abbott chemists used *N*-protected D-aspartic acid **38** as a chiral pool in a seven-step synthesis (*9*) and a six-step approach starting with Fmoc-Asp(O*t*-Bu)-OH (**39**) (*41*). However, when we tried to reproduce these procedures, we experienced difficulties especially during the scale-up. Besides, the relatively high cost of unnatural amino acids could be a burden for a commercial route. This prompted us to investigate new approaches that allow the manufacture of large quantities of sulfonamide **10** at a reasonable cost.

The first evaluated approach relied on the enantioselective alkylation of *t*-butyl glycine analogue **40** by phase transfer catalysis (*42*). When imine **40** was treated with an excess of 1-bromo-2-

chloroethane in the presence of KOH and n-Bu_4NBr, the desired alkylated intermediate **41** was obtained. Other alkylating agents with nitrogen substitution at the 2-position led to degradation. Similarly, attempts to convert **41** to the corresponding dimethylamino compound via reaction with $HNMe_2$ also led to destruction.

A second explored strategy was based on a proline-catalyzed asymmetric α-hydrazination (*43*). 4-Bromobutanal **42** was synthesized starting from the corresponding acid in a two-step sequence (reduction with BH_3•THF complex in THF and subsequent oxidation of the resulting alcohol with TEMPO and bleach). Treatment of **42** with dibenzyl azodicarboxylate (DBAD) in the presence of *L*-proline and subsequent reduction of the aldehyde with $NaBH_4$ provided intermediate **43**. Due to a challenging hydrogenolysis of the N-N bond combined with a disappointing overall yield of 15% starting from the commercially available acid precursor of **42**, this strategy was not further developed.

A third explored approach relied on a Mannich-type reaction with ketone **44**. Reaction of **44** with a mixture of paraformaldehyde and dimethylamine hydrochloride under acidic conditions provided unusable complex product mixtures. Finally, we embarked on developing an alternative synthesis using acid chloride **46** as a starting material to synthesize sulfonamide **10**.

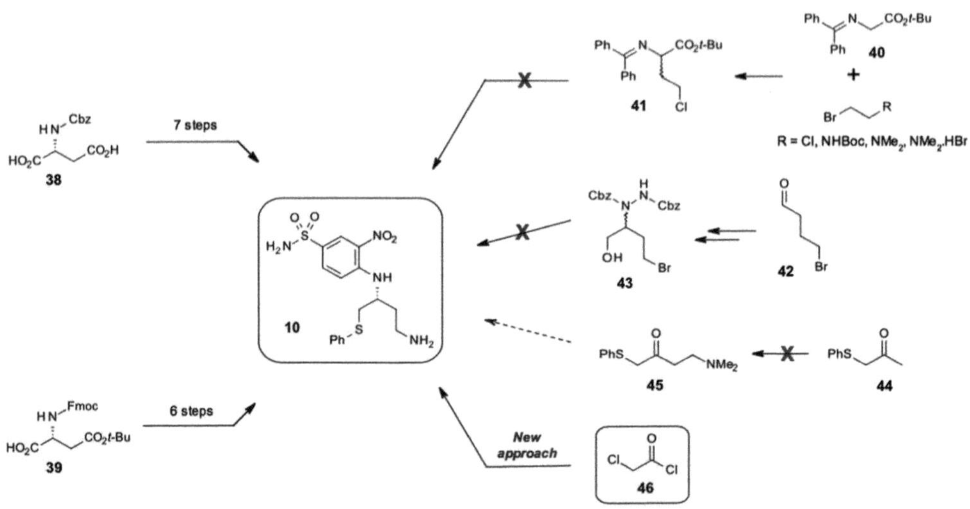

Scheme 7. Existing and alternative approaches to prepare ***10****.*

The synthesis of sulfonamide **10** is depicted in Scheme 8. Chloroacetyl chloride (**46**) was treated with MeONHMe under Schotten-Baumann conditions, to provide the Weinreb amide **47** with a higher purity than the commercially available material. A nucleophilic substitution carried out with sodium thiophenolate in THF provided **48** as an oil after concentration to dryness. The latter was then converted into ketone **50** in a two-step sequence involving a Grignard addition and a subsequent 1,4-addition of dimethylamine on unisolated enone **49**. Maintaining a low reaction temperature was critical to mitigate chemical degradation but despite extensive efforts, both yield and purity remained low. Due to the moderate stability of **50** on storage, it had to be used quickly in the next step.

Reaction of ketone **50** with $MeONH_2$•HCl in 2-MeTHF provided oxime **51** as an *E/Z*: 6/4 mixture of isomers. To avoid any degradation, this step was then telescoped into a BH_3•Me_2S-reduction to obtain racemic amine **52** after concentration to dryness. Isolation of **52** as its difumarate salt **53** proved to be an effective purifying step, and was achieved in a mixture of EtOH and EtOAc.

At this stage, difumarate salt **53** was obtained in 28% yield over three steps and >99 HPLC area % purity (*44*). Basic treatment to break the salt followed by chiral column chromatography finally led to enantiopure diamine **54**. A nucleophilic aromatic substitution on **55** using diamine **54** in DMSO in the presence of DIPEA provided desired sulfonamide **10**.

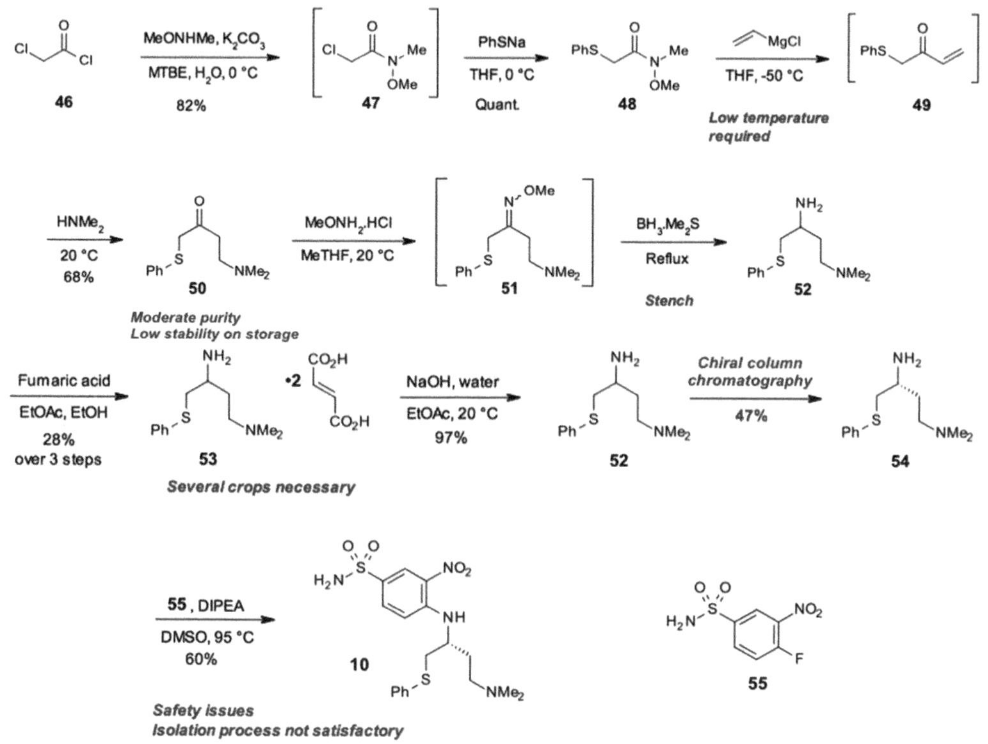

*Scheme 8. First generation synthesis of sulfonamide **10**.*

Although this approach enabled the manufacture of the first quantities of **10**, a few areas were identified as concerns for a future commercial route:

1. Handling the foul smell of $BH_3 \bullet Me_2S$ in the pilot plant was challenging in terms of health and safety.
2. Isolation of difumarate **53** was not straightforward and obtaining several crops were necessary to isolate a significant yield.
3. DSC studies on **55** revealed that a suspected autocatalytic decomposition occurred at 160 °C with a heat flow of 2000 J/g corresponding to a ΔT_{ad} of 1000 °C.
4. The isolation process of **10** proved to be unsuitable on larger scale (appearance of gums) and furnished **10** in only a moderate yield.

To overcome these processing issues, improvements had to be established and we embarked on evaluating alternative conditions.

Improvements to the Original Synthesis of 54

Regarding the conversion of oxime **51** into diamine **52**, a brief survey of reducing agents was carried out using the BH_3•THF complex, a mixture of $NaBH_4/ZrCl_4$ and Dibal-H (45). Despite encouraging results when applied to an *E/Z* mixture of **51**, BH_3•THF complex was not selected for the large scale work because of safety concerns. The mixture of $NaBH_4/ZrCl_4$ was not optimal either (a major exotherm occurred when mixing $NaBH_4$ and $ZrCl_4$ and during the reaction quench).

However, addition of 3.2 equiv of Dibal-H led to a complete conversion to the amine and this process was evaluated in a RC1 reaction calorimeter. Extended times for both the addition of Dibal-H and the quench with aqueous NaOH were recommended to dissipate the released energy at a rate to match with our plant cooling capacities. This slower addition also controlled gas evolution to a realistic degree.

Taking advantage of the improved conversion of the ketone **50** into oxime **51** in toluene, we telescoped three steps: oxime formation with $MeONH_2$•HCl (89 HPLC area % purity instead of 85% in 2-MeTHF), hydride reduction with Dibal-H, and the salt formation (Scheme 9). The last was carried out in MeOH and the best conditions were identified using a Crystal 16 apparatus (65/35 v/v mixture of toluene/MeOH at 25 L/kg of **52**).

To our delight, executing this sequence on a 8 kg scale of ketone **50** of only 62 % purity allowed us to isolate difumarate **53** of 99.5% purity and 47% overall yield without any appreciable material remaining in the mother liquors. Aqueous NaOH treatment provided the racemic free base that was separated using chiral column chromatography to furnish diamine **54** of 99.5% ee.

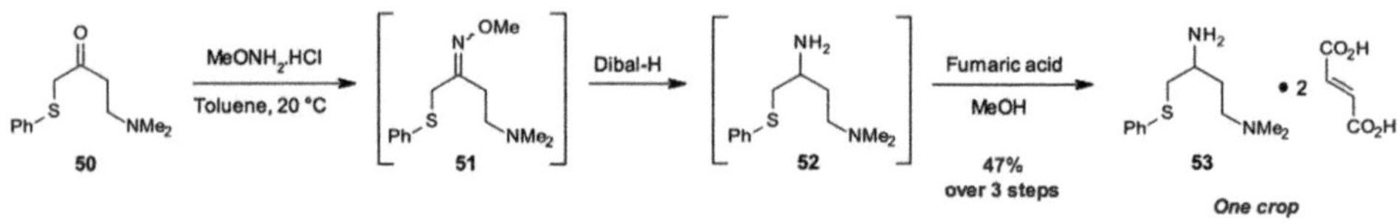

Scheme 9. New telescoped process to manufacture ***53****.*

With enantiopure 1,3-diamine **54** in hand, we then focused on the aromatic nucleophilic substitution (see Table 4). Instead of using fluoro-compound **55** as the substrate, we selected the chloro analogue **56** as DSC studies displayed that decomposition of **56** only started at 225 °C with a maximum reached at 306 °C. In order to ensure process safety, we evaluated the aromatic nucleophilic substitution at a temperature not higher than 90 °C using diisopropylethylamine (DIPEA) in various solvents. The desired product was formed in all solvents, but DMSO led to the most encouraging results (entry 6 and 7).

Lowering the temperature to 60 °C for 40 h resulted in a sluggish reaction (entry 7). We therefore retained 90 °C as a good balance between kinetics and safety (entry 6). Isolation of this step was also revisited since filtration of **10** obtained after a solvent swap from EtOAc to methylcyclohexane (MCH) produced sticky hard spheres that were difficult to drain. Moreover there was a risk of damaging our vessels upon scale-up. A better process resulted when MCH was replaced with *i*-PrOH. Incorporation of a reslurry with MTBE helped to remove traces of remaining **54** and facilitated the drying of the cake. These improvements allowed isolation of **10** on a 3.6 kg scale with a 74% yield and a chemical purity of >99%.

Table 4. Screening of Solvents at Gram Scale to Synthesize 10[a]

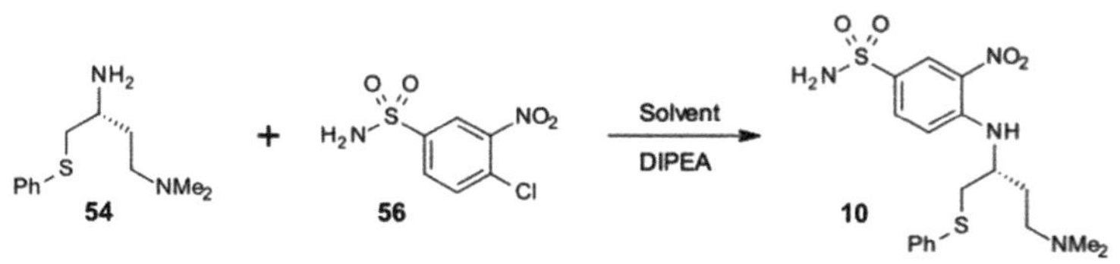

Entry	Solvent	**54** (%)[b]	**56** (%)[b]	**10** (%)[b]
1	AcOEt	11	25.8	57.2
2	*i*-PrOH	14	35	48.2
3	Toluene	11.6	23.1	51.4
4	DMF	44	5	32
5	NMP	3.5	19.2	57.6
6[c]	DMSO	0	3.6	81.7
7[d]	DMSO	4	8	73

[a]Reaction conditions: **54** (1.1 equiv), **56** (1.0 equiv), DIEA (4.5 equiv), 24 h, 90 °C. [b]Measured by HPLC analysis of a sample of reaction mixture. [c]Isolated yield of 56%. [d]40 h, 60 °C.

End-Game Chemistry

It is well known that acylsulfonamides represent an important class of compounds in medicinal chemistry as they serve as carboxylic acid bioisosteres. They can be prepared by condensing sulfonamides and carboxylic acids in the presence of dehydrating agents such as EDC (*N*-(3-dimethylaminopropyl)-*N*'-ethylcarbodiimide), for example. Due to the low nucleophilicity of sulfonamides, an excess of dimethylaminopyridine (DMAP) is usually required to facilitate the conversion. An acidic workup is then necessary to protonate the acylsulfonamide-DMAP salt and release the desired product (*46*). In our case, the original conditions involved 2.5 equiv of EDC and 3 equiv of DMAP in DMF at 20 °C to couple **37** and **10** (Scheme 10) (*47*). However isolation proved to be tricky since acylsulfonamide **57** as barely soluble in common solvents such as MeOH, EtOH, CH_2Cl_2, THF or DMF.

Attempts to isolate **57** by crystallization failed and only furnished gums or gels. Isolation as the hydrochloride salt of **57** was achieved and provided the final product free of any residual solvent. This salt option was eventually not selected since it still required a lyophilization step that promised to be a burden. Encouraging results were obtained when treating the reaction mixture with EtOAc and aqueous NaOH. The sodium salt **57** remained in the organic phase and DMF was removed into the aqueous layer. Isolation by filtration after a solvent switch to *i*-PrOH furnished **57** in 82% yield and 98.5% purity. A polishing filtration in acetone was incorporated and efficiently removed traces of

10. Solvent swap to *i*-PrOH and subsequent filtration provided the desired API with 92% yield and 99.3% HPLC area purity at 5 kg scale.

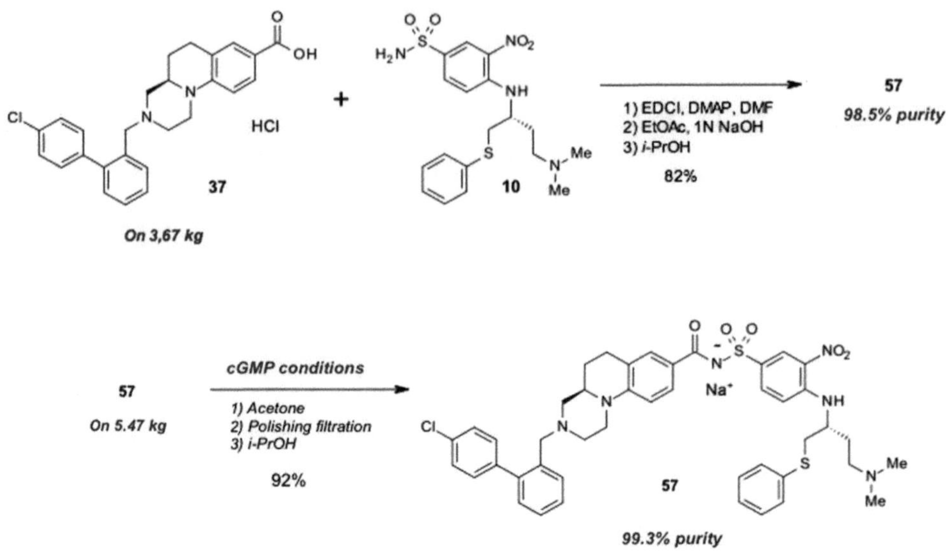

Scheme 10. Multikilogram-scale synthesis of Na salt **57**.

Although this synthesis enabled production of the first batches of API, we identified a few drawbacks that could be detrimental for larger scale implementation:

- Chiral phase chromatographic separations of both racemic mixtures of key intermediates **23** and **52** were carried out to furnish the desired enantiomers.
- A six-step sequence was required to convert tricyclic amine **27** into carboxylic acid **37**, including protecting group installation / removal, a tricky cryogenic carboxylation and *N*-alkylation with mutagenic benzylic chloride derivative **12** (positive Ames test).
- Benzyl chloride derivative **12** was prepared using Suzuki coupling, involving a patented phosphine ligand and required optimization.
- The EDC-mediated final coupling between acid **37** and sulfonamide **10** was not practical for scale-up. Alternative conditions required development to address issues related to:
    - stickiness
    - cost
    - instability on storage of EDC
    - the use of DMF as solvent in the final step
    - a tedious isolation process.

All these factors were subsequently addressed.

Development of Alternative Key Reactions

Toward an Enantioselective Synthesis of Tricyclo 27

Although the undesired enantiomer could be recycled, we sought to develop a synthesis that avoided any chiral column chromatography separation. An approach based on enantioselective hydrogenation of 2-substituted quinoline derivatives was therefore envisioned to access the corresponding enantiomerically enriched 1,2,3,4-tetrahydroquinoline (Scheme 11) (*48*, *49*). Moderate enantiomeric excess was obtained starting with methyl ester derivative **58** upon iridium-catalyzed hydrogenation in toluene using (*R*)-Segphos as a ligand. Gratifyingly, a brief substrate screen showed that Boc-protected tetrahydroquinoline **63** was obtained with 91% ee after a smooth hydrogenation of quinoline **62**. The latter was prepared from ester **58** following a three-step sequence involving a $NaBH_4$ reduction, a PBr_3-mediated bromination and a nucleophilic displacement with di-*tert*-butyliminodicarboxylate in the presence of potassium carbonate.

Although some work needed to be done (absolute configuration required establishment and ready access to the tricyclic amine **27** had to be in place), the proof of concept of this novel enantioselective approach was demonstrated and was a real step toward a more efficient synthesis.

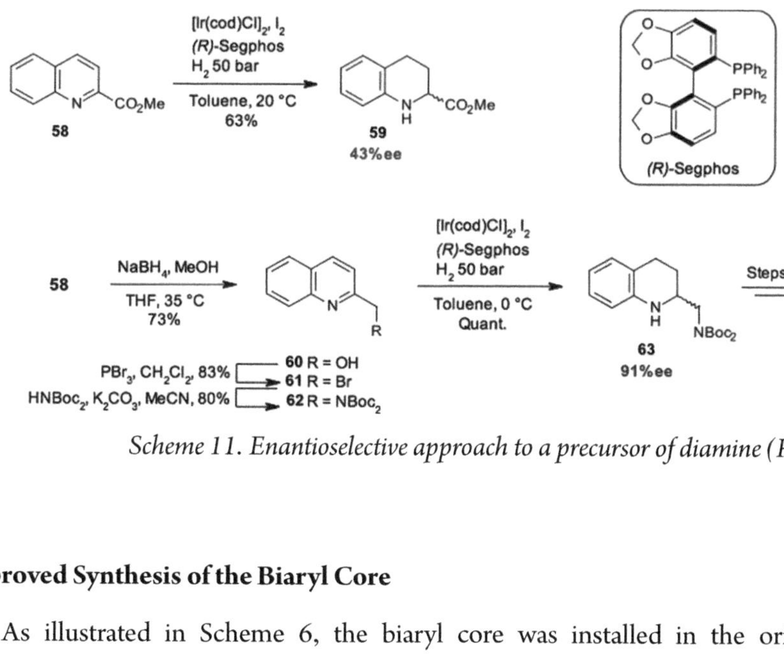

*Scheme 11. Enantioselective approach to a precursor of diamine (R)-**27**.*

Improved Synthesis of the Biaryl Core

As illustrated in Scheme 6, the biaryl core was installed in the original synthesis via a homogeneous Suzuki coupling. However, column chromatography of intermediate **12** was needed to improve the purity. Moreover, because of its mutagenic properties, the initial isolation process involving evaporations to dryness was not suitable in the long run. We therefore explored an alternative catalytic process to both generate the carbon-carbon bond and to keep the chloride derivative in solution in order to avoid handling a powder. Our interest quickly focused on ligandless coupling reactions mediated by palladium supported on charcoal. Pd/C is actually well-known to generate carbon-carbon bonds and is a valuable alternative to homogeneous catalysts (*50*). A first survey was undertaken in 2 dram glass vials and the results are summarized in Table 5.

Table 5. Screening of Conditions for Suzuki Coupling in Water

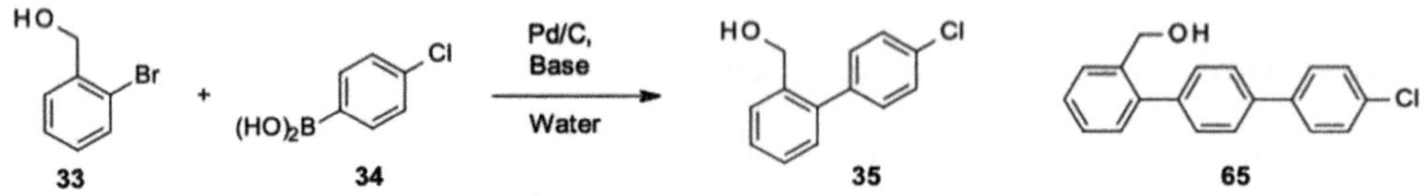

Entry	Reaction conditions[a]	33 (%)[b]	34 (%)[b]	35 (%)[b]	65 (%)[b]
1	0.2 g of **33**, NaOH (2 equiv), CTAB (0.06 equiv), 48 h at 60 °C	4.0	2.4	67	5.0
2	0.2 g of **33**, KF (2 equiv), CTAB (0.06 equiv), 48 h at 60 °C	9.7	25.7	10.5	0.5
3	3 g of **33**, K_2CO_3 (2 equiv), CTAB (0.06 equiv), 22 h at 60 °C	0.1	3.4	94.5	0.5
4	3 g of **33**, K_2CO_3 (2 equiv), 22 h at 60 °C	0.9	4.0	92.9	0.4
5	3 g of **33**, K_2CO_3 (2 equiv), 21 h at 80 °C	0.1	0.9	95.4	1.5

[a]Water 10 mL/g, 10% Pd/C type 91 (0.25 mol %). [b]Determined as HPLC area %.

Heating the mixture was necessary to achieve completion. Of the few bases surveyed, K_2CO_3 gave the best results in terms of conversion (entries 3, 4 and 5). A phase transfer catalyst such as hexadecyltrimethylammonium bromide (CTAB) was deemed not necessary in order to obtain a satisfactory conversion (entry 4). Reaction temperature appeared to be critical since heating at 80 °C enhanced the formation of impurity **65** (isolated and identified by NMR).

Since it has been established that the physical nature of heterogeneous catalysts can affect a reaction, we evaluated several kinds of Pd/C such as: type 91, type 39, type 395M, type 58, type 331, type 338, type 458, type 487, type 38H, type 394, type 87L, 2020/U,W,T2 from Johnson Matthey; Pd/CP ERW and Pd/CP M/R Delink from BASF (*51*, *52*). The properties of some of them are summarized in Table 6. Evaluation in a parallel reaction station at 0.2 g scale suggested that the oxidation state had no impact on the outcome of the reaction (see Figure 6). Regardless of whether Pd/C was reduced or unreduced, all conditions provided desired biaryl **35** in about the same yield. However this screening displayed higher rates of impurity **65** presumably due to different mass transfer. Promising results were obtained with catalysts 394, 2020/U,W,T2, 91, 458 and 39 leading to 78–83% of **35** in the reaction mixture. Further optimization in a 0.5 L cylindrical glass reactor on 20 g scale using Pd/C type 394 as catalyst provided the best results yielding biphenyl **35** of 97% purity after extraction with toluene. The amount of undesired **65** was controlled to NMT 0.7% HPLC area under these conditions.

With biaryl **35** as a solution in toluene, we decided to telescope the Suzuki coupling with the chlorination step using $SOCl_2$ as the chlorination agent. A brief survey of bases was carried out (Table 7). Conversion to benzyl chloride **12** occurred upon heating and depending on the nature of the base added, variations in terms of chemical purity arose. Pyridine (entry 7) provided the best results and chloride **12** was obtained in 77% yield over two steps.

Starting with 20 g of **33**, alcohol **35** was easily prepared and converted into chloride **12** with an overall yield of 83% estimated from an aliquot (Scheme 12). At this stage, we considered that a 97.3% purity was sufficient to be carried through to the next step without any further purification. Since in-house studies have established the mutagenic properties of benzyl chloride derivative **12**, handling this compound in solution appeared to us to be a better option than working with the powder.

Table 6. Physical Properties of Some Pd/C Catalysts in the Coupling Reaction of 33 with 34

Entry	*Catalyst*[a]	*Pd content*	*Water content*	*Oxidation state*
1	5R39	9.1%	60%	Unreduced
2	5R458	5%	47%	Reduced
3	5R394	5%	54%	Unreduced
4	5R38H	5%	55%	Reduced
5	5R338	5%	52%	Reduced
6	ERW	5%	65%	Reduced
7	MR Delink	5%	57%	Reduced

[a] All catalysts have an eggshell-type distribution.

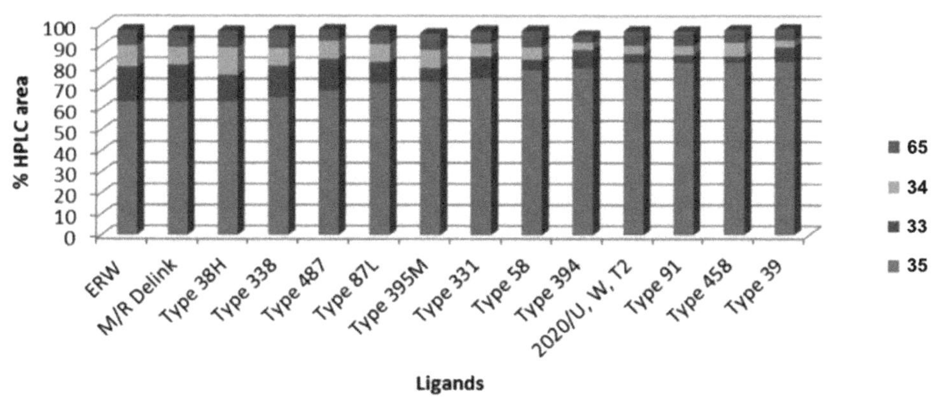

*Figure 6. Suzuki coupling of **33** (0.2 g) and **34** (1.3 equiv) with K_2CO_3 (2 equiv) and Pd/C (0.3 mol %) in water (10 mL/g) at 60 °C.*

Table 7. Chlorination of 35 with $SOCl_2$ and Various Bases

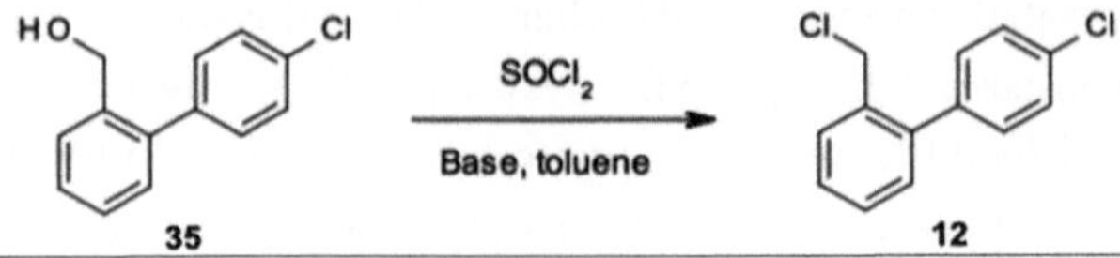

Entry	Reaction conditions[a]	Purity (%)[b]	Yield (%)[c]
1	16 h at 50 °C	75.4	95
2	Et_3N, 6 h at 90 °C	97.2	64
3	DMAP, 4 h at 90 °C	93.3	66
4	DBU, 4 h at 90 °C	81.0	-
5	DABCO, 2 h at 90 °C	86.8	-
6	Imidazole, 4 h at 90 °C	94.2	64
7	Pyridine, 4 h at 90 °C	97.3	77

[a]1 g **35**, $SOCl_2$ (1.2 equiv), base (1.5 equiv), toluene (9 mL/g). [b]Determined as HPLC area % of species. [c]Isolated yield starting from **35**.

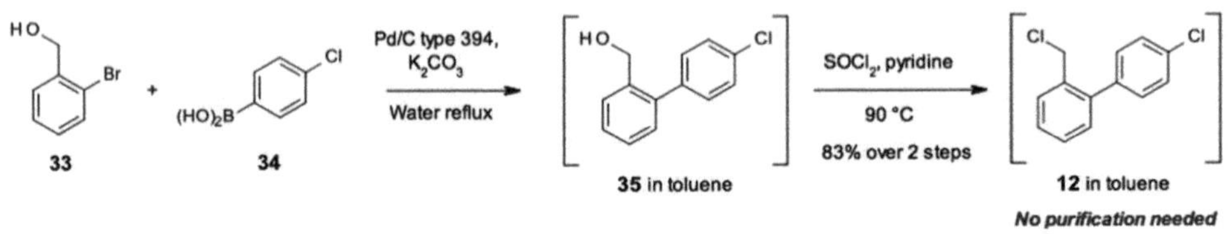

Scheme 12. Improved telescoped process to synthesize dichloride ***12***.

Improved Synthesis of Carboxylic Acid Derivative 31

Conversion of aromatic bromide **30** into its carboxyl analogue via a halogen metal exchange with *n*-BuLi and carboxylation was a challenging step for the synthesis. This reaction proved to be tricky and we experienced batch-to-batch variation attributed to the instability of the organolithium species. In order to mitigate the debromination side reaction (full debromination without carboxylation at -20 °C), a strict low temperature control was required (-78 °C).

To circumvent challenging cryogenic conditions on large scale, we focused on lithium trialkylmagnesates that have been used as metal-halogen exchange reagents for aryl halides. These have the advantage in that the resultant organomagnesiates are generally more thermally stable than the corresponding organolithium species and act successfully as nucleophiles under noncryogenic conditions (53, 54).

We began to evaluate a one-pot reaction with a solution of bromide **30** in THF that was added to a solution of n-Bu_2(i-Pr)MgLi (prepared by treating *n*-BuLi with *i*-PrMgCl) at -5 °C. Slow sparging addition of CO_2 to the reaction mixture was highly exothermic and provided not only the desired **31**, but also up to 20% of side product **66** that was isolated and characterized by NMR (Scheme 13).

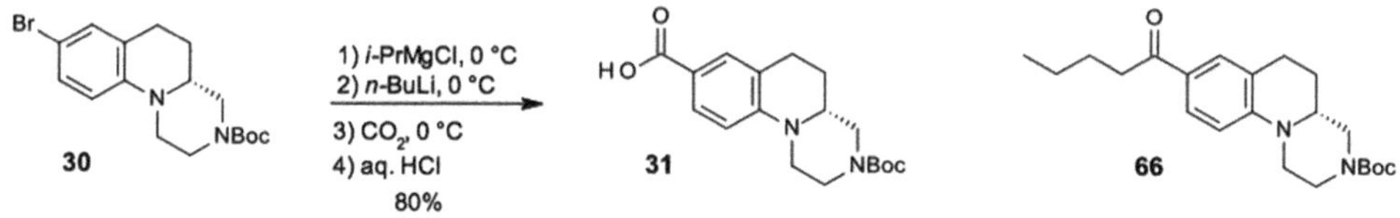

*Scheme 13. Metalation of **30** and subsequent carboxylation under noncryogenic conditions.*

This undesired pathway was fortunately shut down when CO_2 was introduced above the surface of the reaction mixture. This rather simple adjustment also efficiently mitigated the uncontrolled temperature increase. On 730 g scale, this operationally simple procedure provided **31** in 80% yield and 95.9% purity. Although there was some room left to improve the purity, this encouraging result prompted us to implement this approach on a more sophisticated substrate.

Second-Generation Synthesis of 37

As shown before, the synthesis of carboxylic acid **37** required numerous steps including group protection/removal (Me ester and Boc, see Scheme 5 and Scheme 6). In a desire to increase the step economy of the overall process, a straightforward synthesis of **37** was envisioned based on starting at *N*-alkylation of free amine **27** with chloride **12** in toluene solution (Scheme 14). With an excess of Et_3N and addition of catalytic KI, the reaction was completed after six hours at reflux. In the absence of KI, the reaction was sluggish and 23 h were needed to obtain full conversion.

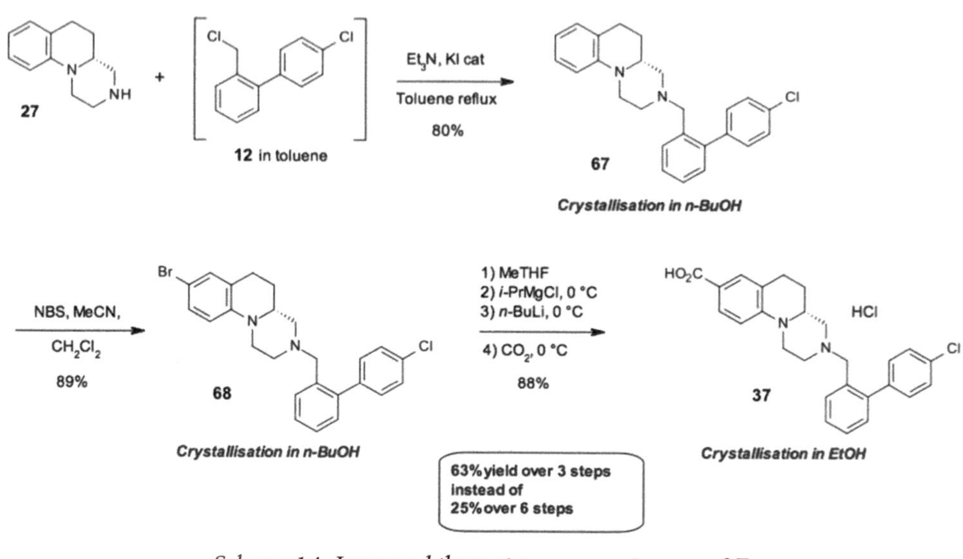

*Scheme 14. Improved three-step sequence to access **37**.*

Filtration of the reaction mixture combined with an aqueous wash of the filtrate was followed by a solvent switch to *n*-BuOH and provided crystalline **67** with a purity of 99% HPLC area after precipitation (55). The yield was improved to 80% when water (1.6 mL/g of **27**) was added to the suspension in *n*-BuOH (16 mL/g of **27**) and the solution cooled to 0 °C.

Regioselective bromination of **67** was successfully accomplished using NBS, followed by a basic workup to eliminate residual succinimide and by a solvent switch to *n*-BuOH. Further optimization demonstrated that seeding at 55 °C was beneficial to crystallize **68** (56). The yield was also improved (75% instead of 50%) when water was added before cooling to 0 °C. Finally, a slight increase of amount of seed (2 wt % instead of 1%) eased the filtration and provided bromide **68** in 89% yield and a chemical purity of 99.4%. Implementation of halogen exchange-carboxylation conditions were then undertaken (Table 8).

Table 8. Preliminary Screening of Solvents and Isolation Conditions for Carboxylation of Compound 68

Entry	*Solvent*	*i-PrMgCl (equiv)*	*n-BuLi (equiv)*	*Hydrolysis and isolation*[a]	*Purity (%)*[b]	*Yield (%)*[c]
1	THF	0.6	1.2	aq NaOH then aq HCl; filtration	93.3	66
2	THF	0.6	1.2	aq HCl then solvent switch to EtOH; filtration	92.9	81
3	Toluene	0.6	1.2	No isolation		
4	Toluene	1.2	2.4	aq NH_4Cl, aq NaOH then aq HCl; filtration	96.8	40
5	2-MeTHF	0.6	1.2	aq HCl		
6	2-MeTHF	0.6	1.2	anhyd HCl in MeOH; filtration	95.4	36
7	2-MeTHF	0.7	1.4	aq NH_4Cl then anhyd HCl in EtOH; filtration	94.6	79
8	2-MeTHF	0.7	1.4	aq NH_4Cl then anhyd HCl in MeOH; filtration	96.9	78
9	2-MeTHF[d]	0.9	1.8	aq NH_4Cl then anhyd HCl in MeOH; filtration	97.3	79

[a] Reactions run between -5 and 0 °C on 1 g of **68**, concentration: 10 mL/g of **68**. [b] Determined as HPLC area % of species. [c] Isolated yield. [d] Concentration: 5 mL/g.

Temperature was critical for the outcome of the reaction. When the reaction was carried out between -5 and 0 °C, we obtained satisfactory conversion rates. On the other hand, the reaction became sluggish at lower temperatures and led to the formation of impurities at room temperature. A quick survey of solvents showed that desired **37** was the predominant product in THF (entry 1 and 2) but an acidic or a basic quench led to a moderate yield and purity. Replacement of THF with toluene resulted in a slow conversion (entry 3) along with the formation of the undesired debromination product. Increasing both amounts of *n*-BuLi and *i*-PrMgCl led to a better conversion but the yield remained low due to a tedious isolation process (entry 4). A switch to 2-MeTHF allowed the smooth formation of the magnesiate intermediate after adding sequentially *i*-PrMgCl and *n*-BuLi to bromide **68** (entry 6 and 9).

Interestingly, separate generation of the trialkylmagnesium ate complex and subsequent addition of **68** provided only a 20% conversion. Adjustments such as increase of the excess of organometallic species led to a better conversion (entry 7, 8 and 9). An acidic quench was not selected since aqueous or anhydrous conditions with it led to an inoperable filtration (entry 5) and low yield (entry 6), respectively. The process was dramatically improved in terms of phase separation efficiency after switching to the use of aqueous NH_4Cl (entry 7, 8 and 9). Formation of the hydrochloride salt was evaluated in anhydrous conditions with EtOH (entry 7) and MeOH (entry 8). Due to a better purity observed in MeOH, the impact of a higher concentration of the reaction mixture combined with an excess of reagents was examined (entry 9). To our delight, crystalline **37** was isolated on a 5 g scale at 89% yield and 97.4% purity. Although there was still some room to improve the purity, this new sequence significantly reduced the number of steps (3 instead of 6) with a significant increase in the yield (65% instead of 25%) (*57*).

Toward a Third-Generation Synthesis of 37

N-Alkylation of primary and secondary amines is a common reaction that can be accomplished in different ways, including reaction of the amine with alkylating agents such as alkyl halides (*58*). Because of the established mutagenic properties of benzyl chloride **12**, it could be seen as a potential burden for future pharmaceutical manufacture. As a consequence, a borrowing hydrogen approach was envisioned involving tricyclic amine **27** and the benzylic alcohol **35**.

Water is the only byproduct generated during the condensation that features excellent atom efficiency (*59*). Iridium and ruthenium-based catalysts were surveyed in refluxing toluene (Table 9). Disappointing results were obtained with iridium, regardless of the nature of the counter-halide (entry 1 and 3). When K_2CO_3 was replaced with $NaHCO_3$, the conversion remained low (entry 1 and 2). Increase of catalyst loading led to a more reliable reaction (entry 4). This trend was also observed with ruthenium (entry 5 and 6). Since the latter provided encouraging results, a control experiment on 2 g scale involving 2.5 mol % of $[Ru(p\text{-cymene})Cl_2]_2$ and 5 mol % of dppf was carried out. After filtration of the crude solution through a pad of silica gel and a solvent switch to EtOH, crystalline **67** was isolated in 70% yield and 98% purity. The proof of concept of this alternative approach was therefore demonstrated and we could circumvent the use of the hazardous chloride **12**.

Toward an Asymmetric Synthesis of 1,3-Diamine 54

Despite its simple chemical structure, access to enantiopure diamine **54** had been one of the toughest challenges of this project. Our efforts ultimately culminated in an asymmetric synthesis relying on a diastereoselective aza-Reformatsky reaction and a chemoselective amination involving Mitsunobu conditions (Scheme 15) (*60*). We took advantage of Ellman's *N*-*tert*-butanesulfonamide **71**, known to be a powerful auxiliary for the synthesis of chiral amines. After choosing commercially available bromide **69** as starting material, conversion to sulfinimine **72** (*61*) was achieved after telescoping nucleophilic displacement with sodium thiophenolate, acetal deprotection and condensation of the resulting aldehyde with *N*-*tert*-butanesulfonamide **71**.

Table 9. Preparation of 67 Using the H-borrowing Approach

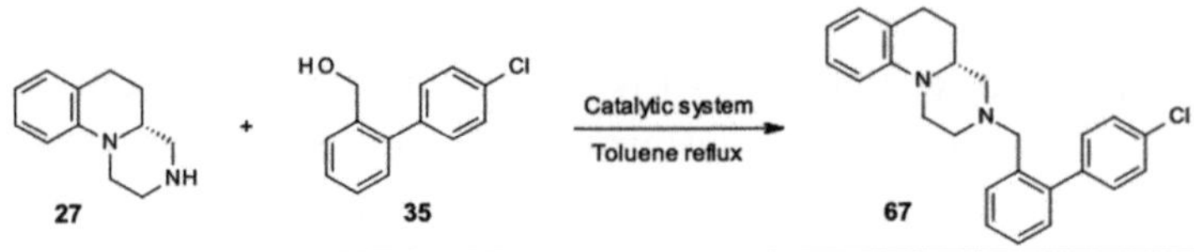

Entry	Reaction conditions[a]	**27** (%)[b]	**35** (%)[b]	**67** (%)[b]
1	[Cp*$IrCl_2$] (2.5 mol %), K_2CO_3 (5 mol %), 25 h	22	56	22
2	[Cp*$IrCl_2$] (2.5 mol %), $NaHCO_3$ (5 mol %), 25 h	25	50	33
3	[Cp*IrI_2] (2.5 mol %), K_2CO_3 (5 mol %), 39 h	23	45	21
4	[Cp*$IrCl_2$] (5 mol %), K_2CO_3 (10 mol %), 65 h	2	7	68
5	[Ru(*p*-cymene)Cl_2]$_2$ (1.25 mol %), dppf (2.5 mol %), 26 h	37	47	6
6	[Ru(*p*-cymene)Cl_2]$_2$ (2.5 mol %), dppf (5 mol %), 20 h	2	3	85

[a]All reactions were conducted on 0.2 g scale in refluxing toluene with an equimolar ratio of **27** and **35**. [b]Measured by HPLC analysis of a sample of reaction mixture.

The aza-Reformatsky reaction was the key step of this approach and appeared to be the main hurdle we encountered. Mixing activated zinc metal and α-bromoester is the common way to generate the Reformatsky reagent and this reaction is known to be highly exothermic with an unpredictable induction time. In order to overcome this issue, we took advantage of a zinc-activation procedure using diisobutylaluminium hydride (Dibal-H) that avoided the need of reflux to activate zinc, and limited the induction period (*62*). A quick survey showed that 1.7 equiv of methyl bromoacetate and three equiv of zinc were sufficient to obtain a satisfactory conversion.

To generate the Reformatsky reagent, zinc was suspended in THF at 30 °C with vigorous stirring before adding 0.3 equiv of methyl bromoacetate and 0.1 equiv of Dibal-H. Based on the results of RC1 experiments, the remaining bromide was added stepwise: 15% over 15 min and then 85% over 105 min (see Figure 7). Addition of the sulfinimine turned out to be easily manageable in terms of safety: 50% of the total charge was added in 30 min at 20 °C followed by 50% in the subsequent 15 min. In order to ensure adequate control during the exothermic quench, a three-step procedure was also developed involving addition of brine followed by aqueous citric acid and again by brine (*63*). Desired **73** was isolated after filtration through silica gel with a diastereomeric ratio of 94:6.

Thanks to the use of toluene as solvent, the reduction of the ester mediated by Red-Al was telescoped with a subsequent Mitsunobu reaction and led to **75** of >99% purity after purification on silica gel. Removal of the chiral auxiliary was achieved using acidic conditions and was telescoped with furamic acid salt formation to provide diamine **76** of >99% enantiomeric purity.

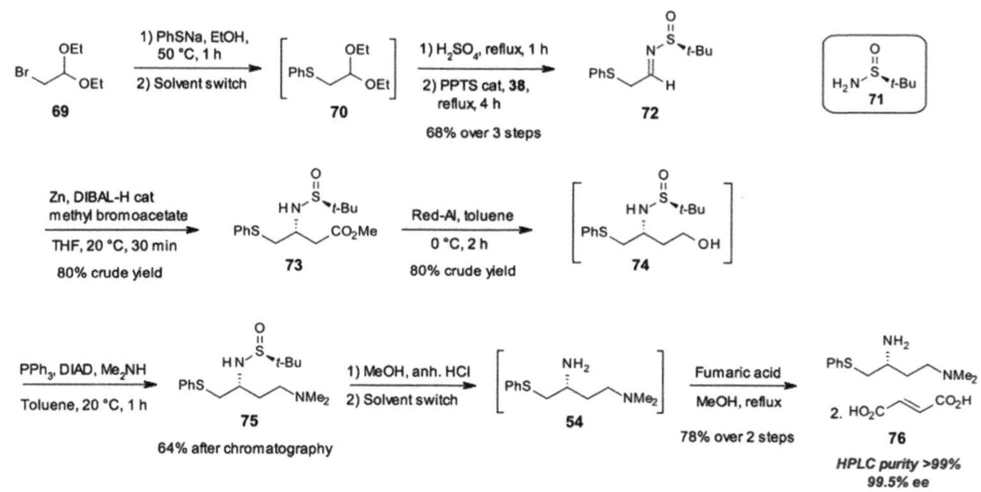

*Scheme 15. Second-generation synthesis of enantiopure diamine **54** isolated as difumarate salt **76**.*

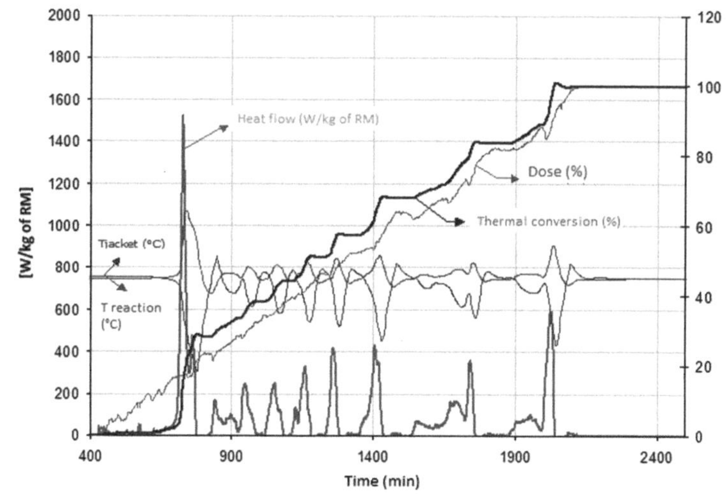

*Figure 7. Heat flow profile for addition of methyl bromoacetate to sulfinimine **72**.*

It is worthwhile to point out that during the acidic removal of the chiral auxiliary, we experienced an unexpected incident in the laboratory (*64*). After opening a bottle of a concentrated methanolic hydrogen chloride solution prepared a few months before, a massive spray suddenly occurred when a pipet sampled the solution. Fortunately, except for spillage inside the fume hood, no injury resulted. Investigations to determine the root cause were undertaken and showed the intrinsic instability of concentrated methanolic hydrogen chloride solutions on storage, especially at room temperature, led to the problem. Ethanolic hydrogen chloride solution avoided this off-gassing, and was subsequently used as it displayed a better stability with negligible variations of HCl concentration when stored for few months at 0– 5 °C.

Final Coupling: Searching for Alternative Conditions

Although the original synthesis allowed the manufacture of the first kilograms of API, there existed issues related to the final coupling step of **37** with **10**, such as the use of DMF as solvent and EDC as a coupling agent (sticky oil, concerns about its stability on storage, expensive). We had also concerns about the practicality of the isolation on larger scale because of the scale-up of what was a tedious process. Second-generation conditions were therefore surveyed employing alternative coupling agents such as thionyl chloride, CDI, propanephosphonic anhydride (T3P) and *N*-(3-(dimethylamino)propyl)-*N*'-ethylcarbodiimide hydrochloride (EDC-HCl). Disappointing results were obtained when we used hydrochloride **37** as a starting material. This prompted us to switch to a two-step procedure involving initial neutralization of the hydrochloride and subsequent treatment with the coupling agent (Table 10). The free amine of **37** was easily obtained in EtOAc after addition of an excess of Et_3N and filtration of the resulting $Et_3N•HCl$.

Table 10. Step-wise Sequence to Couple 37 with 1.1 Equivalents of 10

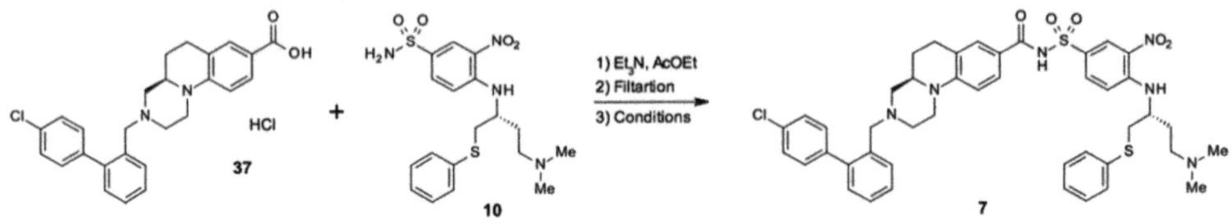

Entry	Reaction conditions	37 (%)[a]	10 (%)[a]	7 (%)[a]	Yield (%)[c]
1	$SOCl_2$ (2.5 equiv), EtOAc, 6 h at reflux	-	-	-	-
2[b]	CDI (1.5 equiv), THF, DBU (1 equiv), 3 h at reflux then **10** (1 equiv), 20 h at reflux	12.2	25.4	10.8	-
3	EtOAc, Et_3N (2 equiv), T3P (5 equiv), 23 h at reflux	3.2	0.9	86.6	-
4[b]	EtOAc, Et_3N (2 equiv), T3P (5 equiv), 23 h at reflux	5.7	0.5	84.4	55
5	EtOAc, Et_3N (2 equiv), T3P (2 equiv), 20 h at reflux	34.4	12.2	12.0	-
6	EtOAc, T3P (5 equiv), 24 h at reflux	2.0	0.6	92.8	60
7	EtOAc, T3P (5 equiv), 30 h at reflux	1.6	0.4	94.0	63

[a]Measured by HPLC analysis of a sample of reaction mixture. [b]1.0 equiv of **10**. [c]Isolated yield.

We had poor results with $SOCl_2$ and CDI, but encouraging coupling results were obtained with T3P. We discovered that the presence of $Et_3N•HCl$ was not detrimental to the coupling reaction whereas the amount of T3P was critical. An extended time was beneficial in terms of conversion since the reaction was sluggish. However basic quench with 6 N NaOH followed by solvent switch to *i*-PrOH and crystallization led to a disappointingly low 96.9% purity. Since no improvement was observed after extra reslurries, *N*-(1-methanesulfonyl)benzotriazole **78** was instead evaluated as a coupling partner. This was better but a superior synthesis of **78** was required.

Synthesis of Coupling Agent 78

Amidation mediated with *N*-acylbenzotriazoles acting as acylating reagents has been described by the Katritzky group (*65*). Their synthesis consisted of treating benzotriazole **77** with methanesulfonyl chloride in the presence of pyridine. Desired *N*-(1-methanesulfonyl)benzotriazole **78** ($BtSO_2Me$) was originally isolated in 89% yield after concentration to dryness and recrystallization from benzene. Since large quantities of this non-commercial reagent were anticipated, its access was revisited as we were searching for a procedure that could fit within industrial constraints.

A brief survey of alternative conditions was carried out (Table 11). As a starting point, we reproduced the original conditions to obtain the desired product in 48% yield after addition of water and concentration of the organic phase followed by recrystallization from toluene (entry 1). Using Et_3N instead of pyridine was beneficial in terms of yield (entry 2) but extended heating at reflux led back to benzotriazole 77.

Table 11. Alternative Conditions to Prepare 78

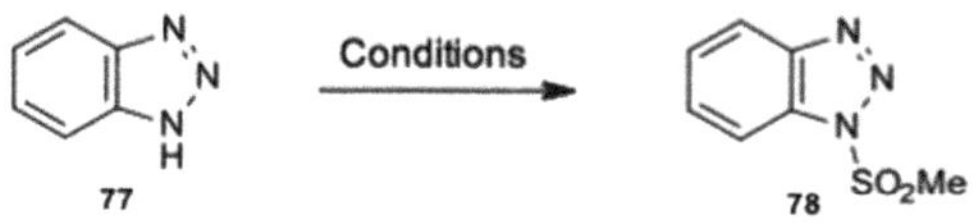

Entry[a]	Solvent, base (equiv), MesCl (equiv)	**77** (%)[g]	**78** (%)[g]	Yield (%)[h]	Purity (%)
1[b]	Toluene, pyridine (1.6), (1.2)	5.8	74.0	48	99.1
2[b]	Toluene, Et_3N (1.6), (1.2)	1.4	89.9	70	99.1
3[c]	2-MeTHF, pyridine (1.5), (1.3)	2.7	98.2	-	86.5
4[d]	2-MeTHF, 2,6-collidine (1.6), (1.4)	0.6	86.0	-	90.9
5[c]	2-MeTHF, Et_3N (1.5), (1.3)	0.1	99.4	-	98.8
6[e]	THF, Et_3N (1.3), (1.2)	0.5	99.2	57	96.6
7[f]	THF, Et_3N (1.5), (1.3)	0.6	99.2	87	99.7

[a]All reactions were run at 20 °C. [b]Addition of water, concentration of the organic phase and recrystallization in toluene. [c]Filtration of the reaction mixture, **78** obtained in solution. [d]Addition of water then phase separation. **78** obtained in solution. [e]Filtration of the reaction mixture, concentration to dryness, recrystallization in toluene. [f]Addition of water, filtration of the slurry. [g]Measured by HPLC analysis of a sample of reaction mixture. [h]Isolated yield.

Upon switching solvents to 2-MeTHF, Et_3N provided the best results with respect to both kinetics (1 h compared to 6 h with 2,6-collidine and 22 h with pyridine) and conversion (entry 3, 4 and 5). However, degradation occurred when **78** was stored in 2-MeTHF (loss of up to 10% purity after 72 h at 4 °C), seeking to telescope this reaction with the coupling with **37** (*66*, *67*). This would have placed constraints on a robust process.

Gratifyingly, the reaction went to completion after one hour in THF and a quench with water in turn led to a smooth crystallization of **78** (entry 7). On 500 g scale, **78** was isolated after filtration in 90% yield and >99% purity as a white powder which was stable on storage (no noticeable degradation after one year). This process was therefore much more practical than filtering off Et_3N•HCl from the reaction mixture before quenching and separating the phases (entry 6). With supply of **78** being secured, we then focused on developing new conditions for the coupling of **37** with **10**.

Condensation of acid **37** with $BtSO_2Me$ (**78**) was evaluated with an excess of Et_3N (3.5 equiv) in THF and 2-MeTHF. Upon reaction completion, the resulting suspension was filtered to remove Et_3N•HCl and the *N*-acetyl-benzotriazole **79** was obtained with a similar chemical purity in both solvents (>90 % with remaining acid <0.5 %). These solutions were subsequently treated with the anion of sulfonamide **10** prepared by adding various bases as described in Table 12. In order to generate the sodium salt of **10**, we focused on bases such as NaH, *t*-BuONa and NaOH (*68*). Deprotonation proceeded instantly at 20 °C with 2.5 equiv of NaH (60% in oil) as indicated by the appearance of a dark red color (*69*). After a slow addition of **79** to the reaction mixture at 40 °C followed by heating 1 h at reflux and quench with water, an organic phase containing 69 area % of **57** was obtained. When the crude was reslurried twice in a mixture of *i*-PrOH:MCH 87.5:12.5, the purity was improved to 98.5% (entry 1).

Table 12. Survey of Coupling Conditions Between 79 and 10 in the Presence of a Base

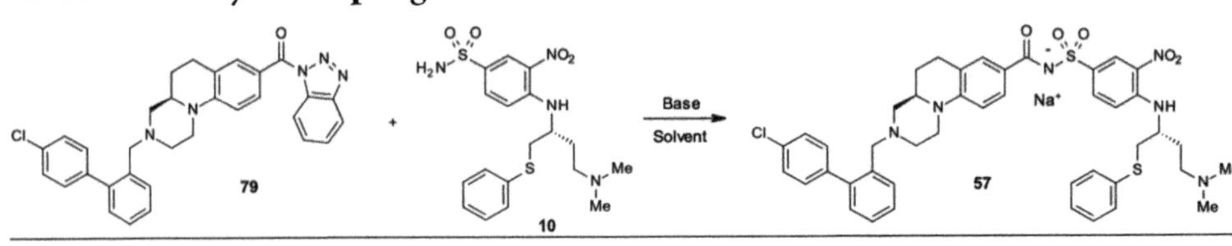

Entry	Reaction conditions[a]	57 (%)[b]	10 (%)[b]	Yield (%)[c]	Purity (%)[d]
1	NaH (2.5 equiv), THF, 1 h	69.4	8.5	70	98.5
2	*t*-BuONa (2.1 equiv), THF, 3 h	58.2	5.9	59	98.9
3[e]	NaOH (1.9 equiv), THF, 22 h	63.0	5.2	65	98.1
4[e]	NaOH (1.9 equiv), 2-MeTHF, 2 h	73.2	1.8	78	98.6
5[f]	NaOH (1.5 equiv), 2-MeTHF, 2 h	75.7	1.0	81	98.8

[a]All reactions run at reflux. [b]Measured by HPLC analysis of a sample of reaction mixture. [c]Isolated yield. [d]After three reslurries in *i*-PrOH:MCH 87.5:12.5. [e]NaOH prills. [f]NaOH flakes.

Due to concerns around the identification and accurate quantification of the residue in the API resulting from the oil in the NaH suspension, the use of NaH was discontinued and the same procedure was applied with *t*-BuONa. In this case, **57** was isolated with a moderate 59% yield and a satisfactory purity (entry 2) (*70*). Unfortunately, batch-to-batch variations were observed resulting in the formation of 10 to 22% of carboxylic acid **37**.

This lack of reproducibility prompted us to try NaOH as a base. Early experiments suggested that the nature of both NaOH and solvent impacted the outcome of the coupling reaction. Switching from THF to 2-MeTHF in the presence of NaOH prills improved the conversion (entry 3 and 4) (*71*). Using NaOH flakes was also beneficial in terms of conversion; further, the use of flakes allowed the reaction to proceed with fewer equivalents of NaOH (1.5 instead of 1.9) probably because of increased surface area (entry 5). Most of the benzotriazole released during the nucleophilic substitution was removed to the aqueous layer after treatment of the reaction mixture with aqueous NaOH. A solvent switch to *i*-PrOH under vacuum at 50 °C followed by subsequent filtration, facilitated by the addition of a small amount of MCH, provided **57** in 81% yield and 98.5% purity. Traces of **37** (0.3 to 0.5%) were purged after solubilization in refluxing acetone followed by filtration at 40 °C (*72*). A solvent swap to *i*-PrOH and addition of MCH provided desired **57** after filtration with 92% yield and 98.9% purity (Scheme 16).

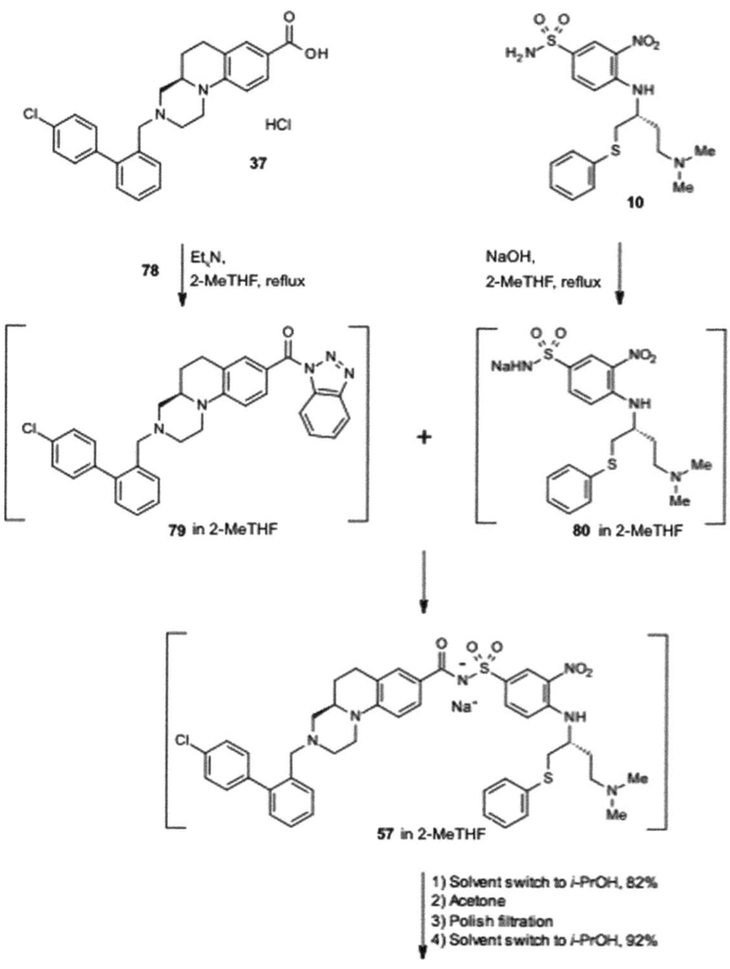

Scheme 16. Coupling with $BtSO_2Me$ and the isolation process.

Regarding the purity profile of the final product, several points are worth noting (Scheme 17). Upon coupling with $BtSO_2Me$, impurity **81**, previously derived from condensation of **37** with EDCI and difficult to purge, was no longer a concern. The formation of compounds **82**, **83** and **84** were attributed to two side reactions that occurred during the synthesis of the biaryl motif: dechlorination mediated by Pd to provide **82**, and multiple Suzuki couplings to provide **83** and **84**. Mono *N*-

methylated analogues **85** and **86** were suspected to arise from the contamination of Me_2NH by traces of $MeNH_2$. Finally, our choice to isolate the API in *i*-PrOH was motivated by the formation of impurity **87,** observed when CH_2Cl_2 was used in the presence of NaOH. Fortunately, the isolation process efficiently purged those impurities to levels <0.10%.

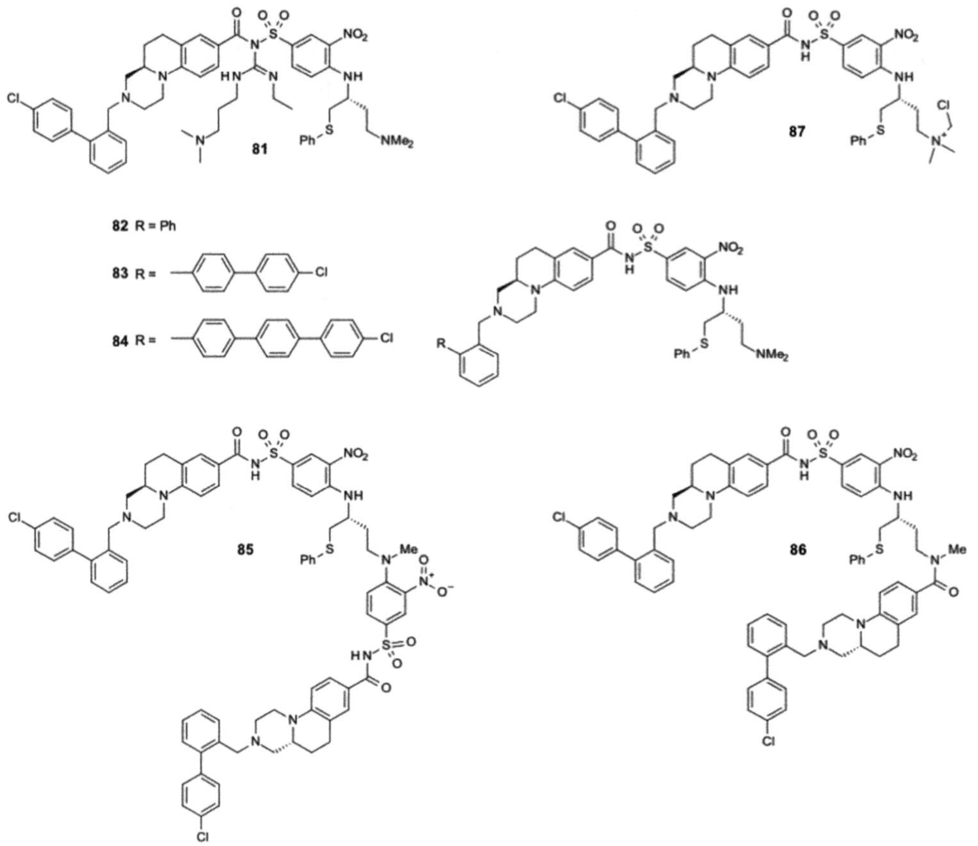

Scheme 17. Process-related impurities in API.

Summary

The discovery of S 44563, a potent inhibitor of anti-apoptotic proteins involved in programmed cell death, has been described. A convergent approach was envisioned to manufacture the API relying on the coupling of key precursors **37** and **10**.

The synthesis of hydrochloride **37** proceeded in 13 steps (longest linear sequence) from quinoline-2-carboxylic acid **20** in an overall yield of 8.4% to deliver batches of up to 5 kg (Scheme 18). To overcome some challenging reaction conditions, we developed alternative procedures featuring an enantioselective hydrogenation of 2-substituted quinoline derivatives, a telescoped Pd/C-mediated Suzuki coupling in water – chlorination – *N*-alkylation sequence and a magnesiate-type intermediate formation to enable incorporation of CO_2 at 0 °C. We also demonstrated that a H-borrowing approach could improve step economy and avoid the formation/isolation of a mutagenic intermediate.

The synthesis of sulfonamide **10** was made possible after designing an approach to manufacture batches of up to 3.6 kg of enantiopure 1,3-diamine **54** starting with chloroacetyl chloride (Scheme

19). In parallel, an enantioselective version was also developed involving an aza-Reformatsky reaction to install the stereocenter coupled with a chemoselective amination.

Finally, the final coupling between **37** and **10** to install the acylsulfonamide motif and isolation of the API proved to be challenging. Implementation of a polishing filtration enabled the manufacture of a first 5 kg clinical batch of API with a satisfactory purity, using EDC as a coupling agent. Alternative conditions involving *N*-(1-methanesulfonyl)benzotriazole were also developed and could be considered as a substitute to EDC.

Unfortunately, because of termination of this project at the beginning of clinical research, we did not have the opportunity to implement those alternative procedures on larger scale. This chapter represents the largest scale achieved.

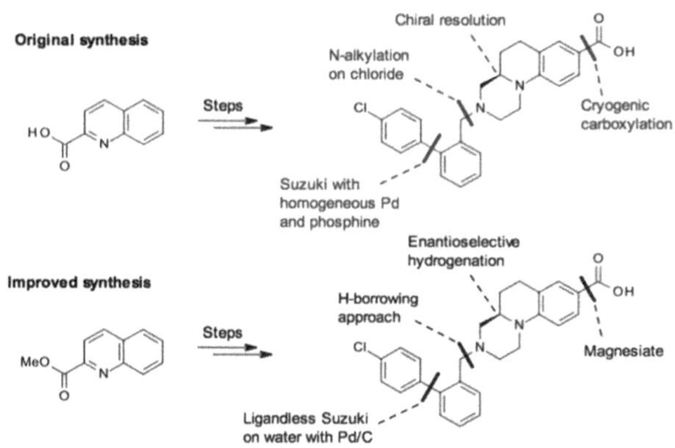

Scheme 18. Original strategy at kilogram scale to manufacture the first batches of carboxylic acid and development of alternative key reactions.

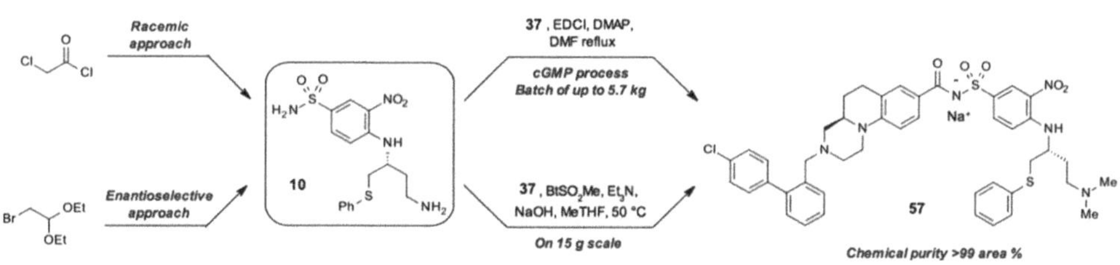

Scheme 19. Racemic approach at kilogram scale to manufacture the first batches of API and development of an enantioselective approach involving new conditions for the final coupling.

Acknowledgments

We thank our many colleagues of the Centre of Excellence Chemistry, Servier Research Institute and of the Industrial Research Centre, Oril Industrie who contributed to this project in addition to our colleagues whose names are included in the references below. Particularly for medicinal chemistry, we would like to acknowledge Patrick Casara and Thierry Le Diguarher for their tireless dedication to the discovery of BH_3 mimetics which has enabled this work.

References

1. Fuchs, Y.; Steller, H. Programmed Cell Death in Animal Development and Disease. *Cell* **2011**, *147*, 742–758.
2. Hanahan, D.; Weinberg, R. A. Hallmarks of Cancer: The Next Generation. *Cell* **2011**, *144*, 646–674.
3. Ziegler, D. S.; Kung, A. L. Therapeutic targeting of apoptosis pathways in cancer. *Curr. Opin. Oncol.* **2008**, *20*, 97–103.
4. Reed, J. C. Apoptosis-based therapies. *Nature Rev. Drug Discovery* **2002**, *1*, 111–121.
5. Danial, N. N.; Korsmeyer, S. Cell death: critical control points. *Cell.* **2004**, *116*, 205–219.
6. Adams, J. M.; Cory, S. The Bcl-2 apoptotic switch in cancer development and therapy. *Oncogene* **2007**, *26*, 1324–1337.
7. Labi, V.; Erlacher, M.; Kiessling, S.; Villunger, A. BH3-only proteins in cell death initiation, malignant disease and anticancer therapy. *Cell Death Differ.* **2006**, *13*, 1325–1338.
8. Youle, R. J.; Strasser, A. The Bcl-2 protein family: opposing activities that mediate cell death. *Nature Rev. Mol. Cell Biol.* **2008**, *9*, 47–59.
9. Oltersdorf, T.; Elmore, S. W.; Shoemaker, A. R.; Armstrong, R. C.; Augeri, D. J.; Belli, B. A.; Bruncko, M.; Deckwerth, T. L.; Dinges, J.; Hajduk, P. J.; Joseph, M. K.; Kitada, S.; Korsmeyer, S. J.; Kunzer, A. R.; Letai, A.; Li, C.; Mitten, M. J.; Nettesheim, D. G.; Ng, S.; Nimmer, P. M.; O'Connor, J. M.; Oleksijew, A.; Petros, A. M.; Reed, J. C.; Shen, W.; Tahir, S. K.; Thompson, C. B.; Tomaselli, K. J.; Wang, B.; Wendt, M. D.; Zhang, H.; Fesik, S. W.; Rosenberg, S. H. An inhibitor of Bcl-2 family proteins induces regression of solid tumours. *Nature* **2005**, *435*, 677–681.
10. Park, C.-M.; Bruncko, M.; Adickes, J.; Bauch, J.; Ding, H.; Kunzer, A.; Marsh, K. C.; Nimmer, P.; Shoemaker, A. R.; Song, X.; Tahir, S. K.; Tse, C.; Wang, X.; Wendt, M. D.; Yang, X.; Zhang, H.; Fesik, S. W.; Rosenberg, S. H.; Elmore, S. W. Discovery of an orally bioavailable small molecule inhibitor of prosurvival B-cell lymphoma 2 proteins. *J. Med. Chem.* **2008**, *51*, 6902–6915.
11. Gandhi, L.; Camidge, D. R.; Ribeiro de Oliveira, M.; Bonomi, P.; Gandara, D.; Khaira, D.; Hann, C. L.; McKeegan, E. M.; Litvinovich, E.; Hemken, P. M.; Dive, C.; Enschede, S. H.; Nolan, C.; Chiu, Y. L.; Busman, T.; Xiong, H.; Krivoshik, A. P.; Humerickhouse, R.; Shapiro, G. I.; Rudin, C. M. Phase I study of Navitoclax (ABT-263), a novel Bcl-2 family inhibitor, in patients with small-cell lung cancer and other solid tumors. *J. Clin. Oncol.* **2011**, *29*, 909–916.
12. Souers, A. J.; Leverson, J. D.; Boghaert, E. R.; Ackler, S. L.; Catron, N. D.; Chen, J.; Dayton, D. B.; Ding, H.; Enschede, S. H.; Fairbrother, W. J.; Huang, D. C.; Hymonwitz, S. G.; Jin, S.; Khaw, S. L.; Kovar, P. J.; Lam, L. T.; Lee, J.; Maecker, H. L.; Marsh, K. C.; Mason, K. D.; Mitten, M. J.; Nimmer, P. J.; Oleksijew, A.; Park, C-H; Phillips, D. C.; Roberts, A. W.; Sampath, D.; Seymour, J. F.; Smith, M. L.; Sullivan, G. M.; Tahir, S. K.; Tse, C.; Wendt, M. D.; Xiao, Y.; Xue, J. C.; Zhang, H.; Humerickhouse, R. A.; Rosenberg, S. H.; Elmore, S. W. ABT-199, a potent and selective Bcl-2 inhibitor, achieves antitumor activity while sparing platelets. *Nat. Med.* **2013**, 202–208.
13. San Feliciano, A.; Gordaliza, M.; Salinero, M. A.; Miguel del Corral, J. M. Abietane acids: sources, biological activities, and therapeutic uses. *Planta Med.* **1993**, *59*, 485–490.

14. Petros, A. M.; Dinges, J.; Augeri, D. J.; Baumeister, S. A.; Betebenner, D. A; Bures, M. G.; Elmore, S. W.; Hajduk, P. J.; Joseph, M. K.; Landis, S. K.; Nettesheim, D. G.; Rosenberg, S. H.; Shen, W.; Thomas, S.; Wang, X.; Zanze, I.; Zhang, H.; Fesik, S. W. Discovery of a Potent Inhibitor of the Antiapoptotic Protein Bcl-xL from NMR and Parallel Synthesis. *J. Med. Chem.* **2006**, *49*, 656–663.
15. Wipf, P.; Skoda, E. M.; Mann, A. *The Practise of Medicinal Chemistry*, 4th ed.; Elsevier, 2015; pp 279–299.
16. Fang, Z.; Song, Y.; Zhang, P.; Zhang, Q.; Lui, X. Conformational restriction of a type II FMS inhibitor leading to discovery of 5-methyl-N-(2-aryl-1H-benzo[d]imidazo-5-yl)isoxazole-4-carboxamide analogues as selective FLT3 inhibitors. *Future Med. Chem.* **2014**, *6*, 885–901.
17. El Bakali, J.; Muccioli, G. G.; Body-Malapel, M.; Djouina, M.; Klupsch, F.; Ghinet, A.; Barczyk, A.; Renault, N.; Chavatte, P.; Desreumaux, P.; Lambert, D. M.; Millet, R. Conformational Restriction Leading to a Selective CB2 Cannabinoid Receptor Agonist Orally Active Against Colitis. *ACS Med. Chem. Lett.* **2015**, *6*, 198–203.
18. Wang, Y.; Liu, J.; Dransfield, P. P.; Zhu, L.; Wang, Z.; Du, X.; Jiao, X.; Su, Y.; Li, A.; Brown, S. P.; Kasparian, A.; Vimolratana, M.; Yu, M.; Pattaropong, V.; Houze, J. B.; Swaminath, G.; Tran, T.; Nguyen, K.; Guo, Q.; Zhang, J.; Zhuang, R.; Li, F.; Miao, L.; Bartberger, M. D.; Correll, T. L.; Chow, D.; Wong, S.; Luo, J.; Lin, D. C.-H.; Medina, J. C. Discovery and optimization of potent GPR40 full agonists containing tricyclic spirocycles. *ACS Med. Chem. Lett.* **2013**, *4*, 551–555.
19. Zhang, B.; Nikolovska-Coleska, Z.; Zhang, Y.; Bai, L.; Qiu, S.; Yang, C.-Y.; Sun, H.; Wang, S.; Wu, Y. Design, synthesis, and evaluation of tricyclic, conformationally constrained small-molecule mimetics of second mitochondria-derived activator of caspases. *J. Med. Chem.* **2008**, *51*, 7352–7355.
20. Loriot, Y.; Mordant, P.; Dugue, D.; Geneste, O.; Gombos, A.; Opolon, P.; Guegan, J.; Perfettini, J.-L.; Pierre, A.; Berthier, L. K.; Kroemer, G.; Soria, J-C.; Depil, S.; Deutsch, E. Radiosensitization by a novel Bcl-2 and Bcl-XL inhibitor S44563 in small-cell lung cancer. *Cell Death Dis.* **2014**, 5, e1423.
21. Bruncko, M.; Ding, H.; Elmore, S.; Kunzer, A. R.; Lynch, C. L.; McClellan, W.; Park, C.-M.; Petros, A.; Song, X.; Wang, X.; Tu, N.; Wendt, M. D. *N-acylsulfonamide apoptosis promoters*. World Patent WO2005049594, Abbott Laboratories, 2005.
22. Martinelli, J. R.; Watson, D. A.; Freckmann, D. M. M.; Barder, T. E.; Buchwald, S. L. Palladium-catalyzed carbonylation reactions of aryl bromides at atmospheric pressure: a general system based on Xantphos. *J. Org. Chem.* **2008**, 73, 7102–7107.
23. McNulty, J.; Nair, J. J.; Capretta, A. A synthesis of sulfonamide analogs of platensimycin employing a palladium-mediated carbonylation strategy. *Tetrahedron Lett.* **2009**, *50*, 4087–4091.
24. Trost, B. M.; Bunt, R. C.; Pulley, S. R. On the use of O-methylmandelic acid for the establishment of absolute configuration of a-chiral primary amines. *J. Org. Chem.* **1994**, *59*, 4202–4205.
25. Casara, P.; Le Diguarher, T.; Geneste, O.; Hickman, J. *New tricyclic derivatives, method of preparing same and pharmaceutical compositions containing them*. European Patent EP1953161A1, 2008.

26. Hardouin, C.; Baillard, S.; Barière, F.; Copin, C.; Craquelin, A.; Janvier, S.; Lemaitre, S.; Le Roux, S.; Russo, O.; Samson, S. Multikilogram synthesis of a potent dual Bcl2/Bcl-x_L antagonist. 1. Manufacture of the acid moiety and development of some key reactions. *Org. Process Res. Dev.* **2020**, *24*, 652–669.
27. Hardouin, C.; Baillard, S.; Barière, F.; Craquelin, A.; Grandjean, M.; Janvier, S.; Le Roux, S.; Penloup, C.; Russo, O. Multikilogram synthesis of a potent dual Bcl2/Bcl-x_L antagonist. 2. Manufacture of the 1,3-diamine moiety and improvement of the final coupling reaction. *Org. Process Res. Dev.* **2020**, *24*, 670–685.
28. An enzymatic resolution was also developed to produce intermediate **24**. Unpublished results.
29. Hanada, S.; Motoyama, Y.; Nagashima, H. Dual Si–H effects in platinum-catalyzed silane reduction of carboxamides leading to a practical synthetic process of tertiary-amines involving self-encapsulation of the catalyst species into the insoluble silicone resin formed. *Tetrahedron Lett.* **2006**, *47*, 6173–6177.
30. Iding, H.; Wirza, B.; Rogers-Evans, M. Chemo-enzymatic preparation of chiral 3-aminopyrrolidine derivatives. *Tetrahedron* **2004**, *60*, 647–653.
31. Igarashi, M.; Fuchikami, T. Transition-metal complex-catalyzed reduction of amides with hydrosilanes: a facile transformation of amides to amines. *Tetrahedron Lett.* **2001**, *42*, 1945–1947.
32. Panunzio, M.; Tamanini, E.; Bandini, E.; Campana, E.; D'Aurizio, A.; Vicennati, P. 5-Phenylthio-1,3-oxazinan-4-ones via hetero Diels–Alder reactions: synthesis of (*R*)- and (*S*)-Duloxetines and Fluoxetines. *Tetrahedron* **2006**, *62*, 12270–12280.
33. Flaniken, J. M.; Collins, C. J.; Lanz, M.; Singaram, B. Aminoborohydrides. 11. Facile reduction of N-alkyl lactams to the corresponding amines using lithium aminoborohydrides. *Org. Lett.* **1999**, *1*, 799–801.
34. Sharp, L. A.; Zard, S. Z. A short total synthesis of (±)-Aspidospermidine. *Org. Lett.* **2006**, *8*, 831–834.
35. Intermediate **27** can be also isolated as tosylate salt; unpublished results.
36. Carboxylation applied to crude **30** gave a much lower yield.
37. Rate of conversion was dramatically improved with a catalytic amount of KI (6 h instead of 24 h).
38. Busscher, G. F.; Rutjes, F. P. J. T.; van Delft, F. L. 2-Deoxystreptamine: central scaffold of aminoglycoside antibiotics. *Chem. Rev.* **2005**, *105*, 775–791.
39. Hems, W. P.; Groarke, M.; Zanotti-Gerosa, A.; Grasa, G. A. [(Bisphosphine) Ru(II) Diamine] complexes in asymmetric hydrogenation: expanding the scope of the diamine ligand. *Acc. Chem. Res.* **2007**, *40*, 1340–1347.
40. Ji, X.; Huang, H. Synthetic methods for 1,3-diamines. *Org. Biomol. Chem.* **2016**, *14*, 10557–10566.
41. Wendt, M. D.; Shen, W.; Kunzer, A.; McClellan, W. J.; Bruncko, M.; Oost, T. K.; Ding, H.; Joseph, M. K.; Zhang, H.; Nimmer, P. M.; Ng, S.-C.; Shoemaker, A. R.; Petros, A. M.; Oleksijew, A.; Marsh, K.; Bauch, J.; Oltersdorf, T.; Belli, B. A.; Martineau, D.; Fesik, S. W.; Rosenberg, S. H.; Elmore, S. W. Discovery and structure–activity relationship of antagonists of B-cell lymphoma 2 family proteins with chemopotentiation activity in vitro and in vivo. *J. Med. Chem.* **2006**, *49*, 1165–1181.

42. O'Donnell, M. J. The enantioselective synthesis of α-amino acids by phase-transfer catalysis with achiral Schiff base esters. *Acc. Chem. Res.* **2004**, *37*, 506–517.
43. List, B. Direct catalytic asymmetric α-amination of aldehydes. *J. Am. Chem. Soc.* **2002**, *124*, 5656–5657.
44. Contrary to difumarate salt, isolation of the hydrochloride was not robust.
45. Other reducing agents (Red-Al, $NaBH(OAc)_3$, $LiAlH_4$, PMHS and $Pd(OAc)_2$ in EtOH, $NaBH_4$ and MeOH) were evaluated but either degradation or no reaction was observed.
46. Sturino, C. F.; Labelle, M. A convenient method for the preparation of acylsulfonamide libraries. *Tetrahedron Lett.* **1998**, *39*, 5891–5894.
47. Rate of conversion: 24 h in DMF, 48 h in THF and 48 h in CH_2Cl_2.
48. Maj, A. M.; Suisse, I.; Hardouin, C.; Agbossou-Niedercorn, F. Synthesis of new chiral 2-functionalized-1,2,3,4-tetrahydroquinoline derivatives via asymmetric hydrogenation of substituted quinolines. *Tetrahedron* **2013**, *69*, 9322–9328.
49. Maj, A. M.; Suisse, I.; Méliet, C.; Hardouin, C.; Agbossou-Niedercorn, F. Highly enantioselective hydrogenation of new 2-functionalized quinolone derivatives. *Tetrahedron Lett.* **2012**, *53*, 4747–4750.
50. Felpin, F. -X.; Ayad, T.; Mitra, S. Pd/C: an old catalyst for new applications – Its use for the Suzuki-Miyura reaction. *Eur. J. Org. Chem.* **2006**, 2679–2690.
51. Komaromi, A.; Szabo, F.; Novak, Z. Activity of palladium on charcoal catalysts in cross-coupling reactions. *Tetrahedron Lett.* **2010**, *51*, 5411–5414.
52. Liu, X.; Astruc, D. Development of the applications of palladium on charcoal in organic synthesis. *Adv. Synth. Catal.* **2018**, *360*, 3426–3459.
53. Gallou, F.; Haenggi, R.; Hirt, H.; Marterer, W.; Schaefer, F.; Seeger-Weibel, M. A practical non-cryogenic process for the selective functionalization of bromoaryls. *Tetrahedron Lett.* **2008**, *49*, 5024–5027.
54. Li, H.; Balsells, J. Highly selective and efficient conversion of aryl bromides to *t*-butyl benzoates with di-*t*-butyl dicarbonate. *Tetrahedron Lett.* **2008**, *49*, 2034–2037.
55. Other alcohols and solvents were evaluated (MeOH, EtOH, MTBE, acetonitrile, 2-MeTHF). *n*-BuOH was the only one that gave satisfactory results.
56. Crystallization of **68** in toluene after addition of solvents such as MeOH, EtOH, *i*-PrOH, MTBE or acetonitrile failed. No crystallization was observed in neat solvents such as MeOH, EtOH, *i*-PrOH, MTBE, acetonitrile or 2-MeTHF.
57. Coupling of **12** in toluene and **32** was also evaluated. After 5 h in refluxing toluene with Et_3N (4 equiv) and KI (4 mol %), completion of the reaction was observed. A solvent switch to *i*-PrOH provided **37** with 82% yield and 98.1% purity.
58. Constable, D. J. C.; Dunn, P. J.; Hayler, J. D.; Humphrey, G. R.; Leazer, J. L., Jr.; Linderman, R. J.; Lorenz, K.; Manley, J.; Pearlman, B. A.; Wells, A.; Zhang, T. Y. Key green chemistry research areas – a perspective from pharmaceutical manufacturers. *Green Chem.* **2007**, *9*, 411–420.
59. Leonard, J.; Blacker, A. J.; Marsden, S. P.; Jones, M. F.; Mulholland, K. R.; Newton, R. A survey of the borrowing hydrogen approach to the synthesis of some pharmaceutically relevant intermediates. *Org. Process Res. Dev.* **2015**, *19*, 1400–1410.

60. Laclef, S.; Taillier, C.; Penloup, C.; Viger, A.; Brière, J. -F.; Hardouin, C.; Levacher, V. Scalable asymmetric synthesis of a key fragment of Bcl-2/Bcl-xL inhibitors. *RSC Adv.* **2014**, *4*, 39817–39821.
61. DeCroos, P.; Han, Z. S.; Sidhu, K.; Lorenz, J.; Nummy, L.; Byrne, D.; Qu, B.; Xu, Y.; Wu, L.; Lee, H.; Roschangar, F.; Song, J. J.; Senanayake, C. H. Development of a large-scale asymmetric process for *tert*-butanesulfinamide. *Org. Process Res. Dev.* **2019**, *23*, 263–268.
62. Girgis, M. J.; Liang, J. K.; Du, Z.; Slade, J.; Prasad, K. A scalable zinc activation procedure using Dibal-H in a Reformatsky reaction. *Org. Process Res. Dev.* **2009**, *13*, 1094–1099.
63. Diluted aqueous citric acid was used to preserve the chiral auxiliary.
64. Hardouin, C.; Lemaitre, S. Safety case study. Intrinsic instability of concentrated solutions of alcoholic hydrogen chloride: potential hazards associated with methanol. *Org. Process Res. Dev.* **2020**, *24*, 867–871.
65. Katritzky, A. R.; He, H.-Y.; Suzuki, K. *N*-acylbenzotriazoles: neutral acylating reagents for the preparation of primary, secondary, and tertiary amides. *J. Org. Chem.* **2000**, *65*, 8210–8213.
66. When $BtSO_2Me$ was isolated after concentration to dryness and was reacted, an incomplete conversion was observed. When a crystallized $BtSO_2Me$ was employed, conversion was complete.
67. The solution of *N*-(arylacetyl)-benzotriazole **78** in THF or in 2-MeTHF can be stored for 24 h at 5 °C. To avoid any degradation, we decided to telescope the formation of **79** and its condensation with the anion of **10**.
68. A coupling reaction mediated with *n*-BuLi furnished the corresponding lithium salt of **57** and switching to the sodium analogue proved to be impraticable.
69. When 1.5 equiv was used, only 27% of **57** resulted in the reaction mixture.
70. When 1.4 equiv was used, only 26% of **57** resulted in the reaction mixture.
71. Water released during formation of **80** is less soluble in 2-MeTHF, limiting its hydrolysis or hydrolysis of **79**.
72. Amounts of <4% of remaining carboxylic acid led to levels of <0.1% after polishing filtration and a reslurry in *i*-PrOH/MCH.

Chapter 7

Discovery and Chemical Development of Tesirine: An Antitumor Pyrrolobenzodiazepine Antibody-Drug Conjugate Drug-Linker

Arnaud C. Tiberghien[*,1] and Jeremy S. Parker[2]

[1]AstraZeneca Oncology R&D, Spirogen, QMB Innovation Centre, 42 New Road, E1 2AX London, United Kingdom

[2]Early Chemical Development, Pharmaceutical Sciences, R&D, AstraZeneca, SK10 2NA Macclesfield, United Kingdom

[*]Email: arnaud.tiberghien@astrazeneca.com

For the past ten years, antibody-drug conjugates have been recognized as a viable approach for the treatment of cancer. The field initially relied on a limited number of active small molecules (also called payloads) to be delivered in a targeted approach. As the search for new categories of payloads expanded, a set of molecules based on naturally occurring antitumor antibiotics, the pyrrolobenzodiazepines (PBDs), caught the attention of the field. In particular, one PBD drug-linker, tesirine, was conjugated to a variety of tumor-targeting antibodies and entered multiple clinical trials. In this chapter, we detail the path that led to the discovery of tesirine. Factors guiding the design and synthetic route choice are explained. We discuss the rationale for the choice of the Discovery route to provide material on scale for Phase I clinical trials, and the challenges that were encountered in the process. Initial Development campaigns focused on minimizing the number of high-containment steps, affording a significant cost reduction. Subsequent campaigns introduced new chemistry to reduce the overall step count.

Introduction

Tesirine (**1**, Figure 1) is a pyrrolobenzodiazepine (PBD) drug-linker that has been conjugated to a large number of tumor-targeting antibodies such as anti-CD19, CD25, DLL3, 5T4, PSMA (prostate-specific membrane antigen), BCMA (B-cell maturation antigen), and c-Met (mesenchymal-epithelial transition factor), to provide oncological antibody-drug conjugates (ADCs) (*1–7*). After binding to the cell membrane target antigen, internalization and cleavage of the linker, the active cytotoxic drug SG3199 (**2**) is released intracellularly. The PBD dimer SG3199 is a highly efficient DNA crosslinker that can block replication and trigger cell death (*8*).

Many of the tesirine-ADCs have progressed into the clinic, reaching various phases in their development from Phase I through to Phase III. As the drug-linker component of these clinical ADCs is the same, clinicians and ADC scientists are now able to draw conclusions on the impact of other factors affecting the outcome of the trials, such as antigen target and antibody design.

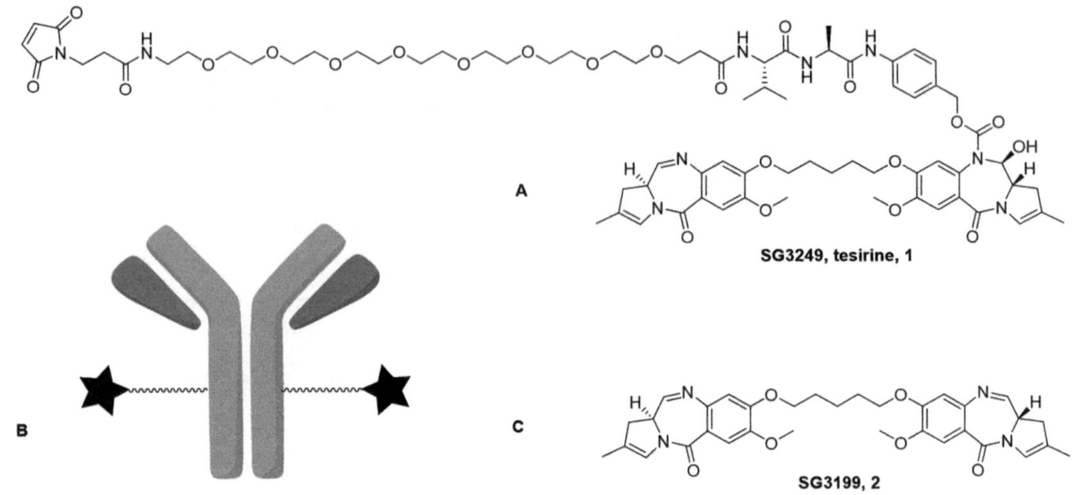

*Figure 1. (A) Structure of drug-linker tesirine (**1**). The terminal maleimide is conjugated in a Michael addition to two cysteines thiols on the antibody, either at random or at specific sites; (B) Schematic depiction of an antibody-drug conjugate bearing two drug molecules (DAR2); (C) SG3199 (**2**), the active drug or "warhead" released by tesirine cleavage in the presence of peptidases. (Adapted with permission from reference (9). Copyright 2018, American Chemical Society).*

Over the years, these multiple clinical trials have created a continuous and increasing demand for tesirine, and Process Development chemists joined forces with the initial Discovery team to face scale-up challenges and analytical development. Many issues were presented, such as synthesis length, number of intermediates requiring chromatographic purifications, and cost of the high-containment facilities required in the latter stages of the synthesis for potent intermediates. The way in which these challenges were addressed are described in this chapter.

Antibody-Drug Conjugates: Considerations Leading to the Selection of Tesirine

The reasons why tesirine is the drug-linker component of so many ADCs in clinical trials are numerous, including business, intellectual property, manufacturing and pharmacological motivations. To understand these drivers, we must first examine what makes a good antibody-drug conjugate.

An ADC is essentially a targeted delivery vehicle, carrying its payload to cells displaying the relevant antigen target. As the payload is typically cytotoxic, it goes without saying that the target must be highly expressed on cancer cells, as homogeneously as possible throughout the tumor, with low or little expression on healthy cells. The antigen expression ratio between cancer and healthy cells can determine whether an ADC will be successful or not. Another safety precaution arises from the biochemistry of the linker between the antibody and the drug. The linker must be stable in circulation to prevent any premature drug release, but should be cleavable under specific conditions (i.e., lysosomal uptake, drop in pH, glutathione sensitivity) after internalization by the targeted cell (Figure 2).

Finally, the drug (or warhead) must be highly potent. Although many factors influence targeted drug delivery via an ADC (drug-to-antibody ratio, antigen target number, expression profile, internalization and processing rates) (*10*), it was empirically determined that traditional cytotoxic agents (with IC_{50}s in the micromolar range such as methotrexate) (*11*) were not potent enough to be delivered by an antibody. As an antibody typically weighs 150 000 Da, the payload only makes up about 1 to 2% of the total ADC mass depending on the loading and molecular weights. In addition, only a small percentage of the injected dose is ultimately absorbed by the tumor. As a result, more active compounds with nanomolar or sub-nanomolar activity were required.

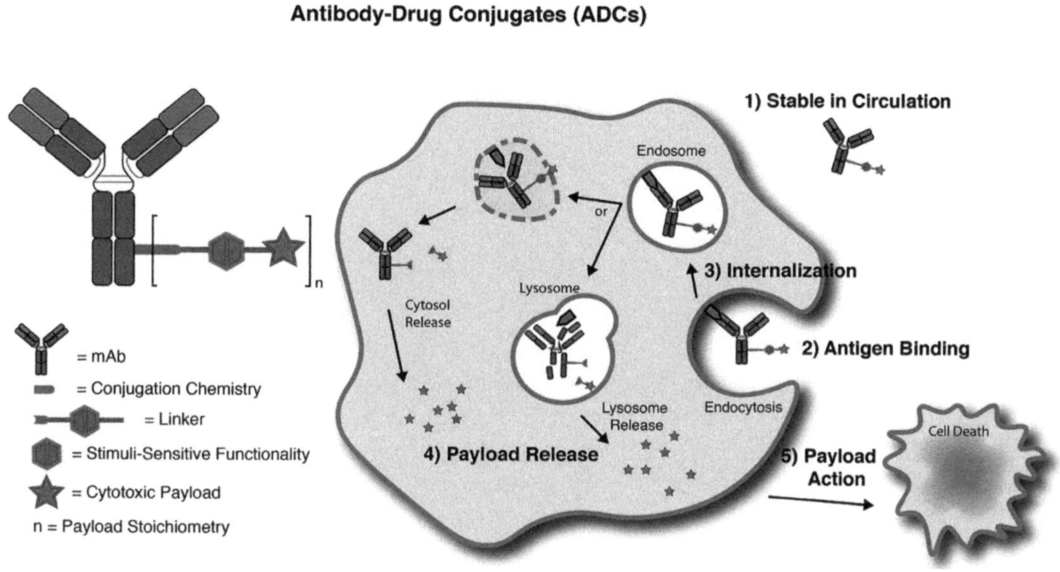

Figure 2. Antibody-drug conjugate components and mode of action. (Reproduced with permission from reference (12). Copyright 2015, American Chemical Society).

In the 1990s, the ADC field started to look for molecules which would be potent enough to match these criteria and came back with two categories: anti-mitotic agents targeting the tubulin system, or DNA damaging agents. Of these molecules, four gave rise to marketed ADCs: anti-tubulin auristatin MMAE (monomethyl auristatin E, Seattle Genetics) which is a component of Adcetris®, Polivy® and Padcev®, maytansinoid DM1 also inhibits the microtubular system and is a component in of Kadcyla® (Genentech), and DNA damaging calicheamicin derivatives used in Besponsa® and Mylotarg® (Pfizer). Finally, the topoisomerase inhibitor deruxtecan is used as the payload moiety in recently approved trastuzumab deruxtecan (Enhertu®) (*13*) by Daichii Sankyo and AstraZeneca (Figure 3).

All four warheads are highly cytotoxic, with the calicheamicin in the pM range, and the auristatins, maytansinoid and topoisomerase molecules in the nM range (*14*).

Despite the clinical success of these four warheads, the ADC field has continued its search for further molecular diversity. In addition to nanomolar potency, other criteria were added to the list, such as multi-drug resistance (MDR) evasion (MDR happens when cancer cells evolve resistance to a variety of drugs by over-expressing efflux systems such as P-glycoprotein; this became an issue evident from MMAE ADCs clinical data (*15*)), the ability to inhibit slowly dividing cancer stem cells (CSCs), as well as synthetic flexibility and scalability. In the 1990s, and in parallel to the burgeoning field of ADCs, medicinal chemistry efforts developed anticancer DNA binding agents based on naturally occurring pyrrolobenzodiazepines that were discovered in several Streptomyces microorganisms (Figure 4).

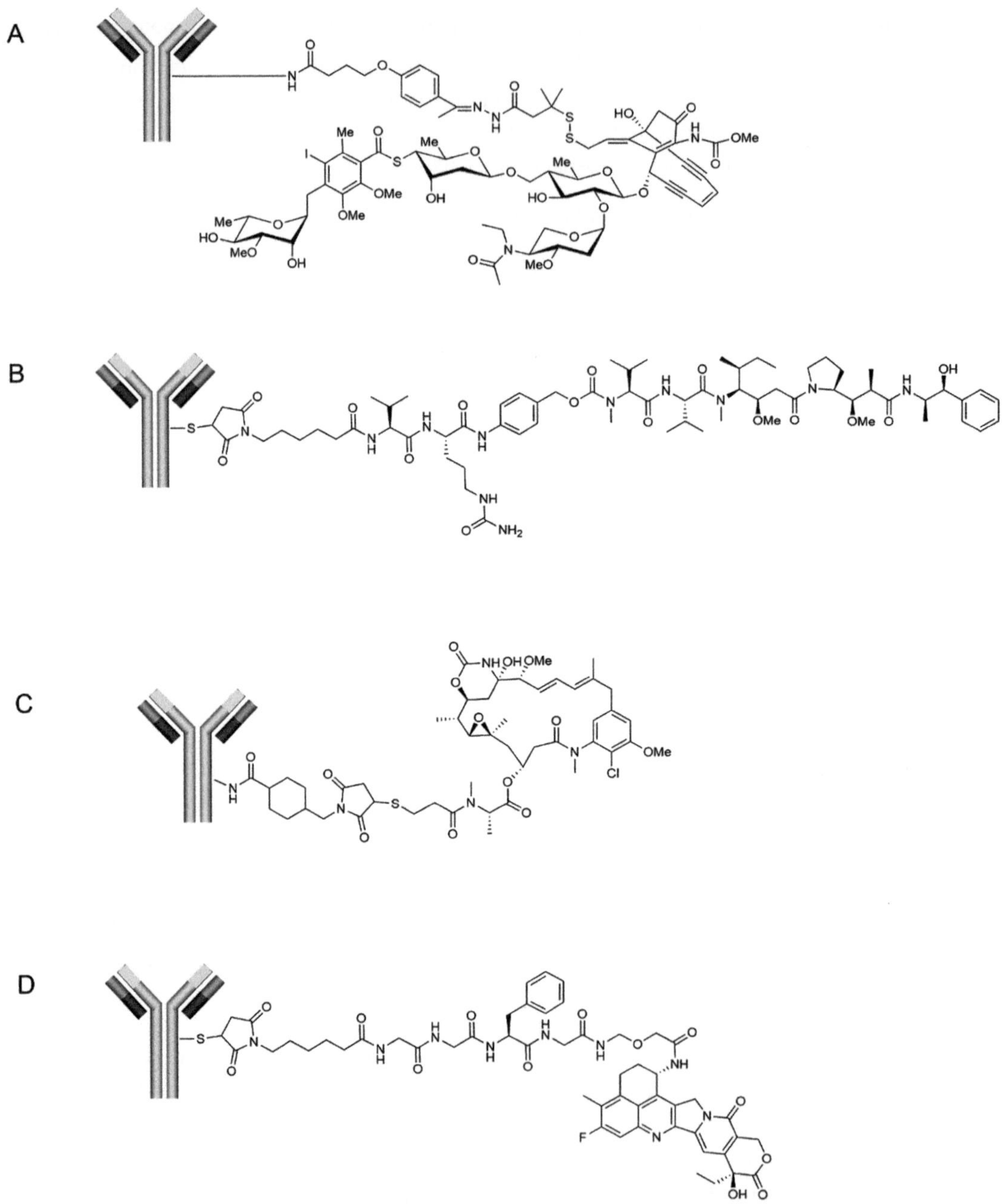

Figure 3. Representative approved ADCs and payloads. (A) Mylotarg® (and structurally related Besponsa®), featuring the DNA double strand cleavage agent calicheamicin. (B) Adcetris®, with the microtubule-binding agent MMAE, a family of compounds isolated from a sub-species of sea hares. (C) Kadcyla®, with the microtubule-binding Maytansinoid DM1. Maytansinoids were discovered in the Ethiopian shrub Maytenus Ovatus. (D) Enhertu®, with deruxtecan, a derivative of topoisomerase inhibitor exatecan.

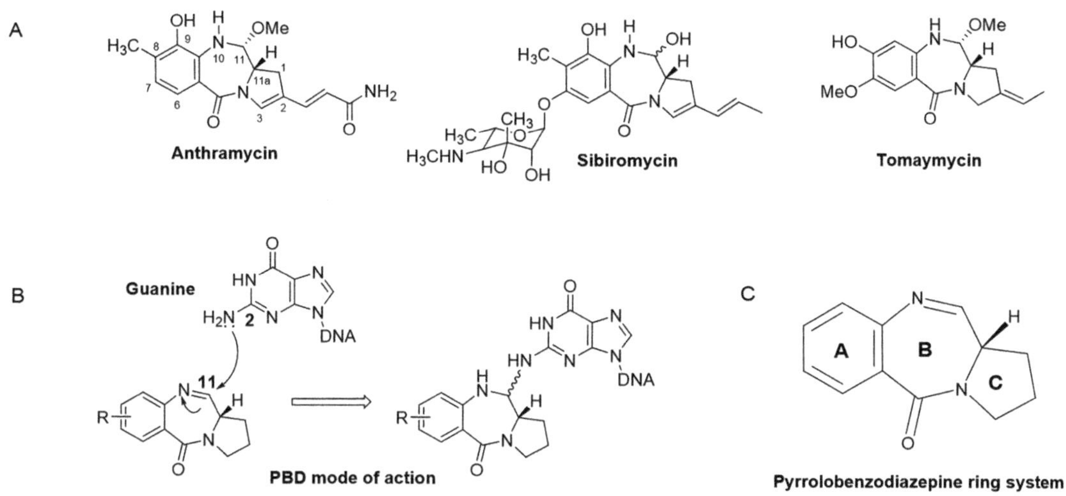

Figure 4. (A) Selected, potent, naturally occurring PBDs. (B) Mode of action: electrophilic guanine alkylation stabilized by DNA minor-groove interactions. C. Pyrrolobenzodiazepine ring system. The imine, methoxyaminal and carbinolamines are interchangeable and solvent-dependent.

Testing a range of naturally occurring PBDs and their analogues helped progress the understanding of the structure activity relationship (SAR) of the molecules. Key structural features required for activity were an alkylating moiety located at N10-C11 (either an imine or solvent adduct equivalent), and an electron-rich aromatic A-ring (Figure 4). Investigation of the mode of action determined electrophilic addition of the N10-C11 imine to the exocyclic 2-amino group of guanine, in the minor groove of the DNA. Under laboratory conditions, the PBDs are reversible alkylators of nucleophiles, but once stabilized in the DNA minor groove, the binding is long-lasting. A flat sp^2 carbon in the C2 position further improved potency by projecting lipophilic side-chains alongside the minor groove, picking-up Van der Waals interactions (*16*).

Mapping of DNA binding sequences (footprinting) additionally uncovered a preference for purine-guanine-purine sequences (*17*). It was hypothesized that two PBDs may be attached together side-by-side to target a longer DNA sequence. After some early attempts linking C7 to C7′, PBD dimer DSB-120 (**4**) composed of two PBD monomeric units DC-81 (**3**) linked C8 to C8′ was highly potent in cells, with an IC_{50} of 0.5 nM in ADJ/PC6 (Figure 5). The mode of action was also changed from simple alkylation, to bivalent binding. Interstrand crosslinks (*18*) were one of the main explanations for the exceptional potency of the C8/C8′ PBD dimers by stalling the replication fork during cell division and triggering apoptosis.

The Thurston group continued their investigation of the C8/C8′ PBD dimer and hypothesized that including a C2 unsaturation, as in highly potent monomer tomaymycin, would provide a superior molecule. Obtaining the ethylidene in pure, stereo-controlled fashion (*E*) was challenging, but the methylidene analogues were more easily accessible, and gave rise to the exo-unsaturated PBD dimer SJG-136, or SG2000 (**5**, Figure 6). In their paper (*19*), Gregson and co-workers reported that the C2 unsaturation not only changed the shape of the C-ring to obtain a better fit with the minor groove, but also reduced overall electrophilicity of the N10-C11 imine. This in turns helped prevent rapid deactivation reaction with sulfur nucleophiles such as glutathione. Together with an excellent NCI (National Cancer Institute) hollow fiber score of 54, this feature of "soft electrophilicity" helped SG2000 progress up to Phase II clinical trials as a standalone agent (*20–22*).

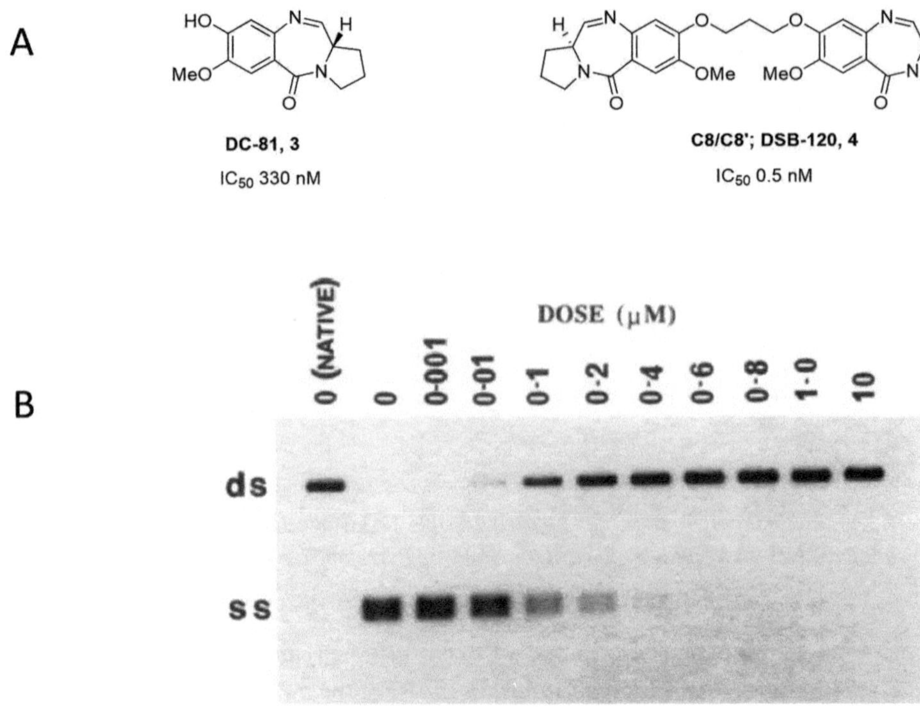

*Figure 5. (A) DSB-120 (**4**), an early PBD dimer designed by joining two monomeric DC-81 (**3**) units back to back through their phenolic hydroxy moiety. IC_{50} in ADJ/PC6 after 4 days incubation. (B) DNA crosslinking assay showing crosslinked double stranded DNA with 0.1 μM of DSB-120 under denaturing conditions. (Reproduced with permission from reference (18). Copyright 1992, American Chemical Society).*

The compound did not advance further in development due to lack of efficacy at non-toxic doses, but provided important insights into the PK/PD behavior of PBDs. It also informed scientists of the toxicology profile of these type of ADC warheads when administered on their own (i.e.: in what could be considered as a background, or control, data set). With this information at hand, one can deduce whether a new toxicity is due to the ADC being taken up in the wrong tissues, or simply due to the warhead alone. To this day, SG2000 remains an excellent PBD warhead with good overall properties and has been included in a number of ADC programs (23, 24).

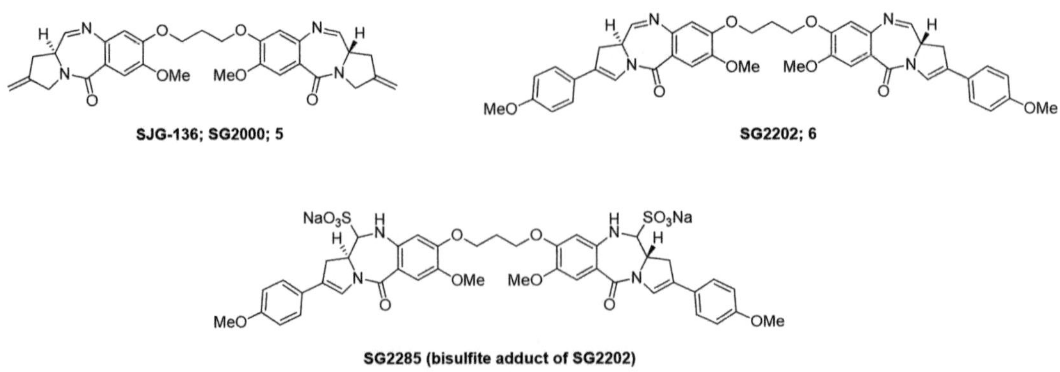

Figure 6. Clinical candidates SJG-136 (SG2000), SG2202 and its water-soluble bisulfite adduct SG2285.

Whilst SG2000 was being developed, synthetic resources were invested in exploring the C2 side-chain potential of the PBDs, with the aim of replicating the *endo-exo* unsaturation system of potent natural products anthramycin and sibiromycin. Insertion of an acrylamide side-chain via Heck chemistry was challenging and gave inconsistent yields (*25*). The Stille reaction to install the side-chain was limited in scope (*26*), but the Suzuki–Miyaura cross-coupling of a C2-triflate proved remarkably efficient, even at room temperature (Scheme 1) (*27, 28*).

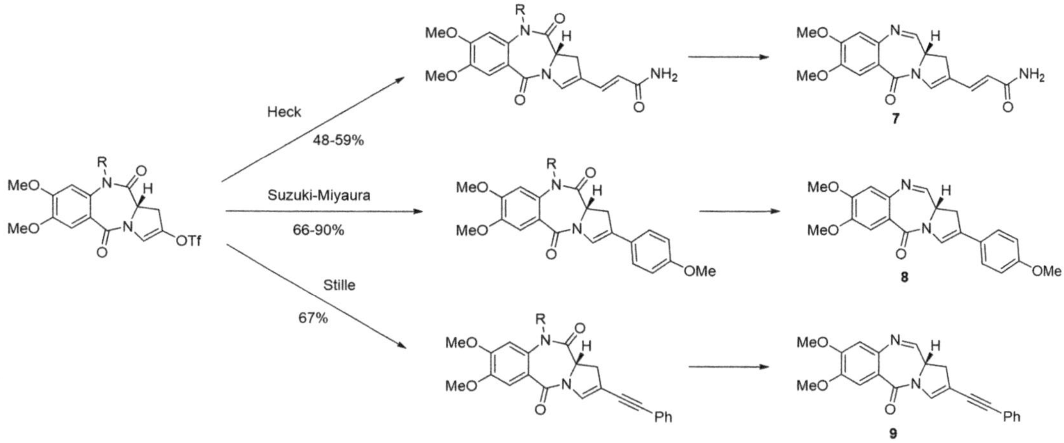

Scheme 1. Heck, Suzuki-Miyaura, and Stille couplings on a PBD C2-triflate. The Suzuki-Miyaura coupling proved remarkably efficient, even at room temperature.

Biological assessment of the resulting C2-aryl PBD monomer DRH-417 (**8**) revealed a surprisingly high cytotoxic activity (*29*). This was rationalized by the ability of the additional aromatic group to fit snuggly into the minor groove, although later models contest the usefulness of the lower part of the ring which projects somewhat outside the groove – (Figure 7).

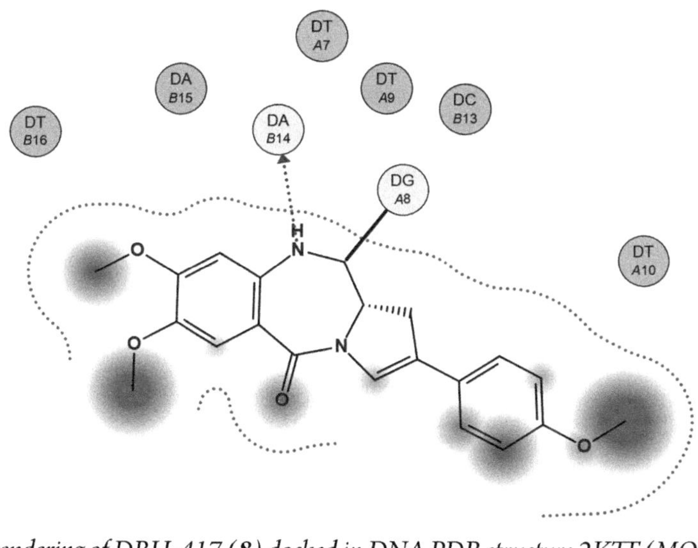

*Figure 7. 2D rendering of DRH-417 (**8**) docked in DNA PDB structure 2KTT (MOE 2018.0101) Solvent-exposed positions in grey. The upper part of the aromatic ring is making tight contacts with the minor groove (similar to the side-chain of sibiromicin). The lower part has an inferior lipophilic efficiency.*

Following the success of the dimerization of DC-81 (**3**) to give rise to DSB-120 (**4**) and SJG-136 (**5**, SG2000), the C2-aryl monomer DRH-417 was dimerized to yield SG2202 (**6**, Figure 6). Both SG2202 and its water-soluble bisulfite adduct prodrug SG2285 (Figure 6) were highly efficacious (GI_{50} = 13 pM for SG2202 in K562 cell line (*30*), as opposed to 419 pM for SG2000) (*28, 31*) but were not pursued in the clinic due to a lack of therapeutic index in preclinical models (*32*). A number of hypotheses were suggested for this, such as the ultra-high potency of the molecules being better suited to targeted prodrug approaches. Another hypothesis was simply the physico-chemical properties of SG2202 with its large number of aromatic groups. Indeed, the solubility forecast index (SFI = ClogD + number of aromatic rings) of SG2202 is equal to 8.79. Compounds with SFI > 7 tend to have low solubility and poorer overall pharmaceutical profiles. In any case, the idea of selectively targeting cancer cells with an antibody-drug conjugate loaded with a PBD payload as potent as SG2202 became attractive.

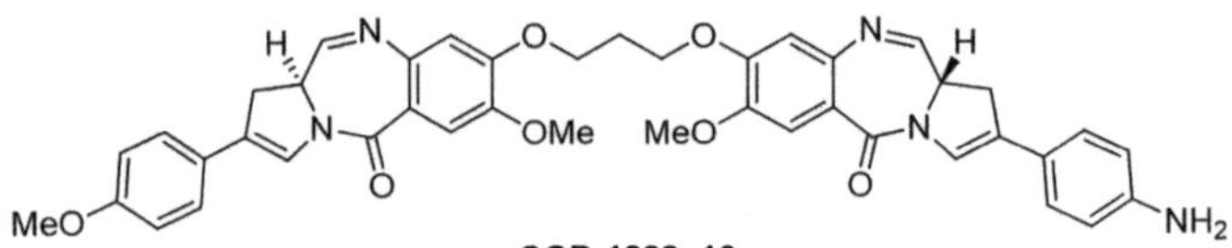

SGD-1882, 10

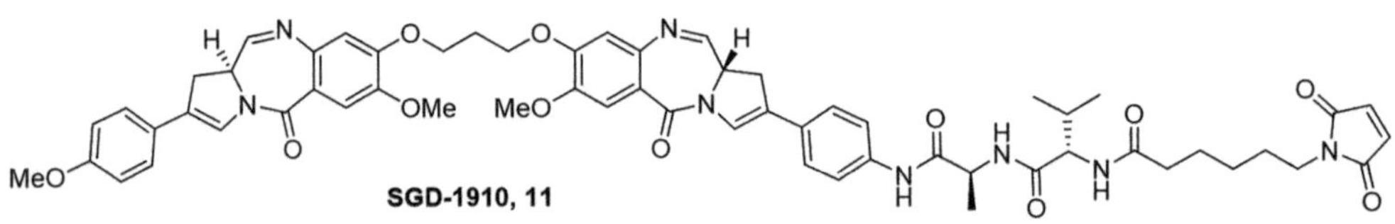

*Figure 8. ADC warhead (SGD-1882, **10**) and ADC warhead with drug-linker (SGD-1910, **11**).*

The simplest way to turn SG2202 into an ADC warhead was to swap one of the C2-*p*-methoxyphenyl groups for a C2-*p*-anilino to provide a handle on which to link the antibody (**10**, SGD-1882, Figure 8). Additionally the use of a valine-alanine dipeptide, as demonstrated by Seattle Genetics (*33*), resulted in a linker stable in systemic circulation, but that hydrolyses back to the aniline group after lysosomal uptake to reveal the warhead. Finally, a maleimide unit provided a suitable thiol conjugation partner to conjugate the drug-linker to the available antibody cysteines (**11**, SGD-1910, Figure 8).

Drug-linker SGD-1910 (**11**) was conjugated at position 239c to an anti-CD33 antibody developed by Seattle Genetics (*34*). On the basis of strong efficacy data and an acceptable safety profile, the conjugate was evaluated up to Phase III clinical trials for acute myeloid leukemia (AML) as a monotherapy, or in combination with hypomethylating agents (*35*). High rates of complete responses (undetectable levels of cancer) were recorded (CR + CRi up to 70%) demonstrating the ADC's efficacy. However, the target CD33 is also present on healthy hematopoietic progenitor stem cells (HSCs – the parent blood cells). This had the disadvantage that the conjugate-drug was also likely to kill these (an argument discussed by Zheng and co-workers (*36*) to justify CLL-1 as a superior target to CD33 for the treatment of AML). This potentially left patients without an effective immune system for a limited period of time, and vulnerable to infections (febrile neutropenia). Only specialist care centers were equipped to protect these patients appropriately and the trial was discontinued as a result of higher death rates in the treatment arm.

SGD-1910 (**11**) was conjugated to an engineered position such as 239c, as it is shielded from the external aqueous solvent. (Many companies choose to deliberately introduce non-natural cysteine(s) in the antibody sequence by genetic engineering, to provide a clean conjugation site for the drug-linker. This can be done by mutation, or insertion. The position(s) of these new cysteine(s) on the antibody is critical for solvent exposure and stability). When SGD-1910 (**11**) is conjugated in a non-engineered fashion in the hinge region (the antibody central region, where up to eight naturally-occurring cysteines are accessible for conjugation), the solvent exposure of the pendant C2-aryl group can trigger a certain level of aggregation (13% for a drug antibody ratio of 2 (DAR2) and 45% for DAR4) (*37*). As the business model of Spirogen at the time (before 2013) was based on licensing its products to a variety of companies - most of these without access to antibody engineering technology - it was imperative to design a new drug-linker which would be compatible to stochastic (natural, semi-random conjugation on of the eight available cysteines) hinge region conjugation, with low levels of aggregations.

As aromatic groups are widely regarded in the pharmaceutical industry as a cause of lower solubility and increased promiscuity (*38*), we decided to remove the lower part of the ring and keep the ligand-efficient top section, as in the naturally occurring potent PBD monomer sibiromycin. To this end, the warhead SG2219 (**12**) was selected (Figure 9), and a self-immolative drug-linker attached at its N10 position (*39*). This had the added advantage of protecting the electrophilic N10-C11 imine in a carbamate prodrug approach. To further enhance the solubility of this construct, a discrete PEG8 chain was inserted between the maleimide and the Val-Ala dipeptide (Figure 10). However, the molecule was found to be unstable, with the maleimide reacting rapidly in a Diels-Alder fashion with the *endo-exo*-unsaturated system. Instead, an iodoacetamide moiety (an alkylating group used in bioconjugation) was used as a substitute to the maleimide, affording SG3227 (**13**).

Conjugates of SG3227 were active and did not suffer deconjugation issues and, as such, could have made good candidates for preclinical development. But the thiol-iodoacetamide conjugation reaction was not as fast and straightforward as the thiol-maleimide (K_d maleimide 50-500 $M^{-1}.s^{-1}$; K_d iodoacetamide = 1.2 $M^{-1}.s^{-1}$) (*40, 41*) and iodoacetamides are better reserved for conjugation to selenocysteines instead (*42*). For this reason, a new warhead compatible with maleimide chemistry was envisaged. The endo-exo system was cut further to simply leave a C2-methyl, and following the structure-activity relationship (SAR) of PBD dimers, two carbons were inserted in the central tether to make up for the expected loss of potency which gave SG3199 (**2**) and the drug-linker SG3249 (**1**, Figures 1, 10, and 11).

An interesting twist to this story is the parallel development of drug-linkers bearing SG2057 (**14**), an isomeric analogue of SG3199 (**2**, Figure 9). Whilst SG3199 was developed through minimalistic evolution of SG2202 (**6**), only retaining the most important C2 carbon atoms, SG2057 (**14**) was designed simply by inserting a further two carbons in the linker of SG2000 (**5**, SJG-136). The warhead was licensed exclusively to Genentech who incorporated SG2057 in a number of disulfide-based drug-linkers (*36, 43*). The resulting antibody-drug conjugates were active and Genentech is now evaluating PBD-ADCs, with a focus on Acute Myeloid Leukemia (AML) (*44*). A most interesting study by Ma, Zhang and co-workers (*45*) systematically measured the exposure of every organ after a single injection in mouse. Zhang and Ma were able to conclude that a cancer cell typically requires 12,000 PBD crosslinks (or 41 PBD crosslinks per million DNA base-pair) to die. These figures may also be applied to SG3199 as the molecules are so close in structure.

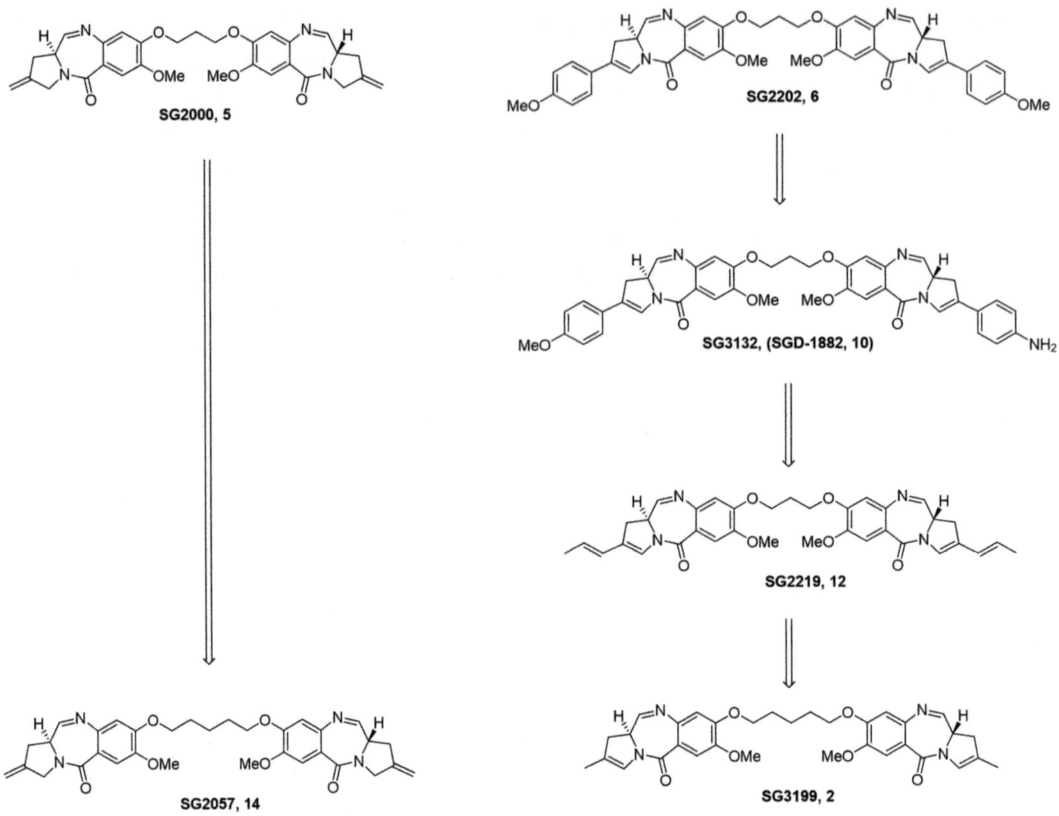

*Figure 9. Lineage and evolution of isomeric ADC warheads SG2057 (**14**) and SG3199 (**2**).*

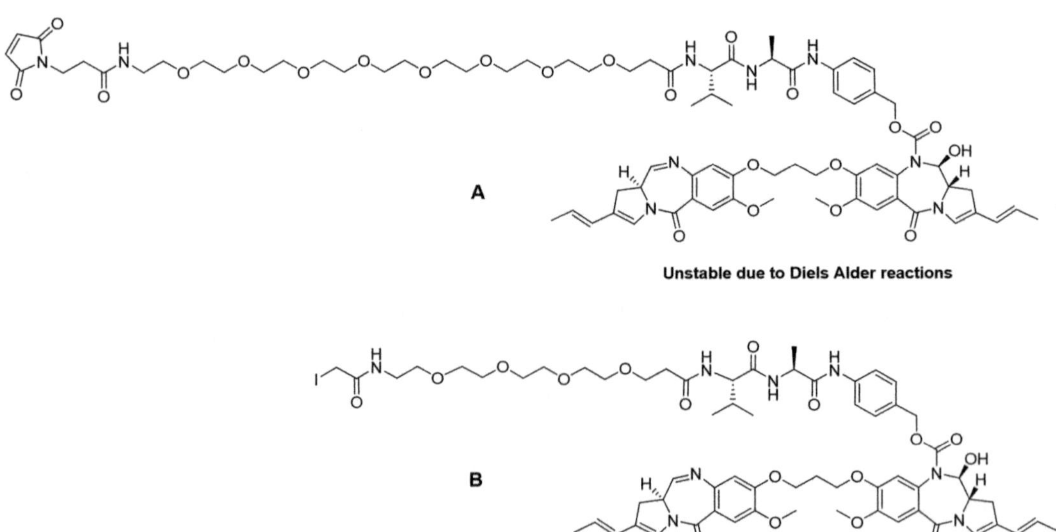

Figure 10. (A) Original design based on sibiromicin side-chains. (B) SG3227, an iodoacetamide drug-linker compatible with unsaturated diene systems.

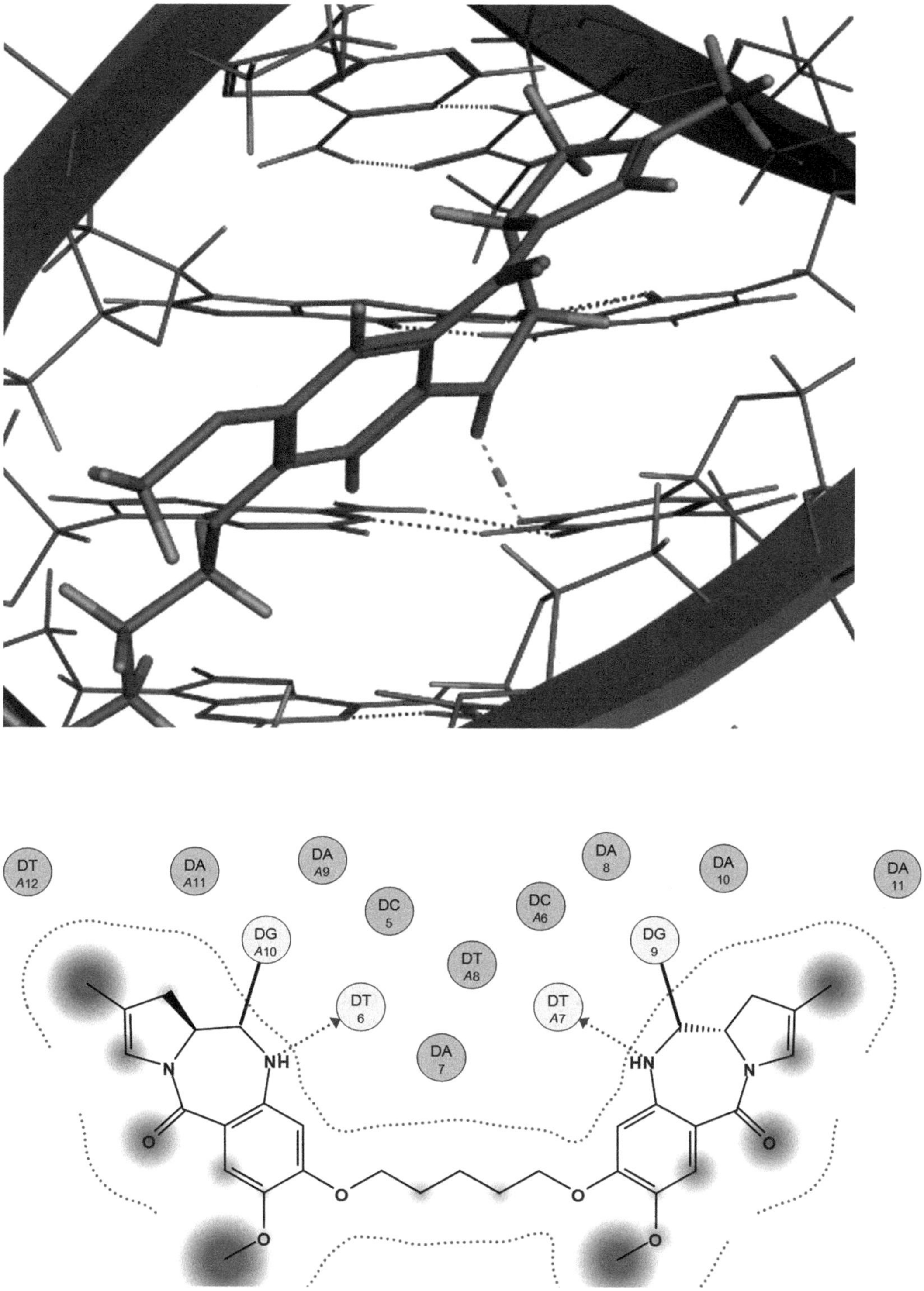

*Figure 11. Molecular model of SG3199 (**2**) docked on a TCTTAGA DNA sequence. (MOE 2018.0101, sequence constructed using the DNA builder option). The covalent guanine-NH-PBD bonds can be observed, as well as the H-bonds with adjacent thymines.*

SG3249 (**1**) fulfilled the necessary requirements for licensing to a broad variety of collaborators, such as good solubility, fast conjugation rate, little aggregation at DAR2, and high potency. A model conjugate of SG3249 with trastuzumab (Herceptin) (*46*) and a drug antibody ratio of 2 (DAR2) showed-single figure ng/mL activity in vitro (Figure 12). A rodent efficacy study with an engineered antibody carrying two SG3249 molecules and targeting a 5T4 positive tumor model (*47*) showed prolonged tumor reduction at 0.3 mg /kg (Figure 13).

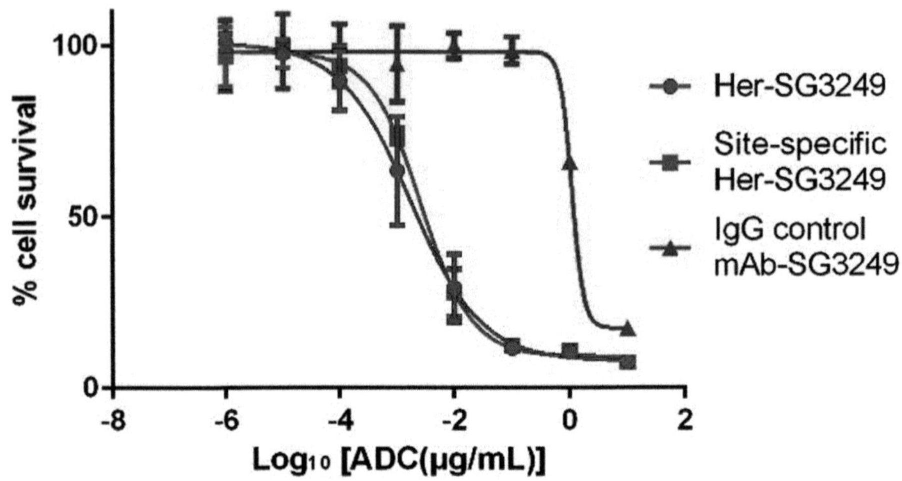

Figure 12. Antibody-targeted cytotoxic activity of SG3249 in HER2 positive SKBR3 cells. Very little activity is observed with non-targeted ADC control. (Reproduced with permission from reference (46). Copyright 2016, American Chemical Society).

Similar levels of activities were observed with other conjugates of SG3249, and typical murine in vivo results gave complete responses (cures) at 1 mg/kg of SG3249-ADC, and stable disease (non-growing tumor) at 0.3 mg/kg. Maximum tolerated levels in rat were observed at 1.5 mg/kg for an anti-EphA2-SG3249 ADC, with a 3 × 0.5 mg/kg regimen better tolerated than a single injection (*23*). It is worth noting that tolerated doses tend to be lower when testing an ADC in an animal species expressing the human antigen target (i.e.: a cross-reactive species), than in a species devoid of the target altogether. Indeed, the target expression profile (how much target antigen is expressed on the tumor cells surface compared to healthy tissues) and the antibody design factors play a large part in the overall toxicity response (*48*). Due to the high level of complete responses in mice and the potential for a therapeutic window, Spirogen licensees (including AstraZeneca/Medimmune who acquired the company in 2013) were keen to evaluate their SG3249-ADCs in clinical settings, and a large number of trials were initiated over the course of five years. This created a demand for material, initially provided by the Spirogen team (now AstraZeneca Early Oncology) through the Discovery route which was optimized to reduce the number of chromatographic steps. The robustness and purity profile also had to be improved considerably to deliver enough clinical-grade material to run multiple Phase I studies.

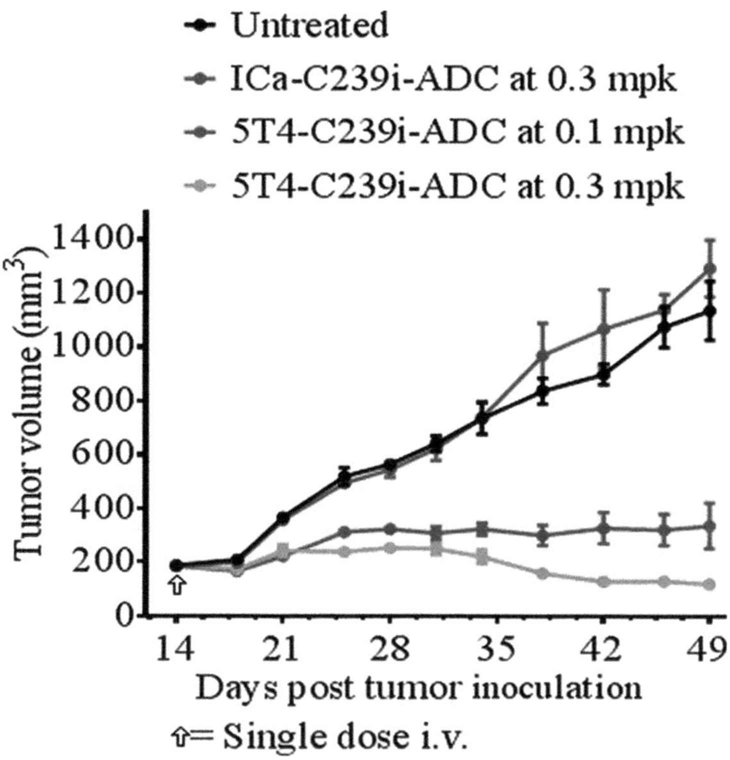

Figure 13. In vivo activity of 5T4-C239i-SG3249 (engineered DAR2) in 5T4 positive MDA-MB-361 tumor model. Typical SG3249 dose responses are observed. Complete responses observed at 1mg/kg (not shown). (Reproduced with permission from reference (47). Copyright 2017, American Chemical Society).

Discovery Route and Initial Scale-Up of Tesirine

Early synthesis of PBD dimers, including SG2000 (**5**) and SG2202 (**6**), started with simplified dimeric structures where chemistry transformations could be completed simultaneously on both halves of the molecules (*19*). This approach was successful for these symmetrical molecules, but for the non-symmetrical tesirine (**1**), it was decided that separate construction of the two halves of the dimer would be the most efficient strategy (*46*). On this basis, a synthetic strategy based around benzopyrrolidine intermediate **15** was devised (Figure 14). From this molecule, separate preparation of the two halves of the tesirine could be completed, followed by dimerization and finalization of the synthesis. This approach was effective in delivering a Discovery synthesis of tesirine (**1**), and the route was judged to be robust enough for scale-up to provide early clinical supplies. Further optimization of the route was completed through a series of campaigns, ultimately affording 169 g of tesirine (**1**) (*9*).

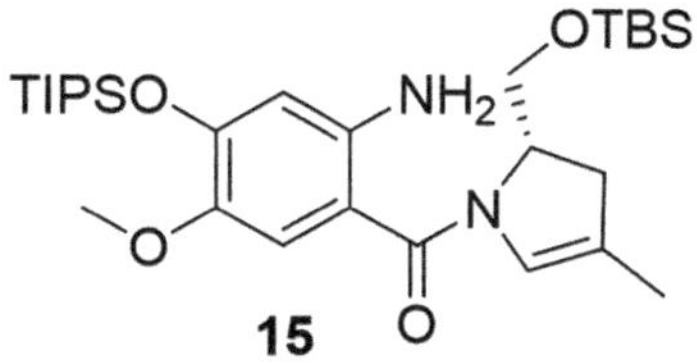

15

*Figure 14. Benzopyrrolidine intermediate **15**.*

This section outlines the successful application of this strategy to afford a Discovery synthesis of tesirine (**1**) and development work completed on the scale-up of this synthetic route.

Preparation of A-Ring Benzoic Acid 21

Preparation of aniline **15** required separate preparation of functionalized benzoic acid (Scheme 2) and pyrrolidine rings (Scheme 3) followed by a number of functional group interconversions. Preparation of the A-ring benzoic acid **21** started with nitration of benzyl vanillin (**17**). This starting material was easy to source on a small-scale, but for scale up we benzylated vanillin (**16**) which was achieved using benzyl bromide and potassium carbonate. The reaction was run initially in DMF, but during development improved reaction rates occurred in NMP thus this solvent was used for large scale preparation. The nitration of vanillin could be achieved with good selectivity at the 6-position by maintaining the temperature at 12 °C, minimizing byproduct formation of which nitro-deformylation was the most significant. Direct nitration of vanillin was investigated, but the benzyl group proved to be critical for delivering the required regioselectivity requiring the protecting group intermediary.

Upon scale-up, the exothermicity of the nitration caused concern, thus a solution of benzyl vanillin (**17**) in acetic acid was added to nitric acid at 22 °C ensuring minimal accumulation of starting material whilst keeping impurities at manageable levels. The crude material was precipitated upon addition to water. The necessary purity was achieved by recrystallisation of crude nitro-product **18** from ethyl acetate. It is important to protect the nitro-product **18** from light to avoid its photochemical degradation.

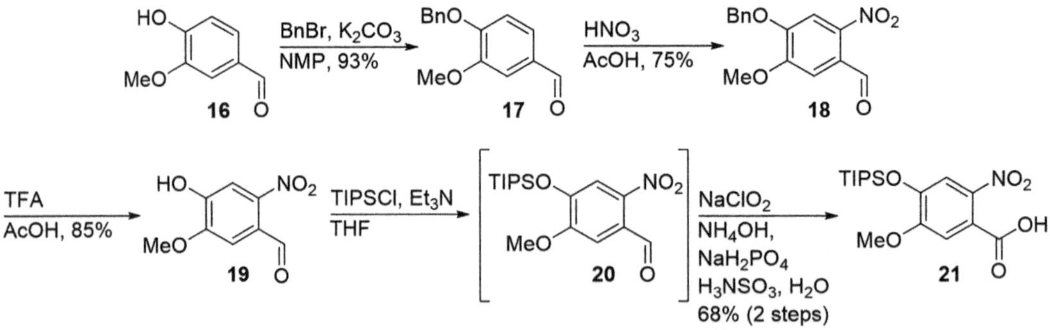

*Scheme 2. Preparation of A-Ring Benzoic Acid **21**.*

Removal of the benzyl protecting group was initially achieved using 1.5 volumes of TFA. For scale-up, the deprotection solution was changed to TFA:AcOH (2:1 v/v, 3 volumes) to enable effective stirring. The phenol **19** was precipitated by addition to *n*-heptane and was isolated by filtration in good purity and improved yield. The TIPS group was selected for protection of the phenol **19** due to its good chemical stability under both the acidic and basic conditions planned in the synthesis. This group was initially introduced using solvent free conditions, melting **19**, imidazole and TIPSCl to facilitate the reaction, followed by purification by flash column chromatography. This procedure was not considered practical or reliable enough to be scaled-up, thus a reaction screen was completed identifying triethylamine and THF as suitable base and solvent, respectively, for TIPS incorporation. On completion of the reaction using these conditions, the precipitated triethylamine hydrochloride could be removed by filtration and the solution of the TIPS phenol **20** telescoped directly into the next reaction.

The final step in the preparation of **21** was the oxidation of the aldehyde to the carboxylic acid. On a Discovery scale this was achieved using standard Pinnick oxidation procedure followed by aqueous work-up and isolation. Upon scale-up, improved conditions using sulfamic acid as chlorine scavenger were utilized and following an aqueous work-up; the carboxylic acid **21** could be simply isolated by slurrying with *n*-hexane.

Preparation of C-Ring Pyrrolidine 27

The steps for the preparation of the C-ring pyrrolidine **27** had been reported previously (*49*), and this chemistry was used directly for the initial synthetic work to prepare tesirine (**1**). For a new scale-up synthesis, development of each of the steps was investigated.

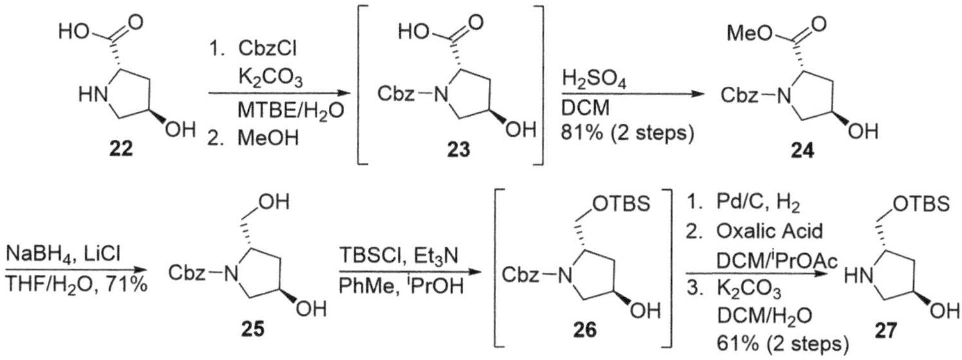

Scheme 3. Preparation of C-Ring Pyrrolidine **27**.

The solvent system for the Cbz-protection of hydroxyproline (**22**) was changed from toluene/water to MTBE/water which produced an improved reaction profile. Following an aqueous work-up and solvent swap, a solution of Cbz-hydroxyproline (**23**) in methanol was used directly in the next stage. Work-up of the direct acid-catalyzed esterification of the methanol solution of ester **24** proved to be challenging on scale. Thus, the solvent system was modified to include DCM affording a simplified work-up with $NaHCO_3$, following which the ester **24** could be isolated by evaporation of the solvent.

Reduction of the ester group of **24** had originally been achieved using $LiBH_4$, but for scale-up the use of the more stable $NaBH_4$ was investigated. The reaction was slower with this reagent, but the addition of LiCl improved reaction rates and matched those in the initial process.

The selective TBS-protection of the primary alcohol **25** was improved from the previous procedure by changing the base from DBU to Et_3N. Although the reaction was slower, improved selectivity for the desired TBS-protected alcohol **26** resulted and the crude product could be telescoped as an *i*-PrOH solution into the next stage, avoiding the need for chromatography. Finally, the Cbz-group was removed by hydrogenolysis and the amine **27** was purified by crystallization as the oxalate salt. A salt-break followed by a solvent swap to *n*-heptane provided the pyrrolidine free-base **27** in preparation for the coupling with the A-ring benzoic acid **21**.

Coupling A-Ring Benzoic Acid 21 and C-Ring Pyrrolidine 27, and Elaboration to Benzopyrrolidine 15

Classic' Discovery conditions were initially applied to couple acid **21** and amine **27** to afford amide **28**, using DCC and HOBt followed by flash column chromatography (Scheme 4). For scale-up, removal of the DCC byproduct proved to be difficult, plus the use of HOBt is not recommended on scale due to Process Safety issues (*50*). Thus, an alternative amide coupling was developed using EDCI and HOPO (2-hydroxypyridine-*N*-oxide), and this procedure was successfully scaled-up delivering high purity **28** following recrystallisation from EtOH/H_2O.

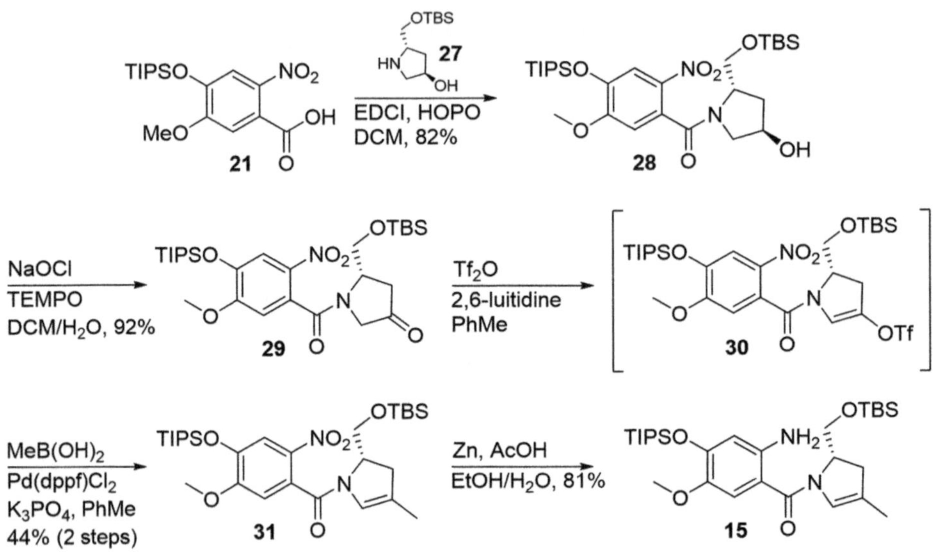

*Scheme 4. Coupling A-ring benzoic acid **21** and C-ring pyrrolidine **27** and elaboration to benzopyrrolidine **15**.*

Oxidation of the hydroxy group of **28** was the next transformation required, and on a Discovery scale this could be achieved using trichloroisocyanuric acid/TEMPO. For scale-up, these conditions were avoided due to possible chlorination of the aromatic ring, and for initial batches Dess-Martin Periodinane (DMP) was used to deliver ketone **29**. DMP has known issues with cost and Process Safety, thus for further scale-up the oxidation was changed to Anelli-Montanari conditions (biphasic buffered NaOCl/TEMPO) which afforded ketone **29** cleanly and in good yield.

Introduction of the methyl group was completed by enol-triflate formation followed by a Suzuki–Miyaura coupling. The thermodynamic enol-triflate **30** was formed using triflic anhydride/ 2,6-lutidine in DCM, and on a small-scale, the material was purified by flash column chromatography. For scale-up, the solvent was changed to toluene enabling a telescope into the Suzuki–Miyaura coupling. Careful optimization of reagent equivalents minimized the amount of 2,6-lutidine triflate formed which poisoned the catalyst of the following coupling reaction.

The Suzuki–Miyaura coupling proved to be a challenging reaction to establish and develop. Initial reactions employed activated conditions using silver oxide and triphenylarsine followed by flash column chromatography, but protodetriflation remained an issue with this byproduct proving challenging to separate from the desired methyl product **31**. For scale-up, an extensive screen of methylating reagent, base, catalyst, solvent and reaction conditions was completed. This work identified $Pd(dppf)Cl_2{\bullet}CH_2Cl_2$ as a better catalyst compared to the original $Pd(PhCN)_2Cl_2$. This allowed discontinuation of the silver oxide and triphenylarsine; potassium phosphate and methyl boronic acid were retained as the base and methylating agent, respectively. The improved conditions were successfully scaled-up to afford 2.1 kg of product, although it was necessary to retain the flash column chromatography purification to ensure the protodetriflation byproduct was controlled below 0.7%. Removal of related impurities in downstream intermediates was not possible, thus this limit was set to ensure final product quality.

Reduction of nitrobenzene **31** to afford the required key aniline intermediate **15** was achieved using zinc as the reductant. This transformation was initially completed in formic acid/ethanol followed by flash column chromatography, but on scale-up this was changed to acetic acid/ethanol/water (removing the requirement for pre-activation of the zinc) and the substrate **31** was added to the reduction mixture to allow control of the exotherm. 1 kg of the aniline **15** was isolated by evaporation of the solvent and used promptly in subsequent steps due to its instability.

Preparation of PBD with Five-Carbon Tether 37

Benzopyrrolidine intermediate **15** was the starting point for the separate preparation of the two halves of the tesirine. Largely equivalent chemistry was required for the two halves, allowing learning to be transferred between the two synthetic streams. The use of *N*-Alloc protection in both streams also provided the opportunity for a single deprotection step late in the synthesis.

Preparation of the PBD with five-carbon tether **37** started with *N*-Alloc protection of **15**, which was completed using allyl chloroformate and pyridine affording *N*-Alloc-aniline **32** (Scheme 5), and this compound could be used without purification both on a small and large (1.4 kg) scale in the subsequent deprotection reaction. Removal of the TBS-protecting group was achieved on a small-scale using a mixture of AcOH, MeOH, THF and water. On scale-up, removal of the AcOH proved problematic during the work-up, thus the conditions were adapted to the use of pTSA in THF and water which simplified the work-up. Purification by flash column chromatography was required for both Discovery and scale-up (1.0 kg) batches to ensure suitable quality alcohol **33** for the oxidative ring closure.

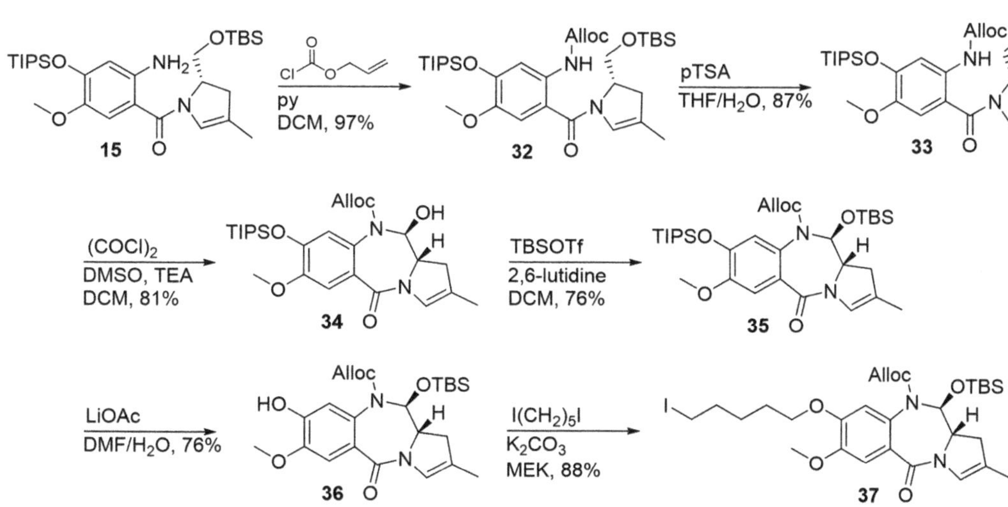

*Scheme 5. Preparation of PBD with five-carbon tether **37**.*

The oxidative ring closure to form the B-ring of the PBD structure is a key synthetic step, both in terms of molecular construction and establishment of DNA binding activity. Any intermediate with completed A-, B- and C-rings have the potential to be cytotoxic and thus preparation and scale-up required high-containment facilities. In vitro cytotoxicity evaluation of all intermediates following the oxidative ring closure was completed, and those compounds showing sub-micromolar activity were handled in a dedicated high-containment facility. Typically, monomers capped by a non-biologically cleavable protected group (such as Alloc) were non-active, although it is worth mentioning iodopentanyl **37** is an alkylating agent. Monomers capped by an enzymatic trigger (such as Val-Ala used in tesirine) were found to be active and were handled under high-containment.

The required stereoselective oxidative ring closure of alcohol **33** was achieved using a Swern oxidation. As is typical for this reaction, achieving the necessary control of stoichiometry and water is challenging on a small-scale, and the reaction proved to be easier to operate upon scale-up. On larger scale the amount of oxalyl chloride could be carefully controlled both in terms of delivering the optimal number of equivalents and drying the reaction mixture to avoid hydrolysis of the reagent. This allowed delivery of A,B,C tricyclo product **34** with the necessary purity to be used directly in the next step, minimizing the amount of over-oxidized byproduct present. It is worth noting the stereocenter at C11 is controlled by the stereochemistry of C11a, as the protected amine attacks the newly formed aldehyde from a single direction. Thus product **34** is obtained as a pure stereoisomer.

Protection of the 11-hydroxy group was now required to preserve the stereochemistry at this center. If unprotected, reversible ring opening to the amino aldehyde, followed by epimerization of the center α to the aldehyde under basic conditions, would degrade the chiral purity. This required the reaction of tricyclo **34** with the reactive TBS-triflate and 2,6-lutidine to afford TBS ether **35**. On a Discovery scale the material was purified by flash column chromatography, but on scale-up the material could be used directly in the subsequent step.

Removal of the TIPS-protecting group was now required to allow attachment of the five-carbon tether. This could be achieved using LiOAc in DMF/water affording phenol **36**. On a small-scale the material was purified by flash column chromatography, but on scale-up the product could be purified by slurrying in *n*-heptane then ethyl acetate.

Finally, the five-carbon tether was introduced in a high-containment facility using a Williamson ether synthesis. On a small-scale this was achieved using 1,5-diiodopentane (5.0 equiv.) and potassium carbonate in acetone, followed by flash column chromatography. For scale-up, the acetone was changed to the higher boiling MEK (methyl ethyl ketone) and fine mesh potassium carbonate was used to improve reaction kinetics. As much as 5.0 equiv. of 1,5-diiodopentane were still required to drive the reaction kinetics and to minimize homodimer formation. Purification was achieved using preparative-scale high pressure chromatography to afford iodopentanyl **37** which was obtained as a solution in EtOAc. It could be used directly in the following dimerization step.

Preparation of PBD with Val-Ala-PABC Trigger 42

Preparation of the second half of tesirine could also make use of **15** as a starting material (Scheme 6), with the first step being the attachment of the protected Val-Ala-PAB sidechain. This motif functions as the linker-payload 'trigger' allowing release of the PBD dimer following internalization of the ADC (Figure 2).

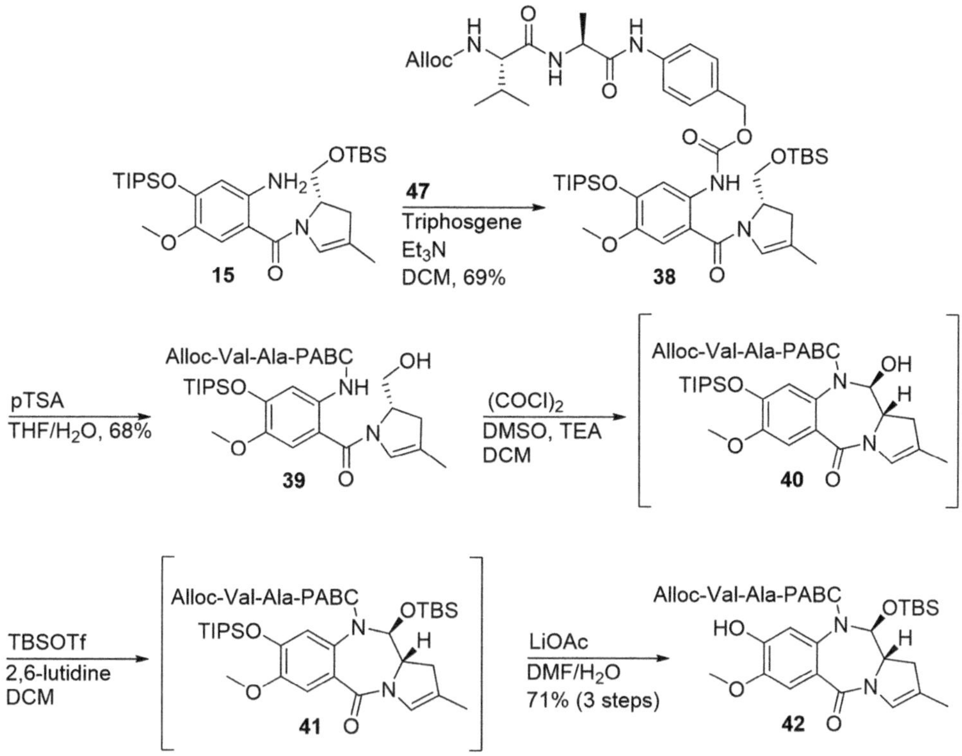

*Scheme 6. Preparation of PBD with Val-Ala-PABC trigger (**42**).*

Preparation of Alloc-Val-Ala-PABOH **47** started with valine (**43**), which following Alloc protection to produce protected amine **44**, was activated as the succinate ester **45** and coupled with alanine to afford dipeptide **46**. This was then coupled with 4-aminobenzyl alcohol to afford dipeptide **47** (Scheme 7) for reaction with **15** to yield coupled intermediate **38**. As part of the work to scale up this route, a base screen was conducted for the Alloc protection step. We were looking to minimize the amount of double Alloc and dimeric species formed and identified a combination of sodium hydroxide (1.76 eq) and sodium carbonate (0.55 eq) as the best conditions for the amine protection. Otherwise minimal changes were made as the chemistry was scaled-up.

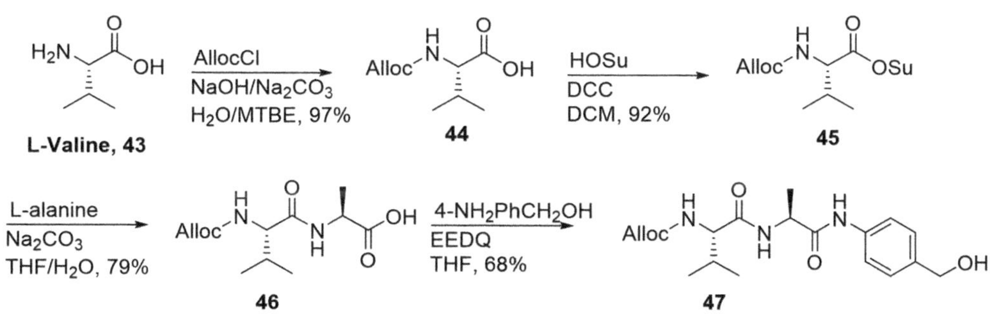

*Scheme 7. Preparation of Val-Ala-PABC trigger **47**.*

For the small-scale coupling of **15** and **47**, the aniline group of the benzopyrrolidine **15** was treated with triphosgene to afford an isocyanate which was reactive enough to couple directly with the dipeptide **47** to produce the carbamate **38**. For scale-up of this reaction, alternatives to the toxic reagent triphosgene were investigated unsuccessfully, thus development work changed the reaction solvent from THF to DCM which mitigated the hygroscopicity of the original conditions. The reduction in the adventitious moisture carried by the solvent allowed reduction in the number of equivalents of **47** from 1.50 eq to 1.05 eq, which in turn simplified the work-up and flash column chromatography purification.

The TBS-deprotection of **38** was achieved in an analogous manner as the deprotection of **32**, with the transformation being completed with AcOH in MeOH, THF and water on small-scale, and pTSA in THF and water during scale-up (affording 1.28 kg of alcohol **39**). The Swern oxidation of alcohol **39** was also scaled-up in a directly analogous manner to the oxidation of **33**, although for this intermediate the reaction needed to be completed in a high-containment facility due to the assumed toxicity of the tricyclo product **40**. On a small-scale, standard isolation and purification were acceptable. But for scale-up, the isolation of large amounts of cytotoxic solids is typically avoided, thus the final product was held as a solution in DCM prior to telescoping into the subsequent reaction.

Protection of the 11-hydroxy group of **40** was again achieved using TBS triflate with 2,6-lutidine to provide **41**. However, for this substrate, when using the same conditions as for the conversion of **34** to **35**, significant impurity formation ensued from cleavage of the PABC carbamate group. This could be avoided by reducing the temperature from 25 °C to -15 °C but required an increase in the equivalents of TBS triflate and 2,6-lutidine to obtain a reasonable reaction rate (<24 h).

On a small-scale the product was purified by flash column chromatography, but upon scale-up the crude product could be directly telescoped into the following reaction to avoid isolation of any solids.

Finally, removal of the TIPS group of **41** was accomplished with lithium acetate in wet DMF. The reaction was performed at ambient temperature on a small-scale, but following development work, was operated at 40 °C on scale-up (568 g of **41**) without any adverse effect on the product purity. The PBD with Val-Ala-PABC trigger **42** was purified by chromatography at all scales, and isolated by evaporation of solvent (small-scale) or precipitation from an ethyl acetate solution using heptane (large scale [338g], preferred when operating in a high-containment facility).

Final Assembly of Tesirine (1)

With the two discrete halves prepared, the synthesis was completed by coupling followed by final functionalization (Scheme 8). On a small-scale, 1.5 equiv of **37** was treated with **42** in a Williamson ether synthesis and the adduct was purified by flash column chromatography. Upon scale up, the stoichiometry of the less expensive **37** was reduced to 1.2 equiv, and the crude product could be solvent-swapped into THF and used directly in the next step.

The next step involved the removal of the two TBS-protecting groups, which could be realized with TBAF, yielding diol **49**. Work at small-scale displayed that acid buffering of the TBAF solution (pH 6 using acetic acid) was necessary to avoid base-mediated epimerization at the C11a center. The product was purified by flash column chromatography.

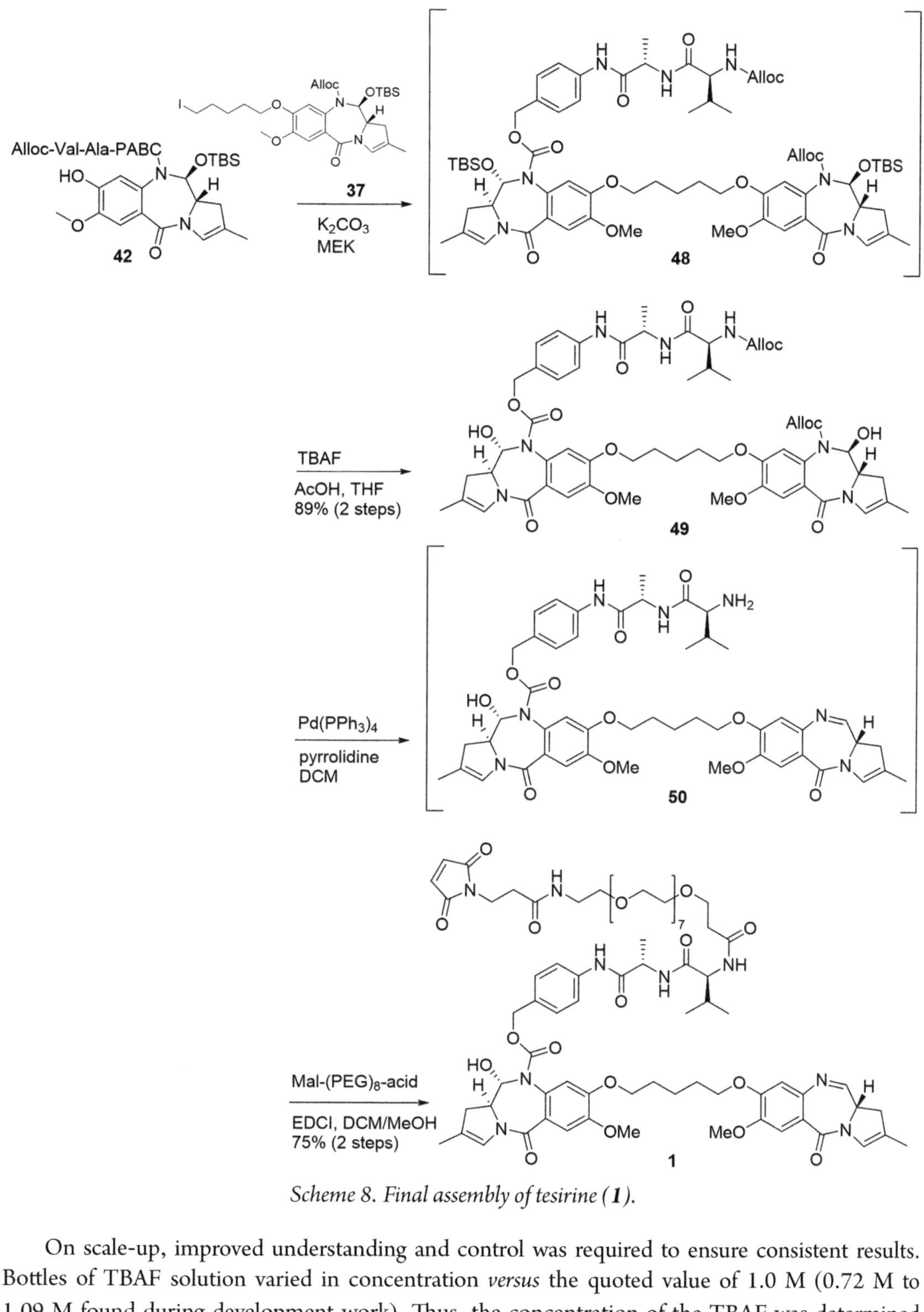

*Scheme 8. Final assembly of tesirine (**1**).*

On scale-up, improved understanding and control was required to ensure consistent results. Bottles of TBAF solution varied in concentration *versus* the quoted value of 1.0 M (0.72 M to 1.09 M found during development work). Thus, the concentration of the TBAF was determined using anion-exchange chromatography or titration prior to use, and following Design of Experiments work, the optimum equivalents of TBAF (3.3 equiv) and AcOH (5.2 equiv) could be charged to the reaction ensuring clean deprotection and minimal epimerization, albeit at a long reaction time (48 h).

Upon scale-up (418 g of **48**), the diol **49** was purified by preparative-scale high-performance liquid chromatography.

Next the *N*-Alloc-protecting groups were removed using palladium catalysis providing imine-amine **50**. Small-scale work indicated that the inclusion of pyrrolidine as an allyl scavenger was critical to avoid the formation of N10-allylated impurities. The crude product was used directly in the subsequent step. Upon scale-up, the amount of *tetrakis*(triphenylphosphine)palladium (0) was reduced from 0.06 equiv to 0.02 equiv, and the crude product was telescoped into the subsequent step as a solution in DCM/MeOH. The low solubility of **50** in pure DCM required the addition of the methanol.

For conducting the final coupling of the sidechain Mal-$(PEG)_8$-acid, we discovered that upon standing (as a solution in DCM/MeOH) the starting material **50** would slowly cyclize to form a macrocycle **51** (Figure 15). On small-scale, both **50** and **51** were shown to react with Mal-$(PEG)_8$-acid to form the product **1** with the use of EDCI. For larger campaigns, process work revealed that mildly acidic conditions promoted the ring-opening of the macrocycle. Thus a small excess of the Mal-$(PEG)_8$-acid was used to ensure an efficient coupling reaction. Control of the quality of the Mal-$(PEG)_8$-acid was important as several commercial batches of this material contained significant amounts of shorter PEG chains (e.g. Mal-$(PEG)_7$-acid). The final product was subsequently purified by flash column chromatography on a small-scale and preparative-scale HPLC on a large scale. Following which evaporation afforded tesirine **1** as a solid.

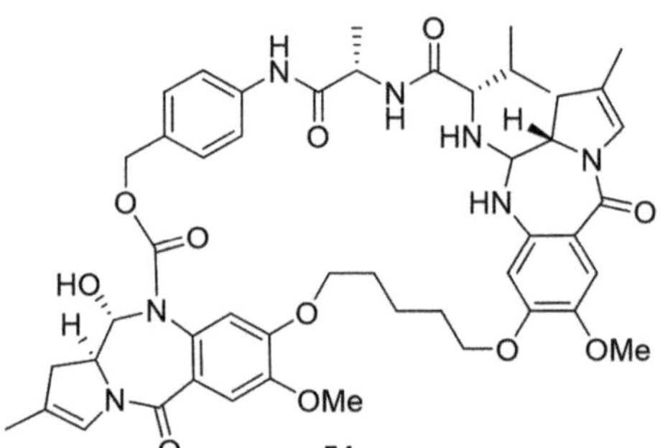

51

*Figure 15. Cyclization impurity of imine-amine **50**.*

This synthesis was efficient in supplying material for a range of toxicology and clinical studies, with improvements measured over the course of development and scale up (Table 1).

The introduction of more robust, scalable Process Chemistry conditions afforded a reduction on overall yield compared to the Discovery Chemistry conditions (2.19% compared to 5.19%). However, it should be noted that the average yield for each step of the longest linear sequence only diminished from 86% to 83%, and the improved conditions delivered an increase in the yield of the steps undertaken in the high-containment facilities as well as a reduction in the number of steps requiring chromatography. Overall a robust and effective synthesis was established that supported a range of ADC programs. A schematic of the route with containment requirements is shown in Figure 16.

Table 1. Route Metrics for the Initial Discovery Synthesis and the Largest Development Synthesis Using the Discovery Route

Parameter	*Initial Discovery Synthesis*	*Largest Development Synthesis*
Amount of tesirine produced	0.86 g	169 g
Number of Steps	34	34
Longest linear sequence (steps)	20	20
Overall Yield (longest linear sequence)	5.19%	2.19%
Average Yield (longest linear sequence)	86%	83%
Yield of high-containment steps (**39** to **1**)	45.2%	58.7%
Number of steps requiring chromatography	17	8
Final Purity	85%	97%

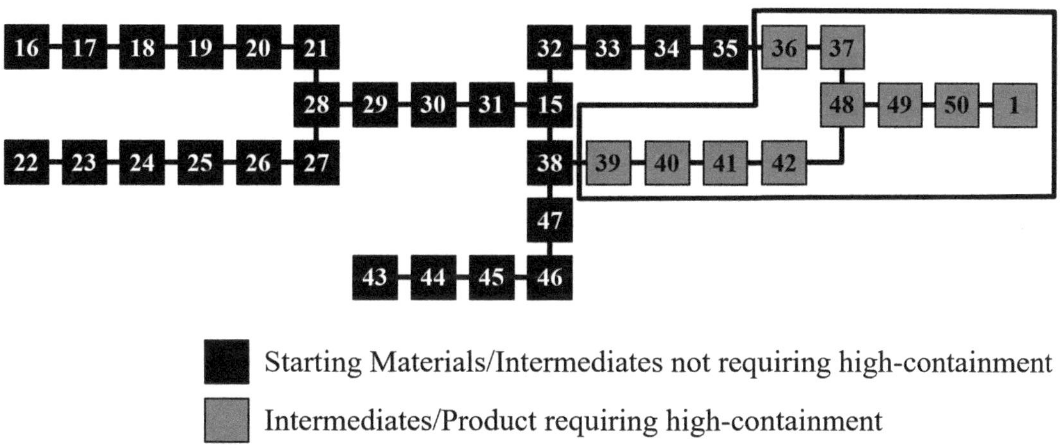

Figure 16. Schematic of discovery Route (each block representing a separate intermediate) showing portion conducted in high-containment (in boxed region).

Route Design to Reduce High-Containment Steps

While the original Discovery and initial scale-up route had been shown to be effective for delivering material for early clinical studies, a detailed Process Chemistry evaluation of the route was undertaken to look for opportunities for long-term improvement. For standard drug molecules, the application of Route Design to reduce the number of steps brings benefits in terms of shortened delivery times and reduced costs. Whilst considering this approach for tesirine, we noted that the majority (90%) of the cost of the campaign was associated with the eight high-containment steps (**39** to **1**), whereas the 26 non-high-containment steps accounted for only 10% of the cost (Figure 16). This is in marked contrast to the manufacture of a standard drug molecule, where most of the manufacturing cost is associated with the earlier stages of the campaign as larger scale equipment is required.

In light of this data and analysis, we decided that an initial focus for process improvement would be to reduce the number of high-containment steps in the synthesis. We expected a significant reduction in the overall cost of goods (CoGs) to result (*51*). To execute this, the number of cytotoxic compounds in the synthetic route would need to be reduced, thus alternative non-cytotoxic intermediates were considered.

The structural features that need to be present in PBD molecules to impart cytotoxicity are intact A-, B- and C-rings. As the B-ring is constructed as part of the synthetic route using an oxidative cyclisation, it was reasoned that moving this step to as late as possible in the synthesis would reduce the number of steps requiring high-containment. Finishing the synthesis with a double oxidative cyclisation was considered, but this chemistry is not compatible with the presence of reactive maleimide group. Thus, the previously used final steps (**49** to **1**) were retained, and work undertaken looking to complete the oxidative cyclisation as the antepenultimate step of the synthesis.

Demonstration of the Closure of the B-rings as the Antepenultimate Step

To prepare a suitable substrate for the oxidative cyclisation investigation, the previously prepared mono-Alloc-protected dimer **52** was coupled with Alloc-Val-Ala-PABOH **47** using triphosgene to afford sidechain-dimer **53** (Scheme 9). The TBS-protecting groups were then removed using standard pTSA in THF and water conditions to create diol **54**. Oxidative cyclisation of the B-rings of this molecule was investigated using a range of reagents which gave varying results; the best conversions were observed with TEMPO/BAIB or Stahl conditions (which also utilizes TEMPO). On detailed examination, the former gave variability in the reaction and the purity of the product, thus the Stahl conditions were developed. With careful optimization of equivalents and air-sparging, excellent conversion (>95%) to the required product **49** could be achieved (Scheme 9).

With the key oxidative cyclisation demonstrated, work was initiated to include the revised synthetic sequence in a manufacturing campaign. The mono-Alloc-protected dimer **52** that was used as part of the demonstration sequence had been prepared by mono-protection of the bis-aniline. This reaction was non-selective, affording a 1:2:1 ratio of bis-aniline/mono-Alloc-protected dimer **52**/bis-Alloc-protected dimer. Extensive development work failed to improve this ratio. We thus decided to prepare the two halves of the dimer separately to fabricate **53**.

Preparation of Ring-Opened PBD with Five-Carbon Tether (56)

Previously prepared aniline **15** was used as the starting point for the preparation of the two ring-opened halves of **1**. For the preparation of the Alloc-protected half **56** (Scheme 10), the conditions for the initial introduction of the Alloc-protecting group was changed from pyridine/DCM to the more process-friendly sodium hydrogen carbonate in aqueous THF. As previously, the Alloc-aniline **32** could be used directly in the following step without purification. The TIPS-protecting group was now removed using the previously demonstrated lithium acetate/DMF conditions, although the transformation could also be achieved using the potassium carbonate/methanol, albeit with lower product purity. The obtained phenol **55** was purified using a silica plug. Finally, the five-carbon linker was introduced using 1,5-dibromopentane using conditions optimized by factorial experimental design, generating **56** with 98% Area HPLC purity following flash column chromatography. Although less reactive than 1,5-diiodopentane, 1,5-dibromopropane is more suitable for large scale synthesis. It is approximately ten times cheaper (Sigma Aldrich catalogue: £20 per 100g, instead of £225 for diiodopentane), more stable, and can activated in situ with TBAI.

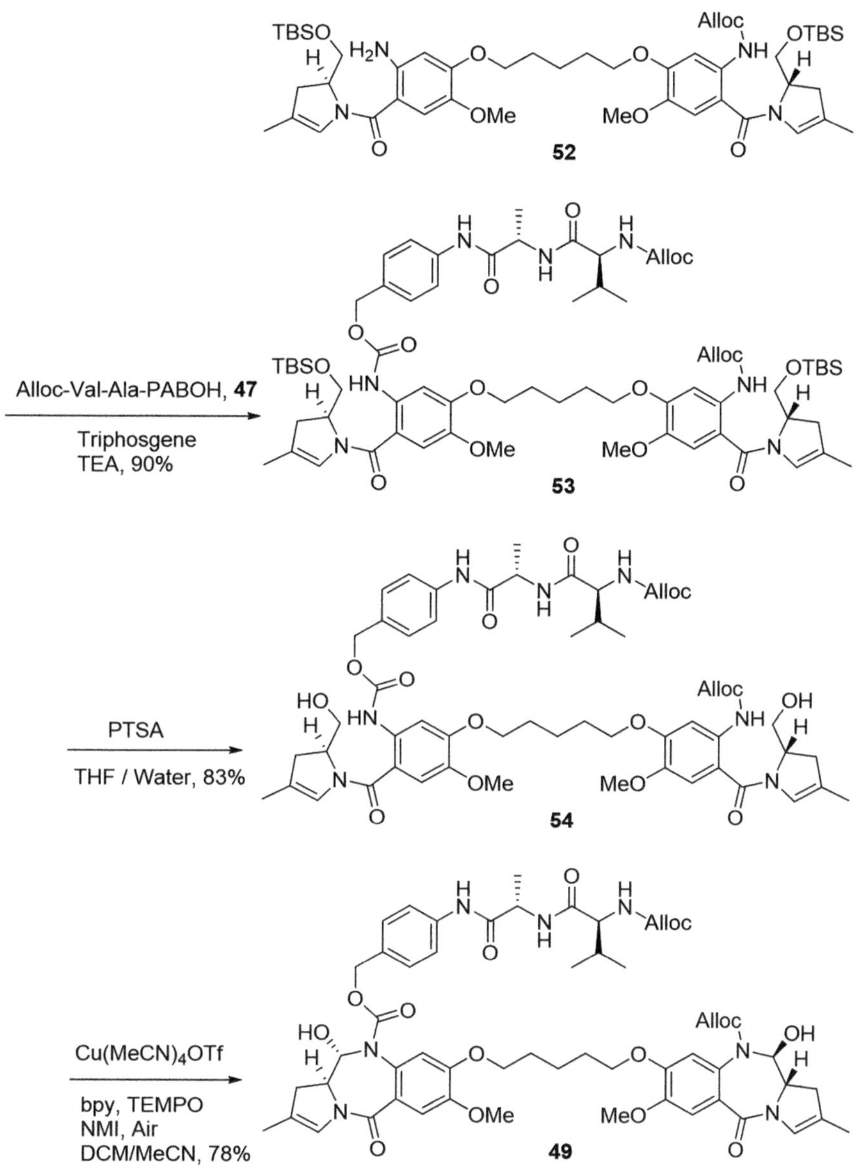

Scheme 9. Demonstration of the late oxidative cyclisation.

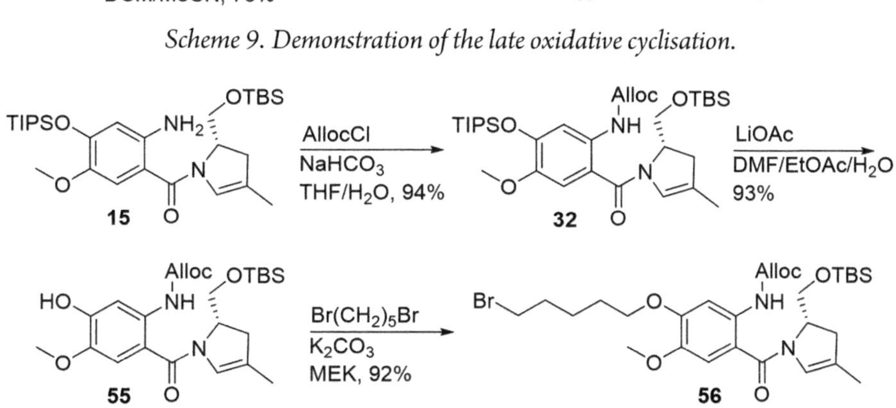

*Scheme 10. Preparation of ring-opened PBD with five-carbon tether (**56**).*

Preparation of Ring-Opened PBD with Val-Ala-PABC Trigger (57) and Coupling of Two Halves

For the preparation of the Val-Ala-PABC trigger half of the molecule **57**, the coupling of aniline **15** with Alloc-Val-Ala-PABOH **47** to deliver **38** was reinvestigated, and we discovered that triphosgene could be replaced by the more practical CDI. The TIPS-protecting group was subsequently removed using the standard lithium acetate/DMF conditions, and the phenol **57** used without further purification. The two halves of the molecule were then coupled together using conditions optimized by factorial experimental design, which included the addition of TBAI (0.96 eq) along with potassium carbonate (1.55 eq) and MEK (10 volumes). Extended reaction times were required, but good conversion could be obtained and the bis-TBS ether product **53** isolated in good purity following flash column chromatography (Scheme 11).

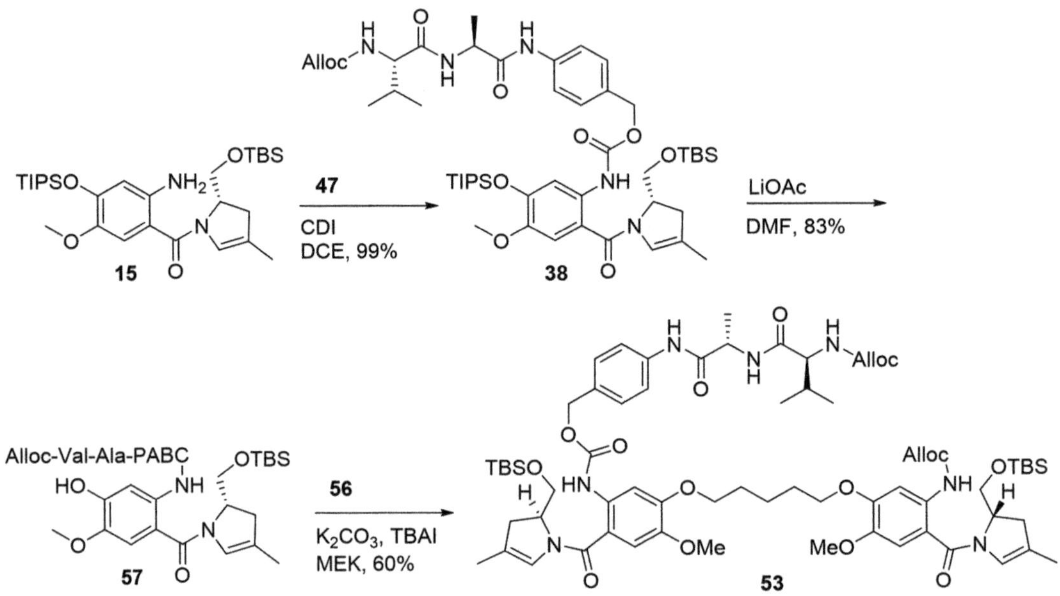

Scheme 11. Preparation of ring-opened PBD with Val-Ala-PABC trigger (57) and coupling of two halves.

Final Assembly of Tesirine (1) Using Late Double Oxidative Cyclisation

The revised synthesis was completed with the previously demonstrated deprotection of the TBS-groups and oxidative cyclisation (Scheme 9). The deprotection of **53** was still completed using pTSA in aqueous THF, but on scale up the diol **54** was obtained as a solid by precipitation from 2-methyl-THF and *n*-heptane. The oxidative cyclisation was achieved using the previously developed Stahl conditions and following purification by reversed-phase HPLC, **49** could be obtained intercepting the original route to tesirine. The final two steps of the synthesis used the methodology developed for the previous campaigns (Scheme 8) including the use of reversed-phase HPLC purification, but the final isolation of tesirine (**1**) was switched to a precipitation from THF/*n*-heptane which was an operation with lower exposure risk than the previously used evaporation.

The revised synthesis was used to supply material for a further set of clinical studies. This route offered a number of advantages compared to the Discovery route (Table 2). Specifically, it resulted in a decrease in the overall number of steps (including the longest linear sequence as well as the high-containment steps), leading to a 60% reduction in CoGs. However, these modifications did not provide a significant increase in the overall yield.

Table 2. Metrics for the Discovery and Reduced High-Containment Routes

Parameter	*Discovery Route*	*Reduced High Containment Route*
Amount of tesirine produced	169 g	151 g
Number of Steps	34	29
Number of High-Containment Steps	8	3
Longest linear sequence (steps)	20	18
Overall Yield (longest linear sequence)	2.19%	2.21%
Average Yield (longest linear sequence)	83%	81%
Number of steps requiring chromatography	8	5
Final Purity	97%	99%

A schematic of the reduced high-containment route is shown in Figure 17.

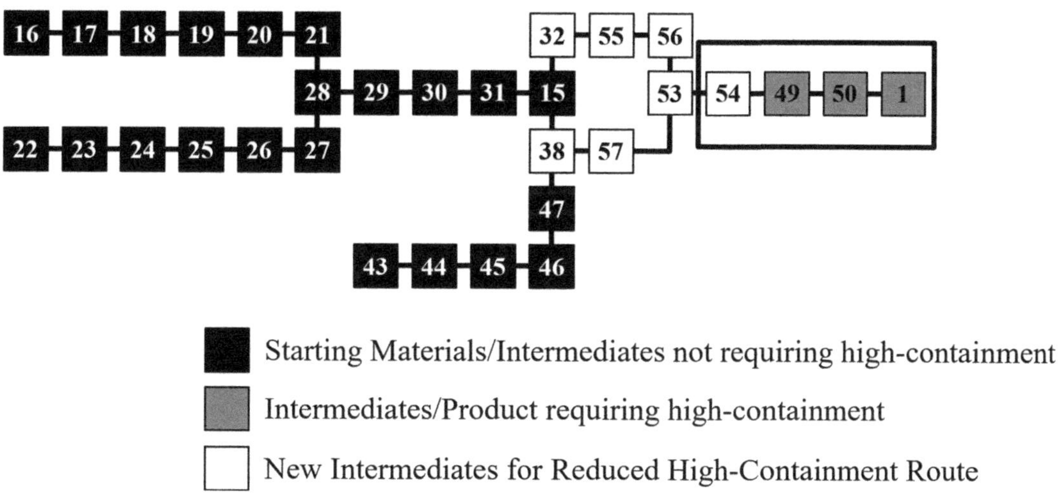

Figure 17. Schematic of reduced high-containment route (each block representing a separate intermediate) showing portion conducted in high-containment (in boxed region).

Route Design Using Isomerization to Reduce Total Number of Steps

Following the successful application of Route Design to reduce the number of high-containment steps, a further evaluation was completed endeavoring to reduce the overall step count and increase the efficiency of the synthesis. This work identified the 9-step synthesis of the nitrobenzene **31** from hydroxyproline (**22**) as an area where a shorter route could be introduced. The initial route (Schemes 3 and 4) included three protection/deprotection steps and three steps to introduce a single carbon atom on the pyrrolidine ring.

As part of this evaluation, (*S*)-4-methylenepyrrolidine-2-carboxylic acid (**58**) was identified as a viable, commercially-available starting material. This molecule has the correct number of carbon atoms and the correct oxidation level to be used as the pyrrolidine component and would only require isomerization of the *exo*-methylene double bond into the ring to afford the required C-ring (Scheme 12) (52).

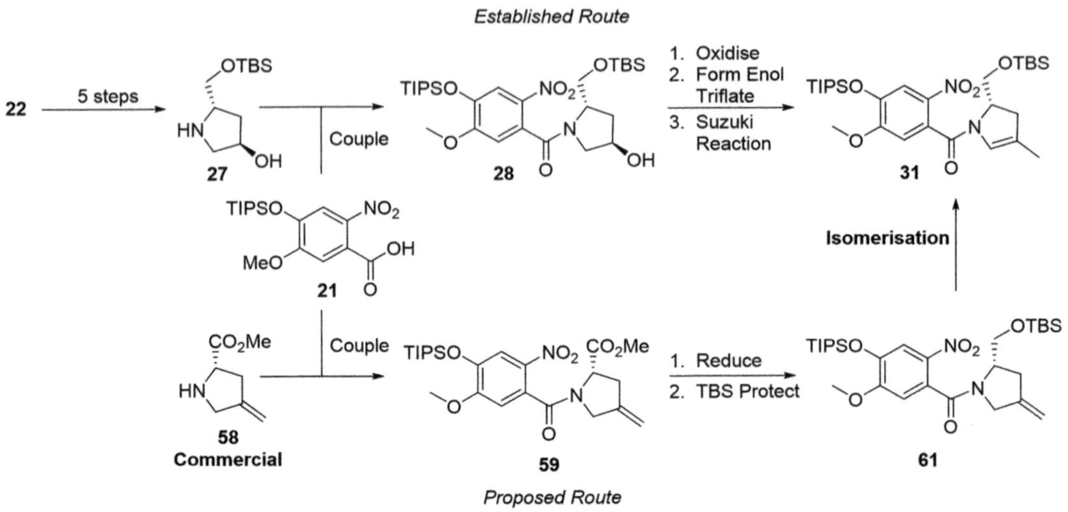

Scheme 12. Original and isomerisation approaches to ***31****.*

The isomerization of *N*-allyl amides to enamides is a known process driven by favorable thermodynamics and can be achieved using a range of transition metal catalysts. For more complex systems, analogous to the desired transformation (**61** to **31**), there are only limited reports using super-stoichiometric amounts of $Fe(CO)_5$ (53). Thus, an isomerization screening program was initiated using nitrobenzene **61**, which was available from a different PBD program.

This screening work identified di-μ-bromobis(tri-*tert*-butylphosphine)dipalladium(I) as an effective catalyst for the transformation (54). Detailed mechanistic studies indicated that the isomerization proceeded via the 4-methyl-2,5-dihydropyrrole **62** (kinetic product), which was then further isomerized to the required 4-methyl-2.3-dihydropyrrole **31** (thermodynamic product) (Scheme 13).

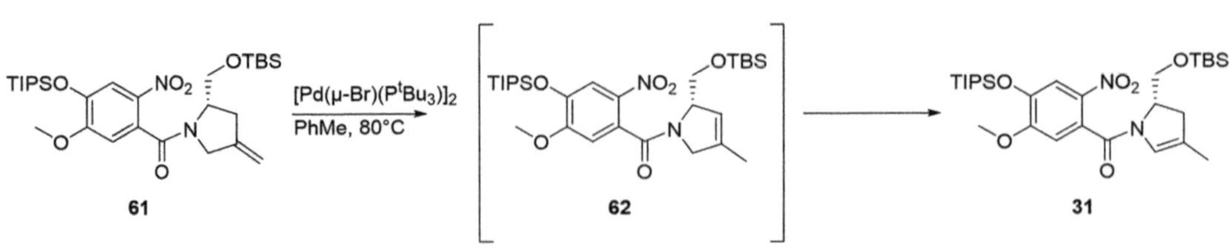

Scheme 13. Isomerization pathway.

This methodology was scaled-up in the laboratory (1.00 g), providing a new formal 24-step route to tesirine. The route provides a reduction in overall number of steps and an improvement in overall yield for the longest linear sequence (Table 3). Scale-up of the route is planned for further deliveries of tesirine (**1**).

Table 3. Metrics for the Reduced High-Containment and Isomerization Routes

Parameter	*Reduced High Containment Route*	*Isomerization Route*
Number of Steps	29	24
Number of High-Containment Steps	3	3
Longest linear sequence (steps)	18	18
Overall Yield (longest linear sequence)	2.21%	3.61%
Average Yield (longest linear sequence)	81%	83%

A schematic of the isomerization route is shown in Figure 18.

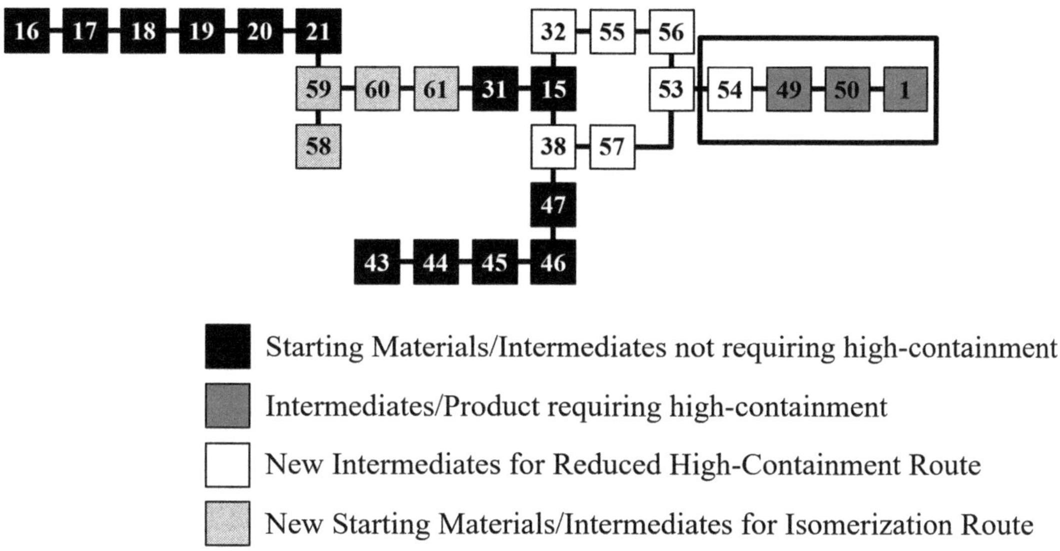

Figure 18. Schematic of isomerization route (each block representing a separate intermediate) showing portion conducted in high-containment (in boxed region).

Tesirine Clinical Landscape in 2020. Clinical Learnings

Since tesirine can be conjugated to any therapeutic antibody, and because of its wide availability to AstraZeneca/Spirogen licensees, a large number of clinical trials has been conducted. Table 4 shows the clinical landscape at the time of writing (1H 2020).

Three companies are currently testing tesirine in the clinic, on different antibodies: AstraZeneca, ADC Therapeutics (ADCT), and Mitsubishi Tanabe Pharma. Additionally, Stemcentrx/AbbVie have tested tesirine on rovalpituzumab (rovalpituzumab tesirine) on small-cell lung cancer (SCLC) up to Phase III. The original responses rates from the Phase I trial (18% in all patients, and 38% in DLL3-high patient set) (55) were not reproducible in a larger Phase III setting and the development of the drug was terminated (56). While the companies using tesirine in their ADCs are keen to understand the reasons behind this failure, they remain aware of the therapeutic potential of tesirine-ADCs observed in preclinical models.

Table 4. Tesirine Clinical Landscape in 1H 2020 (of interest are the two registrational phase II trials of ADCT: loncastuximab tesirine and camidanlumab tesirine in lymphomas)

COMPANY	*AGENT*	*INDICATION*	*TARGET*	*IDENTIFIER*	*PHASE*	*CONJUGATION*
AstraZeneca	*MEDI2228*	*Multiple Myeloma*	*BCMA*	*NCT03489525*	*I*	*Site Specific*
ADCT	Lonca-T (ADCT-402)	B-NHL	CD19	NCT03589469	II	Stochastic
ADCT	Lonca-T + Ibrutinib	B-NHL	CD19	NCT03684694	I	Stochastic
ADCT	Lonca-T + Durvalumab	B-NHL	CD19	NCT03685344	I	Stochastic
ADCT	Lonca-T (ADCT-402)	B-ALL	CD19	NCT02669264	I	Stochastic
ADCT	Cami-T (ADCT-301)	Hodgkin Lymphoma	CD25	NCT04052997	II	Stochastic
ADCT	Cami-T (ADCT-301)	AML/ALL	CD25	NCT02588092	I	Stochastic
ADCT	Cami-T (ADCT-301)	Advanced Solid Tumours	CD25	NCT03621982	I	Stochastic
ADCT	ADCT-602	B-ALL	CD22	NCT03698552	I	Stochastic
ADCT	ADCT-601	Solid Tumours	Axl	NCT03700294	I	Site Specific
Tanabe	TR1801	Solid Tumours	c-MET	NCT03859752	I	Site Specific

At AstraZeneca we use a 5R framework (57) to guide us on the correct path towards drug development. The 5Rs are:

- the right target
- the right tissue
- the right patients
- the right safety
- the right commercial potential

The same approach has been used and extended for the development of an ADC. The target is the most important parameter; it needs to be selective for cancer, and if possible, highly expressed and as homogeneously throughout the tumor as possible. Whereas DLL3 (Delta-like ligand 3) is known as a clean target, its expression numbers are actually relatively low (around 10000 proteins per expressing cancer cell, compared with two orders of magnitude higher for Kadcyla® target, HER2) (58). In addition, the percentage of cells expressing DLL3 in a tumor varies from patient to patient, with significant heterogeneity. A highly heterogenous tumor (with only a fraction of the cells expressing the antigen target) is significantly harder to treat than a homogeneous tumor (where all cells express the target). In a 63 SCLC patients study, on average only 33.9% of the cancer cells in a tumor expressed DLL3 (59). These target parameters, taken together with a challenging tissue and disease setting made it particularly difficult to achieve clinical success.

In contrast to Stemcentrx/AbbVie targeting DLL3 in small-cell lung cancer, ADC Therapeutics (ADCT) trials targeting CD19 and CD25 (biomarkers antigen targets in neoplastic B and T cells) in lymphoma appears to follow a more favorable path. We can return to the 5Rs framework to rationalize the results coming out of these trials.

Target

ADCT-402 (loncastuximab tesirine) targets CD19 in B-cell malignancies such as non-Hodgkin lymphomas. In their preclinical work, Zammarchi and colleagues (*3*) demonstrated good target expression (+++) in a range of lymphomas clinical samples. In addition, more than 80% of the cells in the tumor expressed CD19. Representative tumor models such as Ramos typically expressed 100,000 antigens per cell (ten times more than DLL3), thus allowing efficacious delivery of SG3199. Similar numbers were found for CD25, targeted by ADCT-301 (camidanlumab tesirine) (*1*).

Tissue and Safety

SG3199, the warhead of tesirine, has been shown to be extremely effective in hematological settings (*8*), allowing usage of low, non-severely toxic doses of ADCs. Whereas rovalpituzumab tesirine was given at high doses of up to 0.4 mg/kg, loncastuximab tesirine schedule is currently at 0.15 mg/kg with follow-up doses of 75 µg/kg. This dose reduction is even more significant in the case of camidanlumab tesirine at 45 µg/kg. The reader will notice a large difference between the therapeutic doses in human, and the complete responses doses in mouse (1 mg/kg). These discrepancies drive the improvement of our models, to include the role of the immune system (a positive factor) and account for the typically higher resilience of rodents to the toxicity of PBDs.

Another important factor to include in safety preclinical evaluation is half-life measurement. Although an indirect proxy, the half-life gives good indications of the drug fate. As antibodies tend to remain in circulation for long periods of time (weeks), a sharp departure from this trend indicates rapid tissue intake or catabolism. Unless a large tumor burden is present, this can signify uptake by healthy tissues, followed by toxicity. For optimum safety and efficacy, ADCs should aim to have a long half-life in circulation, ranging from 5 to 10 days. camidanlumab tesirine (conjugated antibody) half-life was found to be 9.1 days (*60*) and loncastuximab tesirine half-life was 7.2 days at cycle 1 and 12.5 days after a second cycle (*61*).

Good efficacy results were observed for loncastuximab tesirine, with an overall response rate (ORR) of 53.5% (all doses, Phase I), and a complete response rate of 36% (*61*). The complete response rate matters because those patients have a good chance of longer-term remission. The Phase I results were even more impressive for camidanlumab tesirine, with an ORR of 80.8%, and a complete response rate of 50%.

On the basis of this encouraging data, loncastuximab tesirine is progressing through a pivotal Phase II trial, and ADCT are expected to file for a Biologics License Application (BLA) in 2H 2020 in anticipation of a commercial launch in 1H 2021. (adctherapeutics.com) Camidanlumab tesirine has also begun a Phase II pivotal trial expected to read-out in 2H 2021.

Future Directions

Many of these tesirine-ADCs trials are conducted as a single agent, but clinicians have long pushed for combination trials where the field is expecting more encouraging results. Either the complete response rate can be raised significantly (i.e.: approval of the ADC combination Polivy® in DLBCL (diffuse large B-cell lymphoma) with bendamustine and a rituximab product, giving a 40% complete response rate instead of 18% for Polivy® alone), or the dose can be lowered to tolerable levels such as 0.1 mg/kg for a tesirine-ADC. Rios-Doria and colleagues (*62*) demonstrated the synergy in mice with an immunomodulatory PD-L1 antibody, whereas Zhong and colleagues (*63*) observed a synergy with olaparib in BRCA-deficient tumors. On this basis, ADC Therapeutics is currently conducting two Phase I trials in combination (Table 4): Lonca-T + Ibrutinib (a covalent binder of Bruton's tyrosine kinase), and Lonca-T with immunomodulatory durvalumab. Similarly, AstraZeneca found synergistic interactions with combinations of multiple myeloma tesirine-ADC MEDI2228 with DNA damage response (DDR) inhibitors or bortezomib (*64*).

Effect of Camidanlumab Tesirine on T-regs and Potential in Immunotherapy of Solid Tumors

Finally, a clinical trial of interest is camidanlumab tesirine in selected solid tumors with literature evidence of CD25 Treg content (NCT03621982). In 2018, Christian Schürch noted that depletion of tumor-promoting CD25 positive Treg cells targeted by Cami-T would result in a positive immuno-oncological effect (*65*). It will be particularly encouraging to observe regressions in these tumor models.

Conclusion

Because of its unique history and licensing model, tesirine was evaluated on a large number of antibodies and cancer targets, and underwent a series of clinical trials ranging from Phase I to Phase III. These multiple trials created a drive to improve the lengthy discovery route and provide a robust process on clinical scale. The potency of intermediates in the final stages of the synthesis redefined the priorities of ADC payloads process design, and the highly potent stages were optimized, with a significant reduction in the number of steps (8 to 3).

Because the same payload, tesirine, was used on different antibodies and tumor models, we are now in a position to draw conclusions as to which ADC design factors favor success, or on the contrary, can warn us of poor clinical outcomes.

At the time of writing loncastuximab tesirine is scheduled for commercial launch in the first half of 2021.

References

1. Flynn, M. J.; Zammarchi, F.; Tyrer, P. C.; Akarca, A. U.; Janghra, N.; Britten, C. E.; Havenith, C. E.; Levy, J. N.; Tiberghien, A.; Masterson, L. A.; Barry, C.; D'Hooge, F.; Marafioti, T.; Parren, P. W.; Williams, D. G.; Howard, P. W.; van Berkel, P. H.; Hartley, J. A. ADCT-301, a Pyrrolobenzodiazepine (PBD) Dimer-Containing Antibody-Drug Conjugate (ADC) Targeting CD25-Expressing Hematological Malignancies. *Mol. Cancer Ther.* **2016**, *15*, 2709–2721.
2. Saunders, L. R.; Bankovich, A. J.; Anderson, W. C.; Aujay, M. A.; Bheddah, S.; Black, K.; Desai, R.; Escarpe, P. A.; Hampl, J.; Laysang, A.; Liu, D.; Lopez-Molina, J.; Milton, M.; Park, A.; Pysz, M. A.; Shao, H.; Slingerland, B.; Torgov, M.; Williams, S. A.; Foord, O.; Howard, P.; Jassem, J.; Badzio, A.; Czapiewski, P.; Harpole, D. H.; Dowlati, A.; Massion, P. P.; Travis, W. D.; Pietanza, M. C.; Poirier, J. T.; Rudin, C. M.; Stull, R. A.; Dylla, S. J. A DLL3-targeted

antibody-drug conjugate eradicates high-grade pulmonary neuroendocrine tumor-initiating cells in vivo. *Sci. Trans. Med* **2015**, 7, 302ra136.

3. Zammarchi, F.; Corbett, S.; Adams, L.; Tyrer, P. C.; Kiakos, K.; Janghra, N.; Marafioti, T.; Britten, C. E.; Havenith, C. E. G.; Chivers, S.; D'Hooge, F.; Williams, D. G.; Tiberghien, A.; Howard, P. W.; Hartley, J. A.; van Berkel, P. H. ADCT-402, a PBD dimer-containing antibody drug conjugate targeting CD19-expressing malignancies. *Blood* **2018**, *131*, 1094–1105.
4. Kinneer, K.; Flynn, M.; Thomas, S. B.; Meekin, J.; Varkey, R.; Xiao, X.; Zhong, H.; Breen, S.; Hynes, P. G.; Fleming, R.; Bezabeh, B.; Chen, C. T.; Wetzel, L.; Chen, R.; Dimasi, N.; Tai, Y. T.; Anderson, K. C.; Herbst, R.; Howard, P. W.; Hurt, E. M.; Tice, D. A. Preclinical assessment of an antibody-PBD conjugate that targets BCMA on multiple myeloma and myeloma progenitor cells. *Leukemia* **2019**, *33*, 766–771.
5. Harper, J.; Lloyd, C.; Dimasi, N.; Toader, D.; Marwood, R.; Lewis, L.; Bannister, D.; Jovanovic, J.; Fleming, R.; D'Hooge, F.; Mao, S.; Marrero, A. M.; Korade, M., 3rd; Strout, P.; Xu, L.; Chen, C.; Wetzel, L.; Breen, S.; van Vlerken-Ysla, L.; Jalla, S.; Rebelatto, M.; Zhong, H.; Hurt, E. M.; Hinrichs, M. J.; Huang, K.; Howard, P. W.; Tice, D. A.; Hollingsworth, R. E.; Herbst, R.; Kamal, A. Preclinical Evaluation of MEDI0641, a Pyrrolobenzodiazepine-Conjugated Antibody-Drug Conjugate Targeting 5T4. *Mol. Cancer Ther.* **2017**, *16*, 1576–1587.
6. Cho, S.; Zammarchi, F.; Williams, D. G.; Havenith, C. E. G.; Monks, N. R.; Tyrer, P.; D'Hooge, F.; Fleming, R.; Vashisht, K.; Dimasi, N.; Bertelli, F.; Corbett, S.; Adams, L.; Reinert, H. W.; Dissanayake, S.; Britten, C. E.; King, W.; Dacosta, K.; Tammali, R.; Schifferli, K.; Strout, P.; Korade, M., 3rd; Masson Hinrichs, M. J.; Chivers, S.; Corey, E.; Liu, H.; Kim, S.; Bander, N. H.; Howard, P. W.; Hartley, J. A.; Coats, S.; Tice, D. A.; Herbst, R.; van Berkel, P. H. Antitumor Activity of MEDI3726 (ADCT-401), a Pyrrolobenzodiazepine Antibody-Drug Conjugate Targeting PSMA, in Preclinical Models of Prostate Cancer. *Mol. Cancer Ther.* **2018**, *17*, 2176–2186.
7. Gymnopoulos, M.; Betancourt, O.; Blot, V.; Fujita, R.; Galvan, D.; Lieuw, V.; Nguyen, S.; Snedden, J.; Stewart, C.; Villicana, J.; Wojciak, J.; Wong, E.; Pardo, R.; Patel, N.; D'Hooge, F.; Vijayakrishnan, B.; Barry, C.; Hartley, J. A.; Howard, P. W.; Newman, R.; Coronella, J. TR1801-ADC: a highly potent cMet antibody–drug conjugate with high activity in patient-derived xenograft models of solid tumors. *Mol. Oncol.* **2020**, *14*, 54–68.
8. Hartley, J. A.; Flynn, M. J.; Bingham, J. P.; Corbett, S.; Reinert, H.; Tiberghien, A.; Masterson, L. A.; Antonow, D.; Adams, L.; Chowdhury, S.; Williams, D. G.; Mao, S.; Harper, J.; Havenith, C. E. G.; Zammarchi, F.; Chivers, S.; van Berkel, P. H.; Howard, P. W. Preclinical pharmacology and mechanism of action of SG3199, the pyrrolobenzodiazepine (PBD) dimer warhead component of antibody-drug conjugate (ADC) payload tesirine. *Sci. Rep.* **2018**, *8*, 10479.
9. Tiberghien, A. C.; von Bulow, C.; Barry, C.; Ge, H.; Noti, C.; Collet Leiris, F.; McCormick, M.; Howard, P. W.; Parker, J. S. Scale-up Synthesis of Tesirine. *Org. Process Res. Dev.* **2018**, *22*, 1241–1256.
10. Lucas, A. T.; Price, L. S. L.; Schorzman, A. N.; Storrie, M.; Piscitelli, J. A.; Razo, J.; Zamboni, W. C. Factors Affecting the Pharmacology of Antibody–Drug Conjugates. *Antibodies* **2018**, *7*, 10.

11. Beck, A.; Goetsch, L.; Dumontet, C.; Corvaia, N. Strategies and challenges for the next generation of antibody-drug conjugates. *Nat. Rev. Drug Discovery* **2017**, *16* (5), 315–337.
12. Gordon, M. R.; Canakci, M.; Li, L.; Zhuang, J.; Osborne, B.; Thayumanavan, S. Field Guide to Challenges and Opportunities in Antibody-Drug Conjugates for Chemists. *Bioconjug. Chem.* **2015**, *26*, 2198–215.
13. Ogitani, Y.; Aida, T.; Hagihara, K.; Yamaguchi, J.; Ishii, C.; Harada, N.; Soma, M.; Okamoto, H.; Oitate, M.; Arakawa, S.; Hirai, T.; Atsumi, R.; Nakada, T.; Hayakawa, I.; Abe, Y.; Agatsuma, T. DS-8201a, A Novel HER2-Targeting ADC with a Novel DNA Topoisomerase I Inhibitor, Demonstrates a Promising Antitumor Efficacy with Differentiation from T-DM1. *Clin. Cancer Res.* **2016**, *22*, 5097–5108.
14. M. Gromek, S.; J. Balunas, M. Natural Products as Exquisitely Potent Cytotoxic Payloads for Antibody- Drug Conjugates. *Curr. Trends Med. Chem.* **2014**, *14*, 2822–2834.
15. Liu-Kreyche, P.; Shen, H.; Marino, A. M.; Iyer, R. A.; Humphreys, W. G.; Lai, Y. Lysosomal P-gp-MDR1 Confers Drug Resistance of Brentuximab Vedotin and Its Cytotoxic Payload Monomethyl Auristatin E in Tumor Cells. *Front. Pharmacol.* **2019**, *10*, 749.
16. Antonow, D.; Thurston, D. E. Synthesis of DNA-interactive pyrrolo[2,1-c][1,4]benzodiazepines (PBDs). *Chem. Rev.* **2011**, *111*, 2815–64.
17. Wells, G.; Martin, C. R.; Howard, P. W.; Sands, Z. A.; Laughton, C. A.; Tiberghien, A.; Woo, C. K.; Masterson, L. A.; Stephenson, M. J.; Hartley, J. A.; Jenkins, T. C.; Shnyder, S. D.; Loadman, P. M.; Waring, M. J.; Thurston, D. E. Design, synthesis, and biophysical and biological evaluation of a series of pyrrolobenzodiazepine-poly(N-methylpyrrole) conjugates. *J. Med. Chem.* **2006**, *49*, 5442–61.
18. Bose, D. S.; Thompson, A. S.; Ching, J.; Hartley, J. A.; Berardini, M. D.; Jenkins, T. C.; Neidle, S.; Hurley, L. H.; Thurston, D. E. Rational design of a highly efficient irreversible DNA interstrand cross-linking agent based on the pyrrolobenzodiazepine ring system. *J. Am. Chem. Soc* **1992**, *114*, 4939–4941.
19. Gregson, S. J.; Howard, P. W.; Hartley, J. A.; Brooks, N. A.; Adams, L. J.; Jenkins, T. C.; Kelland, L. R.; Thurston, D. E. Design, Synthesis, and Evaluation of a Novel Pyrrolobenzodiazepine DNA-Interactive Agent with Highly Efficient Cross-Linking Ability and Potent Cytotoxicity. *J. Med. Chem.* **2001**, *44*, 737–748.
20. Hartley, J. A.; Spanswick, V. J.; Brooks, N.; Clingen, P. H.; McHugh, P. J.; Hochhauser, D.; Pedley, R. B.; Kelland, L. R.; Alley, M. C.; Schultz, R.; Hollingshead, M. G.; Schweikart, K. M.; Tomaszewski, J. E.; Sausville, E. A.; Gregson, S. J.; Howard, P. W.; Thurston, D. E. SJG-136 (NSC 694501), a Novel Rationally Designed DNA Minor Groove Interstrand Cross-Linking Agent with Potent and Broad Spectrum Antitumor Activity: Part 1: Cellular Pharmacology, In vitro and Initial In vivo Antitumor Activity. *Cancer Res.* **2004**, *64*, 6693–6699.
21. Hochhauser, D.; Meyer, T.; Spanswick, V. J.; Wu, J.; Clingen, P. H.; Loadman, P.; Cobb, M.; Gumbrell, L.; Begent, R. H.; Hartley, J. A.; Jodrell, D. Phase I study of sequence-selective minor groove DNA binding agent SJG-136 in patients with advanced solid tumors. *Clin. Cancer Res.* **2009**, *15*, 2140–7.
22. Puzanov, I.; Lee, W.; Chen, A. P.; Calcutt, M. W.; Hachey, D. L.; Vermeulen, W. L.; Spanswick, V. J.; Liao, C. Y.; Hartley, J. A.; Berlin, J. D.; Rothenberg, M. L. Phase I pharmacokinetic and pharmacodynamic study of SJG-136, a novel DNA sequence selective

minor groove cross-linking agent, in advanced solid tumors. *Clin. Cancer Res.* **2011**, *17*, 3794–3802.

23. Hinrichs, M. J. M.; Ryan, P. M.; Zheng, B.; Afif-Rider, S.; Yu, X. Q.; Gunsior, M.; Zhong, H.; Harper, J.; Bezabeh, B.; Vashisht, K.; Rebelatto, M.; Reed, M.; Ryan, P. C.; Breen, S.; Patel, N.; Chen, C.; Masterson, L.; Tiberghien, A.; Howard, P. W.; Dimasi, N.; Dixit, R. Fractionated Dosing Improves Preclinical Therapeutic Index of Pyrrolobenzodiazepine-Containing Antibody Drug Conjugates. *Clin. Cancer Res.* **2017**, *23*, 5858–5868.
24. Gregson, S. J.; Barrett, A. M.; Patel, N. V.; Kang, G. D.; Schiavone, D.; Sult, E.; Barry, C. S.; Vijayakrishnan, B.; Adams, L. R.; Masterson, L. A.; D'Hooge, F.; Snaith, M.; Harper, J.; Hartley, J. A.; Howard, P. W. Synthesis and evaluation of pyrrolobenzodiazepine dimer antibody-drug conjugates with dual beta-glucuronide and dipeptide triggers. *Eur. J. Med. Chem.* **2019**, *179*, 591–607.
25. Chen, Z.; Gregson, S. J.; Howard, P. W.; Thurston, D. E. A novel approach to the synthesis of cytotoxic C2-C3 unsaturated pyrrolo[2,1-c]benzodiazepines (PBDs) with conjugated acrylyl C2-substituents. *Bioorg. Med. Chem. Lett.* **2004**, *14*, 1547–9.
26. Tiberghien, A. C.; Hagan, D.; Howard, P. W.; Thurston, D. E. Application of the Stille coupling reaction to the synthesis of C2-substituted endo-exo unsaturated pyrrolo[2,1-c][1,4]benzodiazepines (PBDs). *Bioorg. Med. Chem. Lett.* **2004**, *14*, 5041–5044.
27. Cooper, N.; Hagan, D. R.; Tiberghien, A.; Ademefun, T.; Matthews, C. S.; Howard, P. W.; Thurston, D. E. Synthesis of novel C2-aryl pyrrolobenzodiazepines (PBDs) as potential antitumor agents. *Chem. Commun. (Cambridge, U. K.)* **2002** (16), 1764–1765.
28. Howard, P. W.; Chen, Z.; Gregson, S. J.; Masterson, L. A.; Tiberghien, A. C.; Cooper, N.; Fang, M.; Coffils, M. J.; Klee, S.; Hartley, J. A.; Thurston, D. E. Synthesis of a novel C2/C2′-aryl-substituted pyrrolo[2,1-c][1,4]benzodiazepine dimer prodrug with improved water solubility and reduced DNA reaction rate. *Bioorg. Med. Chem. Lett.* **2009**, *19*, 6463–6466.
29. Burger, A. M.; Loadman, P. M.; Thurston, D. E.; Schultz, R.; Fiebig, H. H.; Bibby, M. C. Preclinical Pharmacology of the Pyrrolobenzodiazepine (PBD) Monomer DRH-417 (NSC 709119). *J. Chemother.* **2007**, *19*, 66–78.
30. Unlike the IC50 which is used to measure the inhibition of a specified individual target, e.g. an enzyme, GI50 is used to measure the inhibition of whole-cell. GI50 is the concentration that cause 50% of maximal inhibition of cell growth (proliferation).
31. Hartley, J. A.; Hamaguchi, A.; Coffils, M.; Martin, C. R. H.; Suggitt, M.; Chen, Z.; Gregson, S. J.; Masterson, L. A.; Tiberghien, A. C.; Hartley, J. M.; Pepper, C.; Lin, T. T.; Fegan, C.; Thurston, D. E.; Howard, P. W. SG2285, a Novel C2-Aryl-Substituted Pyrrolobenzodiazepine Dimer Prodrug That Cross-links DNA and Exerts Highly Potent Antitumor Activity. *Cancer Res.* **2010**, *70*, 6849–6858.
32. Hartley, J. A. The development of pyrrolobenzodiazepines as antitumour agents. *Expert Opin. Invest. Drugs* **2011**, *20*, 733–744.
33. Jeffrey, S. C.; Torgov, M. Y.; Andreyka, J. B.; Boddington, L.; Cerveny, C. G.; Denny, W. A.; Gordon, K. A.; Gustin, D.; Haugen, J.; Kline, T.; Nguyen, M. T.; Senter, P. D. Design, synthesis, and in vitro evaluation of dipeptide-based antibody minor groove binder conjugates. *J. Med. Chem.* **2005**, *48*, 1344–1358.
34. Kung Sutherland, M. S.; Walter, R. B.; Jeffrey, S. C.; Burke, P. J.; Yu, C.; Kostner, H.; Stone, I.; Ryan, M. C.; Sussman, D.; Lyon, R. P.; Zeng, W.; Harrington, K. H.; Klussman, K.;

Westendorf, L.; Meyer, D.; Bernstein, I. D.; Senter, P. D.; Benjamin, D. R.; Drachman, J. G.; McEarchern, J. A. SGN-CD33A: a novel CD33-targeting antibody-drug conjugate using a pyrrolobenzodiazepine dimer is active in models of drug-resistant AML. *Blood* **2013**, *122*, 1455–1463.

35. Fathi, A. T.; Erba, H. P.; Lancet, J. E.; Stein, E. M.; Ravandi, F.; Faderl, S.; Walter, R. B.; Advani, A. S.; DeAngelo, D. J.; Kovacsovics, T. J.; Jillella, A.; Bixby, D.; Levy, M. Y.; O'Meara, M. M.; Ho, P. A.; Voellinger, J.; Stein, A. S. A phase 1 trial of vadastuximab talirine combined with hypomethylating agents in patients with CD33-positive AML. *Blood* **2018**, *132*, 1125–1133.
36. Zheng, B.; Yu, S. F.; Del Rosario, G.; Leong, S. R.; Lee, G. Y.; Vij, R.; Chiu, C.; Liang, W. C.; Wu, Y.; Chalouni, C.; Sadowsky, J.; Clark, V.; Hendricks, A.; Poon, K. A.; Chu, W.; Pillow, T.; Schutten, M. M.; Flygare, J.; Polson, A. G. An Anti-CLL-1 Antibody-Drug Conjugate for the Treatment of Acute Myeloid Leukemia. *Clin. Cancer Res.* **2019**, *25*, 1358–1368.
37. Jeffrey, S. C.; Burke, P. J.; Lyon, R. P.; Meyer, D. W.; Sussman, D.; Anderson, M.; Hunter, J. H.; Leiske, C. I.; Miyamoto, J. B.; Nicholas, N. D.; Okeley, N. M.; Sanderson, R. J.; Stone, I. J.; Zeng, W.; Gregson, S. J.; Masterson, L.; Tiberghien, A. C.; Howard, P. W.; Thurston, D. E.; Law, C. L.; Senter, P. D. A potent anti-CD70 antibody-drug conjugate combining a dimeric pyrrolobenzodiazepine drug with site-specific conjugation technology. *Bioconjug. Chem.* **2013**, *24*, 1256–1263.
38. Ritchie, T. J.; Macdonald, S. J. The impact of aromatic ring count on compound developability--are too many aromatic rings a liability in drug design? *Drug Discovery Today* **2009**, *14* (21-22), 1011–20.
39. Kemp, G. C.; Tiberghien, A. C.; Patel, N. V.; D'Hooge, F.; Nilapwar, S. M.; Adams, L. R.; Corbett, S.; Williams, D. G.; Hartley, J. A.; Howard, P. W. Synthesis and in vitro evaluation of SG3227, a pyrrolobenzodiazepine dimer antibody-drug conjugate payload based on sibiromycin. *Bioorg. Med. Chem. Lett.* **2017**, *27*, 1154–1158.
40. Koniev, O.; Leriche, G.; Nothisen, M.; Remy, J. S.; Strub, J. M.; Schaeffer-Reiss, C.; Van Dorsselaer, A.; Baati, R.; Wagner, A. Selective irreversible chemical tagging of cysteine with 3-arylpropiolonitriles. *Bioconjug. Chem.* **2014**, *25*, 202–206.
41. Saito, F.; Noda, H.; Bode, J. W. Critical evaluation and rate constants of chemoselective ligation reactions for stoichiometric conjugations in water. *ACS Chem. Biol.* **2015**, *10*, 1026–1033.
42. Li, X.; Patterson, J. T.; Sarkar, M.; Pedzisa, L.; Kodadek, T.; Roush, W. R.; Rader, C. Site-Specific Dual Antibody Conjugation via Engineered Cysteine and Selenocysteine Residues. *Bioconjug. Chem.* **2015**, *26*, 2243–2248.
43. Pillow, T. H.; Schutten, M.; Yu, S. F.; Ohri, R.; Sadowsky, J.; Poon, K. A.; Solis, W.; Zhong, F.; Del Rosario, G.; Go, M. A. T.; Lau, J.; Yee, S.; He, J.; Liu, L.; Ng, C.; Xu, K.; Leipold, D. D.; Kamath, A. V.; Zhang, D.; Masterson, L.; Gregson, S. J.; Howard, P. W.; Fang, F.; Chen, J.; Gunzner-Toste, J.; Kozak, K. K.; Spencer, S.; Polakis, P.; Polson, A. G.; Flygare, J. A.; Junutula, J. R. Modulating Therapeutic Activity and Toxicity of Pyrrolobenzodiazepine Antibody-Drug Conjugates with Self-Immolative Disulfide Linkers. *Mol. Cancer Ther.* **2017**, *16*, 871–878.

44. Leipold, D. D.; Figueroa, I.; Masih, S.; Latifi, B.; Yip, V.; Shen, B. Q.; Dere, R. C.; Carrasco-Triguero, M.; Lee, M. V.; Saad, O. M.; Liu, L.; He, J.; Su, D.; Xu, K.; Vuillemenot, B. R.; Laing, S. T.; Schutten, M.; Kozak, K. R.; Zheng, B.; Polson, A. G.; Kamath, A. V. Preclinical pharmacokinetics and pharmacodynamics of DCLL9718A: An antibody-drug conjugate for the treatment of acute myeloid leukemia. *MAbs* **2018**, *10*, 1312–1321.
45. Ma, Y.; Khojasteh, S. C.; Hop, C. E.; Erickson, H. K.; Polson, A.; Pillow, T. H.; Yu, S. F.; Wang, H.; Dragovich, P. S.; Zhang, D. Antibody Drug Conjugates Differentiate Uptake and DNA Alkylation of Pyrrolobenzodiazepines in Tumors from Organs of Xenograft Mice. *Drug Metab. Dispos.* **2016**, *44*, 1958–1962.
46. Tiberghien, A. C.; Levy, J.-N.; Masterson, L. A.; Patel, N. V.; Adams, L. R.; Corbett, S.; Williams, D. G.; Hartley, J. A.; Howard, P. W. Design and Synthesis of Tesirine, a Clinical Antibody-Drug Conjugate Pyrrolobenzodiazepine Dimer Payload. *ACS Med. Chem. Lett.* **2016**, *7*, 983–987.
47. Dimasi, N.; Fleming, R.; Zhong, H.; Bezabeh, B.; Kinneer, K.; Christie, R. J.; Fazenbaker, C.; Wu, H.; Gao, C. Efficient Preparation of Site-Specific Antibody–Drug Conjugates Using Cysteine Insertion. *Mol. Pharmaceutics* **2017**, *14*, 1501–1516.
48. Lucas, A. T.; Robinson, R.; Schorzman, A. N.; Piscitelli, J. A.; Razo, J. F.; Zamboni, W. C. Pharmacologic Considerations in the Disposition of Antibodies and Antibody-Drug Conjugates in Preclinical Models and in Patients. *Antibodies* **2019**, *8*, 3.
49. Gregson, S. J.; Howard, P. W.; Gullick, D. R.; Hamaguchi, A.; Corcoran, K. E.; Brooks, N. A.; Hartley, J. A.; Jenkins, T. C.; Patel, S.; Guille, M. J.; Thurston, D. E. Linker length modulates DNA cross-linking reactivity and cytotoxic potency of C8/C8′ ether-linked C2-exo-unsaturated pyrrolo[2,1-c][1,4]benzodiazepine (PBD) dimers. *J. Med. Chem.* **2004**, *47*, 1161–1174.
50. Wehrstedt, K. D.; Wandrey, P. A.; Heitkamp, D. Explosive properties of 1-hydroxybenzotriazoles. *J. Hazard Mater.* **2005**, *126*, 1–7.
51. Tiberghien, A. C.; Howard, P. W.; Goundry, W. R. F.; McCormick, M.; Parker, J. S. An Alternative Focus for Route Design for the Synthesis of Antibody–Drug Conjugate Payloads. *J. Org. Chem.* **2019**, *84*, 4830–4836.
52. Campbell, A. D.; Tomasi, S.; Tiberghien, A. C.; Parker, J. S. An Isomerization Approach to Tesirine and Pyrrolobenzodiazepines. *Org. Process Res. Dev.* **2019**, *23*, 2543–2548.
53. Sakaine, G.; Smits, G.; Zemribo, R. Late stage Fe(CO)5 promoted double bond migration: total synthesis of limazepines C and D. *Tetrahedron Lett.* **2015**, *56*, 4767–4769.
54. Mamone, P.; Grunberg, M. F.; Fromm, A.; Khan, B. A.; Goossen, L. J. [Pd(mu-Br)(P(t)Bu3)]2 as a highly active isomerization catalyst: synthesis of enol esters from allylic esters. *Org. Lett.* **2012**, *14*, 3716–3719.
55. Rudin, C. M.; Pietanza, M. C.; Bauer, T. M.; Ready, N.; Morgensztern, D.; Glisson, B. S.; Byers, L. A.; Johnson, M. L.; Burris, H. A., 3rd; Robert, F.; Han, T. H.; Bheddah, S.; Theiss, N.; Watson, S.; Mathur, D.; Vennapusa, B.; Zayed, H.; Lally, S.; Strickland, D. K.; Govindan, R.; Dylla, S. J.; Peng, S. L.; Spigel, D. R.; investigators, S. Rovalpituzumab tesirine, a DLL3-targeted antibody-drug conjugate, in recurrent small-cell lung cancer: a first-in-human, first-in-class, open-label, phase 1 study. *Lancet Oncol.* **2017**, *18*, 42–51.

56. Poirier, J. T. IA24 Targeting DLL3 in Small-Cell Lung Cancer with Novel Modalities. *J. Thorac. Oncol.* **2020,** *15*, S8.
57. Cook, D.; Brown, D.; Alexander, R.; March, R.; Morgan, P.; Satterthwaite, G.; Pangalos, M. N. Lessons learned from the fate of AstraZeneca's drug pipeline: a five-dimensional framework. *Nat. Rev. Drug Discovery* **2014,** *13*, 419–431.
58. Sharma, S. K.; Pourat, J.; Abdel-Atti, D.; Carlin, S. D.; Piersigilli, A.; Bankovich, A. J.; Gardner, E. E.; Hamdy, O.; Isse, K.; Bheddah, S.; Sandoval, J.; Cunanan, K. M.; Johansen, E. B.; Allaj, V.; Sisodiya, V.; Liu, D.; Zeglis, B. M.; Rudin, C. M.; Dylla, S. J.; Poirier, J. T.; Lewis, J. S. Noninvasive Interrogation of DLL3 Expression in Metastatic Small Cell Lung Cancer. *Cancer Res.* **2017**, *77*, 3931–3941.
59. Tanaka, K.; Isse, K.; Fujihira, T.; Takenoyama, M.; Saunders, L.; Bheddah, S.; Nakanishi, Y.; Okamoto, I. Prevalence of Delta-like protein 3 expression in patients with small cell lung cancer. *Lung Cancer* **2018,** *115*, 116–120.
60. Hamadani, M.; Collins, G. P.; Samaniego, F.; Spira, A. I.; Davies, A.; Radford, J.; Caimi, P.; Menne, T.; Boni, J.; Cruz, H.; Feingold, J.; He, S.; Wuerthner, J.; Horwitz, S. M. Phase 1 Study of Adct-301 (Camidanlumab Tesirine), a Novel Pyrrolobenzodiazepine-Based Antibody Drug Conjugate, in Relapsed/Refractory Classical Hodgkin Lymphoma. *Blood* **2018**, *132* (Supplement 1), 928–928.
61. Kahl, B. S.; Hamadani, M.; Radford, J.; Carlo-Stella, C.; Caimi, P.; Reid, E.; Feingold, J. M.; Ardeshna, K. M.; Solh, M.; Heffner, L. T.; Ungar, D.; He, S.; Boni, J.; Havenith, K.; O'Connor, O. A. A Phase I Study of ADCT-402 (Loncastuximab Tesirine), a Novel Pyrrolobenzodiazepine-Based Antibody–Drug Conjugate, in Relapsed/Refractory B-Cell Non-Hodgkin Lymphoma. *Clin. Cancer Res.* **2019**, *25*, 6986–6994.
62. Rios-Doria, J.; Harper, J.; Rothstein, R.; Wetzel, L.; Chesebrough, J.; Marrero, A.; Chen, C.; Strout, P.; Mulgrew, K.; McGlinchey, K.; Fleming, R.; Bezabeh, B.; Meekin, J.; Stewart, D.; Kennedy, M.; Martin, P.; Buchanan, A.; Dimasi, N.; Michelotti, E.; Hollingsworth, R. Antibody–Drug Conjugates Bearing Pyrrolobenzodiazepine or Tubulysin Payloads Are Immunomodulatory and Synergize with Multiple Immunotherapies. *Cancer Res.* **2017**, *77*, 2686.
63. Zhong, H.; Chen, C.; Tammali, R.; Breen, S.; Zhang, J.; Fazenbaker, C.; Kennedy, M.; Conway, J.; Higgs, B. W.; Holoweckyj, N.; Raja, R.; Harper, J.; Pierce, A. J.; Herbst, R.; Tice, D. A. Improved Therapeutic Window in BRCA-mutant Tumors with Antibody-linked Pyrrolobenzodiazepine Dimers with and without PARP Inhibition. *Mol. Cancer Ther.* **2019**, *18*, 89–99.
64. Xing, L.; Lin, L.; Yu, T.; Li, Y.; Cho, S.-F.; Liu, J.; Wen, K.; Hsieh, P. A.; Kinneer, K.; Munshi, N.; Anderson, K. C.; Tai, Y.-T. A novel BCMA PBD-ADC with ATM/ATR/WEE1 inhibitors or bortezomib induce synergistic lethality in multiple myeloma. *Leukemia* **2020**, *34*, 2150–2162.
65. Schürch, C. M. Therapeutic Antibodies for Myeloid Neoplasms—Current Developments and Future Directions. *Front. Oncol.* **2018**, *8* (152) https://doi.org/10.3389/fonc.2018.00152

Chapter 8

Discovery and Chemical Development of Relebactam: A Potent β-Lactamase Inhibitor in Combination with Primaxin® for the Treatment of Serious and Antibiotic-Resistant Bacterial Infections

Feng Xu,*[,1] Benjamin D. Sherry,[1] and Timothy A. Blizzard[2,3]

[1]Department of Process Chemistry, Merck & Co., Inc., Rahway, New Jersey 07065, United States

[2]Department of Discovery Chemistry, Merck & Co., Inc., Kenilworth, New Jersey 07033, United States

[3]Current address: Xpharma Consulting, 43 Bertrand Dr., Princeton, New Jersey 08540, United States

*Email: feng_xu@merck.com

Relebactam (MK-7655) is a novel beta-lactamase inhibitor. In combination with imipenem/cilastatin, it was recently approved by the FDA (as Recarbrio®) for the treatment of resistant Gram-negative bacterial infections. The medicinal chemistry program that led to the discovery of relebactam is reviewed herein. An important element in the success of this program was the early involvement of our Process Chemistry department which developed a large-scale synthesis of key intermediates to enable the discovery program. Later, more efficient synthetic routes were developed to support clinical trials. The development of these large-scale syntheses and a manufacturing route for relebactam are discussed in detail. The end result of this effective collaboration between the medicinal and process chemistry groups was a novel antibiotic that is expected to play an important role in the treatment of serious bacterial infections.

Medicinal Chemistry Program: Introduction

Since the first clinical use of penicillin in 1941, numerous beta-lactam antibiotics, including cephalosporins and carbapenems in addition to penicillins, have been developed for the treatment of bacterial infections (*1*).

One of the most potent beta-lactam antibiotics is imipenem (**1**, Figure 1), a member of the carbapenem class of beta-lactams. Imipenem, in combination with the renal dehydropeptidase (DHP) inhibitor cilastatin (**2**), is marketed by Merck & Co., Inc., Kenilworth, NJ, USA (MSD) as Primaxin.® Primaxin® is used primarily for the treatment of serious bacterial infections in a hospital setting (*2*).

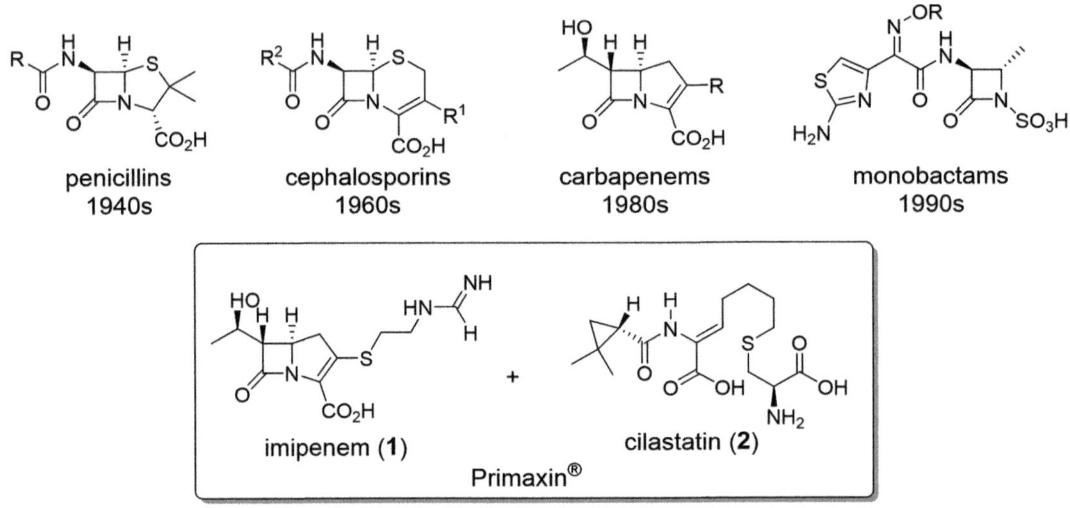

Figure 1. Beta-lactam antibiotics: class structures and Primaxin®.

Beta-lactam antibiotics kill bacteria by inhibiting penicillin binding proteins (PBPs), enzymes that are essential for construction of the bacterial cell wall. In susceptible bacteria, the beta-lactam (e.g. amoxicillin, **3**) acylates a PBP active site serine (e.g. PBP-Ser-OH) thereby inactivating the enzyme. Lacking an active PBP, the bacteria cannot construct a properly cross-linked cell wall and eventually lyse and die (Figure 2). However, some bacteria produce beta-lactamase enzymes (e.g. BLI-Ser-OH)) that can hydrolyze and thus inactivate beta-lactams before they react with the PBPs. These beta-lactamase enzymes thus allow the bacteria to survive treatment with beta-lactam antibiotics.

In recent years, many Gram-negative bacteria, e.g. *Pseudomonas*, *Klebsiella*, and *Acinetobacter*, have developed resistance to imipenem. This resistance is primarily due to the production of a variety of beta-lactamase enzymes, vide supra (2). Loss of porins (outer membrane channels that allow small hydrophilic molecules like imipenem to enter the periplasmic space) also plays a role in Gram-negative resistance to carbapenems.

Beta-lactamase enzymes have been classified into four structure-based groups (class A, B, C, and D) (3). Classes A, C, and D are serine proteases while the class B enzymes are metallo beta-lactamases that work by a fundamentally different mechanism. In Gram-negative bacteria, resistance to imipenem is primarily due to classes A and C although class D enzymes also play a role.

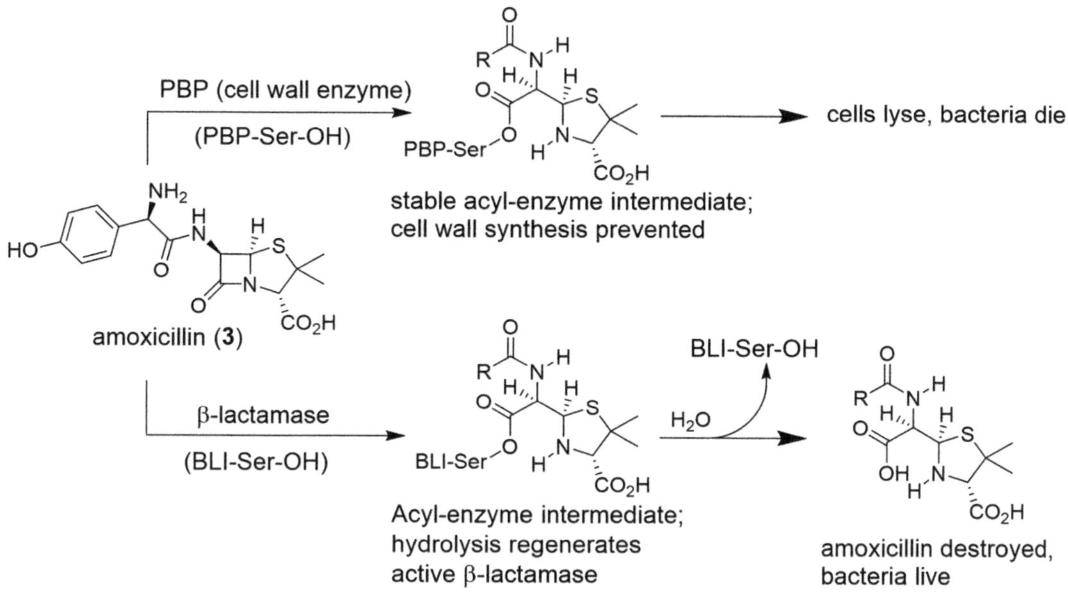

Figure 2. Antibiotic mechanism of action of amoxicillin and inactivation of amoxicillin by beta-lactamase.

Clavulanic acid (**4**) is a beta-lactamase inhibitor (BLI) that effectively inhibits the class A beta-lactamases that are primarily responsible for amoxicillin resistance in Gram-positive bacteria. The combination of amoxicillin with clavulanic acid (marketed as Augmentin®) restores useful clinical activity (Figure 3) (*4*). The success of the amoxicillin/clavulanate combination has prompted extensive research on novel beta-lactamase inhibitors (*5*). Hypothesizing that a similar strategy might successfully overcome beta-lactamase mediated imipenem resistance in Gram-negative bacteria, MSD initiated a medicinal chemistry program to discover a novel BLI for combination with Primaxin.® The outcome of this successful program was the discovery and development of relebactam (MK-7655), as related in the following sections. This novel beta-lactamase inhibitor, in combination with imipenem/cilastatin, was recently approved by the FDA (as Recarbrio™) for the treatment of resistant Gram-negative bacterial infections (*6*).

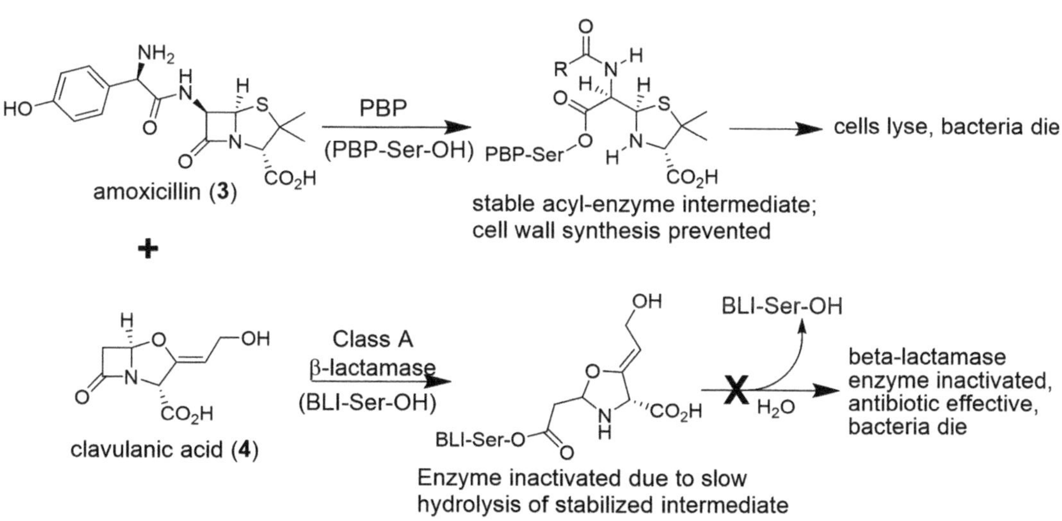

Figure 3. Restoration of antibiotic activity by addition of a beta-lactamase inhibitor.

Discovery of Relebactam

Initially, the program focused on the discovery of a class C BLI. At the time, *Pseudomonas* was the most important resistant bacteria and the main resistance driver was the class C enzyme AmpC. The program began as a collaboration with the company MethylGene, which had discovered that phosphonate **5** was active against a range of class A and C beta-lactamases (Scheme 1). Starting with this lead, an MSD chemistry group worked with a group at MethylGene to discover the project's first preclinical candidate, MK-4698 (**6**), which failed in preclinical safety studies. After the collaboration ended, MethylGene conducted further studies on MK-4698 (as MG96077) but the compound never reached the clinic (*7*).

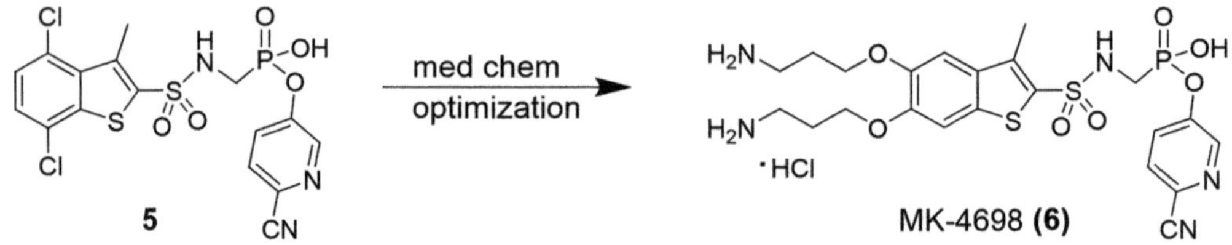

5

MK-4698 (6)

Scheme 1. Merck Research Laboratories and MethylGene collaboration results.

With the demise of MK-4698, the focus of the program shifted to structurally diverse alternatives. Numerous options were considered, including de novo design and modification of existing BLI core structures. Two of these alternatives, the bridged monobactams (e.g. Ro48-1256, **7**) (*8*) originally discovered at Roche and the bicyclic ureas (e.g. avibactam, **8**) (*9*) originally discovered at Aventis, were pursued and both resulted in preclinical candidates for MSD (Figure 4).

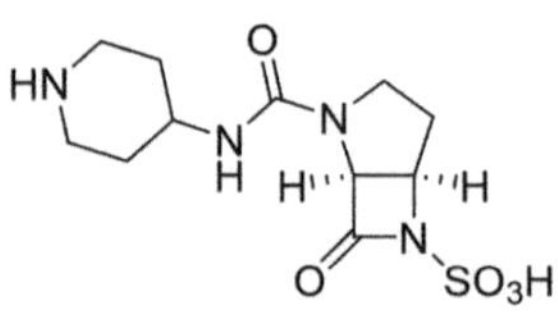

Ro48-1256 (**7**)

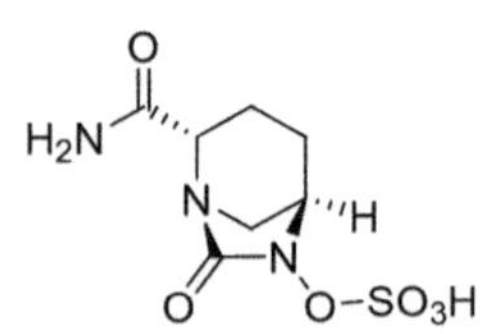

avibactam (**8**)

*Figure 4. Ro48-1256 (**7**) and avibactam (**8**).*

For practical reasons, the initial focus was on the bridged monobactam lead, Ro48-1256. Although this series lacked activity against class A enzymes, carbapenems are inherently resistant to most Class A beta-lactamases and, at this point, the program was primarily focused on class C activity. Furthermore, this series was much more synthetically accessible.

An X-ray crystal structure of Ro48-1256 bound in the active site of AmpC, the key class C beta-lactamase responsible for imipenem resistance in *Pseudomonas,* demonstrated that the piperidine side chain did not fill the binding pocket (Figure 5). Molecular modeling studies suggested that 7-membered (**9**) and even 8-membered (**10**) ring side chains (Figure 6) would fit. Neither of these compounds had been reported in the literature.

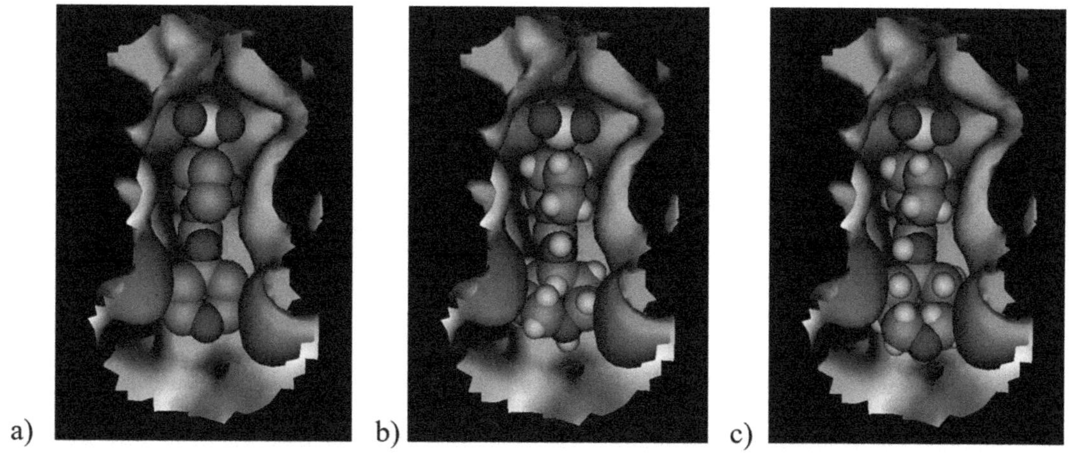

Figure 5. X-ray crystal structure of AmpC active site with Ro48-1256 (a), 7-membered analog (b), and 8-membered analog (c) docked.

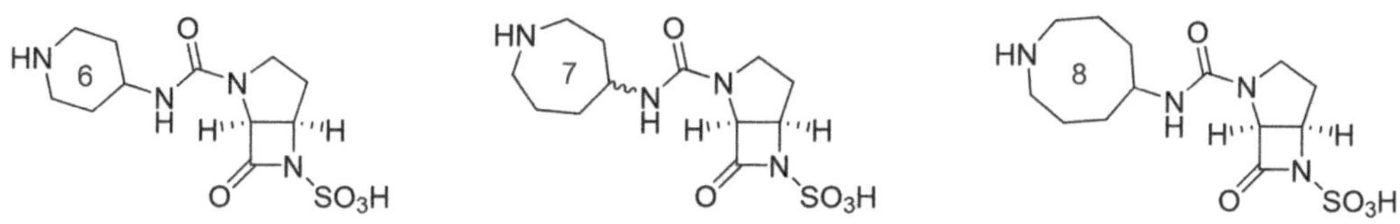

7 **9** **10**

Figure 6. Ro48-1256 (7) and 7- and 8-membered ring analogs.

We therefore undertook the synthesis of the seven-membered ring analog **9** (*10*). The bicyclic core **12** was prepared initially using the published Roche synthesis (starting from **11**), but the MSD Prep Lab quickly developed an improved route from commercially available hydroxy-proline **13** (Scheme 2).

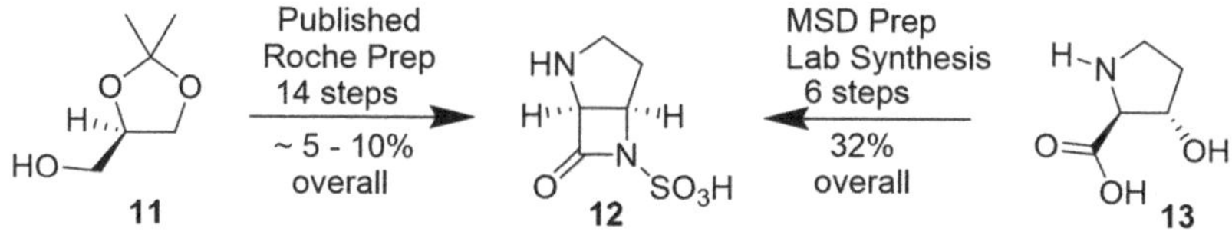

*Scheme 2. Hydroxy-proline derived synthesis of **12**.*

The azepine side chain was initially prepared in racemic form by reductive amination of the azepinone **14** (Scheme 3). The racemic amine **15** was then coupled to the enantiomerically pure core **12** (*8*) to afford a separable mixture of diastereomers **16** and **17** (*10*). Synthetic efforts were complicated somewhat by the reactivity and high water solubility of the two diastereomers.

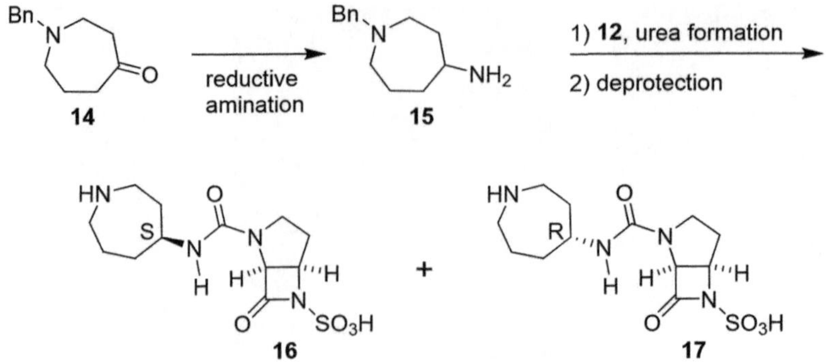

Scheme 3. Coupling of azepine ring to ***12****.*

All compounds were evaluated in an enzyme inhibition assay and in an in vitro synergy assay (*11*). The enzyme assay measured the compound's ability to inhibit the hydrolysis of nitrocefin by beta-lactamase. The synergy assay measured the compound's ability to reduce the minimum inhibitory concentration (MIC) of imipenem against CL5701, a clinically-relevant beta-lactamase producing strain of *Pseudomonas*. Concentrations required to reduce the imipenem MIC by factors of 8 and 16 are shown in Figure 7. For this strain, an 8-fold reduction would be expected to restore clinically useful activity.

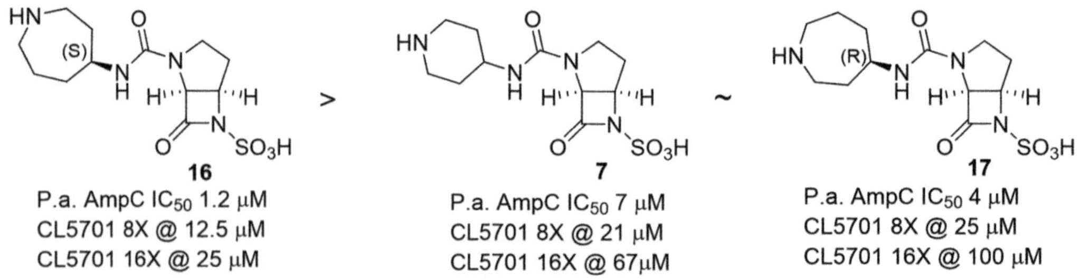

Figure 7. Biological activity of 7-membered ring analogs; CL5701 is a strain of Pseudomonas aeruginosa (P.a); P.a. AmpC is a Class C β-lactamase isolated from P.a.

The stereochemistry of the diastereomers **16** and **17** was determined by X-ray analysis of crystals of the *Pseudomonas* beta-lactamase AmpC with **16** or **17** absorbed (*10*). Since the absolute stereochemistry of the core was known, the X-ray analysis allowed a definitive assignment of the side chain stereochemistry. It is worth noting that molecular modeling, enabled by the existence of the crystal structure of AmpC, predicted that **16** would be the more active diastereomer.

A substantial number of analogs were prepared but **16**, which was one of the first two compounds synthesized, still resulted in the best compound (*10*). To briefly summarize the SAR, a side chain was required for activity and the presence of a basic nitrogen and its location in the side chain were critical. A 7- or 8- membered ring side chain was optimal as the 6- and 9-membered ring analogs were less active.

Selected compounds were evaluated in an in vivo infection model. Mice were infected with imipenem-resistant *Pseudomonas* then treated four times a day for one day with imipenem and the BLI. A 2.35 mg/kg dose of **16** in combination with 2.5 mg/kg of imipenem gave a 4.6 log reduction of bacterial colony forming units. By comparison, the piperidine analog **7** produced only a 1 log reduction at twice the dose (*10*).

The (*S*) azepine analog **16** had superior in vitro and in vivo activity, acceptable PK and physical properties, and there were no significant off-target activities or toxicity in preliminary safety testing. It was nominated for development as MK-8712. Unfortunately, a marginal safety signal ensued in the more extensive safety studies required before clinical trials and development of MK-8712 was terminated.

During the time that the bridged mono-bactam series was being pursued, it had become increasingly clear that class A beta-lactamase mediated resistance, especially from the KPC-2 enzymes expressed by some strains of *Klebsiella*, was becoming too important to ignore. Accordingly, the focus of the program shifted to a next-generation compound with a broader spectrum of activity and included both class A and class C beta-lactamases.

The bridged bicyclic urea BLIs exemplified by avibactam (**8**) were discovered at Sanofi-Aventis in the early 2000s. Compounds in this series are more difficult to synthesize than the bridged monobactams but the series is active against both class A and C enzymes. Based on the SAR of the monobactams, a series of avibactam analogs with basic heterocyclic side chains was targeted and this effort led to the discovery of relebactam (**18**) (*12*).

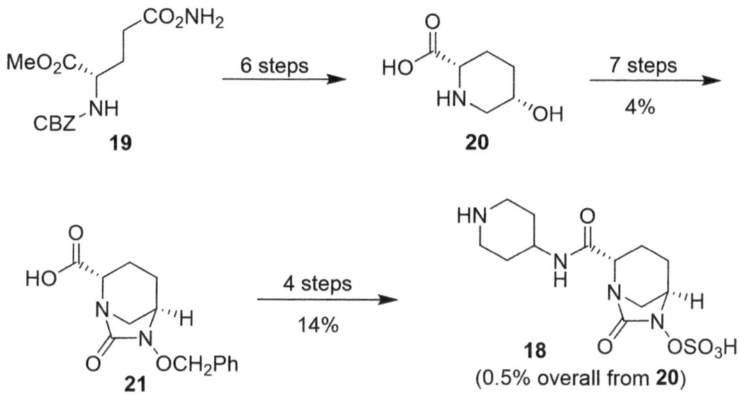

*Scheme 4. Original medicinal chemistry synthesis of relebactam (**18**).*

The original med chem synthesis of **18** is summarized in Scheme 4. Initially, the bicyclic core **21** was prepared using the published Aventis procedure (*13*), which was exceedingly long and possessed several steps with low yields. However, our Process Chemistry group quickly developed an improved procedure (vide infra). This early involvement by the Process Group and the resulting new synthesis of the core was critical to the success of the discovery program.

An X-ray crystal structure of **18** bound in the active site of Pseudomonal AmpC confirmed the structure and showed that the binding was comparable to that of MK-8712 (Figure 8). The key binding elements of MK-8712 (in cyan) and **18** (in yellow) overlap fairly well.

Piperidine **18** was very effective in our in vitro enzyme inhibition assay, inhibiting both KPC-2, a class A enzyme from *Klebsiella*, and AmpC, a class C enzyme from *Pseudomonas* with IC_{50}s in the 200-500 nM range. It was less active against AmpC from *Acinetobacter* and was inactive against Oxa-40, a class D enzyme from *Acinetobacter*. However, the majority of clinical cases are due to *Pseudomonas* or *Klebsiella* so activity against *Acinetobacter* enzymes was considered to be less important. It was also very effective in our in vitro imipenem synergy assay, reducing the imipenem MICs for resistant *Klebsiella* and *Pseudomonas* to the clinical breakpoint of 4 μg/mL at concentrations of 12.5 and 4.7 mM, respectively (*12*).

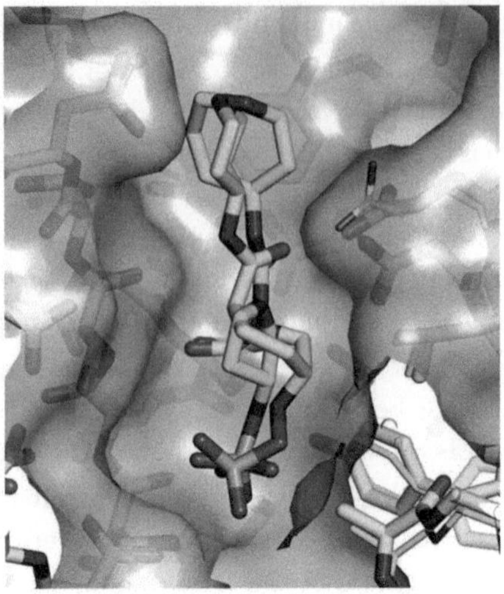

Figure 8. Overlay of MK-8712 & MK-7655.

About 150 bridged bicyclic ureas were synthesized and evaluated. The in vitro SAR is summarized in Figure 9. In contrast to the bridged monobactams, the side chain is not required for activity. However, if a side chain is present, a basic nitrogen is required for maximum synergy. An aromatic side chain increases enzyme inhibition but this increase is not fully reflected in synergy activity, even with a basic nitrogen present. Also, the stereochemistry of the core is critical.

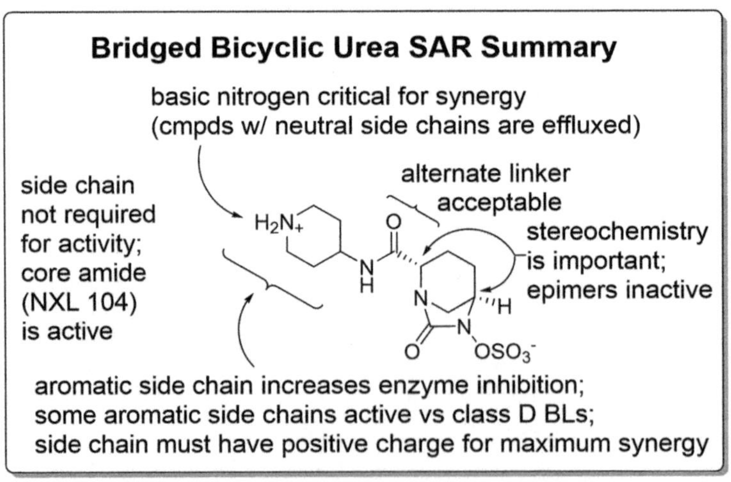

Figure 9. Summary of bicyclic urea BLI series SAR.

The piperidine analog **18** had excellent activity against the key class A and C beta-lactamase enzymes of *Pseudomonas*, *Klebsiella,* and synergized imipenem to its breakpoint at clinically achievable concentrations. Additional in vitro data for **18** have recently been reported (*14*). Pharmacokinetics were appropriate for combination with imipenem. It also had excellent in vivo potency in our mouse spleen and lung infection models for *Pseudomonas* and our mouse spleen infection model for *Klebsiella* (*15*). The lung infection model is especially important since hospital-acquired pneumonias are frequently caused by resistant *Pseudomonas*. It was selected for development as MK-7655 and subsequently assigned the generic name relebactam.

Process Development to Access Relebactam

Introduction

The structure of relebactam presents a host of opportunities for innovation in process development. The strained azabicyclic core, *N*-hydroxy functionality and zwitterionic structure all contribute to the synthetic complexity of the target structure. An early investment in enabling synthetic chemistry was deemed essential to rapidly progress the target in clinical development, which was followed by a more extensive commercial route development program. The results of these investigations are summarized within this section.

First-Generation Process

To enable rapid access to relebactam for early clinical development, a series of strategic disconnections was envisaged that trace back to readily accessible pyroglutamic acid and 4-aminopiperidine as key building blocks (Figure 10) (*16*).

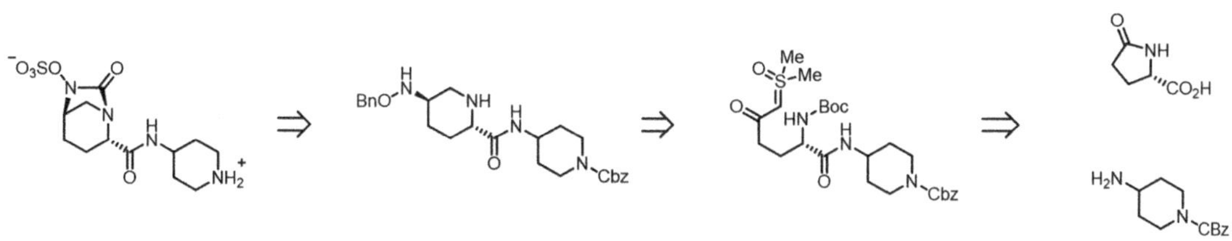

Figure 10. Retrosynthetic analysis for the first-generation relebactam process.

For this approach to succeed, several key enabling transformations were required;

- selective N-H insertion into a carbenoid intermediate
- diastereoselective introduction of the hydroxylamine moiety
- late-stage introduction of the strained azabicyclic scaffold

To address these challenges, the linear carbene precursor was assembled in a series of straightforward steps that were amenable to kilogram-scale processing (Scheme 5). Amide coupling between Cbz-protected 4-aminopiperidine **22** and *L*-pyroglutamic acid was carried out under standard conditions. The product stream after aqueous workup was treated directly with Boc anhydride in the presence of catalytic DMAP. The Boc-derivative **24** was isolated after aqueous workup and crystallization from IPAc. Addition of the sulfur ylide prepared from trimethylsulfoxonium iodide and potassium *tert*-butoxide in DMSO proceeded cleanly and the product was crystallized directly by addition of water to the reaction mixture affording the key carbene precursor **25** in 71% yield.

Access to the sulfur ylide **25** set the stage to investigate two essential reactions, the N-H insertion and diastereoselective reduction (Scheme 6). After screening a number of transition metal catalysts, we discovered that an iridium(I) catalyst was most effective for the desired carbenoid N-H insertion reaction. The ketopiperidine product **26** was subjected directly to diastereoselective reduction using lithium borohydride in the presence of methanol to afford the secondary alcohol **27** in 67% yield as a single diastereomer after crystallization from 1-propanol/toluene. Activation of the secondary alcohol as the *p*-trifluoromethylphenyl sulfonate proceeded cleanly and the activated hydroxyl compound **28** was isolated in 96% yield after crystallization from MTBE/heptanes. The use of a *p*-trifluoromethylphenyl sulfonate activating group in place of more common sulfonate esters offered the dual advantages of improved selectivity for nucleophilic displacement over competing E2 elimination and isolation of the product as a stable crystalline material. Displacement of the activated hydroxyl group with *O*-benzyl-*N*-Boc-hydroxylamine was nevertheless hampered by competitive E2 elimination, but ultimately conditions were developed employing potassium *tert*-butoxide in DMAc which allowed for clean substitution. Deprotection of the Boc group with methanesulfonic acid and isolation of the secondary amine **30** as its *p*-toluenesulfonic acid salt allowed access to the diamine precursor in 53% yield over the final two steps.

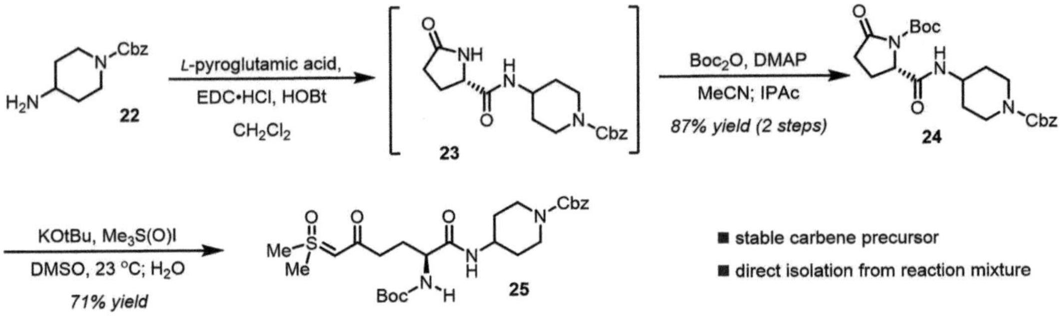

Scheme 5. Process to prepare the carbene precursor ***25****.*

Completion of the synthesis required installation of the strained [3.2.1]-azabicyclic scaffold (Scheme 7). Towards this end, the diamine *p*-TSA salt **30** was treated with triphosgene in the presence of Hünig's base to afford the protected bicyclic product **31** in 87% yield after crystallization from ethanol/*n*-heptane. The benzyl and Cbz groups were removed by catalytic hydrogenolysis and the piperidine nitrogen re-protected in situ with a Boc group to afford **32**. The *N*-hydroxy moiety was sulfated through the action of sulfur trioxide-pyridine complex and after salt metathesis to the *tetra*-butylammonium, the target structure was prepared through Boc cleavage with tetrafluoroboric acid diethyl ether complex in trifluoroethanol. Relebactam **18** was isolated after aqueous workup by crystallization from 2-propanol/water in 68% yield over the final two steps. This process proved amenable to kilogram-scale manufacturing and was used to fund early clinical development.

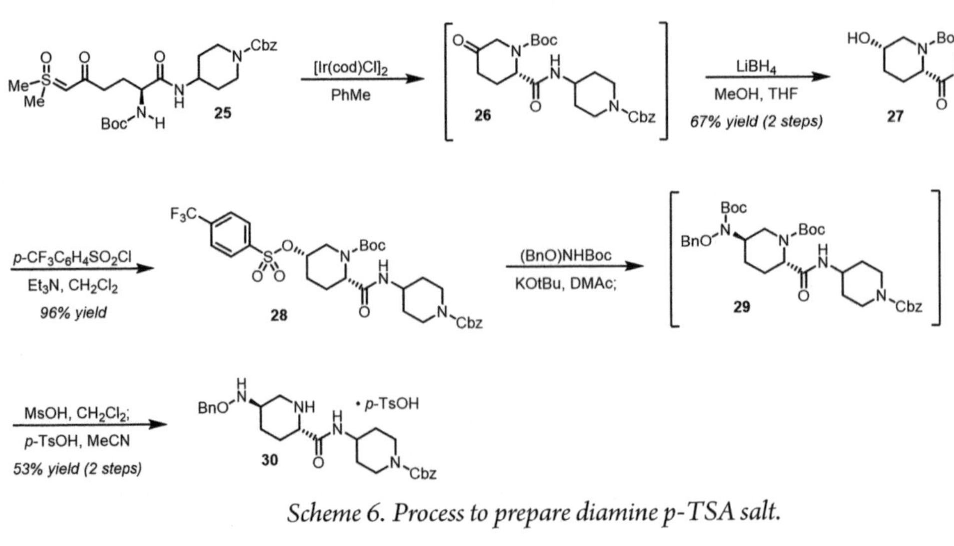

Scheme 6. Process to prepare diamine p-TSA salt.

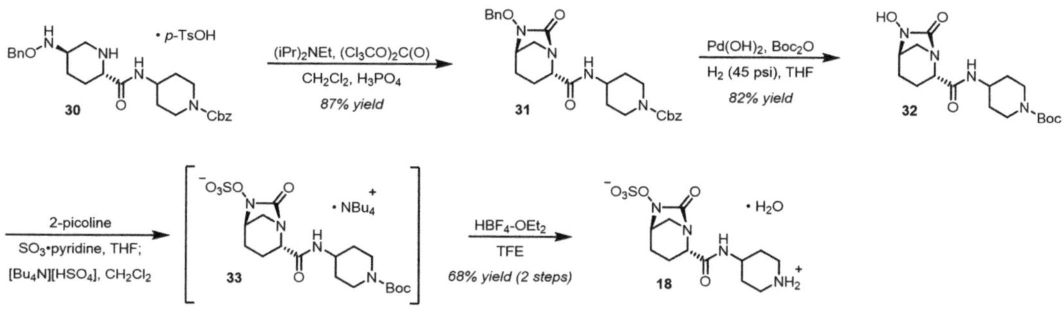

Scheme 7. Completion of the first-generation process to relebactam.

Second-Generation Process

As relebactam continued to progress through clinical development and increasingly larger quantities were necessary, limitations of the first-generation route began to come into focus. Specifically, the use of an expensive transition metal catalyst, multiple protecting group manipulations and the use of a fluorinated reagent and solvent in the final step were all elements that could be improved. Therefore, a second-generation process was conceived, which shared some strategic similarities with respect to strained bicycle formation, but traced back to an alternative readily available material, hydroxypipecolic acid (Figure 11) (*17*).

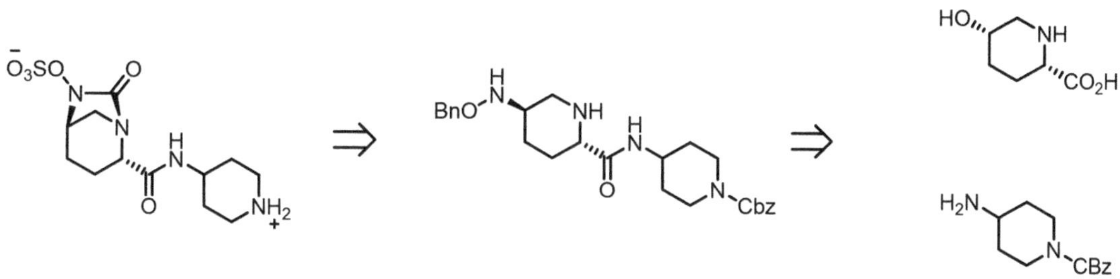

Figure 11. Retrosynthetic analysis of a second-generation relebactam process.

At the outset, successful employment of biosynthetically derived hydroxypipecolic acid would require an efficient procedure to upgrade the purity of this material which was received in 77 wt% purity. An efficient sequence that leveraged lactone formation as a purifying step in the route to the diamine precursor was developed (Scheme 8). Treatment of crude hydroxypipecolic acid **20** with 2-nitrobenzenesulfonyl chloride (2-NsCl), in a water acetone mixture delivered compound **34**, which without isolation was treated with additional 2-NsCl in ethyl acetate and acetone to afford the bridged bicyclic lactone **35** in 79% yield after crystallization from ethyl acetate. This structure was poised for ring opening with Boc protected 4-aminopiperidine **36**. After activation of the secondary hydroxyl group as the 2-nitrobenzenesulfonate ester [O-(2-Ns)], the product **38** was isolated in 98% yield by crystallization from aqueous methanol. Displacement of the 2-nitrobenzenesulfonate ester [O-(2-Ns)] with *O*-benzyl-*N*-(4-nitrobenzenesulfonyl) hydroxylamine in the presence of potassium carbonate and DMAc proceeded smoothly and the nosyl groups were removed using methanolic thioglycolic acid. The desired diamine intermediate **40** was isolated in 70% yield after crystallization from *n*-heptane/toluene.

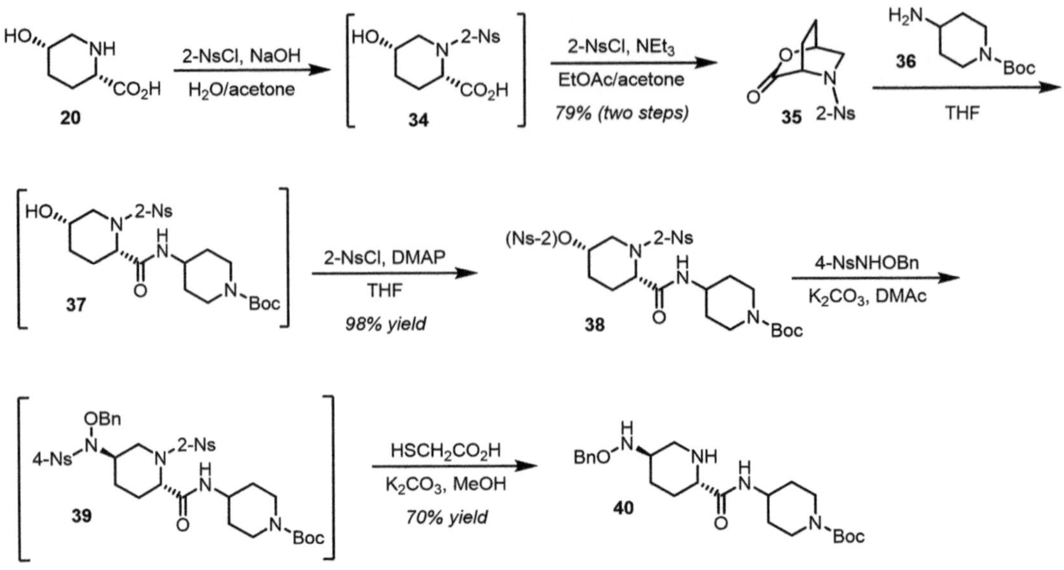

Scheme 8. Preparation of diamine intermediate from hydroxypipecolic acid.

Advancing this intermediate to relebactam followed a strategically similar sequence to the first-generation process, but using reagents and conditions that were targeted for larger-scale deliveries (Scheme 9). The diamine intermediate **30** was converted to the azabicyclic intermediate **41** using triphosgene and the product isolated after crystallization from *n*-heptane/MTBE in 92% yield. Hydrogenolysis removed the *O*-benzyl group providing the desired product **32** in 90% yield after crystallization from ethyl aetate/*n*-hexanes. The hydroxyl group was sulfated with sulfur trioxide pyridine complex and after aqueous workup and salt metathesis to the tetrabutylammonium salt (**33**), the Boc group was removed using trimethylsilyl iodide in acetonitrile, the most robust conditions that were identified after extensive screening. This produced the same polymorphic form of relebactam **18** in 93% yield.

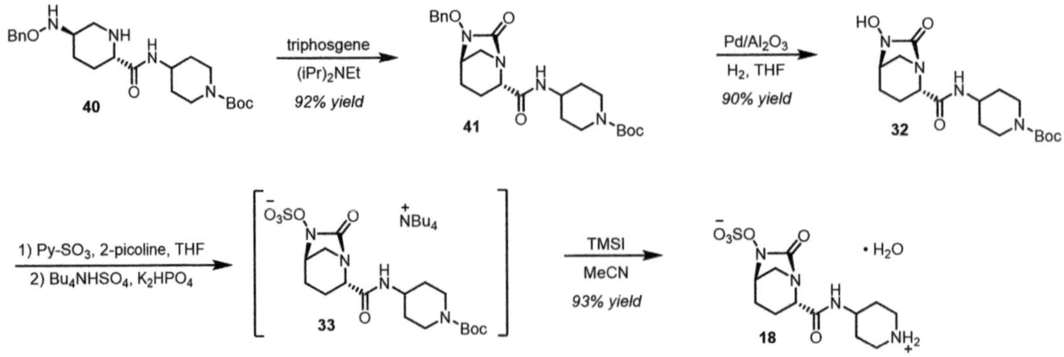

Scheme 9. Completion of the second-generation process to relebactam.

Strategy for the Development of a Manufacturing Process to Relebactam

Up to this point, two generation routes have been developed, which provided a means to enable the preparation of large quantities of relebactem (>50 kg) to support the early safety and clinical studies. In comparison with the first-generation synthesis (*16*), the second-generation route (*17*) has shortened the longest linear synthetic steps (8 steps from 14 steps) and lowered the cost of goods.

In particular, we avoided the use of the expensive iridium catalyst to form the second piperidine skeleton. On the other hand, the second-generation route has introduced a series of protection-deprotection operations to prepare the chiral piperidine intermediate **40**, which allowed applying the synthetic strategy utilized in the first-generation route to construct the rigid bicyclic urea skeleton. However, both routes were deemed unsatisfactory for long-term use due to the steppy (protection-deprotection protocols) synthesis of chiral piperidine amide intermediates **30** and **40**, along with the safety concerns of using hazardous triphosgene on factory scale.

The existing synthetic strategy from piperidine amide **30/40** to relebactam provided a straightforward approach to form the rigid hydroxy urea function group, given the fact that the hydroxy urea intermediate **32** had limited chemical stability (*16*). We would retain this chemistry as part of the ultimate manufacturing route to relebactam, if process robustness, stability and practicality could be realized with further development. Therefore, our development goal for this sector of the chemistry would focus on:

- eliminating triphosgene for the construction of the rigid urea skeleton
- improving the hydrogenolysis efficiency for the preparation of labile hydroxy urea **32** in high purity
- developing a more practical endgame process without using TMSI (*18*).

We were confident that this goal could be realized with further modification of the chemistry. With this strategy in hand, we turned out attention to evaluate alternative preparation of the chiral piperidine intermediate **40**.

Scheme 10 provides an overview of the preparation of the piperidine intermediates **30**, **40** and **44** at the time (*16*, *17*, *19*). All the existing synthetic routes were inefficient, in terms of atom economy, manipulation to effect activation, protection-deprotection protocols, and/or overall yield. In addition, in the AstraZeneca route (*19*) the reduction of the oxime ether **43** resulted in a low diastereoselectivity (3:1 *dr*).

A retrosynthetic analysis resulted in two approaches to access the requisite piperidine **40** (Scheme 11). Option A would rely on a highly chemo- and diastereoselective reduction of oxime **46**. Of concern was that reduction of a corresponding benzyl ester oxime **43** in AstraZeneca route (*19*) was achieved in only 3:1 *dr* (Scheme 10). To improve atom economy and synthetic efficiency without introducing the protection-deprotection operations, we required:

1) a selective and direct coupling of **36** with **45**
2) a direct conversion of commercially available (*S*)-*N*-Boc pyroglutamic acid (**47**) to oxime **46**
3) maintaining the carboxylic acid moiety which we envisioned could function as an additional handle to selectively deliver the hydride for the reduction of the requisite oxime **46** in high diastereoselectivity.

These requirements appeared realistic. For option B, the *N*-benzyloxy stereogenic center in **40** could be established via an S_N2 displacement of the corresponding activated precursor from **48**. Different from the synthetic strategy in the second-generation synthesis (*17*), we proposed a direct and selective coupling of (2*S*,5*S*)-5-hydroxypiperidine-2-carboxylic acid (**20**) with **36**. If we could avoid involving protection-deprotection procedures, it would significantly improve the synthetic efficiency. We envisioned that a practical manufacturing process for the preparation of the key intermediate **40** could arise from the development of these synthetic routes.

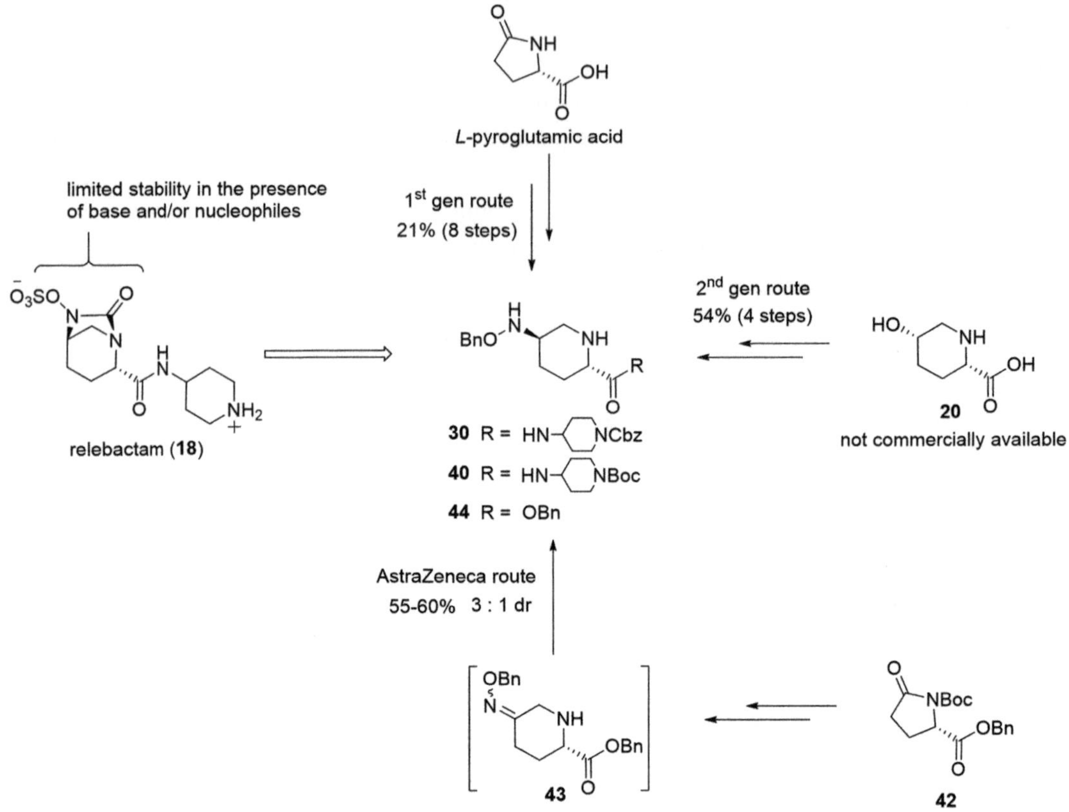

Scheme 10. Overview of previous preparation of piperidine intermediates ***30, 40*** *and* ***44.***

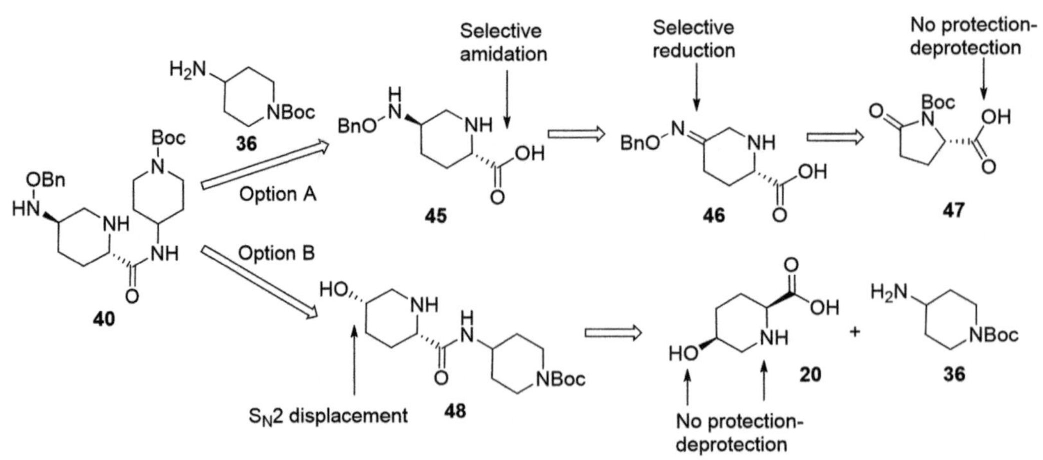

Scheme 11. Retrosynthetic analysis.

Process to Piperidine 40 via Highly Diastereoselective Reduction of Oxime 46

In the previous synthesis (Scheme 10), oxime ester **43** was prepared from pyroglutamate ester via a known one-carbon extension-cyclization sequence (*19*). To streamline the synthesis by eliminating the additional carboxylic acid protection-deprotection steps, (*S*)-*N*-Boc pyroglutamic acid (**47**) became our starting material for the preparation of oxime **46**. With the presence of an additional acidic proton, the one-carbon homologation-cyclization sequence required significant process development. Treatment of **47** with trimethylsulfoxonium iodide and KO*t*Bu in a mixture of THF and DMF did give the desired **49**. However, control experiments revealed that **47** was unstable upon treatment with KO*t*Bu, which could lead to decomposition/polymerization side-reactions. The key to suppress the decomposition was to slowly charge KO*t*Bu over 10 h to a well-agitated reaction mixture of **47** at -20 to -10 °C. As such, the corresponding sulfur ylide could be preferentially generated first, resulting in 90% assay yield of **49** (Scheme 12).

Without workup, the crude reaction stream was directly treated with LiCl, benzyloxy ammonium chloride and cyanoacetic acid to afford the desired chloromethyl oxime **50**. Regarding the choice of acid used in this step, cyanoacetic acid (pKa 2.45) was chosen based on the determined optimum pKa that provided the balanced acidity to promote the desired transformation but suppress side reactions. A major byproduct in this transformation was des-chloro acid **51**, which was most likely formed via an iodide nucleophilic attack (*20*) on the chloride in oxime **50** (Scheme 12). The iodide originated from trimethylsulfoxonium iodide in the previous ylide reaction in the same pot. Indeed, treatment of isolated **50** with NaI in toluene resulted in complete conversion to **51.** After several experiments, we found that the formation of **51** could be suppressed to < 3% when DMSO was introduced (*21*), affording 81% assay yield of **50** over two steps.

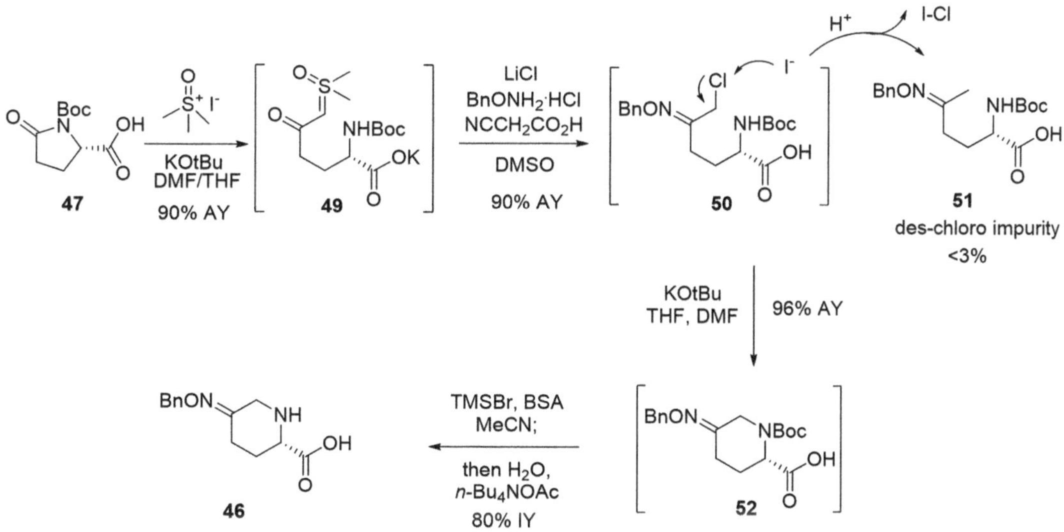

*Scheme 12. Process to oxime **46**.*

Treatment of the crude stream containing **50** with KO*t*Bu in THF-DMF afforded cyclized product **52** in 96% assay yield after work-up. These conditions avoided the precipitation of its carboxylate salt. Without further purification, crude acid **52** was converted to amino acid **46** by treating with TMSBr in the presence of *N*,*O*-bis(trimethylsilyl)-acetamide (BSA) in acetonitrile. The desired product **46** was isolated as a crystalline zwitterion in 80% yield with >97% purity, after quenching the reaction mixture with water and adding Bu_4NOAc as a base to neutralize the HBr generated during the de-Boc reaction.

With oxime **46** in hand, we proceeded to explore the key chemo- and diastereoselective reduction of the oxime ether C=N double bond. A literature survey of reduction of oxime ethers to alkoxyamines revealed that it is generally a capricious reaction due to the existence of oximes as E/Z mixtures, competing N-O bond cleavage, and high sensitivity to the substituents around the oxime ether (22). Therefore, we carried out a broad screening of reduction conditions including metal-catalyzed hydrogenation, modified borohydride treatments, and borane reduction in various solvents. (Table 1).

Table 1. Selected Screening Results of Reduction of Oxime 46[a]

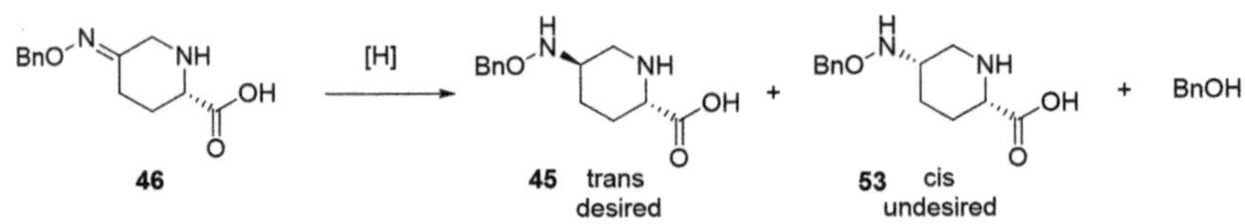

Entry	Catalysts/Reagents	Conditions	Conv.	trans + cis	de (trans)	BnOH
1	0.7 mol%$(NBD)_2RhBF_4$/dcpf, 0.6 equiv HBF_4	500 psi H_2, 10% H_2O/EtOH	96%	92%	98%	4%
2	$NaBH(O_2CH)_3$	H_2SO_4, MeCN	95%	94%	66%	1%
3	$NaBH_4$, $FeCl_3$	EtOH, -25 °C to rt	20%	6%	90%	14%
4	$NaBH_4$, $FeCl_3$	*i*-PrOH, -25 °C to rt	80%	45%	87%	31%
5	$NaBH_4$, $FeCl_2$	EtOH, -25 °C to rt	2.5%	0%	-	2.5%
6	$NaBH_4$, $Fe(OEt)_3$	EtOH, rt	6%	3%	20%	2.5%
7	$NaBH_4$, $FeBr_3$	EtOH, -25 °C to rt	22%	17%	60%	5%
8	$NaBH_4$, FeF_3	EtOH, rt	10%	4%	30%	5%
9	$NaBH_4$, $Fe(NO_3)_3$	EtOH, -25 °C to rt	1%	1%	-	
10	$NaBH_4$, $Fe(acac)_3$	EtOH, rt	6%	5%	70%	
11	$NaBH_4$, LiOTf	EtOH, -25 °C to rt	6%	6%	-20%	0%
12	$NaBH_4$, $Al(OPr)_3$	*i*-PrOH, -25 °C to rt	6%	6%	-20%	0%
13	$NaBH_4$, $Co(OAc)_2$	EtOH, -25 °C to rt	100%	0%	-	100%

[a]All results were obtained by HPLC analysis.

Although Rh-catalyzed hydrogenation (23) afforded a clean reaction with near complete diastereoselectivity (Table 1, entry 1), the high cost of 1% catalyst/ligand contributed significantly to the overall cost of relebactam drug substance. Efforts to reduce the catalyst loading below 0.7 mol% gave low conversions. This option was abandoned due to the high cost of catalyst, the elevated pressure (500 psi H_2) requirement, and the high sensitivity to impurities in the substrate **46**.

The second hit (Table 1, entry 2) which used sodium triformyloxyborohydride (*24, 25*) provided a moderate 5:1 *dr*. Attempts to improve the diastereoselectivity under various conditions with sodium triformyloxyborohydride and its various analogues were unsuccessful (*26*).

An encouraging result arose from the use of $NaBH_4$ and $FeCl_3$ in EtOH (Table 1, entry 3), which allowed a 95:5 *dr*, albeit at a low conversion of only 20%. Reduction of oximes without cleavage of N-O bond in the presence of $NaBH_4$ and $FeCl_3$ is unprecedented. It was intriguing that among the various Lewis acids (>50) screened in MeOH, EtOH, *i*-PrOH and CF_3CH_2OH, $FeCl_3$ was the only one that provided >87% *de*. Not surprisingly, initial laboratory scale-up of the $NaBH_4/FeCl_3$-promoted oxime reduction in ethanol also resulted in a low conversion. We reasoned that the low conversion was mainly due to the difficulty to control local reaction temperature and agitation due to bursts of foam after each solid $NaBH_4$ addition.

These problems were overcome through controlling the concentration of $NaBH_4$ in the reaction mixture and eliminating the foam issue via a slow addition of a triglyme solution of $NaBH_4$ at -25 °C. As such, the reduction could smoothly proceed to >98% conversion without any operational issues. The use of triglyme also improved the overall solubility of the reaction mixture. It is worthy of note that maintaining the reduction temperature below -25 °C was important to achieve a good conversion and high *dr*. At ambient temperature, the predominant reaction was the reduction of Fe(III) to Fe(II) with negligible formation of **45** (Table 1, entry 6).

During studies on this reduction, we observed the reaction variability on conversion (15 – 80%), rate and *dr* (4:1 – 9:1), when anhydrous $FeCl_3$ was used. Further studies revealed that this variation was water-control dependent. Addition of 9 – 18 equiv of water was crucial to achieving a good conversion and high *dr*. Lowering the water to 6 equiv depressed the *dr* to 4 : 1. Therefore, non-hygroscopic $FeCl_3{\cdot}6H_2O$ became our reagent of choice. The optimum conditions were determined as follows: 1.6 equiv $FeCl_3{\cdot}6H_2O$, 1.6 – 2.0 equiv $NaBH_4$ and 8– 12 volumes of ethanol at -35 to -30 °C. Typical end-of-reaction profile after a 16 h addition of a triglyme $NaBH_4$ solution produced a >99% conversion, ~ 95:5 *trans : cis* ratio, and ~5% BnOH, with 87% assay yield of **45**.

Upon completion of the reaction, the bisulfate salt of **45** was directly crystallized from the reaction mixture through an inverse quench into aqueous H_2SO_4 solution at 30 °C. The inorganic salts were fully dissolved and rejected into the mother liquor. The wet cake was washed with 0.05 M H_2SO_4 to provide 1:1 **45**·H_2SO_4 salt in 87% recovery yield with 99.2 : 0.8 *dr* (76% overall yield) (Scheme 13). Washing the wet cake with water led to disproportionation of the bisulfate salt of **45**.

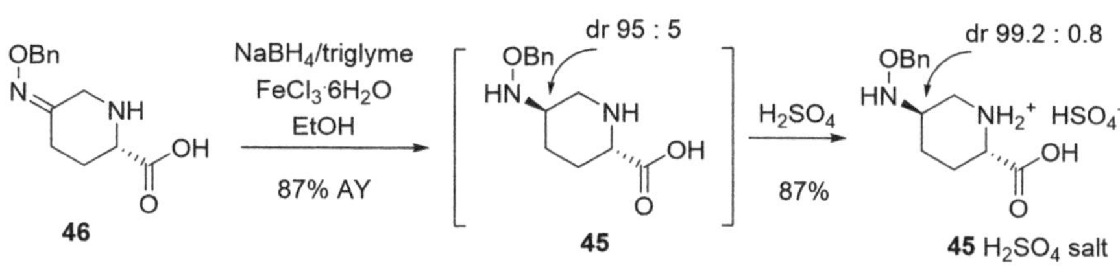

*Scheme 13. Optimized process to intermediate **45** bisulfate salt.*

To support the process development, various techniques were applied to gain the mechanistic understanding. By applying HPLC-ICP-MS technique (ion chromatography coupled with inductively coupled plasma mass spectrometry), we determined that the oxidation state of iron at end of reaction was completely Fe(II). Stoichiometric amount of Fe(III) was reduced during the reduction of oxime **46**. In addition, this result was also consistent with the initial results of Lewis acid screening (Table 1, entry 5), which confirmed that Fe(II) – $NaBH_4$ did not promote the reduction of oxime **46**.

We applied NMR techniques to probe the structure of Fe(III)-complexes in this reaction. Although Fe(III) is paramagnetic, the atom-specific longitudinal (R_1) paramagnetic relaxation enhancements (PREs), which is predominantly through the dipolar interaction between a nuclear spin and the Fe(III)-centered unpaired electrons (*27, 28*), responds in a distance-dependent manner (r^{-6}) between a nuclear spin and Fe(III). To better resolve the Fe(III) binding site, site-specific PREs were quantified by changes in ^{1}H and ^{13}C R_1 relaxations. The paramagnetic R_1 relaxation rates of protons and carbons are summarized in Figure 12. The R_1 relaxation of oxime **46** in MeOH-d_4 in the absence of $FeCl_3$ was very small. In contrast, the observed PRE effects for oxime **46** and its sodium salt in the presence of $FeCl_3$ was more profound. The Fe(III)-ligand interaction underwent fast exchange relative to the NMR time-scale, so the effective ligand relaxation rate was the average result of the free-state and different Fe(III) complexation states. When oxime acid **46** was mixed with $FeCl_3$, the PRE data suggested that iron was bound to the piperidine nitrogen only from the "top" of the ring, as protons pointing "up" had systematically larger R_1's than those pointing "down". But there was no chelation with the free carboxylic acid moiety (**54**, Figure 12). In interesting contrast, when the sodium salt of **46** was mixed with $FeCl_3$, the PRE data supported:

1) the iron complexing to both the nitrogen and the carboxylate oxygen (**55**, Figure 12).
2) in comparison with the free acid-iron complex, the iron flipped from the "top" to the lower surface of the ring, as protons pointing "down" consistently had larger R_1's than those pointing "up".
3) the Fe(III) coordination was further supported by the observed relatively small PRE of the carboxylate carbon in oxime acid **46**-$FeCl_3$ (6.8 s^{-1}) vs. the second largest PRE (72.5 s^{-1}) in oxime **46** sodium salt-$FeCl_3$.

In addition, the significantly broadened water signal in the PRE studies suggested Fe(III) was water-bound.

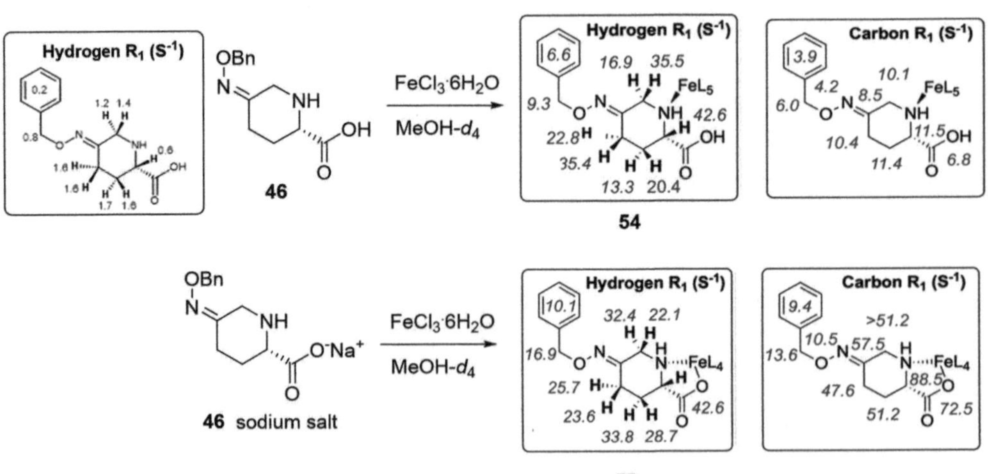

Figure 12. PRE data of oxime acid-$FeCl_3 \cdot 6H_2O$ and oxime acid sodium salt-$FeCl_3 \cdot 6H_2O$.

With the above information in hand, we had a further understanding the reaction species through density-functional theory (DFT) calculations. Our DFT structure of the proposed bicyclic iron complex supported the R_1 PRE data and provided some insight about the role of water for this diastereoselective reduction. In the most stable structure **55a**, a water ligand on the concave face of the bicyclic system was found to hydrogen-bond to the oxime nitrogen, which provided additional 1.1 kcal/mol stability vs the corresponding structure without OH···N hydrogen bonding (Scheme 14).

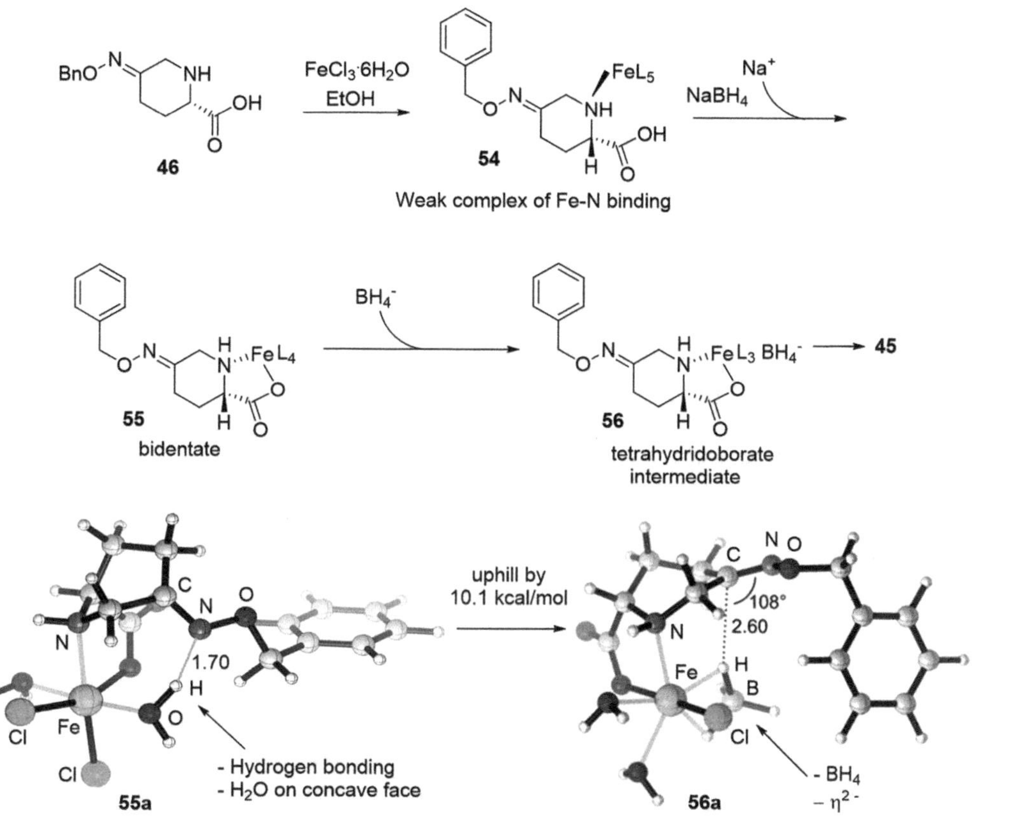

Scheme 14. Plausible reaction pathways.

Based on these results, we proposed that the initial mixing of oxime acid **46** with $FeCl_3·6H_2O$ would form a weak complex of iron binding to the piperidine nitrogen (**54**, Scheme 14). The subsequent addition of $NaBH_4$ would result in the formation of the bidentate **55**. The water molecule played an important role to shape the unique geometry stabilized through a hydrogen bonding. Through a subsequent exchange of a chloro ligand on **55a** with BH_4^-, tetrahydridoborate species **56a** could be thus formed, which was predicted by DFT to be uphill by only 10.1-11.7 kcal/mol depending on the *E*/*Z* configuration of the oxime and the hapticity of the tetrahydridoborate ligand. The geometry of **56a** revealed a plausible explanation for the high *anti*-selectivity in the reduction of oxime **46**. As shown in Scheme 14, a boron-bound hydrogen in **56a** is poised at a stereoelectronically favorable geometry right below the oxime moiety, which allowed the hydride to be perfectly delivered to the concave face of the bicyclic structure, resulting in a high diastereoselective reduction of the C=N bond.

With alkoxyamine **45** in hand, we turned our attention to prepare amide **40** via a selective amidation between **45** and amine **36**, without a protection/deprotection operation. The amidation studies were initially carried out with the available TFA salt of **45** in DMF, before we developed the process to isolate the bisulfate salt of **45** directly from the $FeCl_3$-$NaBH_4$-triglyme reaction mixture. Promising results were observed with the EDC/2-hydroxypyridine 1-oxide (HOPO) combination, yielding **40** in about 80% yield with the byproducts **57** (3%), **58**, and **59** (Scheme 15).

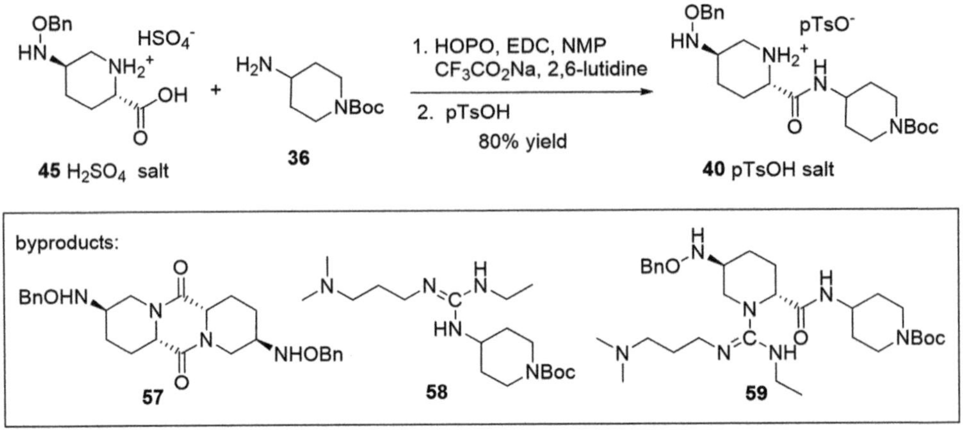

*Scheme 15. Selective amidation of amino acid **45** with amine **36** and byproducts.*

While the low solubility of the bisulfate salt of **45** had greatly benefited its direct isolation after the $NaBH_4/FeCl_3$ reduction of **46**, it presented a problem in the amidation reaction. Attempts to treat the bisulfate salt under the similar conditions that had worked well for the TFA salt gave a thick and unstirrable slurry, resulting in the desired amide **40** in only ~2% yield along with about 50% of undesired dimer **57**. Recalling our positive experiences with the TFA salt of **45** in our initial investigation (vide supra), we decided to carry out a salt metathesis with either sodium or potassium trifluoroacetate to convert the poorly soluble bisulfate salt to the more soluble TFA salt. We were very glad to find that addition of 3 equiv of CF_3CO_2Na completely solubilized the bisulfate salt of **45** in about 10 volumes of NMP. To suppress the dimer formation, we premixed ~50% of **45** (in solution) with the full molar equivalent of amine **36** to neutralize the highly acidic sulfuric acid salt of **45** and therefore formed a buffered weakly acidic solution. This also suppressed the formation of EDC adducts **58** and **59**. Slow addition of the remaining **45** did suppress the formation of dimer **57** to ~1%, however, the reaction stalled at 94% conversion. We reasoned that the incomplete conversion was due to increased acidity of the reaction solution, as the basic amine **36** was gradually converted to amide **40**. Therefore, two equivalents of 2,6-lutidine were added to the reaction mixture to effectively maintain the weak acidity of the reaction solution in the optimal range throughout the reaction. Under these improved conditions, the amidation proceeded to $\geq$ 97% conversion with 93% assay yield of amide **40**, containing < 2% of **57** and almost undetectable levels of **58** and **59**. After aqueous workup, amide **40** was crystallized as its *p*-TSA salt from toluene/acetonitrile in 80% yield and > 99.5% purity (26).

Process to Piperidine 40 from Amino Acid 20

As outlined in Scheme 11, one desired preparation of hydroxy amide **40** would be a direct coupling of unprotected amino acid **20** with amine **36** without introducing any protection-deprotection protocols. This would improve the atom economy and overall synthetic efficiency. The initial studies showed the direct amidation could be realized in 70% yield when amino acid **20** was added relatively quickly (within 30 min) to a mixture of **36**, HOPO and EDC in MeCN at 40 °C (Scheme 16). Further evaluation determined that the amidation suffered from competitive side reactions and was sensitive to reagent charging/reaction time. The amidation yield was also pH dependent as in the absence of HCl, the formation of amide **48** was nearly shut down. The pH of the reaction mixture after HCl charge dropped from ~ 9.1 to ~6.5 at which point the EDC $t_{1/2}$ was about 4 h. As the amidation reaction proceeded, the amine **36** which buffered the pH, was gradually converted to amide **48**. As the reaction pH further decreased to ~5.1, the EDC decomposition accelerated ($t_{1/2}$ to ~8 min) (29). Indeed, the assay yield of amide **48** had decreased from 79% to 74% and 47%, when the addition time of amino acid **20** was extended from 20 min to 1 h and 4 h, respectively.

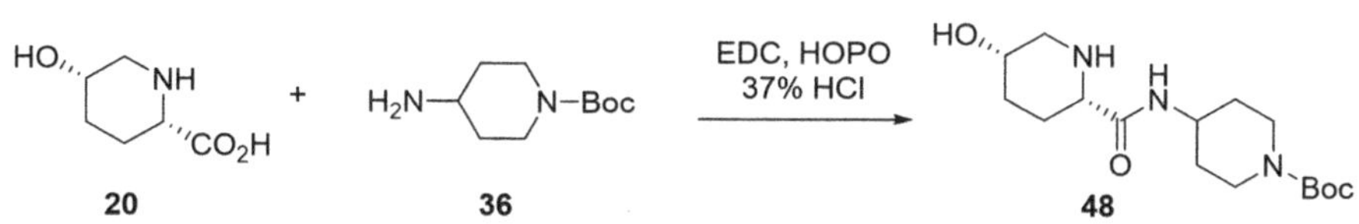

*Scheme 16. Direct amidation of amino acid **20** with amine **36**.*

With this information in hand, it was clear that the most risk-free option with the least sensitivity to time would be to add EDC as the last reagent to a pH buffered reaction mixture. After additional studies, we determined the use of 1.0 equiv conc. HCl was optimum. An undercharge of HCl led to increased amine-EDC adduct **61** (Scheme 17). The use of excess HCl led to increased amino acid dimer **60.** Furthermore, amide **40** could also react with EDC to form byproducts **63**, **65** and **66**.

EDC decomposition was faster than the amide coupling at 0 °C, while more dimer **60** and amide-EDC adduct **62** were formed between 0 and 20 °C. Carrying out the reaction above 40 °C also resulted in fast EDC decomposition in addition to an increase of **61**. Thus, the optimum reaction temperature range was 20 – 40 °C.

It was interesting to note that the amidation slurry consistency varied from a thick paste to a thin slurry, depending on the addition order of the reagents and starting materials. We therefore initiated a detailed study of the physical states of various binary mixtures. Photomicroscopy was useful to define the physical characteristics of the results in acetonitrile:

- A mixture of **36** with aqueous HCl produced a thin, easy-to-stir slurry that consisted of small spherical particles.
- Mixing amine **36** with acidic HOPO produced a thin slurry consisting of short rods with blunt ends. The solids remained well-suspended and very fluid.
- Mixing amine **36** with amino acid **20** produced a thick, needle-like slurry salt, indicating the interaction between **36** and **20** was likely the cause of the thick, difficult-to-stir slurry observed in the amidation.
- The slurry of amino acid **20** and HOPO produced small cuboid solids that were sticky and tended to settle into a thick bed.

- Mixing **20** with HCl produced hard, clumpy pieces of material.

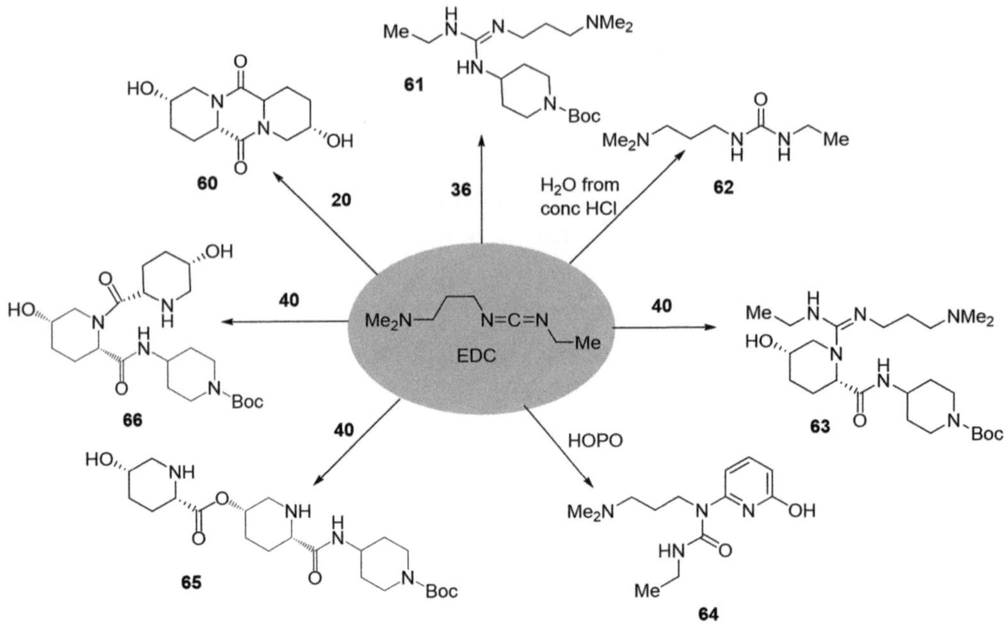

Scheme 17. EDC reactivities.

With this understanding established, an optimal addition mode to eliminate the formation of a thick slurry before EDC addition was developed using XRPD and Raman techniques. Mixing 1.0 equiv of concentrated HCl with amine **36** resulted in a HCl salt slurry. Subsequent addition of HOPO displayed no change in the morphology or behavior of the slurry, indicating that there was no salt metathesis with the **36**-HCl salt. Addition of **20** to this slurry showed little visible change in morphology or behavior of the slurry. As confirmed by Raman spectroscopy, the **36**-HCl salt remained unchanged prior to the addition of EDC, significantly suppressing the salt metathesis and therefore avoiding the formation of a unstirrable slurry. In contrast, switching the addition order of HOPO and **20** resulted in a thick and sticky slurry.

Using the defined optimal protocol, amidation with EDC at 40 °C robustly produced amide **48** within 4 h in 88% assay yield. The reaction was quenched and partitioned between acetonitrile/MTBE and 5 M NaOH to ensure the full rejection of HOPO sodium salt into the aqueous phase. HOPO was not effectively rejected during the subsequent crystallization of amide **48**. It is noteworthy to mention that the 5 M NaOH wash not only led to minor loss of product to the aqueous phase (<1%), but more importantly decreased the residual inorganic salt level in the wet MeCN-MTBE phase to < 0.1 wt%. As such, amide **48** could be isolated from azeotropically dried MeCN-MTBE stream in > 99% purity and 75% yield.

Treatment of amide **48** with 2.2 equiv of 4-nitrobenzenesulfonyl chloride (4-NsCl) and 2.4 equiv of DMAP in MeCN at ambient temperature afforded the desired bis 4-nitrobenzenesulfonyl product **67**, which was directly crystallized upon addition of water in 93% isolated yield and 99% purity to produce **67**-hydrate. Thus, we replaced the bis 2-nosyl amide **38** (Scheme 8) with the corresponding bis 4-nosyl amide **67** (Scheme 18). The bis 4-nosyl amide **67** reacted equally well with 4-NsNHOBn in the presence of K_2CO_3 in DMAc at 50 °C, to introduce the desired the *N*-benzyloxy group effectively without ee erosion. The reaction was quenched with methanol and thioglycolic acid followed by addition of K_2CO_3 at ambient temperature to remove both 4-nosyl groups (>99% conversion) and produce the desired diamine intermediate **40**. The deprotection of the 4-nosyl groups proceeded through the transient mono-protected amine intermediates **69** and **70**. We noted that the amine **70** was slower to deprotect than **69**. After an aqueous workup and addition of *p*-toluenesulfonic acid monohydrate in toluene-acetonitrile, crystalline diamine *p*-TSA salt of **40** was isolated in 78% yield and >99% purity.

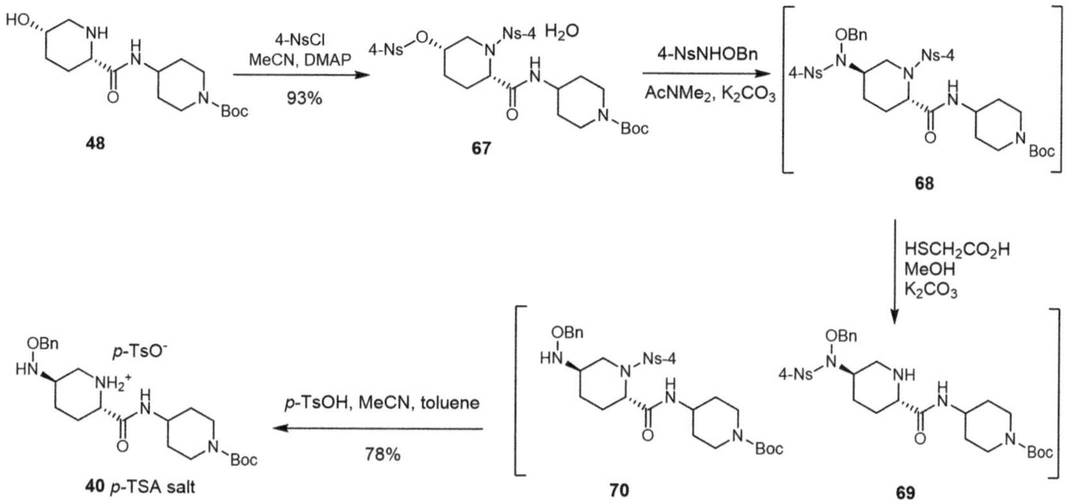

Scheme 18. Process to diamine p-TSA salt ***40****.*

In summary, we have developed two practical processes suitable for the large-scale preparation of the key intermediate **40**. The oxime reduction route (*26*) started from commercially available (*S*)-*N*-Boc-pyroglutamic acid and was accomplished without introducing additional protection-deprotection sequences. The four-step process prepared **40** in 40% overall yield and >99.5% purity. The synthesis is highlighted with an unprecedented and highly diastereoselective reduction of oxime acid **46** with $NaBH_4/FeCl_3{\cdot}6H_2O$. In contrast, the S_N2 displacement route was equally attractive. Starting from custom-synthesized (2*S*,5*S*)-5-hydroxypiperidine-2-carboxylic acid (**20**), amide **48** was prepared via a direct amidation with **36**. Through a subsequent activation/protection – S_N2 displacement/deprotection, the three-step process prepared **40** in 54% overall yield and >99% purity. We chose to implement the S_N2 route for commercialization of relebactam.

Development of Final Manufacturing Process (Intermediate 40 to Relebactam)

Starting from amine *p*-TSA salt **40**, our first late-stage development goal was to eliminate the use of hazardous triphosgene. The use of triphosgene posed serious safety concerns for the preparation of benzyloxy urea **41** at factory scale (Scheme 9). Our approach to overcome this issue started by gaining a better mechanistic understanding of the urea formation with triphosgene. We discovered that triphosgene preferentially reacted with the piperidine nitrogen and the adjacent amide moiety in **40**, presumably via the pathway shown in Scheme 19. When triphosgene was charged in sub-stoichiometric amounts, formation of intermediate **73** was observed. Upon introducing additional triphosgene, intermediate **73** could be further converted to intermediate **74**. The desired benzyloxy urea **41** was not formed until aqueous phosphoric acid was added.

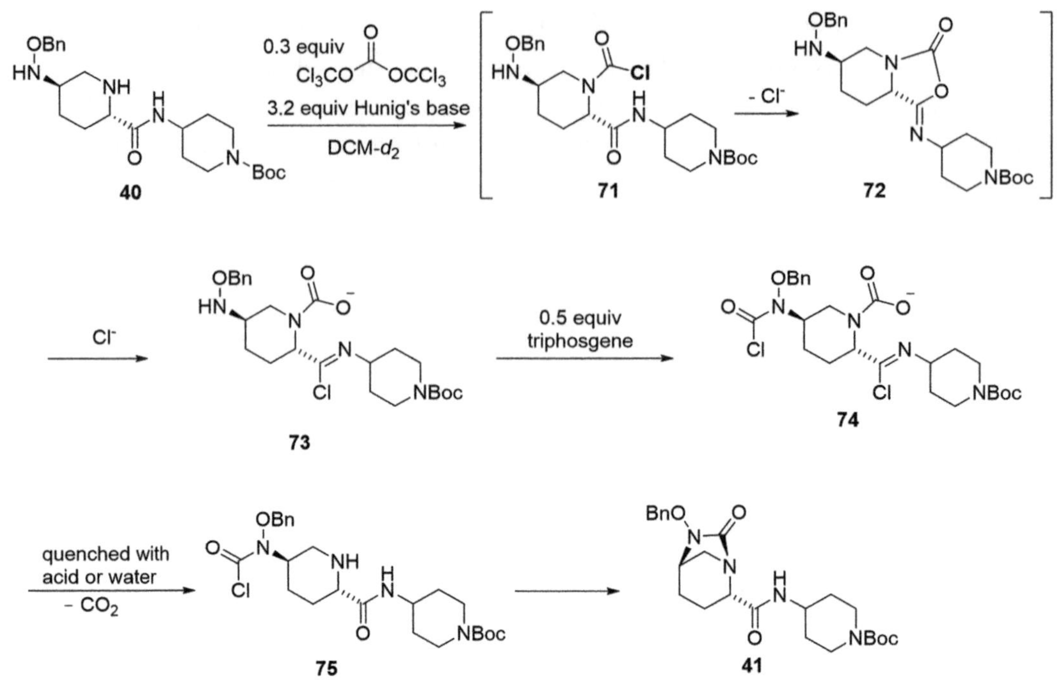

*Scheme 19. NMR studies on the reaction of **40** with triphosgene.*

This result clearly showed the first equivalent of phosgene (from triphosgene) served as a protecting reagent by blocking the more nucleophilic piperidine nitrogen. Subsequently the second equivalent of phosgene formed the carbamoyl chloride **74**. As such, benzyloxy urea **41** could be formed via an intramolecular nucleophilic attack of the piperidine nitrogen in **75**. Guided by this understanding, a robust process to prepare benzyloxy urea **41** via in situ protection of the piperidine nitrogen and carboxylate with Me_2SiCl_2 was developed (Scheme 20).

The diamine *p*-TSA salt **40** was treated with Me_2SiCl_2 in the presence of *i*-Pr_2NEt in MeCN at 0 – 20 °C to afford the in situ protected intermediate **76**. Subsequent treatment with CDI at 45 °C led to the installation of the imidazole urea on the benzyloxy nitrogen, while the silyl group also migrated from O to the adjacent N to form a more thermodynamically stable amide **77**. *i*-PrOH was then introduced to deprotect the dimethylsilyl group and the resulting intermediate **78** spontaneously cyclized to form the desired urea product **41**. After aqueous workup, benzyloxy urea **41** was crystallized from toluene/*n*-heptane in 85% yield and >99% purity.

While the benzyloxy urea **41** showed excellent stability upon storage, the subsequent hydroxy urea **32** was much less stable due to the increased electrophilicity of the bicyclic urea (*16*). Based on the earlier challenging hydrogenolysis experience (lack of reproducibility on reaction conversion and suppression of the decomposition of hydrogenolysis product **32**) of benzyloxy urea **41** at pilot plant scale, we envisioned that an in situ protection of the hydroxy amine group in **32** during the hydrogenolysis would present a corresponding intermediate with increased stability similar to benzyloxy urea **41**. Thus, trimethylsilyl (TMS) protected urea **79** was identified as the optimal species, as the TMS protection and deprotection could be easily carried out under mild conditions. Indeed, in the presence of stoichiometric *N,O*-bis(trimethylsilyl)acetamide (BSA), the TMS protected hydroxyl urea **79** was cleanly formed in nearly quantitative yield immediately upon hydrogenolysis of **41** in the presence of Pd/C in *i*-PrOAc.

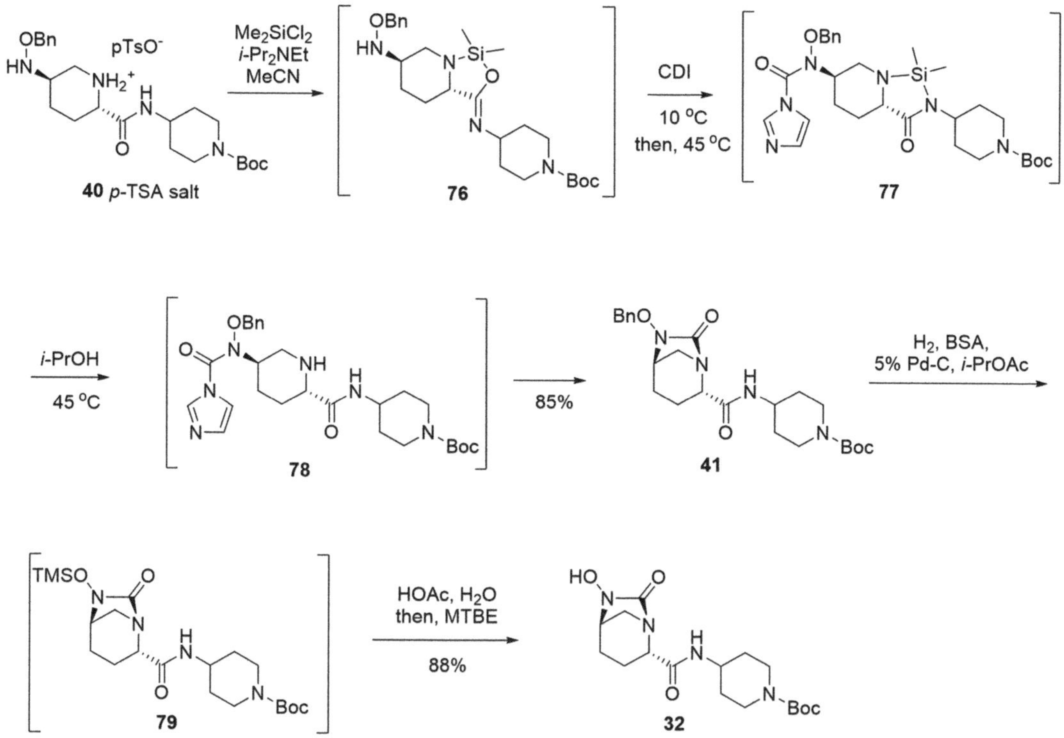

Scheme 20. Preparation of hydroxy urea **32**.

During the development of the hydrogenolysis, however, there was significant variation of the hydrogenolysis rate. This was true even with the same lot of benzyloxy urea **41** under identical process conditions. After further scrutiny, the rate variation was ultimately traced back to the source of BSA. When batches of BSA contaminated with a small amount of triethylamine were used, the hydrogenolysis rate was surprisingly accelerated, in contrast to reported examples of catalyst poisoning in *O*-benzyl ether hydrogenolysis reactions (*30–33*). Thus, various basic additives were screened for this reaction and 1,4-diazabicyclo[2.2. 2]octane (DABCO) proved eminently acceptable. In the presence of DABCO, the hydrogenolysis proceeded much faster and the rate was reproducible (Figure 13). Of note, the impact of DABCO on the reactivity of Pd/C and other solid supported Pd catalysts proved to possess wide application scope (*34*).

Our studies on the stability of **32** revealed that complicated oligomers would be formed under certain conditions via nucleophilic attack on the electrophilic urea. To overcome this issue, we required a process allowing direct crystallization of **32** from the reaction stream along with minimal process operation and under mild conditions. The use of *i*-PrOAc as the hydrogenation solvent made the direct isolation feasible, since the solubility of the desilylated hydroxy urea **32** in *i*-PrOAc was low (2.1 mg/mL). After optimization, hydrogenolysis of **41** in the presence of 2 mol% of DABCO, 1.25 equiv of BSA, and 5 wt% of Pd-C in *i*-PrOAc under H_2 (50 psig) at ambient temperature afforded the desired silylated **79** in >99% conversion. Upon filtration to remove the catalyst and subsequent concentration, a solution of HOAc (0.7 equiv) and water (2.6 equiv) followed by MTBE was added at 15 °C. Intermediate **79** smoothly converted to **32**, which was directly crystallized from the reaction stream in 88% isolated yield (*34*).

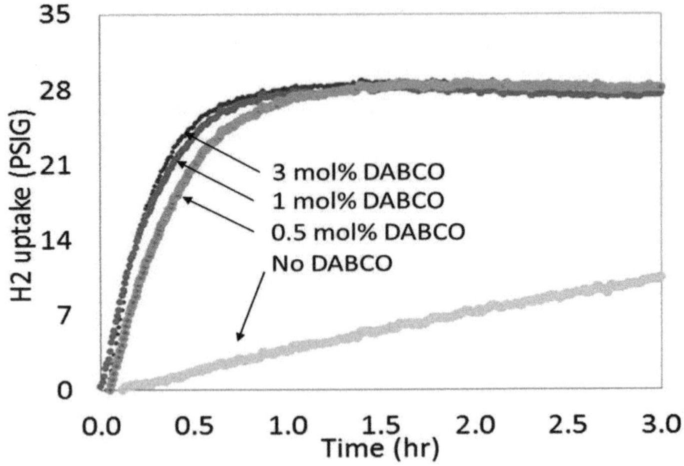

Figure 13. Hydrogen uptake vs. DABCO loadings with 5 wt% Pd-C in i-PrOAc under 50 psig of H_2 at ambient temperature.

Previously, the sulfation of **32** was carried out in the presence of SO_3-pyridine complex in acetonitrile (Scheme 9). While these conditions were effective, the reaction resulted in a thick stirrable slurry. With the hydrogenolysis process established, the quality of the isolated hydroxy urea **32** was improved. In particular, the oligomeric impurities in the isolated **32** could be reproducibly controlled to below not-quantifiable (NQ) level as compared to ~1-2% that resulted from the previous hydrogenolysis process. Interestingly, the sulfation using the hydroxy urea **32** with higher purity prepared under the optimal hydrogenolysis conditions resulted in a much worse and unstirrable thick slurry. This dramatic effect of the oligomeric impurities was unambiguously confirmed by a series of control experiments with various levels of oligomers.

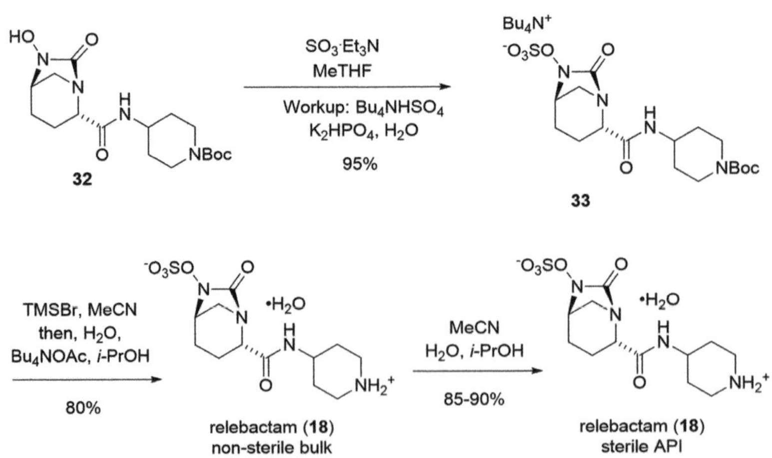

Scheme 21. Process to relebactam.

We postulated that replacing pyridinium ion of product salt generated during the sulfation with a corresponding different amine cation could be beneficial on improving the solubility of the salt product through counter ion effect, allowing the sulfation to be carried out in a homogeneous solution. After screening multiple sulfation reagents under various conditions, SO_3 triethylamine complex became our sulfation reagent of choice. The use of SO_3 triethylamine complex resulted in a near homogeneous reaction solution through counter ion effect (Scheme 21).

With these conditions discovered, a robust sulfation process was consequently developed. Treatment of **32** with $SO_3{\cdot}Et_3N$ in 2-MeTHF at 40 °C produced **33** in >98% conversion. Subsequent addition of an aqueous K_2HPO_4 solution at ambient temperature over several hours allowed the complete quench of unconsumed SO_3. Any remaining SO_3 in the wet organic phase would slowly form sulfuric acid, which caused the decomposition of the sulfate **33** to form decarbonylated impurity **80** (Scheme 22). It was important to control the formation of decarbonylated impurity **80** to low level by ensuring the complete quench of SO_3 as described above, because impurity **80** acted as initiator for oligomer formation under the subsequent de-Boc conditions (For more information, see the discussion about the development of the de-Boc step).

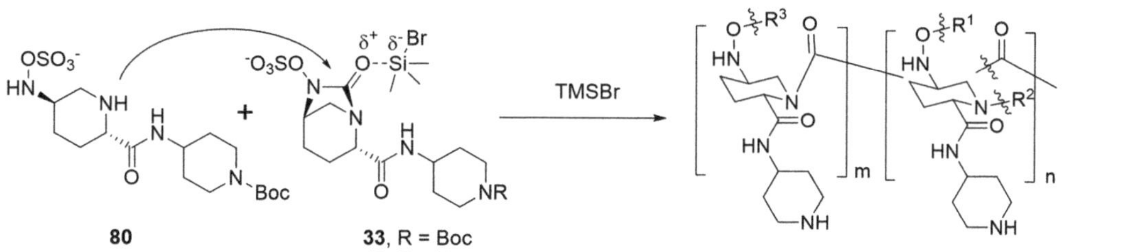

*Scheme 22. Plausible pathway of oligomerization via decarbonylated impurity **80**.*

Bu_4NHSO_4 was then added to the quenched biphasic mixture. The Bu_4N^+ salt of sulfate **33** was extracted into the organic phase, while the other salts were rejected into the aqueous phase. After a brine wash, the crude stream was treated with activated carbon and azeotropically solvent-switched to MeCN. The solution of **33** was directly carried through to the subsequent de-Boc step. This robust process prepared **33** in 95% assay yield, while the decarbonylated impurity **80** was controlled to ~ 0.1%.

An effective de-Boc process using TMSI in MeCN at 0 °C was established during the development of the second-generation synthesis (*17, 35*). This process significantly improved the robustness and simplified the isolation of relebactam zwitterion. However, the use of TMSI at factory scale was deemed non-ideal in terms of handling, storage, shipment and cost (*18*). The use of TMSBr to replace TMSI was examined. It was quickly found that the Boc group could be effectively removed when **33** was treated with TMSBr in MeCN at ambient temperature. This result established a baseline process to relebactam for further optimization.

Although the oligomeric impurities could be controlled to below not-quantifiable (NQ) levels in the isolated **32**, we wished to control the oligomeric impurities to not-detectable (ND) levels in the subsequent isolated crude relebactam before final aseptic recrystallization. We confirmed that the optimized sulfation process did not result in an increase of oligomeric impurities. Based on the early development experience on the observed variation of oligomer levels in crude **18**, we suspected additional oligomeric impurities could be generated from the de-Boc step. To address the capability of rejection of oligomeric impurities during the isolation of relebactam, we developed two process controls. Implementation of a carbon treatment in aqueous MeCN prior to recrystallization of crude **18** ensured effective rejection of oligomers as high as 2% levels, obviating the need for further downstream controls. To provide further confidence for oligomer control, multiple experiments were conducted to demonstrate that oligomers present at 0.66 – 0.75% levels in crude **18** were rejected to ND level through a recrystallization without carbon treatment.

In addition, we scrutinized the fate of every impurity during the de-Boc process. This led us to discover that the decarbonylated impurity **80** increased the level of oligomers. Control experiments clearly displayed **80** as functioning as an initiator for the formation of oligomers under the de-Boc conditions, presumably via intermolecular nucleophilic attack of the free amine nitrogen in **80** onto the electrophilic ureas in **18**/**33** (Scheme 22). With the sulfation process conditions optimized (vide supra), the decarbonylated impurity could be well controlled to ~0.1% in sulfate **33**. This further enhanced the capability to fully reject the oligomers to not-detectable (ND) levels in the isolated crude relabactam.

Under the optimized de-Boc process conditions (Scheme 21), sulfate **33** was treated with 1.35 equiv of TMSBr at ambient temperature. Upon completion of the Boc deprotection, remaining TMSBr was quenched with a solution of 0.25 equiv of Bu_4NOAc mono-HOAc solvate in aqueous MeCN. Bu_4NOAc as a base neutralized the HBr liberated from the reaction. Non-sterile relebactam was crystallized as the desired zwitterion monohydrate from aqueous *i*-PrOH–MeCN in 85% isolated yield and high quality (>99% purity).

Since relebactam is intravenously co-dosed to patients with Primaxin® (imipenem/cilastatin), the non-sterile bulk relebactam was further recrystallized from aqueous *i*-PrOH–MeCN at ambient temperature under sterilized process conditions to produce sterile relebactam for commercial use (Schemes 21 and 23).

In summary, a robust manufacturing process to relebactam has been developed and implemented in production (Scheme 23). Starting from (2*S*,5*S*)-5-hydroxypiperidine-2-carboxylic acid **20** and *tert*-butyl 4-aminopiperidine-1-carboxylate **36**, relebactam was prepared in 29% overall yield over eight process steps including the final aseptic recrystallization. In contrast to the first-generation large-scale synthesis, the final manufacturing process realized through process evolution and innovation significantly improves the process efficiency and robustness.

Approval Status

On July 16, 2019 relebactam, in combination with imipenem/cilastatin, was approved by the FDA (as Recarbrio®) for the treatment of complicated intra-abdominal infections and urinary tract infections caused by certain Gram-negative bacteria (*6*). On June 4, 2020, the FDA announced the approval of Recarbrio® to treat "hospital-acquired bacterial pneumonia and ventilator-associated bacterial pneumonia (HABP/VABP) in patients 18 years of age and older" (*36*).

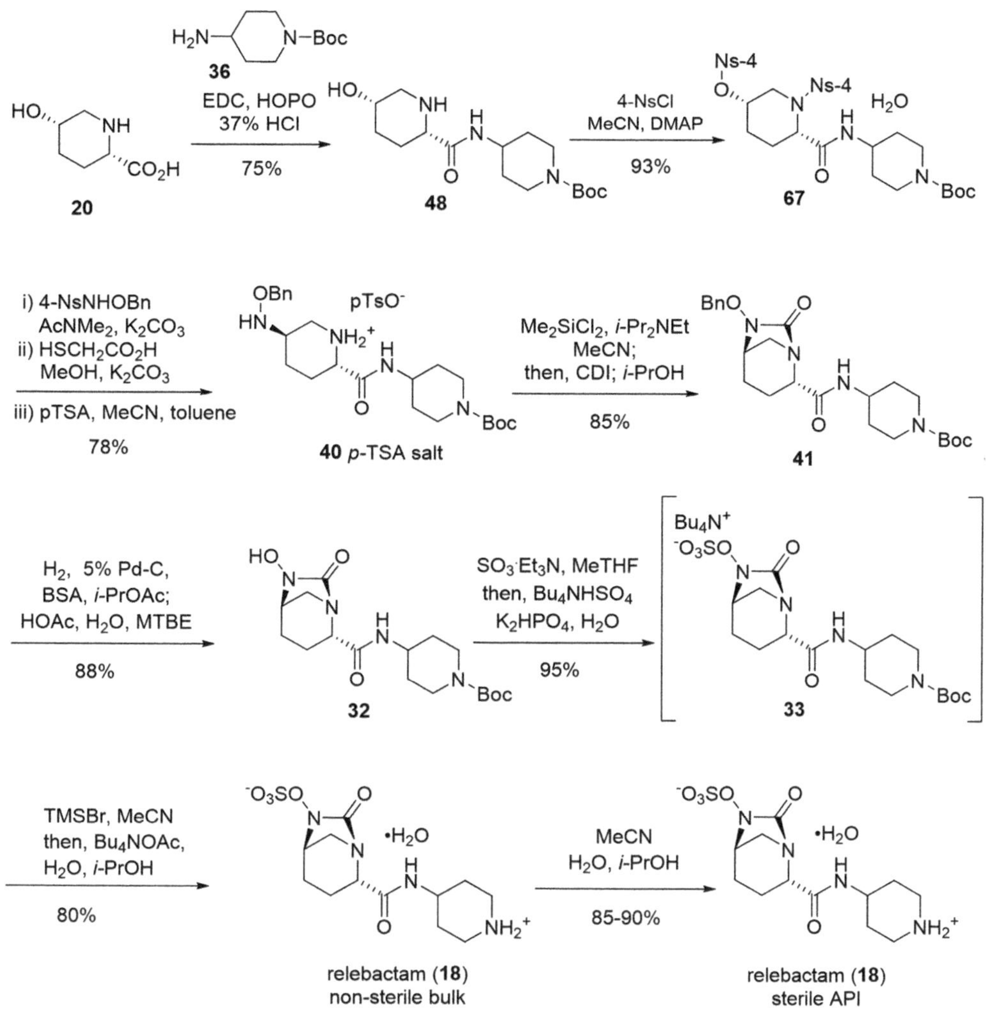

Scheme 23. Manufacturing process to relebactam.

Acknowledgments

The authors would like to thank the countless smart, creative scientists with whom we have had the privilege to work during the discovery and development of relebactam.

References

1. Donowitz, G. R.; Mandell, G. L. Beta-Lactam Antibiotics. *New Engl. J. Med.* **1988**, *318*, 419–426.
2. Rodloff, A. C.; Goldstein, E. J. C.; Torres, A. Two Decades of Imipenem Therapy. *J. Antimicrob. Chemother.* **2006**, *58*, 916–929.
3. Hall, B. G.; Barlow, M. Revised Ambler Classification of β-Lactamases. *J. Antimicrob. Chemother.* **2005**, *55*, 1050–1051.
4. Geddes, A. M.; Klugman, K. P.; Rolinson, G. N. Introduction: Historical Perspective and Development of Amoxicillin/Clavulanate. *Int. J. Antimicrob. Agents* **2007**, *30* (suppl. 2), 109–112.

5. Watkins, R. R.; Papp-Wallace, K. M.; Drawz, S. M.; Bonomo, R. A. Novel β-Lactamase Inhibitors: a Therapeutic Hope against the Scourge of Multidrug Resistance. *Frontiers in Microbiology* **2013**, *4*, article 392 (DOI:10.3389/fmicb.2013.00392.
6. Drug Approval Package: RECARBRIO. https://www.accessdata.fda.gov/drugsatfda_docs/nda/2019/212819Orig1s000TOC.cfm (accessed September 9, 2020).
7. Dininno, F.; Hammond, M. L.; Dykstra, K.; Kim, S.; Tan, Q.; Young, K.; Hermes, J. D.; Chen, H.; Raeppel, S.; Mannion, M.; Gaudette, F.; Vaisburg, A.; Rahil, J.; Georgopapadakou, N.; Zhou, N. Z. *Novel Inhibitors of Beta-Lactamase*. U.S. Patent Application 2010279983 (A1).
8. Heinze-Krauss, I.; Angehrn, P.; Charnas, R. L.; Gubernator, K.; Gutknecht, E.-M.; Hubschwerlen, C.; Kania, M.; Oefner, C.; Page, M. G. P.; Sogabe, S.; Specklin, J.-L.; Winkler, F. Structure-Based Design of β-Lactamase Inhibitors. 1. Synthesis and Evaluation of Bridged Monobactams. *J. Med. Chem.* **1998**, *41*, 3961–3971.
9. Ehmann, D. E.; Jahic, H.; Ross, P. L.; Gu, R. F.; Hu, J.; Kern, G.; Walkup, G. K.; Fisher, S. L. Avibactam is a Covalent, Reversible, Non-β-Lactam β-Lactamase Inhibitor. *Proc. Natl. Acad. Sci. U.S.A.* **2012**, *109*, 11663–11668.
10. Blizzard, T. A.; Chen, H.; Kim, S.; Wu, J.; Young, K.; Park, Y.; Ogawa, A.; Raghoobar, S.; Painter, R. E.; Hairston, N.; Lee, S. H.; Misura, A.; Felcetto, T. J.; Ha, S.; Hickey, E.; Hermes, J.; Hammond, M. Side Chain SAR of Bicyclic β-Lactamase Inhibitors (BLIs). 1. Discovery of a Class C BLI for Combination with Imipenem. *Bioorg. Med. Chem. Lett.* **2010**, *20*, 918–921.
11. Blizzard, T. A.; Chen, H. Y.; Wu, J. Y.; Kim, S.; Ha, S.; Mortko, C.; Variankaval, N.; Chiu, A. *Novel Inhibitors of Beta-Lactamase*. Patent Application WO 2008039420, Example 32.
12. Blizzard, T. A.; Chen, H.; Kim, S.; Wu, J.; Bodner, R.; Gude, C.; Imbriglio, J.; Young, K.; Park, Y.; Ogawa, A.; Raghoobar, S.; Hairston, N.; Painter, R. E.; Wisniewski, D.; Scapin, G.; Fitzgerald, P.; Sharma, N.; Lu, J.; Hermes, J.; Hammond, M. Discovery of MK-7655, a β-Lactamase Inhibitor for Combination with Primaxin. *Bioorg. Med. Chem. Lett.* **2014**, *24*, 780–785.
13. Lampilas, M.; Aszodi, J.; Rowlands, D. A.; Fromentin, C. *Azabicyclic Compounds, Preparation thereof and Use as Medicines, in Particular as Antibacterial Agents*. U.S. Patent 7,112,592, Example 32b.
14. Young, K.; Painter, R. E.; Raghoobar, S. L.; Hairston, N. N.; Racine, F.; Wisniewski, D.; Balibar, C. J.; Villafania, A.; Zhang, R.; Sahm, D. F.; Blizzard, T.; Murgolo, N.; Hammond, M. L.; Motyl, M. R. In vitro Studies Evaluating the Activity of Imipenem in Combination with Relebactam against Pseudomonas Aeruginosa. *BMC Microbiology* **2019**, *19*, 150.
15. Powles, M. A.; Galgoci, A.; Misura, A.; Colwell, L.; Dingley, K.; Tang, W.; Wu, J.; Blizzard, T.; Motyl, M.; Young, K. In Vivo Efficacy of Relebactam (MK-7655) in Combination with Imipenem-Cilastatin in Murine Infection Models. *Antimicrob. Agents Chemother.* **2018**, *62*, eo2577-17.
16. Mangion, I. K.; Ruck, R. T.; Rivera, N.; Huffman, M. A.; Shevlin, M. A Concise Synthesis of a β-Lactamase Inhibitor. *Org. Lett.* **2011**, *13*, 5480–5483.
17. Miller, S. P.; Zhong, Y.-L.; Liu, Z.; Simeone, M.; Yasuda, N.; Limanto, J.; Chen, Z.; Lynch, J.; Capodanno, V. Practical and Cost-Effective Manufacturing Route for the Synthesis of a β-Lactamase Inhibitor. *Org. Lett.* **2014**, *16*, 174–177.

18. The storage and shipment of TMSI requires at controlled temperature with stabilizer. The high moisture sensitivity also is a concern for large-scale use, in particular when accurate stoichiometric charge is required.
19. Ball, M.; Boyd, A.; Ensor, G. J.; Evans, M.; Golden, M.; Linke, S. R.; Milne, D.; Murphy, R.; Telford, A.; Kalyan, Y.; Lawton, G. R.; Racha, S.; Ronsheim, M.; Zhou, S. H. Development of a Manufacturing Route to Avibactam, a β-Lactamase Inhibitor. *Org. Process Res. Dev.* **2016**, *20*, 1799–1805.
20. The iodide attacks of a chloro oxime to form a deschloro oxime has not been reported. For an example of formation of ICl through iodide attack on chloride, see: Davies, I. W.; Taylor, M.; Hughes, D.; Reider, P. J. Hydrogen Iodide-Promoted Reduction of β-Chlorovinamidinium Salts. *Org. Lett.* **2000**, *2*, 3385–3387.
21. Although studied, the mechanism of using DMSO to suppress dechlorination remains unknown.
22. Mohr, J.; Oestreich, M. $B(C_6F_5)_3$-Catalyzed Hydrogenation of Oxime Ethers without Cleavage of the N–O Bond. *Angew. Chem., Int. Ed.* **2014**, *53*, 13278–13281and references therein.
23. Kadyrov, R.; Riermeier, T.; Almena, J.; Monsees, A.; Groß, P.; Rossen, K. *Verfahren zur Herstellung von enantiomerenangereicherten Hydroxylaminderivaten*. European Patent 1862446A2, 2007.
24. Knopf, I.; Cummins, C. C. Revisiting CO_2 Reduction with $NaBH_4$ under Aprotic Conditions: Synthesis and Characterization of Sodium Triformatoborohydride. *Organometallics* **2015**, *34*, 1601–1603.
25. Marchini, P.; Liso, G.; Reho, Antonia; Felice, L.; Micheletti Moracci, F. Sodium Borohydride-Carboxylic Acid Systems. Useful Reagents for the Alkylation of Amines. *J. Org. Chem.* **1975**, *40*, 3453–3456.
26. Chung, J. Y. L.; Meng, D.; Shevlin, M.; Gudipati, V.; Chen, Q.; Liu, Y.; Lam, Y.; Dumas, A.; Scott, J.; Tu, Q.; Xu, F. Diastereoselective $FeCl_3{\cdot}6H_2O/NaBH_4$ Reduction of Oxime Ether for the Synthesis of β-Lactamase Inhibitor Relebactam. *J. Org. Chem.* **2020**, *85*, 994–1000.
27. Bloembergen, N.; Morgan, L. O. Proton Relaxation Times in Paramagnetic Solutions. Effects of Electron Spin Relaxation. *J. Chem. Phys.* **1961**, *34*, 842–850.
28. Guéron, M. Nuclear Relaxation in Macromolecules by Paramagnetic Ions: A Novel Mechanism. *J. Magn. Reson.* **1975**, *19*, 58–66.
29. Lei, Q. P.; Lamb, D. H.; Heller, R. K.; Shannon, A. G.; Ryall, R.; Cash, P. Kinetic Studies on the Rate of Hydrolysis of *N*-Ethyl-*N*'-(dimethylaminopropyl)carbodiimide in Aqueous Solutions Using Mass Spectrometry and Capillary Electrophoresis. *Anal. Biochem.* **2002**, *310*, 122–124.
30. Sajiki, H.; Hirota, K. A Novel Type of Pd/C-catalyzed Hydrogenation Using a Catalyst Poison: Chemoselective Inhibition of the Hydrogenolysis for O-Benzyl Protective Group by the Addition of a Nitrogen-containing Base. *Tetrahedron* **1998**, *54*, 13981–13996.
31. Sajiki, H.; Hattori, K.; Hirota, K. The Formation of a Novel Pd/C–Ethylenediamine Complex Catalyst: Chemoselective Hydrogenation without Deprotection of the O-Benzyl and N-Cbz Groups. *J. Org. Chem.* **1998**, *63*, 7990–7992.

32. Sajiki, H.; Kuno, H.; Hirota, K. Suppression Effect of the Pd/C-Catalyzed Hydrogenolysis of a Phenolic Benzyl Protective Group by the Addition of Nitrogen-Containing Bases. *Tetrahedron Lett.* **1998**, *39*, 7127–7130.
33. Researchers at Abbott Laboratories reported rate enhancement in the hydrogenolysis of a benzyl ester in conjunction with the reduction of Ar–NO_2 to aniline by addition of TEA. See: Seif, L. S.; Hannick, S. M.; Plata, D. J.; Morton, H. E.; Sharma, P. N. A Novel Large Scale Hydrogenation Strategy of *N,N*-Dimethyl 4-(Acetoxy-3-nitrophenylmethyl)-1-H-imidazole-1-sulfonamide, a Key Intermediate in the Synthesis of ABT-866. *Chem. Industries (Dekker)* **2003**, *89*, 633–641.
34. Yin, J.; Weisel, M.; Ji, Y.; Liu, Z.; Liu, J.; Wallace, D. J.; Xu, F.; Sherry, B. D.; Yasuda, N. Improved Preparation of a Key Hydroxylamine Intermediate for Relebactam: Rate Enhancement of Benzyl Ether Hydrogenolysis with DABCO. *Org. Process Res. Dev.* **2018**, *22*, 273–277.
35. Liu, Z.; Yasuda, N.; Simeone, M.; Reamer, R. A. *N*-Boc Deprotection and Isolation Method for Water-Soluble Zwitterionic Compounds. *J. Org. Chem.* **2014**, *79*, 11792–11796.
36. FDA Approves Antibiotic to Treat Hospital-Acquired Bacterial Pneumonia and Ventilator-Associated Bacterial Pneumonia. https://www.fda.gov/news-events/press-announcements/fda-approves-antibiotic-treat-hospital-acquired-bacterial-pneumonia-and-ventilator-associated (accessed September 9, 2020).

Chapter 9

Discovery and Chemical Development of Grazoprevir: An HCV NS3/4a Protease Inhibitor for the Treatment of the Hepatitis C Virus Infection

Feng Xu[*,1] and John A. McCauley[2]

[1]Department of Process Chemistry, Merck & Co., Inc., Rahway, New Jersey 07065, United States

[2]Department of Medicinal Chemistry, Merck & Co., Inc., West Point, Pennsylvania 19486, United States

[*]Email: feng_xu@merck.com

This chapter outlines the discovery and chemical development of the hepatitis C virus NS3/4a protease inhibitor, grazoprevir, approved in combination with the NS5a inhibitor elbasvir as Zepatier® for the treatment of hepatitis C virus (HCV) infection. Highly potent and efficacious HCV NS3/4a protease inhibitors have been a key factor in the recent evolution of all-oral HCV infection therapies. The path to the final manufacturing process of grazoprevir required chemistry evolution with innovative breakthroughs and thorough evaluation of various bond connection strategies. Focusing on problem solving and mechanistic understanding of the chemistry issues, this chapter describes the entire chemical development journey starting from Discovery Chemistry. The development of two generation large-scale syntheses for grazoprevir are discussed in detail. The initial safety concerns and hazardous issues encountered during process scale-up are explained, and the solutions to overcome these issues through staged process evolution and ultimately with new chemical transformation for an ideal process are described. Our chemical development delivers the most effective synthesis of this highly complex molecule with an optimal sequence of bond connection to assemble four building blocks in the final manufacturing process, which has been successfully implemented at factory scale.

Introduction

Grazoprevir (**1**, Figure 1) is a hepatitis C virus (HCV) NS3/4a protease inhibitor approved for the treatment of HCV infection in over 40 countries. Grazoprevir is approved in combination with the NS5a inhibitor elbasvir as the fixed-dose, single-pill combination therapy Zepatier.® In this chapter, we describe the discovery and initial synthesis as well as the conception and implementation of a large-scale manufacturing route for grazoprevir.

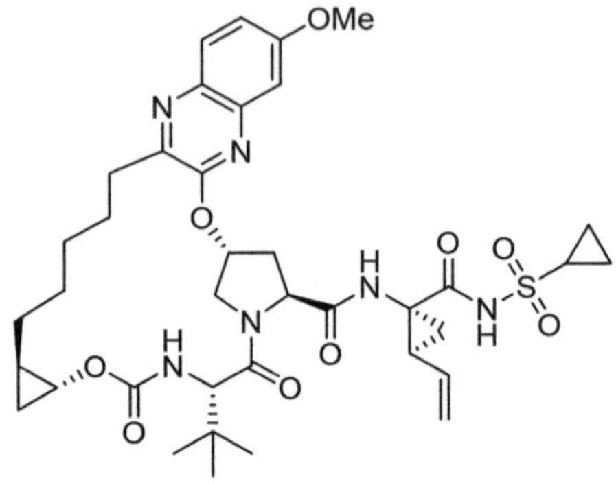

1 (Grazoprevir, MK-5172)

Figure 1. Structure of grazoprevir.

Hepatitis C NS3/4a Protease

Hepatitis C virus (HCV) infection (*1*) continues to represent a major health issue (*2*). While HCV is slow to progress, the disease ultimately leads to liver damage in up to 50% of patients as well as hepatocellular carcinoma in a significant number of cases (*3*). There are multiple genotypes in addition to multiple quasispecies of this positive strand RNA *Flaviviridae* virus (*4*). The modest sustained virologic response (SVR) rates for genotype 1 (gt1), the most common genotype worldwide, coupled with significant adverse events (AEs) and extended treatment duration of the previous standard-of-care therapies (pegylated interferon-alpha/ribavirin; PEG-IFN/RBV) (*5*), sparked a significant effort to identify more effective and better-tolerated treatments. These efforts have focused on direct inhibitors of several key nonstructural proteins essential for viral replication (*6*).

The treatment of HCV infection has undergone a revolution over the last ten years. Initially, the PEG-IFN/RBV standard of care was modified with the addition of NS3/4a protease inhibitors (*7*). This treatment change led to increased SVR rates in both naïve and treatment-experienced patients. More recently, all-oral therapies that exhibit SVR rates of better than 90% have eliminated the need for the less-effective interferon-based therapies that showed high rates of severe side effects and required difficult dosing regimens. The all-oral therapies currently on the market are based on several different combinations of inhibitors of NS5b polymerase, NS3/4a protease, and the NS5a replication complex.

The HCV NS3 protein consists of two domains: a *N*-terminal serine protease and a *C*-terminal RNA helicase (*8*). NS4a is a peptide that binds to NS3 and is a critical co-factor for polyprotein maturation. The NS3/4a complex mediates proteolysis of the HCV polyprotein, cleaving four junctions between the non-structural proteins, and is required for replication (*9*).

HCV NS3/4a protease inhibitors that have entered clinical development can be classified by their modes of interaction with the protease (*10*); ketoamides that participate in slowly reversible covalent binding events with the active site serine (*11, 12*) and typically more rapidly reversible non-covalent inhibitors. The reversible inhibitor structural class is significantly larger than the covalent binder class and can be broken down further into three sub-classes based on structure: P1–P3 macrocyclic, acyclic, and P2–P4 macrocyclic compounds. The prototypical non-covalent inhibitor is a highly potent P1–P3 macrocycle discovered at Boehringer-Ingelheim, ciluprevir (BILN-2061, **2**, Figure 2) (*13*) which was the first HCV NS3/4a inhibitor to demonstrate efficacy in the clinic, showing multi-log drops in viral load during a Phase 1b study (*14*). Following the HCV NS3/4a protease inhibition clinical proof-of-concept established with ciluprevir (*13, 14*), many research groups chose this molecule as a starting point for novel compound/series development (*10*). Crystal structures of ciluprevir revealed that the compound bound to pockets (P1 through P4) in the NS3 protease enzyme active site and this structure became an important design tool for many groups (*13*).

Invention of Grazoprevir

Our novel P2–P4 macrocyclic design hypothesis (see Figure 2) resulted from in silico analysis of ciluprevir (**2**) bound to a model of the full length NS3/4a protein. This model removed the six *C*-terminal residues (DLEVVT) of the helicase domain which lie in the active site of the apo X-ray structure. (*15*). This allowed visualization of ciluprevir bound in the active site of the full-length protein rather than just to the protease domain. Furthermore, overlay of this truncated amino acid sequence in the model with inhibitor bound, revealed that the glutamate sidechain filled a now defined pocket between P2 and P4. We hypothesized that a new linker from P2 to P4 could be employed to fill this space. Initial macrocyclic compounds were synthesized using a ring-closing metathesis-based route, and then optimized to the first generation NS3/4a inhibitor, vaniprevir (**3**, Figure 2) (*16, 17*).

A second-generation program focused on improving activity against the variety of genotypes of HCV and against mutant NS3 enzymes that were appearing in the clinic, while maintaining the favorable pharmacokinetic properties seen with vaniprevir. Initial lead **4** (Figure 2) showed promising activity against the genotype 3 (gt3) enzyme as compared to vaniprevir (IC_{50} = 45 nM vs. 54 nM) and modeling suggested that the P2 region of the molecule would be a key area of focus for optimization (*18*). An effort to rapidly access chemical diversity in P2 led to discovery of unique tricyclic P2 groups exemplified in compound **5** (gt3 IC_{50} = 12 nM). Aiming for further improvements in the potency profile against mutant enzymes, we then simplified the tricycle to a quinoline and lengthened the linker giving compound **6** (*18*). Additional improvements in potency, pharmacokinetics and physical properties resulted in the discovery of **1** which was advanced to development as MK-5172 and later renamed grazoprevir (*18, 19*).

Medicinal Chemistry Synthesis of Grazoprevir

The medicinal chemistry synthesis of grazoprevir initiated at 4-methoxy-phenylenediamine hydrochloride (**7**) which was heated with diethyl oxalate to give dihydoxyquinoxaline **8** in 69% yield (Scheme 1). Regioselective chlorination with thionyl chloride followed by addition to activated cis-hydroxyproline derivative **9** (*20*) and deprotection then led to coupled intermediate **10** in 69% yield. Straightforward amide coupling with **11**, prepared in enantiopure form in seven steps, produced chloroquinoxaline **12**. Next, while vinylation of the ring-chloride proceeded smoothly, the key ring-closing metathesis (RCM) using the Zhan 1B catalyst (*21*), was problematic and the desired macrocycle was isolated in only 25% yield. Reduction of the resulting olefin gave macrocycle **13**, which was readily converted to **1** (grazoprevir, MK-5172) in 89% yield by hydrolysis and coupling to the known cyclopropylamino acyl-sulfonamide **14** (*22*).

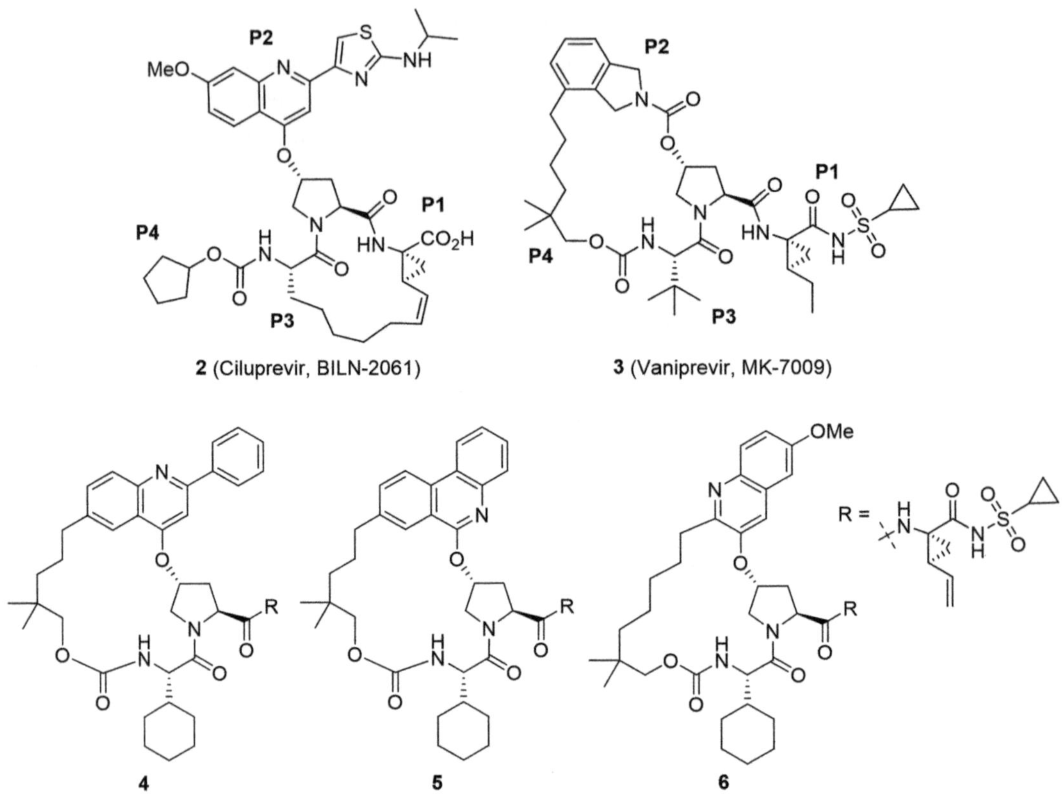

Figure 2. HCV NS3 protease inhibitors.

Reagents and conditions: (a) diethyl oxalate, Et_3N, 150 °C; (b) thionyl chloride, DMF, 110 °C; (c) **9** (*20*) Cs_2CO_3, NMP; (d) HCl, dioxane; (e) **11**, HATU, *i*-Pr_2NEt, DMF; (f) potassium vinyltrifluoroborate, Et_3N, dichloro[1,1-bis(diphenylphosphino) ferrocene] palladium (II) chloride, EtOH; (g) dichloro[1,3-bis(2,4,6-trimethylphenyl)-2-imidazolidinylidene][[5-[(dimethylamino)sulfonyl]-2-(1-methylethoxy-O)phenyl]methylene-C]ruthenium(II) (Zhan 1B catalyst) (*21*), DCE; (h) H_2, 10% Pd/C, MeOH, dioxane; (i) LiOH, THF, water; (j) **14** HCl salt (*22*), TBTU, *i*-Pr_2NEt, DMAP, DMF.

Grazoprevir exhibited an excellent profile regarding broad genotype and mutant enzyme potency, cellular activity, and liver exposure in preclinical species. The compound was approved for clinical development following preclinical toxicology studies. The grazoprevir program ultimately benefited from several chemistry and medicinal chemistry decisions and strategies (*23*). Analysis of bound structures, and specifically in silico methods to analyze structure when no crystal structure was available, was critical in the formulation of the P2–P4 macrocyclization hypothesis. The flexibility and reliability of the RCM reaction allowed for extensive and rapid evaluation of the macrocyclic framework. At several points during the invention of grazoprevir, novel designs were also enabled by the discovery and application of new synthetic routes and availability of key advanced intermediates, which made the exploration of chemical space more facile. Following the invention of grazoprevir, several other P2–P4 macrocyclic HCV NS3/4a protease inhibitors have been more recently approved by the FDA (*24*, *25*).

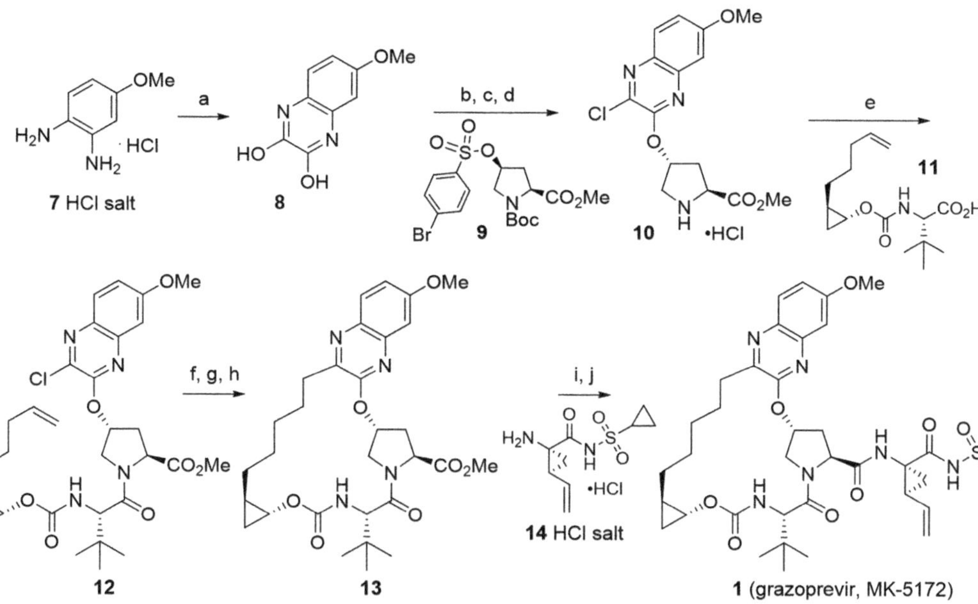

Scheme 1. Medicinal chemistry synthesis of grazoprevir (MK-5172).

Development of Manufacturing Process through Chemistry Evolution

Once a drug candidate is selected for further preclinical and clinical development, the efforts on synthetic chemistry shift to developing a practical route suitable for large-scale preparation. Rather than emphasizing the flexibility of a synthetic route allowing quick access to analogs for SAR studies during drug discovery phase, the chemical development at this stage is target-molecule driven with an ultimate goal of developing the most efficient synthesis of the selected drug candidate. However, the initial process route developed to quickly initiate and support the early safety and clinical studies, which is often achieved under urgent and compressed timeline, usually does not meet all the criteria to be considered as a manufacturing route. From this point of view, the process to develop a satisfactory manufacturing route requires chemistry evolution, which could be effectively realized through innovation to overcome the drawbacks of the existing chemistry and to discover the ultimate route and process. In general, a good manufacturing process evolved through chemical development should possess the core elements illustrated in Figure 3.

Retrosynthetic Analysis

Grazoprevir (**1**) contains seven stereogenic centers and a unique *trans* cyclopropoxy moiety with a carbamate linkage incorporated in the 18-membered macrocyclic ring. Retrosynthetic analysis with various C–C, C–O and C–N bond disconnection strategies logically led to building blocks **14** – **21** (Scheme 2). Among these building blocks, sulfonamide **14**, which can be prepared from amino-acid **20** and cyclopropyl sulfonamide **21**, has been utilized in the pharmaceutical industry at large-scale (*26, 27*). Quinoxaline **15** (X, Y = OH) can be readily accessed by condensation of 3,4-diaminoanisole with oxalate (*28*).

In addition, amino ester **17** and amino acid **19** are commercially available in large quantities. Based on our retrosynthetic assessment, it was attractive to incorporate the hydroxyproline moiety to the grazoprevir skeleton via an S_NAr displacement of dichloroquinoxaline **15** (X, Y = Cl) with Boc hydroxyproline ester **17**, as building blocks **14** and **16** could be installed by amidation.

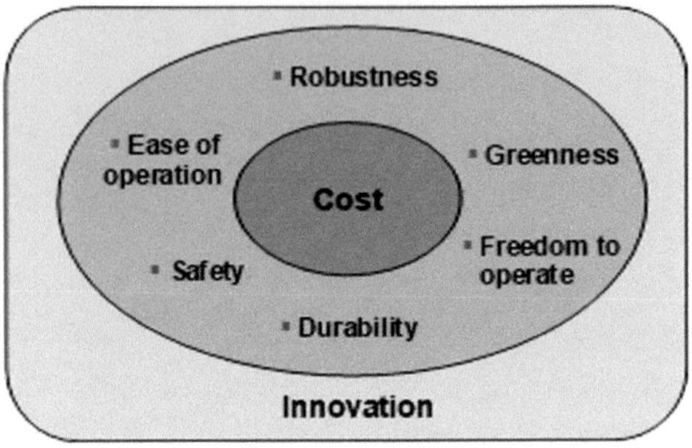

Figure 3. Examples of essential elements of a good manufacturing process.

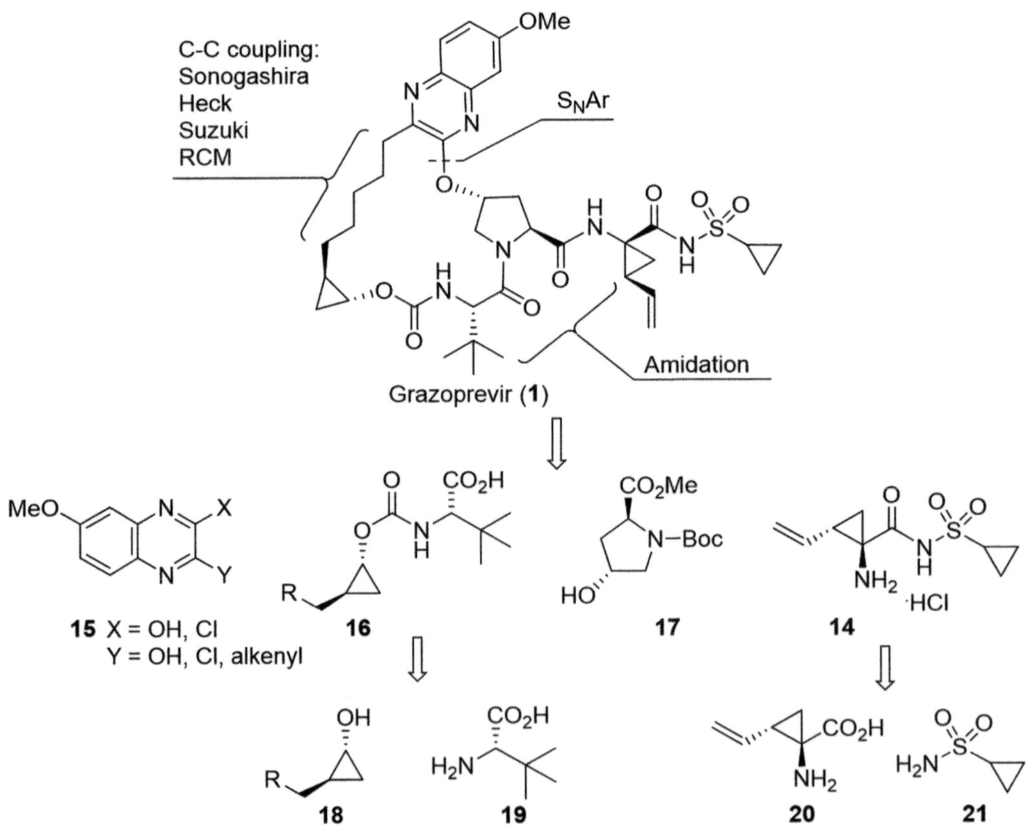

*Scheme 2. Retrosynthetic analysis to **1**.*

The discovery of the function of the unique cyclopropoxy moiety in grazoprevir was achieved through extensive studies based on an understanding the binding interaction of the HCV virus with drug candidates at the level of molecular structure (*18*). The incorporation of the *trans* cyclopropoxy moiety in the macrocycle ring resulted in significant improvement of the potency against the genotype 1-3 NS3/4a protease enzymes as well as clinically relevant mutant enzymes (*18*). But it also raised the chemical challenge of synthesizing grazoprevir, because the methods for practical asymmetric preparation of cyclopropanols, especially 2-substituted cyclopropanols, were lacking at the time (*29–35*). To support the development of an effective synthesis of grazoprevir suitable for large-scale preparation, we required a practical and efficient synthesis of building block **18** containing the *trans* cyclopropanol moiety.

In general, isolation of crystalline solid intermediates on large-scale synthesis, particularly at a late stage, was crucial for rejection of process impurities. This provided a practical means for the requisite process quality control to meet the high-quality standards and agencies' regulations in the pharmaceutical industry. However, cyclopropanol **18** (R = alkene or alkyne) is an oil, and purification and isolation of **18** could be problematic on large-scale. In addition, for the sake of obtaining an isolatable solid intermediate, it was not a preferred practice to derivatize the intermediate. This sacrifices efficiency by introducing additional process steps and operations. As a compromise, we envisioned that purification of carboxylic acid **16**, derived from building blocks **18** and **19**, could be practically realized through isolation of a suitable crystalline salt. Given an overall consideration of the synthetic strategy and convergence, compounds **14** – **17** became our four core building blocks of choice.

Regarding the construction of the macrocycle skeleton, we proposed that several strategies of C–C bond coupling including sp^2–sp^3, sp^2–sp^2, and sp^2–sp couplings, and ring closure metathesis (RCM) with building block **16** should be investigated. The strategy for closure of the grazoprevir macrocycle would be ultimately responsible for the sequence of events and types of reactions employed to introduce building blocks **15**, **16** and **17**. Building block **14** was most reasonably installed through amidation at the last stage of synthesis.

First-Generation Route to the Cyclopropanol-Leucine Building Block

The alkene cyclopropanol-leucine acid **11** utilized in the discovery route (Scheme 1) was prepared over seven steps (Scheme 3) ((*18*), *Supporting Information*). Simmons–Smith cyclopropanation of enol ether **23** afforded racemic cyclopropanol **24**. Enzymatic esterification resolution followed by column chromatography purification and transesterification yielded **25** in 50% ee. Carbamate formation of **25** with leucine **26** gave optically pure ester **27** after column chromatography purification. Finally, hydrolysis of **27** yielded alkene acid **11** in 17% overall yield.

We anticipated that an alkyne-substituted cyclopropanol **18** (R = alkynyl) would be a more suitable building block. This could be easily transformed to the corresponding analogs, providing the requisite synthetic flexibility to explore multiple C–C bond coupling approaches including sp^2–sp^3, sp^2–sp^2, and sp^2–sp couplings, and ring closure metathesis (RCM) (Scheme 2). Thus, we concentrated our efforts on developing a practical synthesis to access cyclopropanol building block **28** (Scheme 4) and analogs with varied length of terminal or internal alkyne substituents.

Considering the state of the art for the asymmetric synthesis of cyclopropanols at the time (*29–35*), the development of a scalable asymmetric synthesis of alkyne acid **28** presented a significant challenge. The first-generation large-scale preparation of alkyne acid **28** relied on an enzymatic resolution to obtain enantiomerically pure cyclopropanol **36**. We were confident that the efficiency of enzymatic resolution, which was encountered in the preparation of alkene cyclopropanol-leucine acid **11** (Scheme 3), could be improved through enzyme choice and process optimization. Thus, we could eliminate impractical column chromatography purification for a large-scale application.

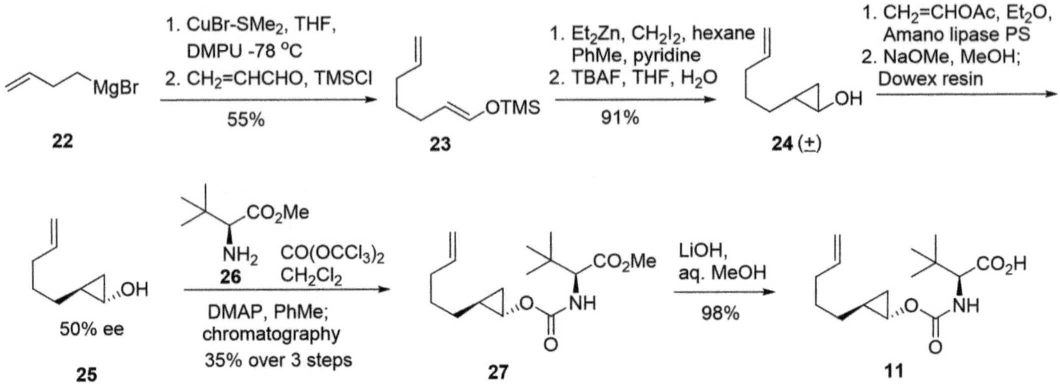

*Scheme 3. Synthesis of alkene acid **11**.*

After evaluation of several approaches, we selected chloropentyne **29** as the starting material for the preparation of **28** (*36*). Note that *E*-vinyl boronate **30** was commercially available, but it was expensive and possessed long lead-times for shipment. Hydroboration (*37*) of **29** with pinacol borane in the presence of a catalytic amount of BH_3•THF (5 mol%) and cyclohexene (10 mol%) prepared **30** in 84% yield. The slow addition of cyclohexene allowed control of the exothermic hydroboration at the desired reaction temperature (0 °C to ambient temperature) and avoided the hazardous accumulation of the reagents. As such, the hydroboration could be safely carried out on large-scale.

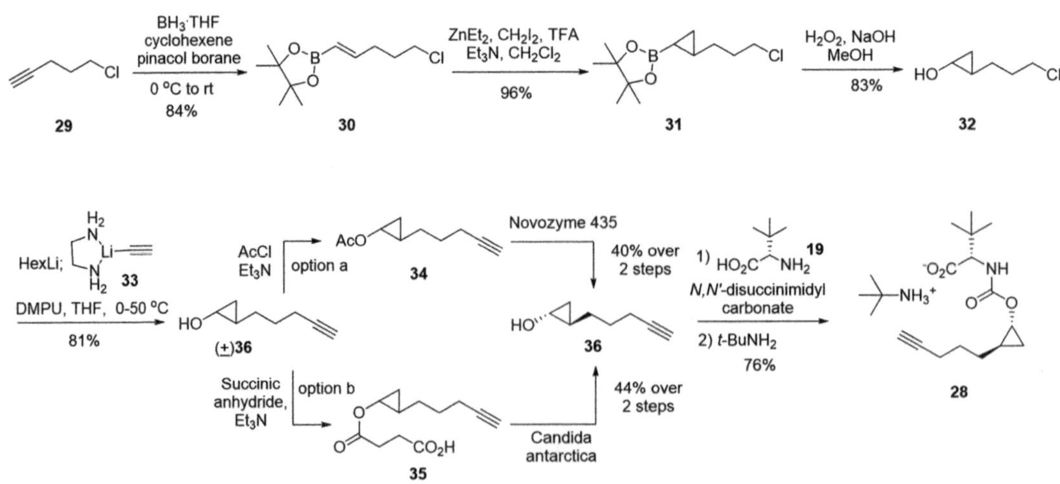

*Scheme 4. First generation of alkyne acid **28**.*

Racemic Simmons-Smith cyclopropanation under Shi's modified conditions (*38*) provided boronate **31** in 96% yield. The high conversion (> 98%) and clean purity profile (> 95% purity) represented a significant improvement over the literature precedent using $Zn(CH_2I)_2$ (*39*). However, large exotherms occurred upon sequential addition of the reagents during cyclopropanation. To address the safety concern for the implementation on pilot plant scale, the cyclopropanation was examined further using differential scanning calorimetry (DSC) and RC-1 calorimetry. DSC exhibited no exotherms in the reaction slurries after additions, quench, and concentration. RC-1 calorimetry was used to examine the heat transfer of the reaction during the reagent additions and to determine a safe addition rate and vessel-jacket temperature (*36*). As a result, at pilot plant scale, the Simmons–Smith cyclopropanation proceeded smoothly over multiple batches.

Oxidation of boronate **31** produced cyclopropanol **32**, which was deprotonated with hexyl lithium followed by coupling with lithium acetylide – ethylene diamine complex **33** in DMPU/THF to afford racemic coupling product (±)**36**. Although lithium acetylide – ethylene diamine complex **33** is a relatively air-stable and commercially available solid, the thermal stability of this complex in the reaction as well as the conversion of the requisite coupling reaction were solvent dependent. Thermal stability studies suggested that DMPU would be a suitable solvent for safe scale-up of the acetylide coupling reaction at 50 °C without thermal decomposition. After further optimization, DMPU/THF became our solvents of choice, which allowed the coupling reaction to be carried out at 50 °C in 81% yield.

Having addressed the issue of reaction solvent, our attention turned to the concern of acetylene generation during the reaction. Acetylene is very sensitive to pressure, high and even moderate temperature, static electricity, and/or mechanical shock (*40*). The explosive limits can range from 2.5 to 81% in air (*41*). In order to understand and quantify the timing and the amount of acetylene produced during this reaction, ReactIR was used to measure the evolution of acetylene gas in the headspace. Suspending the complex **33** in the optimal solvent DMPU resulted in $< 1\%$ acetylene release in the headspace. Under the optimized reaction conditions, low levels of acetylene were released upon addition of the lithium alkoxide solution of **32** to the suspension of lithium acetylide **33** in DMPU and the subsequent heating to 50 °C. In addition, the electrical classification of acetylene is Class 1, Group A in the United States and Group IIC in the EU. Based on these results, safely performing this chemistry on kilogram scale in countries that use the American electrical classification system would require a pilot plant rated for Class 1 Group A gases, and in countries that use the European system would require a pilot plant capable of handling class IIC gases. If the reactor and/or its venting pathways are not rated for acetylene, the amount of acetylene in the headspace should be limited to $\leq 1\%$ for single batch and $\leq 0.6\%$ for regular production, placing significant restrictions on the broad utility of this chemistry at plant scale. In addition, care should be taken when using **33** on any scale to avoid potentially dangerous explosions from the buildup of acetylene.

Enzymatic resolution was initially carried out on acetate **34**, which was prepared by esterification of racemic alcohol **36** with AcCl and Et_3N. Selective enzymatic hydrolysis of the desired enantiomer of the ester gave 96% *ee* of **36** in crude. However, the *ee* of isolated **36** (40% yield over two steps) decreased to 92% due to the partial hydrolysis of the undesired acetate enantiomer during silica gel column chromatography. To overcome this *ee* erosion issue and to remove the need of column purification for rejection of the undesired acetate, a practical enzymatic resolution process was developed. Treatment of racemic **36** with succinic acid anhydride–Et_3N afforded acid **35**. Enzymatic resolution of **35** with *Candida antarctica* delivered cyclopropanol **36** in 44% yield over two steps with >97% ee, while the undesired enantiomer was rejected through basic aqueous workup without the need for column chromatography purification. Treatment of **36** with (*L*)-*tert*-leucine **19** in the presence of *N,N'*-disuccinimidyl carbonate afforded carbamate **28**, which was then isolated as its crystalline *t*-$BuNH_2$ salt (*42*).

This first-generation route provided a means to prepare kilogram quantities of alkyne acid **28** to support the early safety and clinical studies of grazoprevir. While effective, this process was deemed unsuitable for larger-scale use due to inefficient steps together with the safety concerns for:

- use of acetylide
- the control of the exotherm of hydroboration
- Simmons–Smith cyclopropanation.

Manufacturing Route to the Cyclopropanol-Leucine Building Block

Despite the availability of various methodologies to prepare substituted/functionalized cyclopropanes, at the inception of this work literature precedent for the asymmetric synthesis of a cyclopropanol was still lacking (*29–35*). To tackle this problem, we explored numerous approaches for the direct preparation of cyclopropanols including Walsh's cyclopropanation via methylene di-iodozinc (*32*), asymmetric Simmons–Smith cyclopropanation of enolized substrates (*33*), the Kulinkovich reaction (*34, 35*), and 3-exo-tet ring-closing cyclopropanation (*43*). However, results from each of these strategies were unsatisfactory in terms of diastereo- and/or enantio-selectivity, yield, and practicality for large-scale production.

Rather than focusing on direct cyclopropanation for the preparation of cyclopropanols, we considered two different transformations for the second-generation route to carbamate **28**: first asymmetric cyclopropanation through a [1,3]-phosphorus-Brook rearrangement (*44*) and then a stereospecific Baeyer–Villiger oxidation of a cyclopropyl ketone, (Scheme 5). In this approach, introducing the acetylene functional group into cyclopropropanone **38** before Baeyer–Villiger oxidation was unattractive and cost-ineffective, because a multiple step synthesis would be required to prepare chiral alkyne epoxide precursor **41** (*45*). Alternative strategies using stepwise installation of the terminal alkyne group through carbon-carbon coupling also did not improve the synthetic efficiency (*46*). This might require a hazardous acetylide coupling process, which had been used in the first-generation route and was also deemed unfit for large-scale preparation. Thus, readily accessible alkene epoxide **40** became our precursor of choice. We envisioned the alkene functionality could be subsequently transformed to the requisite alkyne group in building block **28**.

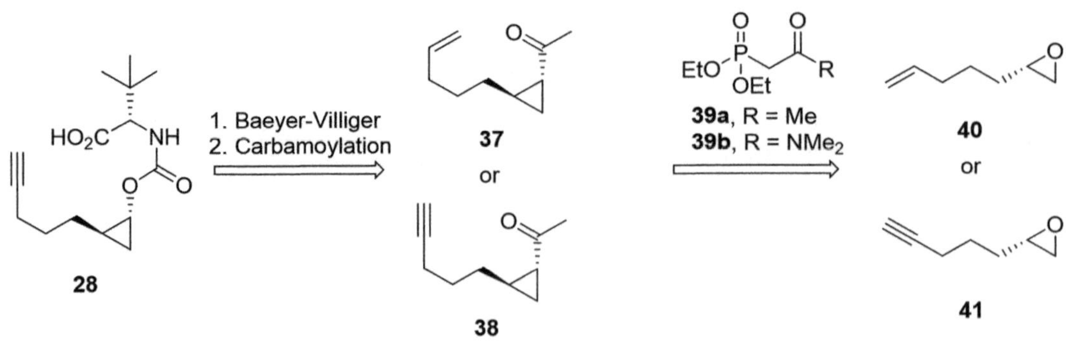

Scheme 5. Second-generation retrosynthetic analysis of ***28****.*

The second-generation route (*47*) initiated with an addition of commercially available 3-butenylmagnesium bromide to (*S*)-epichlorohydrin (**42**) in the presence of 1 mol % CuI to provide chlorohydrin **43** in 94% assay yield with near perfect regioselectivity (*48*). Subsequent treatment of **43** with methyl ketone phosphonate **39a** did give cyclopropyl methyl ketone **37** directly, albeit in low yield due to the competitive side reactions of the corresponding methyl ketone intermediates under the basic reaction conditions. To overcome this incompatibility, we developed a practical preparation of methyl ketone **37** by using amide phosphonate **39b** to suppress side reactions during cyclopropanation and to afford the desired cyclopropyl amide **44** in 87% assay yield.

In addition, introducing the sterically bulky dimethyl amide group proved beneficial in the diastereoselective intramolecular ring closure of **46**, affording the desired cyclopropane **44** in a >200:1 *trans*:*cis* ratio. Treatment of the crude cyclopropyl amide **44** with MeMgCl in MeTHF at 55 °C afforded a corresponding stable magnesium hemiaminal intermediate, which upon aqueous workup gave the desired methyl ketone **37** in 95% assay yield without any epimerization (Scheme 6).

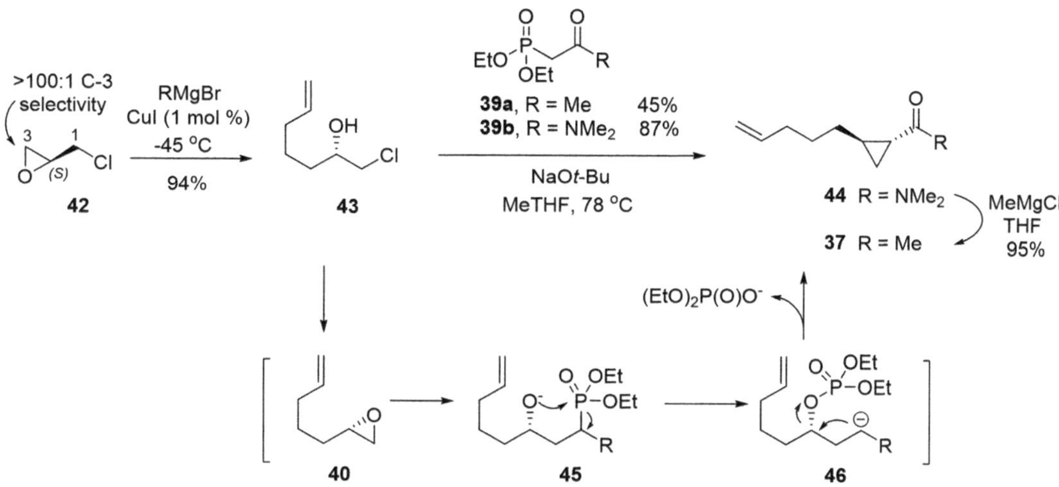

Scheme 6. Cyclopropanation via phosphorus-Brook rearrangement.

The terminal double bond in **37** was then converted to the corresponding dibromide **47**, which eliminated the incompatible reactivity of the unsaturated C–C bond under subsequent Baeyer-Villiger oxidation conditions. Furthermore, the dibromide served as a masked acetylene precursor, which allowed transforming alkene **37** to the desired alkyne functionality through a bromination-elimination sequence. Thus, treatment of **37** with Py•HBr_3 in MeCN afforded the desired dibromo ketone **47** in 96% assay yield (Scheme 7). Treatment of **47** with the strong oxidant, CF_3CO_3H, which was generated in situ by treatment of $(CF_3CO_2)_2O$ with urea-H_2O_2 complex (UHP), afforded optimal reactivity in the Baeyer-Villiger oxidation and resulted in a mixture of the alcohol and acylated products **48a-c** in 91% assay yield. The ratio of the mixture of components of **48a-c** did not impact the performance of the subsequent synthesis. The use of a stable UHP complex together with proper nitrogen sweep to control the level of oxygen in the headspace permitted carrying out the Baeyer-Villiger oxidation safely at >200 kg scale.

Treatment of the mixture of bromides **48a-c** under zipper reaction conditions with *N*-lithium diaminopropane (Li-DAP) yielded the desired alkyne **36** in ~80% assay yield. However, the elimination resulted in a thick triphasic slurry (solid phase and two organic layers), which we viewed as a robustness risk toward implementation on large-scale. Indeed, about 5 - 10% of the corresponding bromoalkene intermediates were not consumed under the heterogeneous Li-DAP conditions.

Since the discovery of alkyne zipper reaction in 1975 (*49*), it has been widely applied to prepare terminal alkynes (*50*). However, the evolution of the zipper reaction from a laboratory reaction toward a more practical process has received comparatively little attention, and Li-DAP remained the most commonly used base at the time (*51–53*). The use of other bases such as *t*-butoxide, $LiNH_2$, LDA, LiHMDS, and DBU failed to provide improved results. In addition, we found that the use of sodium or potassium bases led to significant decomposition of cyclopropanol **36** via a rearrangement to the corresponding aldehyde **49**.

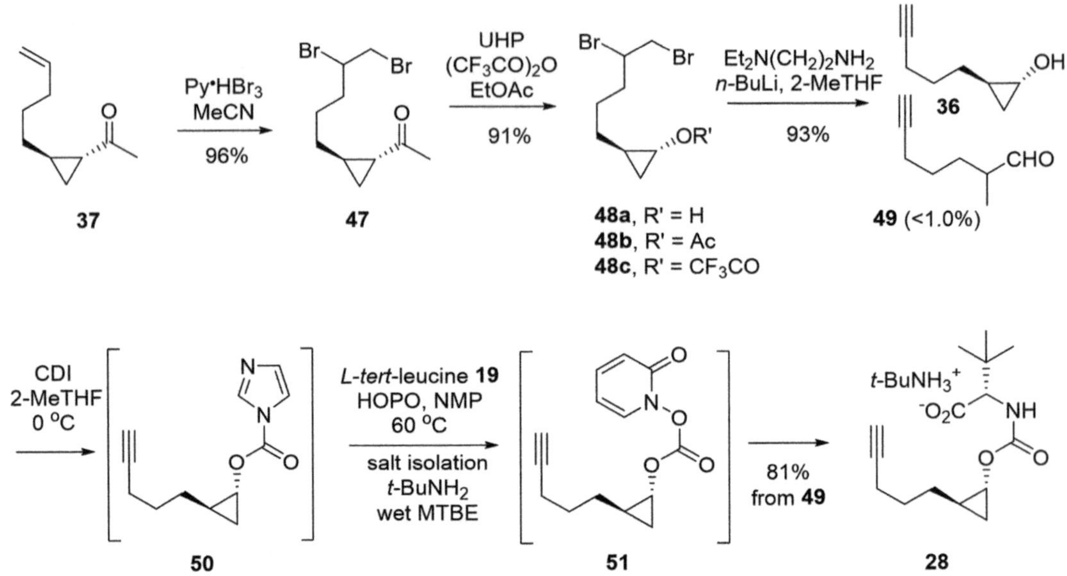

Scheme 7. Preparation of alkyne acid ***28****.*

To overcome the heterogeneity issue, we explored the reactivity and physical properties of $LiNH(CH_2)_nNR_2$. We postulated that modification of the *N*-alkyl groups of these Li-DAP analogs could be beneficial, allowing the elimination to be performed in a homogeneous solution (*54*). Unlike the use of Li-DAP, the elimination of **48** with $LiNH(CH_2)_2NEt_2$ or $LiNH(CH_2)_3NEt_2$ was then carried out in a homogeneous solution in 2-MeTHF or THF at -15 °C, which provided a robust control of the reaction performance and resulted in near complete consumption of the corresponding bromoalkene intermediates (<1%).

After an aqueous workup at 0 - 5 °C, which suppressed the formation of byproduct **49** to <1.0%, the desired alkyne alcohol **36** was obtained in 93% assay yield (Scheme 7). The observed *ee* of **36** confirmed that the Baeyer-Villiger oxidation gave 100% retention of the stereochemistry. To the best of our knowledge, the use of $LiNH(CH_2)_2NEt_2$ or similar lithium amides for the zipper/elimination reaction has not been reported previously (*45*).

We also explored the scope of using $LiNH(CH_2)_2NEt_2$ for elimination, which seemed to be broad. Substrates bearing a variety of functional groups, including silyl ethers, THP, silane, and halo-substituted aromatic ring, were well-tolerated, providing the corresponding alkyne products in high yields (*47*).

The classical conditions for carbamate formation utilized in the first-generation route only allowed a modest yield of compound **28** (Scheme 4) (*36*). To address this issue, we developed an efficient process to form carbamate **28** with the hindered (L)-*tert*-leucine (**19**). Under the optimal process conditions intermediate **50** was converted to activated pyridone intermediate **51** in the presence of a catalytic amount of 2-hydroxypyridine *N*-oxide (HOPO). As such, carbamate formation with (L)-*tert*-leucine was significantly improved and afforded **28** in 90% assay yield (Scheme 7). To the best of our knowledge, the use of HOPO to promote carbamate formation has not been reported previously. After an acid-base aqueous workup, acid **28** was isolated as its *t*-$BuNH_2$ salt from wet MTBE (~1% water) in 81% isolated yield and ≥97 HPLC area% purity with excellent impurity rejection.

In summary, we have developed an asymmetric and practical synthesis of *trans*-cyclopropoxy building block **28**. Starting from (*S*)-epichlorohydrin, the seven-step through-process is carried out in its entirety using the crude stream with non-isolatable solid intermediates, building block **28** *t*-$BuNH_2$ salt is efficiently prepared in 51% overall yield and ≥97% purity without recrystallization. In comparison with the first-generation route (Table 1), the asymmetric second-generation synthesis also addresses process safety and hazardous concerns, and possesses the practical elements required for manufacturing process. The second-generation process has been successfully implemented at manufacturing scale.

Table 1. Comparison of First and Second Generation Syntheses of 28

	1st Gen. Synthesis	*2nd Gen Synthesis*
Synthesis	Racemic with resolution	Asymmetric
Robustness	Scale dependent	Yes
Safety/hazardous concerns	Hydroboration, Simmons–Smith reaction, acetylide coupling	None
Overall yield	9 - 11%	51%

Early Route Scouting to Construct the Grazoprevir Macrocyclic Skeleton

The strategy to construct the grazoprevir macrocycle is ultimately responsible for the sequence of events and types of reactions employed to assemble building blocks **15**, **16** and **17** (Scheme 2). With the second-generation route to building block **28** in hand, cyclopropanol building block **16** with varied R substituents could then be effectively accessed by applying similar chemistry. Thus, we were in a good position to explore various coupling strategies outlined in the retrosynthetic plan (Scheme 2) and to identify the most efficient chemistry to construct the macrocyclic skeleton.

The macrocyclic skeleton in the discovery route was realized through a ring closure metathesis (RCM) reaction, albeit in low yield (25%, Scheme 1) (*18*). During the further efforts to improve the RCM reaction, we discovered that the inefficiency of the RCM reaction was due to the binding affinity of the catalyst with substrate to form a stable chelated Ru complex (*55*). The yield of macrocyclization depended greatly on the relative length of olefin side chains. We found that the catalyst sequestering could be overcame by using a substrate with an extended, pendant alkene chain off the quinoxaline moiety, affording a significant improvement of the RCM yield (71%, Figure 4) (*55*). Given an overall consideration of cost and freedom-to-operate, however, the application of RCM for the preparation of grazoprevir did not provide the necessary benefits suitable for large-scale preparation. Thus, we began our search elsewhere.

Attempts to form the grazoprevir macrolactam ring via an intramolecular Heck or Sonogashira coupling resulted in unsatisfactory results in terms of selectivity, yield, and/or cost of ligands (Figure 4) (*56*). The feasibility of constructing the backbone of the 18-membered macrocycle via intramolecular sp^2-sp^3 coupling was also thoroughly investigated. Under Suzuki-Miyaura coupling conditions, macrocycle **13** was obtained in 22% yield (57). Alternatively, Molander's photoredox coupling gave **13** in only 10% yield (57). Attempts to improve the cyclization yield by either pathway through extensive screening of Pd and Ni sources and ligands under numerous reaction conditions proved unsuccessful.

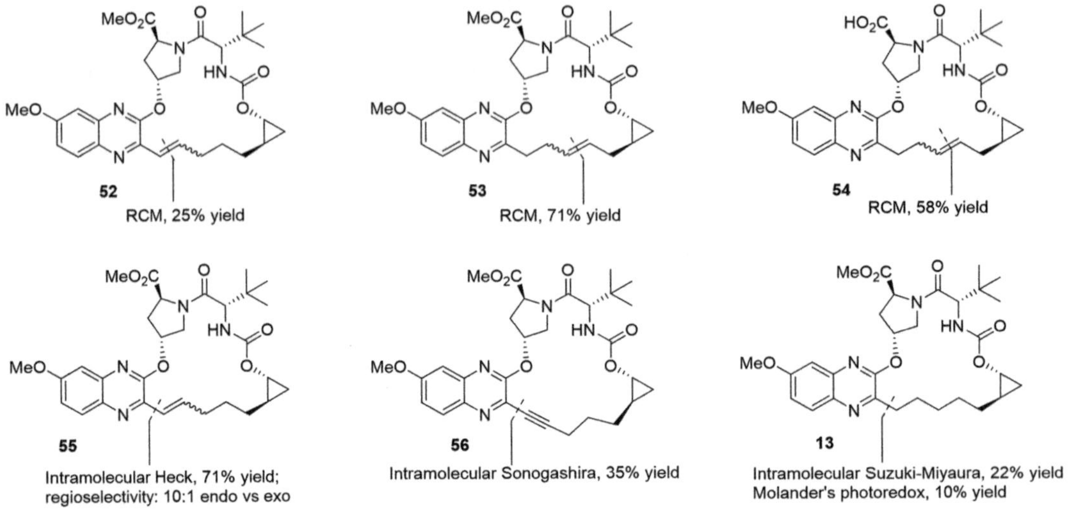

Figure 4. Summary of some early attempts to prepare the macrocyclic skeleton corresponding to **1**.

Given the inefficient syntheses obtained with RCM, sp^2-sp^3 couplings and intramolecular sp^2-sp^2 and sp-sp^2 couplings, we turned our attention to applying an intermolecular sp^2-sp coupling to install building block **28** with quinoxaline building block **15**. The success of this strategy led to two generations of syntheses of grazoprevir: one was applied to support the early drug development; and the other has been implemented for commercialization.

First-Generation Large-Scale Synthesis of Macrocyclic Acid 57

Starting from dichloroquinoxaline **58**, an S_NAr reaction with *N*-Boc-4-hydroxproline methyl ester (**17**) in the presence of Cs_2CO_3 in DMF at room temperature was applied to form the desired ether linkage. Although the initial results showed an encouraging regioselectivity in favoring the desired product (**59** : the corresponding regioisomer = 95:5), the significant amount of double-addition byproduct **60** formed under these conditions was problematic during the isolation without using chromatography. Subsequent optimization indicated the formation of impurity **60** could be suppressed to 2% by carrying out the S_NAr reaction in the presence of DBU in DMAc at 50 °C in 88% assay yield. After workup, crystalline **59** was isolated from MTBE/*n*-heptane in 71% yield (Scheme 8).

The strategy applied in the first-generation route to form the macrocyclic ring called for the use of a Sonogashira reaction to couple building block **28** and chloroquinoxaline **59** followed eventually by an intramolecular amidation with the proline moiety to form the ring (Scheme 8). We applied high throughput experimentation to optimize the Sonogashira reaction catalyst/ligand combinations, solvents, temperature, and bases. The optimal conditions involved addition of **59** and **28** to a slurry of $Pd(OAc)_2$ (3 mol %), $P(t\text{-}Bu)_3HBF_4$ (6 mol %), and K_2CO_3 (2.5 equiv) in a solvent mixture of CPME/MeCN (2.4:1) and heating at 85 °C for 2.5 h. After an aqueous workup, unisolated product **61** was obtained in 98% assay yield. Hydrogenation of crude **61** over Pd/C in *i*-PrOAc/MeOH furnished the reduced product **62** in 89% assay yield. Both Sonogashira product **61** and hydrogenated intermediate **62** were amorphous solids. Without further purification, the Boc group was removed in the presence of $PhSO_3H$ in acetonitrile, resulting in benzenesulfonate salt **63**. Attempts to crystallize this salt were unsuccessful. Therefore, upon neutralization with Hünig's base, the crude stream was slowly charged to a solution of (1-[bis(dimethylamino)methylene]-1H-1,2,3-triazolo[4,5-b]pyridinium 3-oxide hexafluorophosphate (HATU) in MeCN to provide cyclized **13**, which was finally isolated as a crystalline solid in 55% yield over four steps. Hydrolysis of ester **13** with LiOH in biphasic aqueous THF gave the desired acid **57**. After acidic aqueous workup, crystalline anhydrate macrocyclic acid **57** was obtained from EtOAc – hexanes in 98% yield.

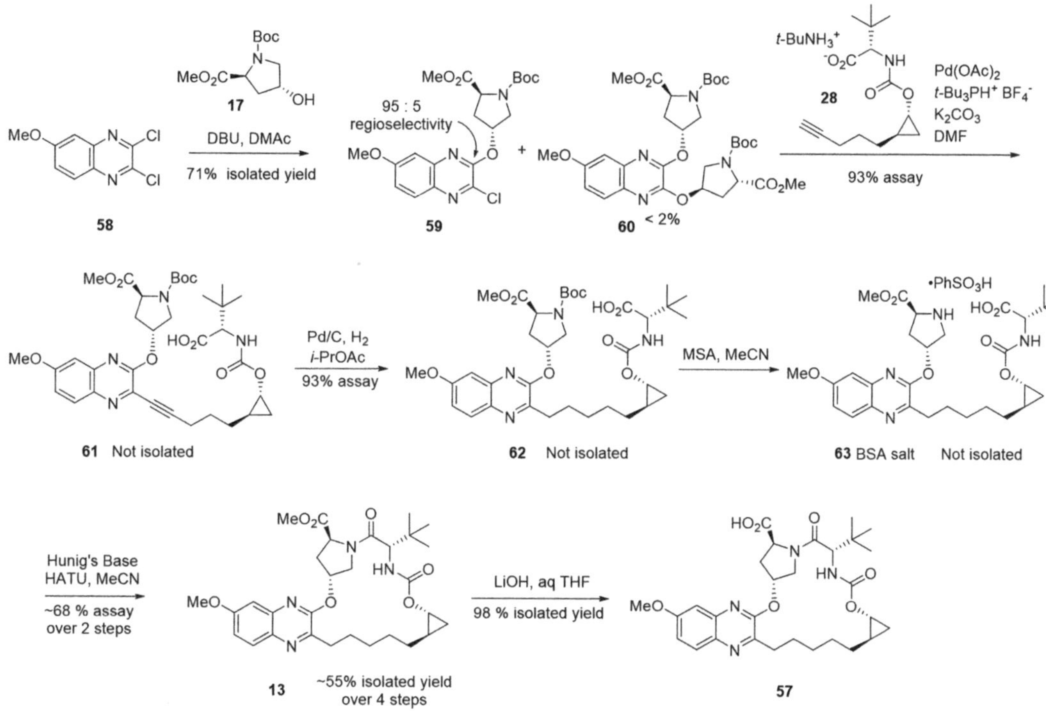

Scheme 8. First-generation synthesis of macrocyclic acid **57**.

The first-generation route to the penultimate macrocyclic acid **57** provided a means to prepare large quantities (>100 kg) of grazoprevir to support the early safety and clinical studies. However, the strategy to construct the grazoprevir macrocycle was still deemed unsatisfactory. While the initial preparation of intermediate **59** through a Sonogashira coupling followed by hydrogenation proved to be effective, the main liability of the first-generation route was the incompatible reactivity of the electron-rich quinoxaline ring in **62** under various conditions attempted for the requisite removal of the Boc group. Numerous impurities were formed through reaction of the quinoxaline moiety with the cationic species generated under the deprotection conditions, adversely impacting the synthetic efficiency, yield, and overall process robustness. In line with this explanation of quinoxaline electron-richness, attempts to change the reaction sequence order to perform the Boc deprotection followed by hydrogenation, as depicted in Scheme 9, did not suppress the byproduct formation during the Boc deprotection.

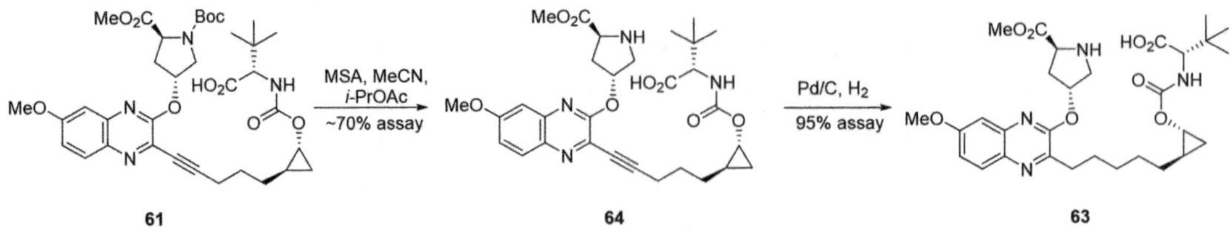

Scheme 9. Alternative approach by de-Bocing first.

Final Manufacturing Route to Macrocyclic Acid 57

This competitive side reaction issue was problematic when intermediates containing the electron-rich quinoxaline moiety were subjected to acidic Boc deprotection conditions. We reasoned that it could be more effective to remove the Boc group (carried from commercially available proline ester **17**) earlier in the synthesis, while the quinoxaline moiety in intermediate **59** maintained its electron-deficient nature.

By carrying out this early Boc deprotection approach, we initially encountered another new set of challenges. The remaining chloride in the electron-deficient quinoxaline ring led to a competitive S_NAr reactivity. The formation of the corresponding quinoxalinol or ether **66** via S_NAr displacement of the chloride by nucleophilic solvents (water or alcohols) became significant, when the Boc deprotection of **59** was carried out under acidic aqueous conditions or in alcohol solvents. It was believed that the protonation of the chloroquinoxaline nitrogen increased its electron-deficient nature and sufficiently lowered the barrier for S_NAr reactivity. For the same reason, acidic aqueous workup of the Boc deprotection reaction proved undesirable.

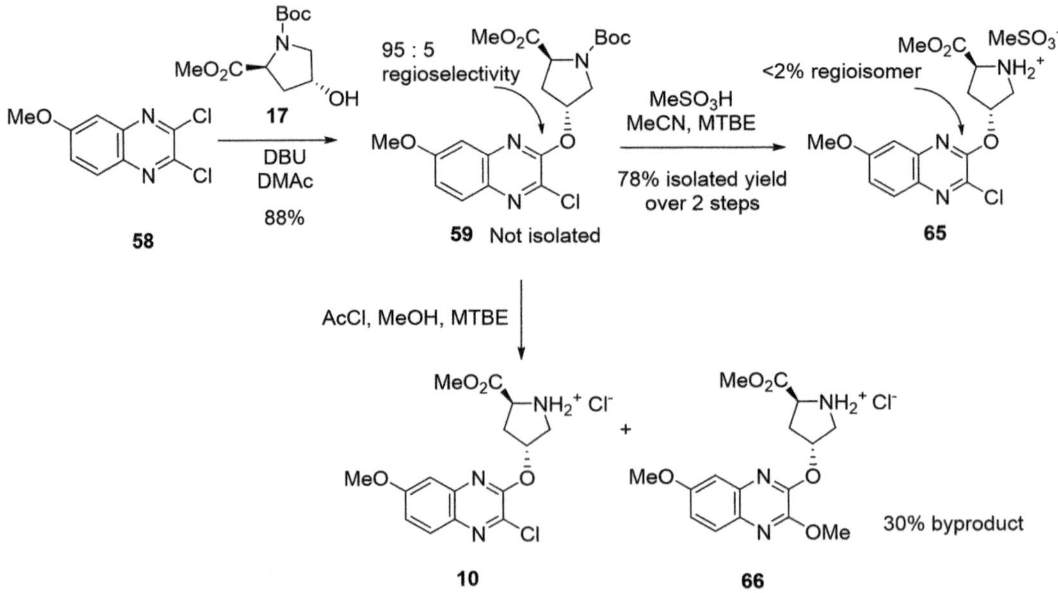

*Scheme 10. Boc deprotection and the through-process to proline chloroquinoxaline MSA salt **65**.*

To address this incompatibility, we studied various acidic Boc deprotection conditions in non-nucleophilic solvents, in an attempt to identify an isolable salt without any aqueous workup. Ultimately, a combination of $MeSO_3H$ together with acetonitrile as a non-nucleophilic solvent proved crucial to suppressing the S_NAr side reactions, resulting in a high-yielding Boc deprotection. Furthermore, the Boc-deprotected methanesulfonate salt **65** directly crystallized from the reaction mixture as an acetonitrile solvate. As such, the chloro-substituted, electron-deficient quinoxaline ring remained intact. With this key result in hand, a robust, two-step through-process of a S_NAr coupling followed by Boc deprotection was developed. The isolation of **59**, which was obtained at the cost of a ~20% yield loss to the mother liquor in the first-generation route (Scheme 8), was avoided. Boc deprotection of a crude stream of **59** resulted in **65** in 78% isolated yield over two steps with a good rejection of the regioisomer (98:2) via a simple filtration.

Under the basic conditions required for sp^2-sp coupling to incorporate building block **28** in the quinoxaline ring, the free base of **65** was generated in situ. With the nitrogen in the proline moiety unprotected, the thermostability of free base of **65** proved concerning. Attempts to apply the Sonogashira conditions developed for the preparation of Boc protected proline **61** in the first-generation route (Scheme 8) resulted in complete decomposition of **65**. However, we were glad to discover that a mild and efficient Sonogashira coupling reaction required to overcome the thermo-instability of the free base could be carried out in an alcohol solvent. Presumably the reduction capability of alcohol solvent improved the catalyst stability and reactivity (58, 59), which was evidenced by the significantly increased catalytic turnover number at mild temperature.

In contrast, carrying out the Sonogashira coupling in alternative solvents under various conditions led to unsatisfactory yields and/or poor conversions. With methanol as solvent, the Sonogashira reaction could be effectively carried out at 35 °C, the catalyst loading was reduced to as low as 0.05 mol% $(Ph_3P)_2PdCl_2$ together with 0.1 mol% CuI. It is noteworthy that air-stable $(Ph_3P)_2PdCl_2$ is one of the least expensive, commercially available palladium catalysts. Under the optimized conditions, the Sonogashira coupling of **65** and **28** in the presence of Hünig's base afforded the desired product **67** in 98% assay yield.

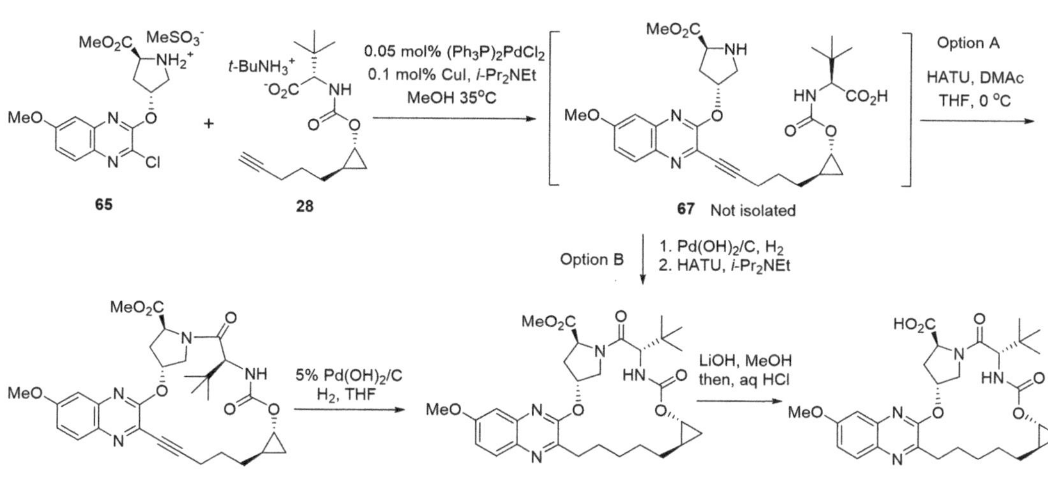

Scheme 11. Manufacturing route to macrocyclic acid.

Competitive intermolecular amidation during macrolactamization could lead to the formation of corresponding byproducts such as dimers, trimers and even oligomers. Under routine HPLC conditions, the oligomers could not be easily detected due to the mobility behavior resulting from their high molecular weight. The oligomeric impurities did not affect the performance of the downstream chemistry. However, we needed to develop an oligomer-sensitive analytical method in order to monitor the purge of oligomers, and to ensure appropriate product quality. With a reliable analytical method using a size-exclusion column specifically designed for analyzing large molecules, optimal lactamization conditions were derived, resulting in significant suppression of the formation of oligomer byproducts. The remaining oligomeric byproducts could be rejected to non-detectable levels in the subsequent isolated intermediates. In practice, without isolation of alkyne acid intermediate **67**, the Sonogashira reaction crude stream was solvent-switched to THF/DMAc and subsequently treated with HATU to afford the desired macrolactam **56** in 84% isolated yield and > 97% purity without any epimerization. In addition, the regioisomer corresponding to **65** (Scheme 10) was further rejected to < 0.2%.

Hydrogenation of alkyne **56** performed smoothly in the presence of 5% $Pd(OH)_2/C$ in THF at 50 psig, giving the desired **13** in > 99% conversion. The high conversion was required due to poor rejection of the corresponding partially-reduced alkene intermediate. Upon removal of the Pd catalyst and concentration, macrocycle **13** was isolated from aqueous *i*-PrOH in 87% yield (Scheme 11, Option A). To identify the best sequence to synthesize macrocycle **13**, we also explored an alternative approach (Scheme 11, Option B), which featured the reverse sequence of hydrogenation followed by macrolactamization. Since intermediate **67** was not isolable as a solid, charcoal treatment of the crude stream **67** prior to hydrogenation was essential to removing impurities poisoning the Pd catalyst and achieving the required high conversion (> 99%). Although the subsequent macrolactamization proceeded well, Option A provided a more robust process and was selected as the macrocyclization process en route to grazoprevir.

Instead of carrying out the hydrolysis of **13** in a basic biphasic solution developed in the first-generation route (Scheme 8), which required an extraction workup before isolation of the free acid **57**, an operationally simpler process by carrying out the hydrolysis with LiOH in aqueous methanol was developed. The combination of LiOH with MeOH was chosen based on the solubility of LiCl in the HCl quenched hydrolysis mixture. As such, upon addition of aqueous HCl to the reaction mixture, the acid **57** hemihydrate was directly isolated in 95% yield through filtration with excellent rejection of impurities including LiCl.

The second-generation route (*57*) provides a practical manufacturing process to penultimate macrocyclic acid **57**, which has been implemented at commercial scale for the preparation of grazoprevir. Starting from (*S*)-epichlorohydrin, the overall yield to macrocyclic acid penultimate has been improved to 36%, vs. 5% yield of the first-generation route. This robust and productive process possesses an efficient sequence as well as bond connection strategy to assemble building blocks **17**, **28**, and **58**.

Mechanistically Driven Endgame Development

In the discovery chemistry route (*18*), the preparation of grazoprevir via coupling of macrocyclic acid **57** with sulfonamide amine HCl salt **14** was accomplished with the expensive TBTU and catalytic DMAP in the presence of *i*-Pr_2NEt in dichloromethane (Scheme 1). During the development of the first-generation endgame process (*42*), we discovered that the use of less expensive *N*-(3-Dimethylaminopropyl)-*N'*-ethylcarbodiimide (EDC) alone could also accomplish the requisite coupling with high conversion, although epimerization of the stereogenic center adjacent to the proline carboxylic acid led to a significant amount of **68** (up to 10%). Rejection of **68** to acceptable levels in isolated grazoprevir for pharmaceutical use proved to be challenging. Optimization of the EDC coupling using additives, therefore, focused on suppression of the formation of **68**. After examining several solvents (DCM, EtOAc, MeCN, DMF, DMAc, DMSO, and NMP) and additives (1-hydroxybenzotriazole, 2-hydroxypyridine *N*-oxide, *N*-hydroxysuccinamide, and $CuCl_2$) and the effect of base loading, we found that EDC coupling in MeCN with an excess of pyridine could yield grazoprevir with < 0.2% of isomer **68**. After an acidic quench, grazoprevir was obtained in 93% yield and with > 99.7% purity. This process (Scheme 12) has been scaled up in pilot plant to supply large quantities of grazoprevir for safety and clinical development.

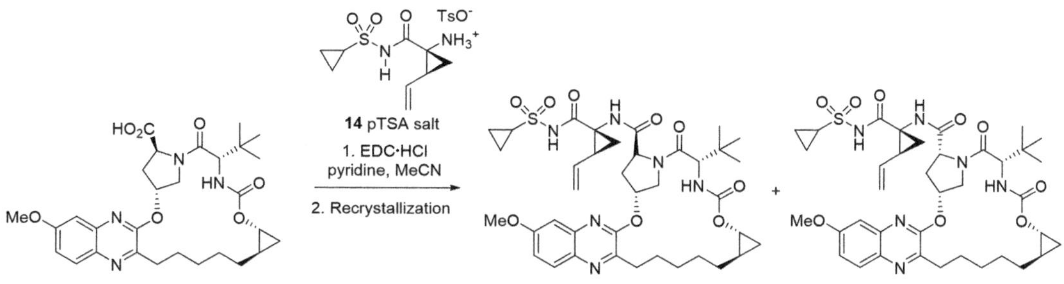

Scheme 12. First generation endgame process.

However, this process lacked operational robustness and impurity control due to a series of significant competitive side reactions. To ensure the quality of isolated grazoprevir, a large excess of EDC•HCl (1.65 equiv) was required to achieve > 99% conversion, because of the poor rejection of unconsumed **57** under the isolation conditions. While the use of excess EDC•HCl resulted in high conversions, these conditions led to the rapid formation of several byproducts that were also poorly rejected during the isolation. It was therefore crucial to quench the reaction as soon as the desired conversion had been achieved. Failing to do so led to a subsequent reaction of product **1** with EDC that unacceptably increased the levels of the byproducts and resulted in poor quality grazoprevir. This delicate balance between the side reactions vs. rejection of impurities and conversion vs. reaction quench timing rendered the control of product quality challenging. Our approach to overcome these issues and therefore to develop a robust endgame coupling process relied on obtaining the understanding of degradation pathways which would both compromise the overall yield and form impurities.

Our studies on the competitive side reactions in the EDC coupling reaction led to identification of byproducts **69**, **72a,b**, and **73a,b**. A plausible pathway for the formation of **72a,b** and **73a,b** via a novel rearrangement is depicted in Scheme 13. Aging the reaction mixture overnight resulted in formation of the EDC adduct guanidine **69** in ~30% yield. Guanidine **69** could be converted to intermediate **70** via a [1,3]-carbonyl migration (*60*). Subsequent intramolecular nucleophilic attack would afford the electron-deficient guanidine **71**, which could undergo fragmentation via a nucleophilic amine attack to afford 7**2a,b** and **73a,b**.

We further reasoned that the acyl sulfonamide moiety in grazoprevir could be acidic, which could react like a carboxylic acid with EDC to form nucleophilic adduct **69** under suitable reaction conditions (*61*). To verify the acidity of the acyl sulfonamide moiety, we measured the dissociation constant of grazoprevir by potentiometric titrations with spectrophotometric analysis. Grazoprevir was titrated in 30% methanol in ionic strength adjusted water (0.15 M KCl in water), in the pH range of 2 to 10, with 0.5 M potassium hydroxide. The measured pKa of the acyl sulfonamide proton in grazoprevir was 5.1. It is known that solvent polarity, dielectric constant and ability to preferentially solvate and thus stabilize certain species can impact compound acidity/basicity and the nucleophilicity of a nucleophile (*62*). In addition, the nucleophilic addition of grazoprevir to EDC could also be promoted by the addition of pyridine (pKa = 5.23 in water) (*63*) in the EDC coupling reaction. Therefore, we initiated studies on utilizing the solvent effect to modify the nucleophilicity and acidity of grazoprevir in the reaction mixture, and consequently suppress the corresponding nucleophilic addition to EDC.

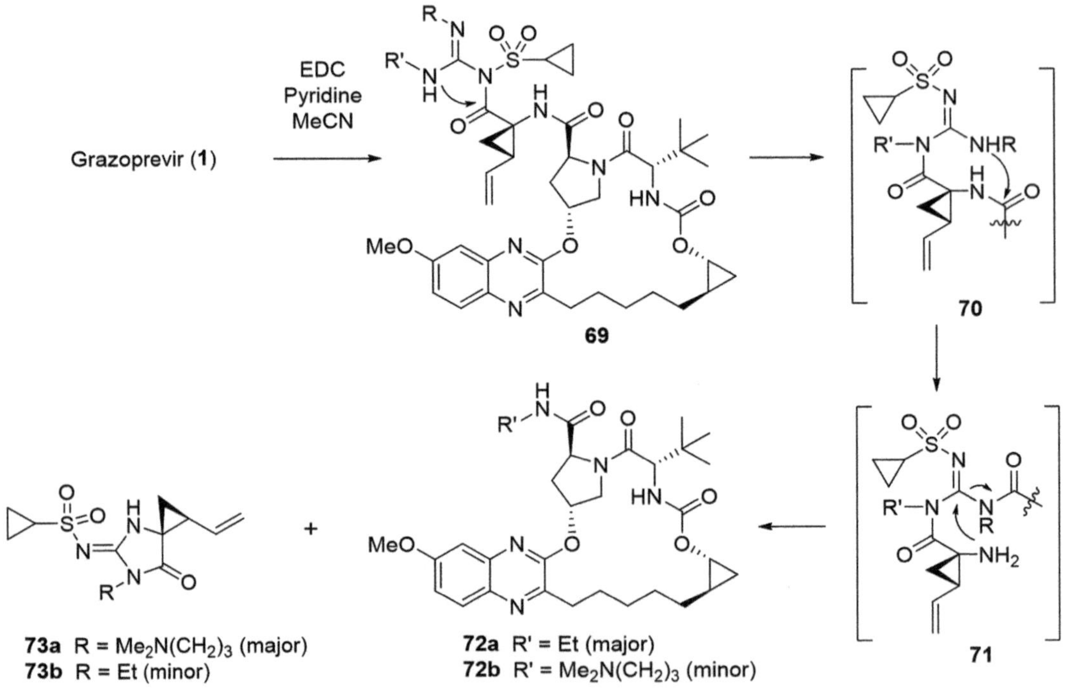

Scheme 13. Plausible pathways to byproducts in EDC coupling.

The solvent effect was studied by carrying out the EDC coupling of acid **57** with amine salt **14** in the presence of pyridine. The results obtained in MeCN, THF, DMAc and THF–DMAc systems are summarized in Table 2. The EDC couplings in THF and/or DMAc were carried out in a homogeneous solution, while the reaction in MeCN was a slurry. The solvent polarity and/or dielectric constant could be also adjusted by mixing DMAc with THF (Table 2, entries 3 and 4). The results showed that the formation of adduct **69** could be significantly suppressed, as the polarity/dielectric constant of the solvent system were decreased, presumably the acidity and/or nucleophilicity of the acyl sulfonamide could be decreased in a less polar solvent, resulting in the suppression of a weak base pyridine promoted the corresponding nucleophilic addition to EDC.

The mechanistic studies resulted in a simple solution to suppress the formation of **69** and subsequently **72a,b** and **73a,b**. By carrying out the coupling in the much less polar solvent THF compared to acetonitrile, the nucleophilic addition to form adduct **69** was significantly suppressed (Table 2, entry 2). Indeed, even aging the EDC reaction mixture in THF at ambient temperature for one week resulted in < 2% of impurities **69**, **72a,b** and **73a,b**.

Further optimization of the EDC coupling conditions was focused on addressing the stability/reactivity of sulfonamide **14**. Sulfonamide **14** has been previously used for amidation on large-scale (*26, 27*); however, studies on its stability/reactivity were lacking. Unlike the stable *p*TSA salt of **14**, we found that the free base **14** was labile at ambient temperature, with ~40% decomposition observed overnight (Scheme 14). The decomposition byproducts **74** and **75** were unambiguously characterized by NMR and LC/MS techniques. The plausible pathways for the formation of these byproducts are depicted in Scheme 14. Ammonia was also formed during the decomposition of **74** and was captured by macrocyclic acid **57** to form byproduct **80** under the endgame EDC coupling conditions.

Table 2. Selected Results of the Solvent Effect on Formation of EDC Adduct 69 at Ambient Temperature

Entry	*Solvents*	*Time*	*Adduct* ***69***
1	DMAc	2.5 h	5%
2	THF	2.5 h one week	< 0.05% < 2%
3	14% DMAc in THF	9.5 h 3 days	0.8% 4.7%
4	18% DMAc in THF	2.5 h 21 h	0.2% 1.4%
5	MeCN	3 h overnight 2 days	6% 30% 61%

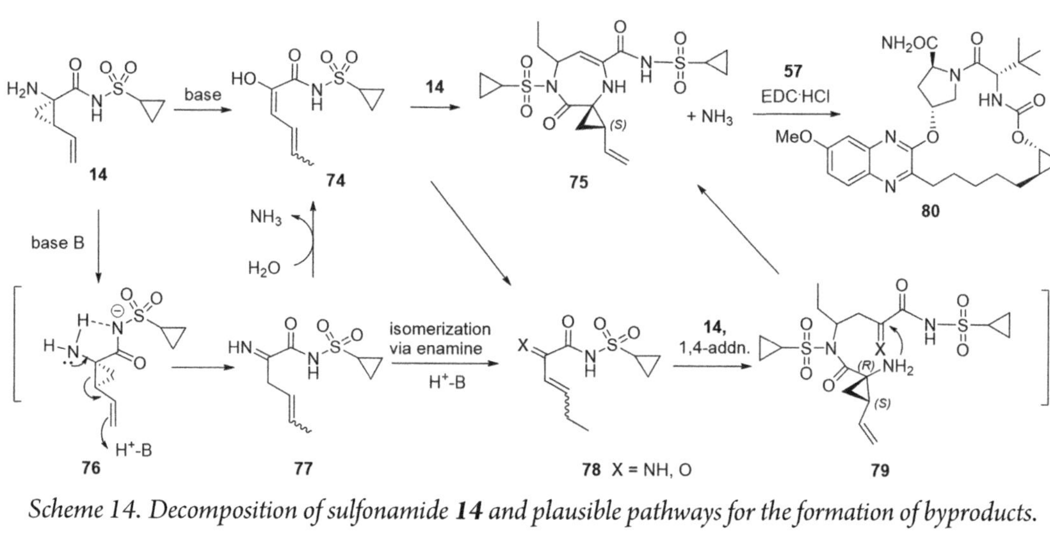

Scheme 14. Decomposition of sulfonamide ***14*** *and plausible pathways for the formation of byproducts.*

We overcame the instability of the free base **14** by lowering the reaction temperature to 0 - 5 °C and by charging pyridine to a slurry of azeotropically dried acid **57** and the *p*TSA salt of **14** in

THF. The resulting homogeneous solution was then treated with EDC to provide grazoprevir (**1**) with an excellent purity profile, as such the decomposition of **14** was suppressed to a minimal level (< 1%). It is noteworthy that this endgame process on large-scale operates with extreme ease and without further reaction concerns. Upon aqueous citric acid workup, grazoprevir monohydrate was crystallized from aqueous acetone in 94% yield and > 99.8% purity with no impurities > 0.1%.

In conclusion, an in-depth mechanistic understanding of the reaction pathways that led to the desired and undesired products played a crucial role to the development of an efficient and robust endgame, which has been successfully implemented for commercialization (57).

Summary

This chapter has outlined the discovery and chemical development of the hepatitis C virus NS3/4a protease inhibitor, grazoprevir, which was approved in combination with NS5a inhibitor, elbasvir as ZEPATIER® for the treatment of HCV infection. The discovery of grazoprevir resulted from careful analysis of bound structures and generation of the P2–P4 macrocyclization hypothesis using structure-based drug design. Dramatic improvements in activity against the variety of genotypes of HCV and against mutant NS3 enzymes that were appearing in the clinic were made, while maintaining favorable pharmacokinetic properties. Chemistry played a major role in the discovery phase of this program as the ring-closing metathesis reaction was critical for the extensive and rapid evaluation of the macrocyclic framework. At several points during the invention of grazoprevir, synthesis-inspired designs were enabled by the discovery and application of new synthetic routes and availability of key advanced intermediates, which made the exploration of the chemical space more facile.

In order to develop an efficient synthesis of grazoprevir, we first developed an asymmetric synthesis of cyclopropanol building block **28**. Starting from (*S*)-epichlorohydrin and without isolation of any intermediates or recrystallization, the cyclopropoxy building block **28** was prepared in >97% purity and 51% isolated yield. After thorough exploration of various strategies to assemble the backbone of grazoprevir, four building blocks have been installed in the most efficient and optimal sequence. In-depth mechanistic understanding successfully guided us to solve the incompatible reactivity issue of Boc deprotection and the competitive side reactions in the endgame. Starting from dichloroquinoxaline **58**, grazoprevir was prepared in 51% overall yield and > 99.8% purity. Combining all the key elements required for a manufacturing process and minimizing the environmental burden from process waste, this convergent route has been implemented for commercialization of grazoprevir (Scheme 15). In contrast to the first-generation large-scale synthesis, the waste output of the final manufacturing process has been reduced by 65%.

Our development of the manufacturing process for grazoprevir is a journey of evolution of chemistry and process. As the existing issues and obstacles were overcome with breakthroughs, the evolved process possessed improved practicality, robustness, and cost-effectiveness. Of importance, mechanistic studies played a crucial role for understanding of the origin of process issues and therefore allowed us to navigate through the process evolution and development. As such, a robust manufacturing process has been realized.

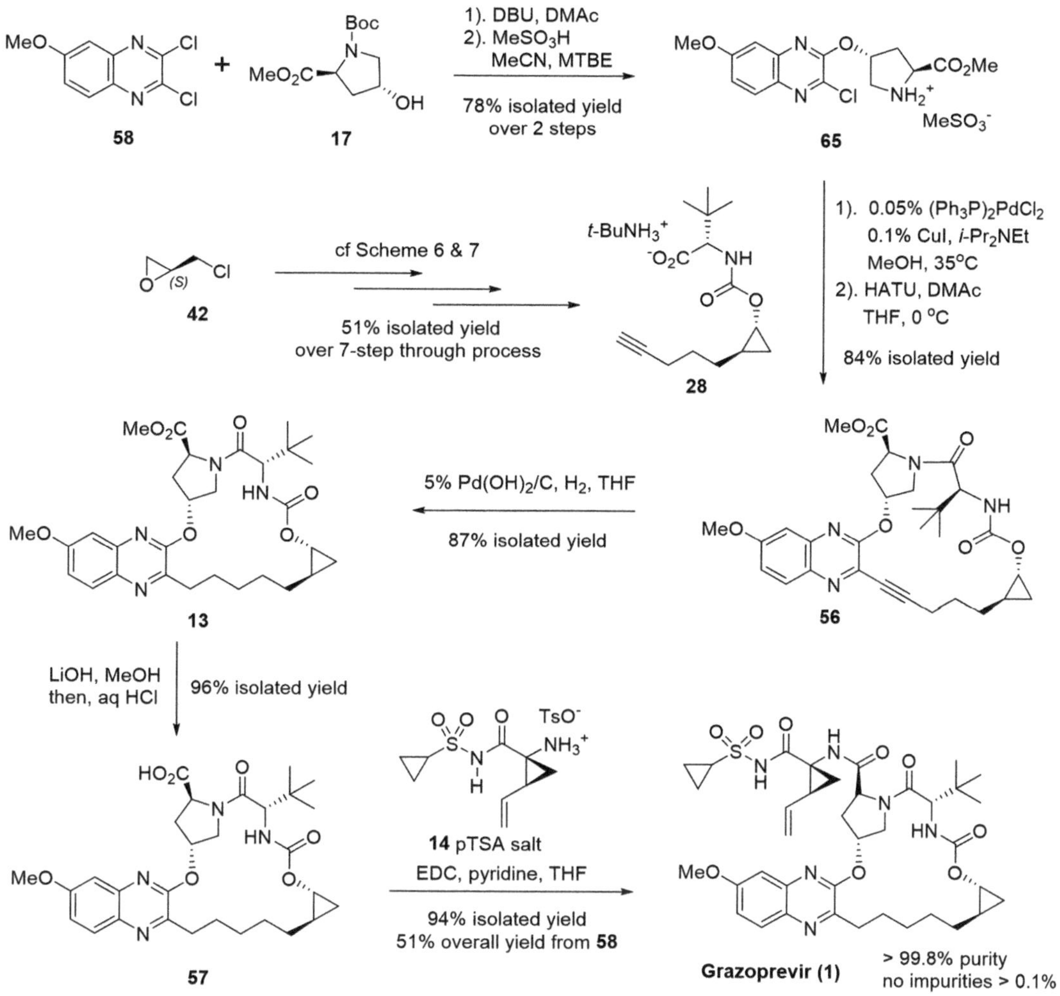

Scheme 15. Summary of the manufacturing route to grazoprevir with isolated intermediates.

Acknowledgments

This chapter represents the outstanding contributions of many scientists within Merck & Co., Inc. The authors hereby express their sincere gratitude to the countless smart, creative scientists with whom we have had the privilege to work during the discovery and development of grazoprevir.

References

1. Choo, Q. L.; Kuo, G.; Weiner, A. J.; Overby, L. R.; Bradley, D. W.; Houghton, M. Isolation of a cDNA Clone Derived from a Blood-Borne Non-A, Non-B Viral Hepatitis Genome. *Science* **1989**, *244*, 359–362.
2. Lavanchy, D. The Global Burden of Hepatitis C. *Liver Int.* **2009**, *29* (Suppl. 1), 74–81.
3. Marcellin, P.; Asselah, T.; Boyer, N. *Hepatology* **2002**, *36*, S47–56.
4. Simmonds, P. Genetic Diversity and Evolution of Hepatitis C Virus – 15 Years on. *J. Gen. Virol.* **2004**, *85*, 3173–3188.

5. Zeuzem, S.; Berg, T.; Moeller, B.; Hinrichsen, H.; Mauss, S.; Wedemeyer, H.; Sarrazin, C.; Hueppe, D.; Zehnter, E.; Manns, M. P. Expert Opinion on the Treatment of Patients with Chronic Hepatitis C. *J. Viral Hepat.* **2009**, *16*, 75–90.
6. Lindenbach, B. D.; Rice, C. M. Unravelling hepatitis C virus replication from genome to function. *Nature* **2005**, *436*, 933–938.
7. Coilly, A.; Dumortier, J.; Botta-Fridlund, D.; Latournerie, M.; Leroy, V.; Pageaux, G.-P.; Agostini, H.; Giostra, E.; Moreno, C.; Roche, B.; Antonini, T. M.; Guillaud, O.; Lebray, P.; Radenne, S.; Saouli, A.-C.; Calmus, Y.; Alric, L.; Debette-Gratien, M.; De Ledinghen, V.; Durand, F.; Duvoux, C.; Samuel, D.; Duclos-Vallée, J.-C. Multicenter Experience with Boceprevir or Telaprevir to Treat Hepatitis C Recurrence after Liver Transplantation: When Present Becomes Past, What Lessons for Future. *PLoS One* **2015**, *10*, e0138091.
8. Kim, J. L.; Morgenstern, K. A.; Lin, C.; Fox, T.; Dwyer, M. D.; Landro, J. A.; Chambers, S. P.; Markland, W.; Lepre, C. A.; O'Malley, E. T.; Harbeson, S. L.; Rice, C. M.; Murcko, M. A.; Caron, P. R.; Thomson, J. A. Crystal Structure of the Hepatitis C Virus NS3 Protease Domain Complexed with a Synthetic NS4A Cofactor Peptide. *Cell* **1996**, *87*, 343–355.
9. Grakoui, A.; McCourt, D. W.; Wychowski, C.; Feinstone, S. M.; Rice, C. M. Characterization of the Hepatitis C Virus-Encoded Serine Proteinase: Determination of Proteinase-Dependent Polyprotein Cleavage Sites. *J. Virol.* **1993**, *67*, 2832–2843.
10. McCauley, J. A.; Rudd, M. T. Hepatitis C Virus NS3/4a Protease Inhibitors. *Curr. Opin. Pharmacol.* **2016**, *30*, 84–92.
11. Venkatraman, S.; Bogen, S. L.; Arasappan, A.; Bennett, F.; Chen, K.; Jao, E.; Liu, Y.-T.; Lovey, R.; Hendrata, S.; Huang, Y.; Pan, W.; Parekh, T.; Pinto, P.; Popov, V.; Pike, R.; Ruan, S.; Santhanam, B.; Vibulbhan, B.; Wu, W.; Yang, W.; Kong, J.; Liang, X.; Wong, J.; Liu, R.; Butkiewicz, N.; Chase, R.; Hart, A.; Agrawal, S.; Ingravallo, P.; Pichardo, J.; Kong, R.; Baroudy, B.; Malcolm, B.; Guo, Z.; Prongay, A.; Madison, V.; Broske, L.; Cui, X.; Cheng, K.-C.; Hsieh, T. Y.; Brisson, J.-M.; Prelusky, D.; Korfmacher, W.; White, R.; Bogdanowich-Knipp, S.; Pavlovsky, A.; Bradley, P.; Saksena, A. K.; Ganguly, A.; Piwinski, J.; Girijavallabhan, V.; Njoroge, F. G. Discovery of (1R,5S)-N-[3-Amino-1-(cyclobutylmethyl)-2,3-dioxopropyl]3-[2(S)-[[[(1,1-dimethylethyl)amino]carbonyl]amino]-3,3-dimethyl-1-oxobutyl]6,6-dimethyl-3-azabicyclo[3.1.0]hexan-2(S)-carboxamide (SCH 503034), a Selective, Potent, Orally Bioavailable Hepatitis C Virus NS3 Protease Inhibitor: A Potential Therapeutic Agent for the Treatment of Hepatitis C Infection. *J. Med. Chem.* **2006**, *49*, 6074–6086.
12. Lin, C.; Kwong, A. D.; Perni, R. B. *Infect. Disord.: Drug Targets* **2006**, *6*, 3–16.
13. Llinas-Brunet, M.; Bailey, M. D.; Bolger, G.; Brochu, C.; Faucher, A.-M.; Ferland, J. M.; Garneau, M.; Ghiro, E.; Gorys, V.; Grand-Maitre, C.; Halmos, T.; Lapeyre-Paquette, N.; Liard, F.; Poirier, M.; Rheaume, M.; Tsantrizos, Y. S.; Lamarre, D. Structure-Activity Study on a Novel Series of Macrocyclic Inhibitors of the Hepatitis C Virus NS3 Protease Leading to the Discovery of BILN 2061. *J. Med. Chem.* **2004**, *47*, 1605–1608.
14. Lamarre, D.; Anderson, P. C.; Bailey, M.; Beaulieu, P.; Bolger, G.; Bonneau, P.; Bos, M.; Cameron, D. R.; Cartier, M.; Cordingley, M. G.; Faucher, A.-M.; Goudreau, N.; Kawai, S. H.; Kukolj, G.; Lagace, L.; LaPlante, S. R.; Narjes, H.; Poupart, M.-A.; Rancourt, J.; Sentjens, R. E.; St George, R.; Simoneau, B.; Steinmann, G.; Thibeault, D.; Tsantrizos, Y. S.; Weldon,

S. M.; Yong, C.-L.; Llinas-Brunet, M. An NS3 Protease Inhibitor with Antiviral Effects in Humans Infected with Hepatitis C Virus. *Nature* **2003**, *426*, 186–189.

15. Liverton, N. J.; Holloway, M. K.; McCauley, J. A.; Rudd, M. T.; Butcher, J. W.; Carroll, S. S.; DiMuzio, J.; Fandozzi, C.; Gilbert, K. F.; Mao, S.-S.; McIntyre, C. J.; Nguyen, K. T.; Romano, J. J.; Stahlhut, M.; Wan, B.-L.; Olsen, D. B.; Vacca, J. P. Molecular Modeling Based Approach to Potent P2–P4 Macrocyclic Inhibitors of Hepatitis C NS3/4A Protease. *J. Am. Chem. Soc.* **2008**, *130*, 4607–4609.
16. Liverton, N. J.; Carroll, S. S.; Di Muzio, J.; Fandozzi, C.; Graham, D. J.; Hazuda, D.; Holloway, M. K.; Ludmerer, S. W.; McCauley, J. A.; McIntyre, C. J.; Olsen, D. B.; Rudd, M. T.; Stahlhut, M.; Vacca, J. P. MK-7009, a Potent and Selective Inhibitor of Hepatitis C Virus NS3/4A Protease. *Antimicrob. Agents Chemother.* **2009**, *54*, 305–311.
17. McCauley, J. A.; McIntyre, C. J.; Rudd, M. T.; Nguyen, K. T.; Romano, J. J.; Butcher, J. W.; Gilbert, K. F.; Bush, K. J.; Holloway, M. K.; Swestock, J.; Wan, B.-L.; Carroll, S. S.; DiMuzio, J. M.; Graham, D. J.; Ludmerer, S. W.; Mao, S.-S.; Stahlhut, M. W.; Fandozzi, C. M.; Trainor, N.; Olsen, D. B.; Vacca, J. P.; Liverton, N. J. Discovery of Vaniprevir (MK-7009), a Macrocyclic Hepatitis C Virus NS3/4a Protease Inhibitor. *J. Med. Chem.* **2010**, *53*, 2443–2463.
18. Harper, S.; McCauley, J. A.; Rudd, M. T.; Ferrara, M.; DiFilippo, M.; Crescenzi, B.; Koch, U.; Petrocchi, A.; Holloway, M. K.; Butcher, J. W.; Romano, J. J.; Bush, K. J.; Gilbert, K. F.; McIntyre, C. J.; Nguyen, K. T.; Nizi, E.; Carroll, S. S.; Ludmerer, S. W.; Burlein, C.; DiMuzio, J. M.; Graham, D. J.; McHale, C. M.; Stahlhut, M. W.; Olsen, D. B.; Monteagudo, E.; Cianetti, S.; Giuliano, C.; Pucci, V.; Trainor, N.; Fandozzi, C. M.; Rowley, M.; Coleman, P. J.; Vacca, J. P.; Summa, V.; Liverton, N. J. Discovery of MK-5172, a Macrocyclic Hepatitis C Virus NS3/4a Protease Inhibitor. *ACS Med. Chem. Lett.* **2012**, *3*, 332–336.
19. Summa, V.; Ludmerer, S. W.; McCauley, J. A.; Fandozzi, C.; Burlein, C.; Claudio, G.; Coleman, P. J.; DiMuzio, J. M.; Ferrara, M.; Di Filippo, M.; Gates, A. T.; Graham, D. J.; Harper, S.; Hazuda, D. J.; McHale, C.; Monteagudo, E.; Pucci, V.; Rowley, M.; Rudd, M. T.; Soriano, A.; Stahlhut, M. W.; Vacca, J. P.; Olsen, D. B.; Liverton, N. J.; Carroll, S. S. MK-5172, a Selective Inhibitor of Hepatitis C Virus NS3/4a Protease with Broad Activity across Genotypes and Resistant Variants. *Antimicrob. Agents Chemother.* **2012**, *56*, 4161–4167.
20. Arasappan, A.; Chen, K. X.; Njoroge, F. G.; Parekh, T. N.; Girijavallabhan, V. Novel Dipeptide Macrocycles from 4-Oxo, -Thio, and -Amino-Substituted Proline Derivatives. *J. Org. Chem.* **2002**, *67*, 3923–3926.
21. CAS# 918870-76-5; 1,3-bis(2,4,6-trimethylphenyl)-4,5-dihydroimidazol-2-ylidene[2-(*i*-propoxy)-5-(*N*,*N*-dimethylaminosulfonyl)phenyl] methyleneruthenium (II) dichloride.
22. Wang, X. A.; Sun, L.-Q.; Sit, S.-Y.; Ny, S.; Scola, P. M.; Hewawasam, P.; Good, A. C.; Chen, Y.; Campbell, J. A. *Hepatitis C Virus Inhibitors*. U.S. Patent 6,995,174, 2006.
23. McCauley, J. A.; Rudd, M. T. *HCV-The Journey from Discovery to a Cure*, Vol. I, Sofia, M. J., Ed.; Springer: Switzerland, 2019.
24. Lawitz, E. J.; O'Riordan, W. D.; Asatryan, A.; Freilich, B. L.; Box, T. D.; Overcash, J. S.; Lovell, S.; Ng, T. I.; Liu, W.; Campbell, A.; Lin, C.-W.; Yao, B.; Kort, J. Potent Antiviral Activities of the Direct-Acting Antivirals ABT-493 and ABT-530 with Three-Day Monotherapy

for Hepatitis C Virus Genotype 1 Infection. *Antimicrob. Agents Chemother.* **2016**, *60*, 1546–1555.

25. Rodriguez-Torres, M.; Glass, S.; Hill, J.; Freilich, B.; Hassman, D.; Di Bisceglie, A. M.; Taylor, J. G.; Kirby, B. J.; Dvory-Sobol, H.; Yang, J. C.; An, D.; Stamm, L. M.; Brainard, D. M.; Kim, S.; Krefetz, D.; Smith, W.; Marbury, T.; Lawitz, E. GS-9857 in Patients with Chronic Hepatitis C Virus Genotype1–4 Infection: a Randomized, Double-blind, Dose-ranging Phase 1 Study. *J. Viral. Hepat.* **2016**, *23*, 614–622.
26. Flick, A. C.; Ding, H. X.; Leverett, C. A.; Kyne, R. E., Jr.; Liu, K. K.-C.; Fink, S. J.; O'Donnell, C. Synthetic Approaches to the 2014 New Drugs. *J. Bioorg. Med. Chem.* **2016**, *24*, 1937–1938.
27. Song, Z. J.; Tellers, D. M.; Journet, M.; Kuethe, J. T.; Lieberman, D.; Humphrey, G.; Zhang, F.; Peng, Z.; Waters, M. S.; Zewge, D.; Nolting, A.; Zhao, D.; Reamer, R. A.; Dormer, P. G.; Belyk, K. M.; Davies, I. W.; Devine, P. N.; Tschaen, D. M. Synthesis of Vaniprevir (MK-7009): Lactamization To Prepare a 22-Membered Macrocycle. *J. Org. Chem.* **2011**, *76*, 7804–7815.
28. Sarges, R.; Howard, H. R.; Browne, R. G.; Lebel, L. A.; Seymour, P. A.; Koe, B. K. 4-Amino[l,2,4]triazolo[4,3-a]quinoxalines. A Novel Class of Potent Adenosine Receptor Antagonists and Potential Rapid-Onset Antidepressants. *J. Med. Chem.* **1990**, *33*, 2240–2254.
29. For a review, see: Lebel, H.; Marcoux, J.-F.; Molinaro, C.; Charette, A. B. Stereoselective Cyclopropanation Reactions. *Chem. Rev.* **2003**, *103*, 977–1050.
30. Benoit, G.; Charette, A. B. Diastereoselective Borocyclopropanation of Allylic Ethers Using a Boromethylzinc Carbenoid. *J. Am. Chem. Soc.* **2017**, *139*, 1364–1367.
31. Wang, Y.; Wen, X.; Cui, X.; Wojtas, L.; Zhang, X. P. *J. Am. Chem. Soc.* **2017**, *139*, 1049–1052.
32. Cheng, K.; Carroll, P. J.; Walsh, P. J. Diasteroselective Preparation of Cyclopropanols Using Methylene Bis(iodozinc). *Org. Lett.* **2011**, *13*, 2346–2349.
33. Du, H.; Long, J.; Shi, Y. Catalytic Asymmetric Simmons–Smith Cyclopropanation of Silyl Enol Ethers. Efficient Synthesis of Optically Active Cyclopropanol Derivatives. *Org. Lett.* **2006**, *8*, 2827–2829.
34. Quan, L. G.; Kim, S.-H.; Lee, J. C.; Cha, J. K. Diastereoselective Synthesis of *trans*-1,2-Dialkylcyclopropanols by the Kulinkovich Hydroxycyclopropanation of Homoallylic Alcohols. *Angew. Chem., Int. Ed.* **2002**, *41*, 2160–2162.
35. Corey, E. J.; RaoMark, S. A.; Noe, C. Catalytic Diastereoselective Synthesis of *Cis*-l,2-Disubstituted Cyclopropanols from Esters Using a Vicinal Dicarbanion Equivalent. *J. Am. Chem. Soc.* **1994**, *116*, 9345–9346.
36. Bassan, E. M.; Baxter, C. A.; Beutner, G. L.; Emerson, K. M.; Fleitz, F. J.; Johnson, S.; Keen, S.; Kim, M. M.; Kuethe, J. T.; Leonard, W. R.; Mullens, P. R.; Muzzio, D. J.; Roberge, C.; Yasuda, N. Multikilogram-Scale Synthesis of a Chiral Cyclopropanol and an Investigation of the Safe Use of Lithium Acetylide–Ethylene Diamine Complex. *Org. Process Res. Dev.* **2012**, *16*, 87–95.
37. Shirakawa, K.; Arase, A.; Hoshi, M. Preparation of (E)-1-Alkenylboronic Acid Pinacol Esters via Transfer of Alkenyl Group from Boron to Boron. *Synthesis* **2004**, 1814–1820.
38. Lorenz, J. C.; Long, J.; Yang, Z.; Xue, S.; Xie, Y.; Shi, Y. A Novel Class of Tunable Zinc Reagents ($RXZnCH_2Y$) for Efficient Cyclopropanation of Olefins. *J. Org. Chem.* **2004**, *69*, 327–334.

39. Fontani, P.; Carboni, B.; Vaultier, M.; Maas, G. Synthesis of Cyclopropylboronic Acid Esters by Carbene Transfer to 1-Alkenylboronic Acid Esters. *Synthesis* **1991**, 605–609.
40. Päassler, P., Hefner, W., Buckl, K., Meinass, H., Meiswinkel, A., Wernicke, H.-J., Ebersberg, G., Müller, R., Bässler, J., Behringer, H., Mayer, D. Acetylene. In *Ullmann's Encyclopedia of Industrial Chemistry*; Wiley-VCH: Weinheim, 2008.
41. http://www.airgas.com/content/productChart.aspx?chart_type=2 (accessed 20 Oct 2011).
42. Kuethe, J.; Zhong, Y.-L.; Yasuda, N.; Beutner, G.; Linn, K.; Kim, M.; Marcune, B.; Dreher, S. D.; Humphrey, G.; Pei, T. Development of a Practical, Asymmetric Synthesis of the Tepatitis C Virus Protease Inhibitor MK-5172. *Org. Lett.* **2013**, *15*, 4174–4177.
43. Militzer, H.-C.; Schömenauer, S.; Otte, C.; Puls, C.; Hain, J.; Bräse, S.; de Meijere, A. Versatile Syntheses of Alkynyl- and Substituted Alkynylcyclopropanes: 2-Alkoxyethynylcyclopropanes for the Anellation of Bicyclo[3.3.0]octane Fragments. *Synthesis* **1993**, 998–1012.
44. For a recent application, see: Sokolsky, A.; Smith, A. B., III Validation of Diethoxyphosphonate as an Effective Agent for Charge Transfer in Anion Relay Chemistry (ARC). *Org. Lett.* **2012**, *14*, 4470–4473.
45. Harris, M. R.; Konev, M. O.; Jarvo, E. R. Enantiospecific Intramolecular Heck Reactions of Secondary Benzylic Ethers. *J. Am. Chem. Soc.* **2014**, *136*, 7825–7828.
46. 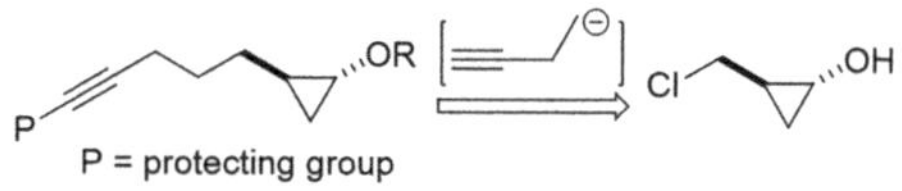

For example, the approach shown to the left suffered from the instability of the cyclopropanol intermediate.
47. Xu, F.; Zhong, Y.-L.; Li, H.; Qi, J.; Desmond, R.; Song, Z. J.; Park, J.; Wang, T.; Truppo, M.; Humphrey, G. R.; Ruck, R. T. Asymmetric Synthesis of Functionalized *trans*-Cyclopropoxy Building Block for Grazoprevir. *Org. Lett.* **2017**, *19*, 5880–5883.
48. Kubizna, P.; Špánik, I.; Kožíšek, J.; Szolcsányi, P. Synthesis of 2, 6-Disubstituted Piperidine Alkaloids from Ladybird Beetles Calvia 10-Guttata and Calvia 14-Guttata. *Tetrahedron* **2010**, *66*, 2351–2355.
49. Brown, C. A.; Yamashita, A. The Acetylene Zipper. An Exceptionally Facile "Contrathermodynamic" Multipositional Isomerization of Alkynes with Potassium 3-Aminopropylamide. *J. Am. Chem. Soc.* **1975**, *97*, 891–892.
50. Avocetien, K.; Li, Y.; O'Doherty, G. A. The Alkyne Zipper Reaction in Asymmetric Synthesis. In *Modern Alkyne Chemistry*; Trost, B. M., Li, C.-J., Eds.; Wiley-VCH: Weinheim, 2014; pp 365–394.
51. Avocetien, K. F.; Li, J. J.; Liu, X.; Wang, Y.; Xing, Y.; O'Doherty, G. A. *De Novo* Asymmetric Synthesis of Phoracantholide J. *Org. Lett.* **2016**, *18*, 4970–4973.
52. Das, S.; Kuilya, T. K.; Goswami, R. K. Asymmetric Total Synthesis of Bioactive Natural Lipid Mycalol. *J. Org. Chem.* **2015**, *80*, 6467–6489.
53. Abrams, S. R.; Shaw, A. C. Triple-bond Isomerizations: 2- to 9-Decyn-1-ol. *Org. Synth.* **1988**, *66*, 127 –129.
54. Kamienski, C. W.; Lewis, D. H. Relationship between Structure and Solubility of Organic Lithium Compounds. *J. Org. Chem.* **1965**, *30*, 3498–3504.

55. Williams, M. J.; Kong, J.; Chung, C. K.; Brunskill, A.; Campeau, L.-C.; McLaughlin, M. The Discovery of Quinoxaline-Based Metathesis Catalysts from Synthesis of Grazoprevir (MK-5172). *Org. Lett.* **2016**, *18*, 1952–1955.
56. Kong, J.; McLaughlin, M.; Williams, M.; Song, Z. J.; Chen, Y.; Jin, Y. *Methods and Intermediates for the Preparation of Macrolactams*. PCT Int. Appl. WO 2015095437, 2015.
57. Xu, F.; Kim, J.; Waldman, J.; Wang, T.; Devine, P. Synthesis of Grazoprevir, a Potent NS3/4a Protease Inhibitor for the Treatment of Hepatitis C Virus. *Org. Lett.* **2018**, *20*, 7261–7265.
58. Xu, F.; Kosjek, B.; Cabirol, F. L.; Chen, H.; Desmond, R.; Park, J.; Gohel, A. P.; Collier, S. J.; Smith, D. J.; Liu, Z.; Janey, J. M.; Chung, J. Y. L.; Alvizo, O. Synthesis of Vibegron Enabled by a Ketoreductase Rationally Designed for High pH Dynamic Kinetic Reduction. *Angew. Chem., Int. Ed.* **2018**, *57*, 6863–6867.
59. Zacuto, M. J.; Xu, F. One-Step $RhCl_3$-Catalyzed Deprotection of Acyclic *N*-Allyl Amides. *J. Org. Chem.* **2007**, *72*, 6298–6300.
60. Graubaum, H.; Martin, D.; Csunderlik, C.; Glatt, H.-H.; Bacaloglu, R.; Malurea-Munteanu, M. Darstellung und Umlagerung von 1-Acetyl-2-amino-benzimidazolen. *Z. Chem.* **1984**, 57–58.
61. Williams, A.; Ibrahim, I. T. Carbodiimide Chemistry: Recent Advances. *Chem. Rev.* **1981**, *81*, 589–636.
62. Reichardt, C.; Welton, T. *Solvent Effects in Org. Chem.*, 4th ed.; Wiley-VCH: Weinheim, 2011.
63. Lõkov, M.; Tshepelevitsh, S.; Heering, A.; Plieger, P. G.; Vianello, R.; Leito, I. On the Basicity of Conjugated Nitrogen Heterocycles in Different Media. *Eur. J. Org. Chem.* **2017**, 4475–4489.

Chapter 10

Discovery and Process Development of BIIB068: A Reversible Bruton's Tyrosine Kinase (BTK) Inhibitor for the Treatment of Autoimmune Diseases

Bin Ma,*[,1] Chaomin Li,*[,2] Brian T. Hopkins,[1] Lloyd Franklin,[2] Robbie Chen,[2] Tamera Mack,[2] Daniel Patience,[2] William F. Kiesman,[2] and Erin M. O'Brien[2]

[1]Department of Medicinal Chemistry, Biogen, Inc., 225 Binney Street, Cambridge, Massachusetts 02142, United States

[2]Department of Chemical Process Development, Biogen, Inc., 225 Binney Street, Cambridge, Massachusetts 02142, United States

*Emails: bin.ma@biogen.com (B.M.); chaomin.li@biogen.com (C.L.)

We describe the discovery and chemical process development of BIIB068, a highly selective, reversible Bruton's tyrosine kinase (BTK) inhibitor for the treatment of autoimmune diseases including systemic lupus erythematosus. The brief overview of the medicinal chemistry effort that led to the discovery of BIIB068 from a proprietary hit is presented in the first part, as well as the first synthetic route used in the discovery stage. In the second part, we discuss chemical process development efforts that led to multi- kilogram production of BIIB068 hemi-adipate. Process optimization resulted in: 1) removal of late transition metal from the reaction, 2) streamlined process and significantly improved overall yield, and 3) appropriate impurity (including PMI) control which enabled the delivery of material to support toxicology studies and clinical trials.

Introduction

Bruton's tyrosine kinase (BTK) is a non-receptor tyrosine kinase that is essential for downstream signaling events including phospholipase C γ (PLCγ) activation, the Ca^{2+} response, and NF-kB (nuclear factor kappa-light-chain-enhancer of activated B cells) activation via the Fc Receptors (FcR) in myeloid cells (*1*) and B cell receptor (BCR) pathways in B cells (*2*). Thus, given the roles of myeloid and B cells playing in both the adaptive immunity and innate immune system, it was important to identify a selective BTK inhibitor that can block:

- antigen presentation in B cells
- cytokine production by monocytes (*3*)
- NETosis by neutrophils (*4*)

This may provide therapeutic value to patients suffering from autoantibody-driven immune disorders. As in systemic lupus erythematosus (SLE), autoreactive B cell-derived antibodies form immune-complexes, bind Fc receptors on myeloid cells and induce proinflammatory cytokine production by monocytes and NETosis (neutrophil extracellular traps) by neutrophils (*5*). One preclinical evidence supporting this hypothesis is that BTK deficient *xid* mice were found to be resistant to acquiring lupus and collagen induced arthritis (*6*), while targeting BTK with small molecule inhibitors has been demonstrated to be efficacious in autoimmune disease models (*7, 8*).

Since the approval of ibrutinib (*9*), a first generation non-selective covalent BTK inhibitor in 2013, numerous second-generation covalent inhibitors (*10*)with improved kinome selectivity have been discovered and advanced into the clinic. This led to the recent approval of acalabrutinib and zanubrutinib for the treatment of B-cell malignancy such as chronic lymphocytic leukemia (CLL) and mantle cell lymphoma (MCL) (*11, 12*).

However, the development of covalent BTK inhibitors for autoimmune indications has been more arduous which resulted a number of failed Phase II studies in rheumatoid arthritis (RA) (*13–15*). Although selective covalent BTK inhibitors such as evobrutinib have demonstrated efficacy in Phase II for the treatment of multiple sclerosis (*16*), based on the mechanism of action (MOA), there is a potential risk for idiosyncratic toxicity (*17*) leading to unfavorable tolerability profiles. Thus, there exists an unmet medical need to develop a selective, non-covalent BTK inhibitor which may afford superior safety and efficacy for the treatment of autoimmune indications.

To date three reversible BTK inhibitors (fenebrutinib, BMS-986142, and AS-1763 (*18–20*)) have advanced into the clinic for the treatments of RA. The most advanced compound, fenebrutinib, was reported to achieve its primary end point in a Phase II proof of concept (POC) study (*21*) validating BTK inhibition as a viable treatment for RA. While fenebrutinib has failed to achieve POC in SLE in a second Phase II study, patients treated with fenebrutinib at doses anticipated to fully inhibit BTK exhibited modulation of pathway-relevant biomarkers (*22*). Here we describe the discovery and chemical process development of BIIB068 (**1**, Figure 1), a highly selective, reversible BTK inhibitor advanced into early clinical trials for the treatment of lupus (*23, 24*).

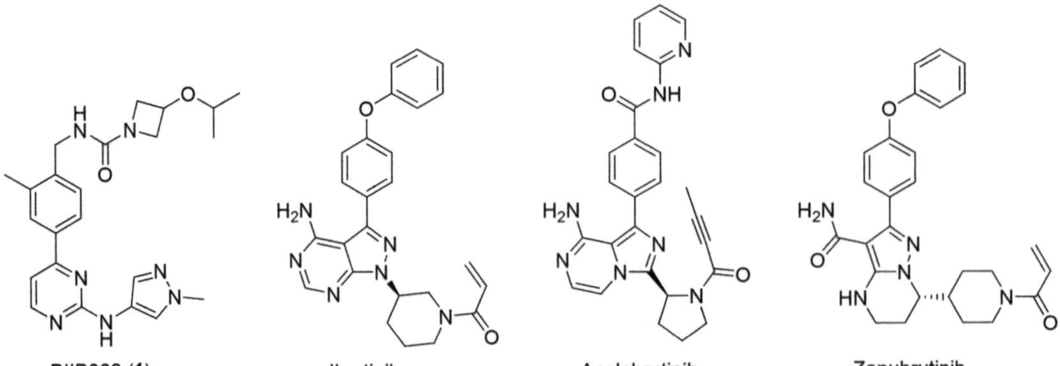

*Figure 1. BIIB068 (**1**), a reversible BTK inhibitor in clinical development, and 3 covalent BTK inhibitors (ibrutinib, acalabrutinib and zanubrutinib) approved for market.*

Discovery of BIIB068

Screening Biogen's proprietary collection led to the discovery of the initial pyrrolopyrimidine hit **2** (BTK IC_{50} = 3700 nM, Figure 2). Upon obtaining an X-ray co-crystal structure with BTK, **2** exhibited a binding pose different from the covalent type I binders such as ibrutinib. Although the pyrrolopyrimidine moiety maintained the key interactions with the hinge region, the benzyl amide moiety was orientated into the "H3" pocket. This region of the protein was unique for BTK, as there were two non-conversed residues Leu-542 and Ser-543 which were considered important for influencing selectivity over other Tec family members.

In addition, it had been reported that introducing lipophilic groups into this region of the protein resulted in significant improvement in binding affinity (25). Thus, analog **3** containing the *p-tert*-butyl moiety was synthesized and resulted in >20-fold improvement in biochemical potency (BTK IC_{50} = 142 nM). A co-crystal structure of **3** with BTK confirmed the orientation of the *tert*-butyl group towards the Tyr-551 (Figure 3) which was hypothesized to stabilize the protein inactive conformation (26). An analysis of the co-crystal structures suggested that 2-substitution on the central phenyl core could reach towards the glycine rich P-loop and gain additional hydrophobic interactions by occupying a small pocket formed by lipophilic amino acid side chains. This led to methyl compound **4** which displayed an additional 30-fold increase in biochemical potency (BTK IC_{50} = 5 nM) and demonstrated good kinome selectivity (4/403 kinase >80% @ 10 μM, FLT3, KIT, PDGFRB and YSK4, with K_d = 82, 150, 53 and 160 nM respectively).

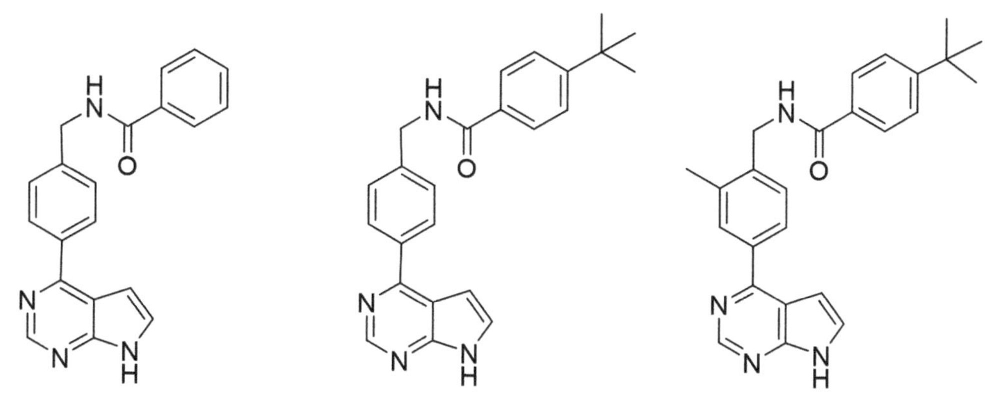

2 (hit)
BTK IC_{50} = 3700 nM

3
BTK IC_{50} = 142 nM

4
BTK IC_{50} = 5 nM

*Figure 2. Initial BTK hit **2** and analogs **3** and **4**.*

Having identified the potential lead **4**, which exhibited good biochemical potency and excellent kinome selectivity, our attention switched to optimizing the properties of the molecule with the main objectives to:

- improve cellular activity in the human whole blood (WB) assay (pBTK IC_{50} = 7.8 μM)
- reducing the clearance (in vivo rat CL $\%Q_h$ > 90)
- increasing the aqueous solubility (< 1 μg/mL at pH 6.8)

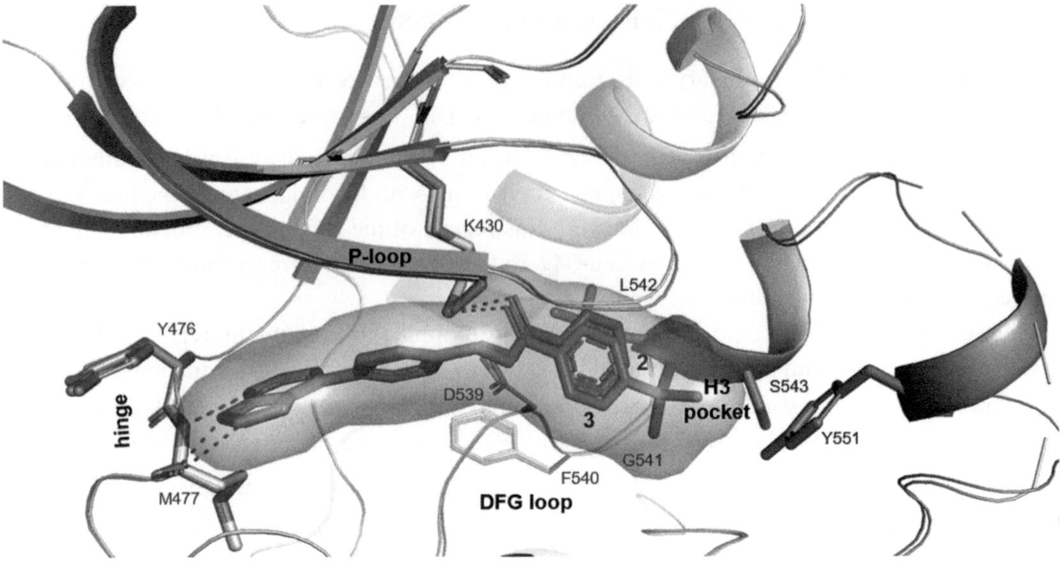

*Figure 3. Overlay of BTK co-crystal structures of hit **2** and analog **3**.*

It was surmised that the physicochemical properties (Elog D = 5.18) contributed to the poor drug-like-properties for this class of molecules, so reducing the lipophilicity was considered the first objective. Replacing the hydrophobic 4-*tert*-butyl phenyl amide with various substituted 5- and 6- membered heteroaryl moieties led to the discovery of the isothiazole **5** (Table 1), which exhibited comparable biochemical potency (BTK IC_{50} = 3 nM), but had improved the lipophilic ligand efficiency (LLE = 3.9 for **5** versus 2.9 for **4**). In addition, **5** exhibited improved aqueous solubility (10 μg/mL at pH 6.8) and lower clearance (in vivo rat CL $\%Q_h$ = 43) compared to **4**.

While focusing on improving the absorption, distribution, metabolism, and excretion (ADME) by modulating the physicochemical properties, we discovered that introducing either the 4- or 2-aminopyrimidine hinge binder afforded compounds **6** and **7** which demonstrated low clearance in vivo (**6** CL $\%Q_h$ = 6 and **7** CL $\%Q_h$ = 11) when dosed IV in a rat PK study. Although this finding was not anticipated based on the in vitro rat liver microsomal data (RLM CL $\%Q_h$ = 72 and 65, respectively), the apparent in-vivo in-vitro correlation (IVIVC) disconnect was resolved when the in vitro clearance values were corrected for the microsomal binding (**6** RLM CL $\%Q_h$ = 10 and **7** RLM CL $\%Q_h$ = 4).

To further improve the WB cellular activity and drug-like-properties, we continued to focus on increasing the biochemical potency while identifying various regions of the molecule which accommodated the presence of polar functionality without being detrimental to the potency. Utilizing the BTK co-crystal structure and the published literature, we hypothesized that introducing substitutions on the hinge binder, which extended towards the solvent front, had the potential to further improve the potency. Initial attempts to access the solvent front utilizing various alkyl substituents proved unsuccessful. Fortunately, introducing a phenyl group led to *N*-phenylpyrimidin-2-amine **8** which demonstrated a modest improvement in potency (3-fold) compared to 2-aminopyrimidine **7.** However, there was no improvement in either the cellular activity or solubility. Although analogs containing various 5- or 6- membered hetero-aryl groups (**9-14** in Table 2) demonstrated potency in the biochemical assay which was consistent with our docking studies, only the 1,4-disubstituted pyrazole **15** afforded improved biochemical potency (BTK IC_{50} = 0.2 nM) which translated into blocking phosphorylation of the BTK in human WB (pBTK IC_{50} = 0.43 μM). The identification of the pyrazole moiety was a critical breakthrough for the program as this molecule also conveyed good in vivo properties (CL %Q_h = 6, IV 1 mg/kg, and %F = 24 in CMC/Tween) when dosed in a rat PK study.

Table 1. Novel Hinge Binder BTK Inhibitors 5, 6 and 7

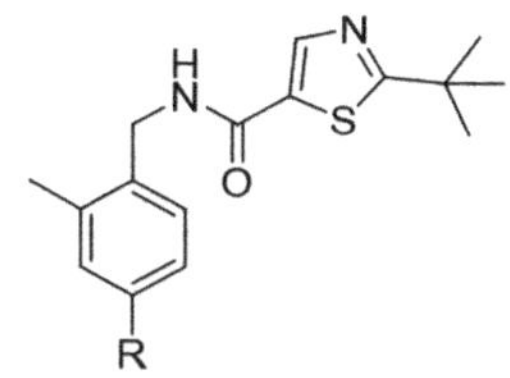

Compound	**5**	**6**	**7**
R	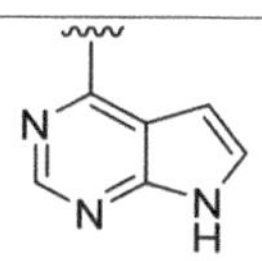	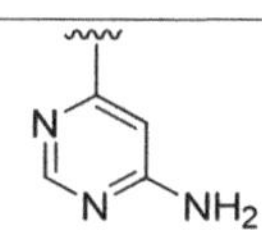	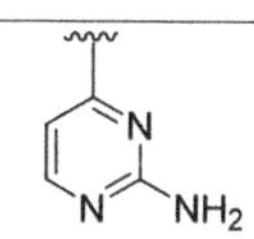
BTK IC_{50} (nM)[a]	3	9	6
LLE	3.9	4.1	4.4
Log D[b]	4.35	3.87	4.04
Solubility pH 6.8 (μg/mL)	10	6	13
Human PPB fu %	<1	3	<1
HLM/RLM (CL %Q_h)	64/91	53/72	56/65
Rat PK (CL %Q_h)	43	6	11

[a]IC50 valaues were determined using non-phosphorylated BTK protein in a FRET assay in duplicates.; [b]Measured using Analiza ELogD assay.

Table 2. SAR of Aminopyrimidine Analogs of 7

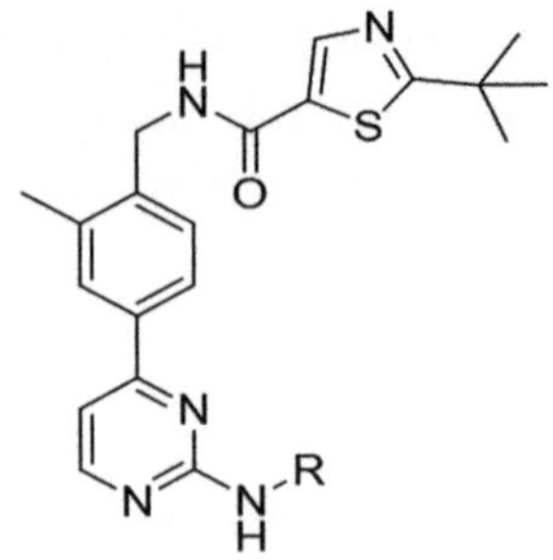

Compound	R	BTK IC_{50} (nM)[a]	WB pBTK IC_{50} (μM)	Log D[a]	Solubility pH 6.8 μg/mL	LLE
8	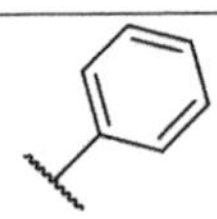	1.8	>10	3.44	<0.1	2.2
9	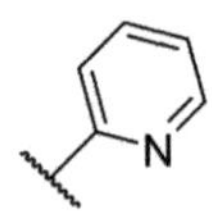	32	ND	2.17	0.23	2.4
10	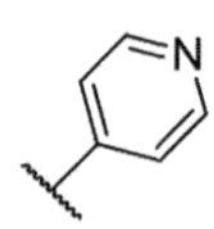	0.4	5.3	2.94	<0.1	4.2
11	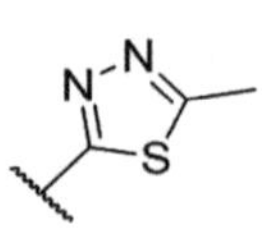	40	ND	2.07	ND	2.8
12	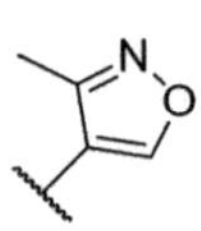	1.7	>10	2.97	0.12	3.7
13	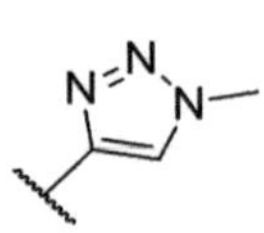	10	ND	2.79	ND	4.1
14	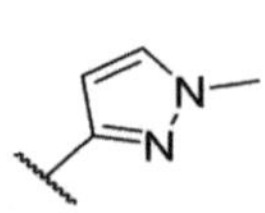	2.1	4.7	3.47	<0.1	4.0
15	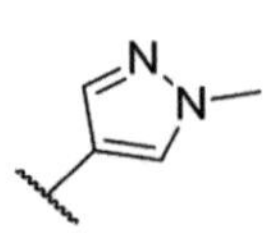	0. 2	0.43	4.76	2.7	4.9

[a]See notes from Table 1.

To enable further development of this chemical series, it was necessary to refocus on modulating the physicochemical properties which were assumed contributing to the low solubility (kinetic pH 6.8 <10 μg/mL) and high plasma protein binding (>99.5%). Based on the co-crystal structure of compound **3** and in silico docking, we determined that further optimization of the distal heteroaryl moiety may be a viable strategy to enhance the drug-like properties for the lead series (Table 3). Replacing the thiazole motif with the thiadiazole **16** helped to reduce the lipophilicity (Log D = 3) and to improve the solubility, but unfortunately this also resulted in a loss of cellular potency.

Table 3. SAR on the Amide Region of Compound 15

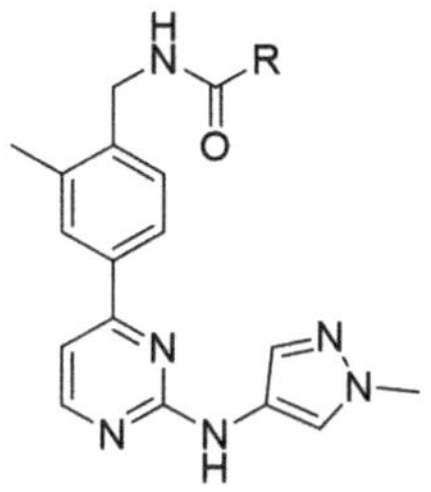

Compound	R	BTK IC_{50} (nM)[a]	LLE	LogD [a]	WB pBTK IC_{50} (μM)	Unbind WB pBTK IC_{50} (nM)[b]	Human PPB fu%	Corr. RLM, HLM CL %Q_h	IV Rat % CL Q_h	Solubility pH 1.5, 6.8 (μg/mL)
15	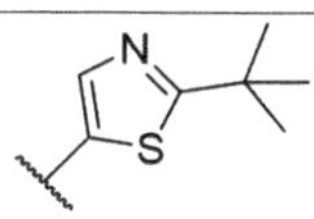	0.2	4.9	4.8	0.43	1.3	0.3	1, 2	6	<10, <10
16	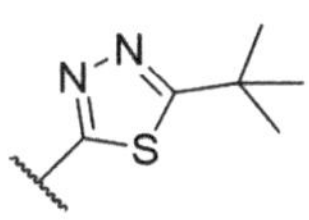	2	4.8	3.0	1.4	9.8	0.7	1, 3	18	>50, <10
17	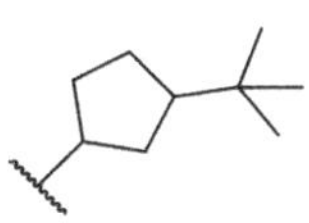	2	3.5	4.6	>10	7	0.07	3, 14	31	<10, <10
18	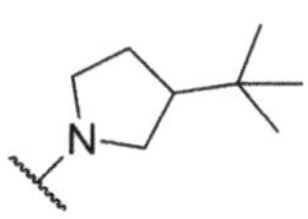	0.2	5.0	3.7	0.35	0.7	0.2	1, 19	4	<10, <10
19	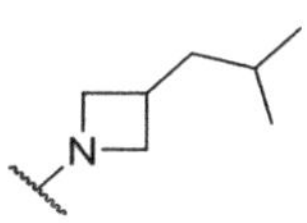	0.4	5.0	3.3	0.65	5.9	0.9	1, 8	17	<10, <10
1	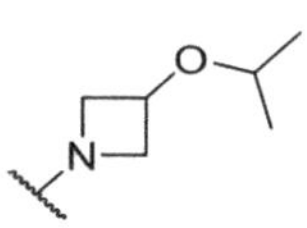	1	5.0	2.3	0.12	7.1	5.9	3, 9	1	>50, >10

[a]See notes from Table 1. [b]Unbound WB pBTK IC50 = WB pBTK IC50 x Human PPB fu.

Switching to a substituted carbocyclic such as cyclopentyl **17** confirmed that non-aromatic motifs were tolerated and maintained the desirable biochemical potency while increasing the sp^3 character of the molecule. Furthermore, replacing the hydrophobic cyclopentane motif with a substituted pyrrolidine **18** reduced the lipophilicity (**18** LogD = 3.7 versus **17** LogD = 4.8) and led to a significant improvement in the cellular potency (human WB pBTK IC_{50} = 0.35 μM).

Having validated the hypothesis that modulating the physicochemical properties was a viable strategy for improving the WB cellular potency, a homologous series of analogs was synthesized, leading to the identification of **19** containing the 3-isobutylazetidine moiety. This lowered the lipophilicity (Log D = 3.3) while maintaining excellent biological activity (BTK IC_{50} = 0.4 nM and human WB pBTK IC_{50} = 0.65 μM).

Finally, introducing an oxygen linkage at the 3-position of the azetidine led to the discovery of compound **1** (BTK IC_{50} = 1 nM). This simple structural modification reduced lipophilicity (LogD = 2.3), increased the aqueous solubility, and improved the whole blood cellular potency (human pBTK IC_{50} = 0.12 μM). Furthermore, compound **1** demonstrated low in vivo clearance (CL %Qh = 6) and acceptable oral bioavailability (%F = 48) when dosed in a rat PK study.

Pharmacology and In Vivo Assessment of BIIB068

BIIB068 (**1**) demonstrated excellent selectivity when screened against a panel of 395 kinases binding only 10 kinases with >70% inhibition at 1 μM. The overall kinase selectivity was based on the K_d values determined against the 10 kinases identified from the kinome screen (AAK1 = 430 nM, BMP2K = 1100 nM, BTK = 0.3 nM, CDKL2 = 420 nM, JAK2= 460 nM, PIP5K1A = >10000 nM, RIOK1= 1200 nM, ROIK2 = 2400 nM, STK1 = 880 nM, YSK4 = 130 nM), BIIB068 afforded more than 400-fold selectivity over other kinases and more than 1000-fold selectivity over the Tec family members.

To further interrogate the overall kinome selectivity in cells, BIIB068 was screened in the BioMAP system, an innovated platform comprising 12 human primary cell-based assays developed to assess both on-target and off-target activity in a cellular system using pathway and disease relevant biomarkers. Consistent with the biological profile anticipated for a selective BTK inhibitor (27), BIIB068 demonstrated dose dependent activity (30, 10, 3.3, and 1.1 μM) which was restricted to the BT culture systems (Figure 4). In addition, BIIB068 inhibited phosphorylation of PLCγ2 with an IC_{50} = 0.4 μM in Ramos cells. Furthermore, in human peripheral blood mononuclear cells (PBMCs), BIIB068 blocked both anti-IgD and anti-IgM BCR-induced B cell activation (IC_{50} = 0.11 μM and 0.21 μM, respectively) which taken together, is consistent with the biological profile of a selective BTK inhibitor.

BIIB068 demonstrated good in vitro ADME properties, high permeability with moderate efflux potential in the Caco-2 cell system and moderate to high plasma protein binding across all preclinical species and in humans, with (fu_{rat} (1.28%) < fu_{cyno} (3.62%) < fu_{human} (5.95%) < fu_{dog} (11.3%)). The potential risk for BIIB068 to cause a drug-drug interaction (DDI) was deemed low as there was no inhibition (reversible or time dependent induction (TDI)) of any CYP isoforms (IC_{50} = >25 μM) in human liver microsomes nor was there induction of CYP activity in the human hepatocyte assay when tested up to 50 μM. The major pathway for metabolism was via CYP 3A4 (66%), followed by minor contributions from CYP1A2, 2C19 and 2D6. The in vitro metabolism of BIIB068 confirmed that no unique human metabolites, glutathione (GSH) or CN adducts were formed after BIIB068 was incubated in the presence of human microsomes and hepatocytes.

BIIB068 exhibited low clearance in rat, dog and cynomolgus monkey, and achieved moderate oral bioavailability (26-32% in dogs and cynomolgus monkeys, and 48% in rats) when dosed as suspension. The non-clinical in vitro cardiovascular safety assessment showed no hERG inhibition (IC_{50} > 50 μM) at the maximum concentration tested or in vitro cytotoxicity (IC_{50} = >20 μM) when tested in the HepG2 cell assay. This molecule was considered as a low risk for genetic toxicity as it was negative for clastogenic and mutagenic responses in vitro and negative in the Ames assay.

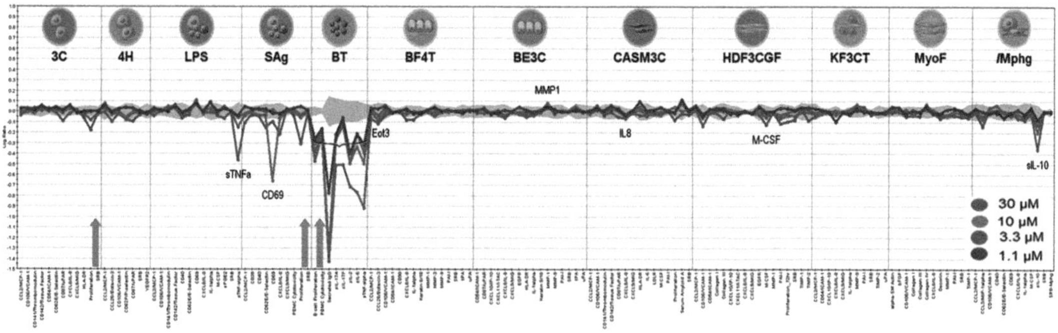

*Figure 4. Phenotypic BioSeek study data of BIIB068 (**1**) at 1.1, 3.3, 10 and 30 μM. The gray area shows the 95% confidence area of each biomarker response in the absence of drug. Data points that fall outside of the gray area signify inhibition (below) or augmentation (above) of the biomarker response at the indicated concentration of BIIB068 (**1**).*

Table 4. In Vitro Biological Profiling of BIIB068 (1)

Assay (measured output, stimulus, cell type)	*IC50 ± SD (μM)*
B cell activation (CD69, anti-IgD/BCR, PBMC)	0.111 ± 0.087 (N = 10)
ROS generation (ROS, SLE IC/FcγR, neutrophils)	0.054 ± 0.057 (N = 4)
Monocyte (PB IgG, IL-6, IL-1b, TNFα/FcγR, PBMC)	0.063 ± 0.034 (N = 4)
pBTK inhibition (pTyr, n/a, human WB)	0.12 ± 0.15 (N = 51)
pBTK inhibition (pTyr, n/a, mouse WB)	0.17 ± 0.100 (N = 5)
B cell activation (CD69, IL-4/anti-CD40/CD40, PBMC)	>10
T cell activation (CD69, anti-CD3/CD28/TCR, PBMC)	>10

BIIB068 selectively inhibited B cell, neutrophil and monocyte activation in vitro and did not inhibit T cell responses in vitro (Table 4). Although inhibition of the BCR-mediated B cell activation with anti-IgD stimulation (CD69 IC_{50} = 0.111 μM) was observed in PBMCs, BIIB068 did not inhibit induction of CD69 on CD19+ B cells (IC_{50} >10 μM) upon treatment with anti-CD40 and IL-4 in PBMC's. This clearly demonstrated that not all B cell activity was blocked by inhibition of BTK, unlike B-cell depletion therapies. Further, when human neutrophils from healthy donors were activated with immune-complexes (IC) from SLE patients (SLE IC), BIIB068 inhibited FcγR-mediated ROS production (IC_{50} = 0.054 μM).

To evaluate the translatability of the in vitro biological potency to in vivo efficacy, BIIB068 was dosed in a thymus-independent type 2 (TI-2) antigen mouse model. As reported (28), TI-2 antigens rely heavily on BCR signaling and require BTK expression to generate a productive antibody response in vivo, thus the TI-2 model was considered a suitable model for demonstrating on-target proof of biology preclinically. BIIB068 was dosed BID for ten consecutive days in mice, after which serum, plasma and WB were collected, and the antigen-specific antibody responses, plasma concentration and target engagement in WB PD were measured (Figure 5). A dose dependent reduction of anti-NP IgM antibody titers (93%, 90%, 63%, 22% and -23%) was observed at 50, 30, 10, 3 and 1 mg/kg/dose with a calculated ED_{50} of 4.4 mg/kg/dose. Similarly, a dose dependent PK/PD response was observed for the inhibition of pBTK at the 50, 30, 10, 3 and 1 mg/kg/dose corresponding to an in vivo EC_{50} = 0.16 μM, consistent with the in vitro mouse WB potency. Analysis of the PK/PD/efficacy data demonstrated that the 30 mg/kg/dose was sufficient to achieve a ~90% reduction in anti-NP IgM antibody titers which suggested that full target coverage throughout the entire dosing period was not required to drive high efficacy in the preclinical TI-2 mouse model. Having established a PK/PD and efficacy relationship and demonstrating excellent in vitro safety, BIIB068 was advanced into tolerability studies.

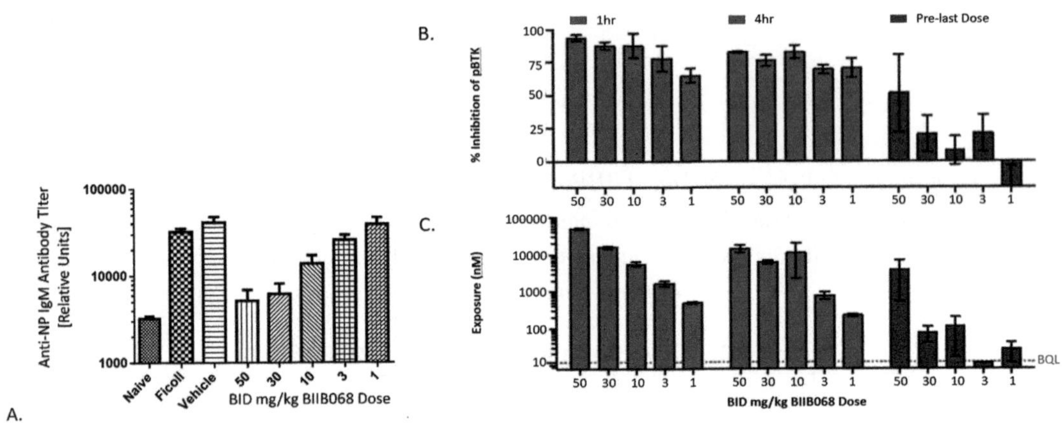

*Figure 5. PK and PD, efficacy analysis of BIIB068 (**1**) in the TI-2 efficacy model. (A). Antigen-specific antibody response to a TI-2 antigen. (B). pBTK % inhibition at 1 h, 4 h and pre-last dose. (C). Total drug concentration at 1 h, 4 h and pre-last dose.*

BIIB068 was dosed in a rodent (Sprague-Dawley rat) and non-rodent (cynomolgus monkey) GLP toxicity studies for 13 weeks and was shown be safe and well tolerated. There was no compound related adverse effects such as changes in body weight, food consumption, ECG, clinical pathology, organ weight or gross pathology/histopathology observed in the studies. The No Observed Adverse Effect Level (NOAEL) was determined at 50 mg/kg/day for rat and 100 mg/kg/day for monkey which corresponded to a daily rat Auc_{inf} 376 μg*h/mL and monkey Auc_{inf} 129 μg*h/mL.

BIIB068 was dosed in a Phase I SAD study in healthy human volunteers to evaluate the PK and tolerability. The starting dose of 10 mg was chosen to afford a safety margin > 10-fold based on the NOAEL determined from the GLP preclinical toxicity studies. BIIB068 was well tolerated when dosed in healthy volunteers, and dose linear PK was observed in 10 mg, 30 mg and 60 mg cohorts although less than dose portion increase in C_{max} or Auc was observed in the 120 mg cohorts. The peak plasma concentration (C_{max}) occurred at ~0.9-2 h after dosing and median $t_{1/2}$ ranged from 6-9 h across all cohorts using an immediate release formulation (29). BIIB068 exhibited a dose dependent PK/PD response as measured by the inhibition of constitutive BTK phosphorylation in WB lysate (Figure 6). The 60 mg cohort achieved maximal inhibition of BTK phosphorylation (>

90%) for up to 8 h post-dosing, and more than 70% inhibition of BTK phosphorylation for up to 12 h post-dosing was achieved at or above 30 mg single doses.

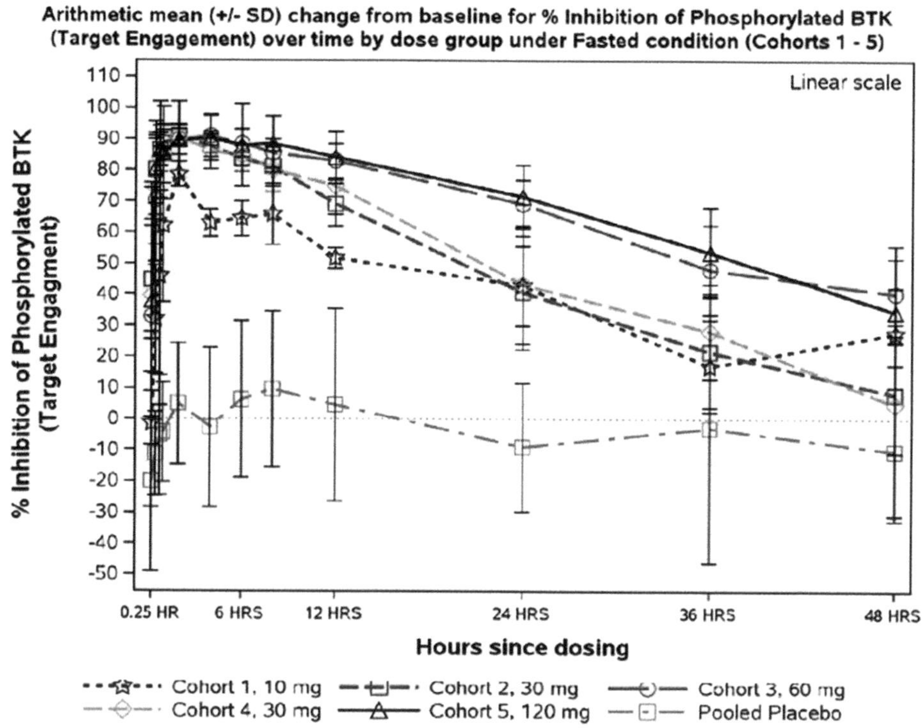

A.

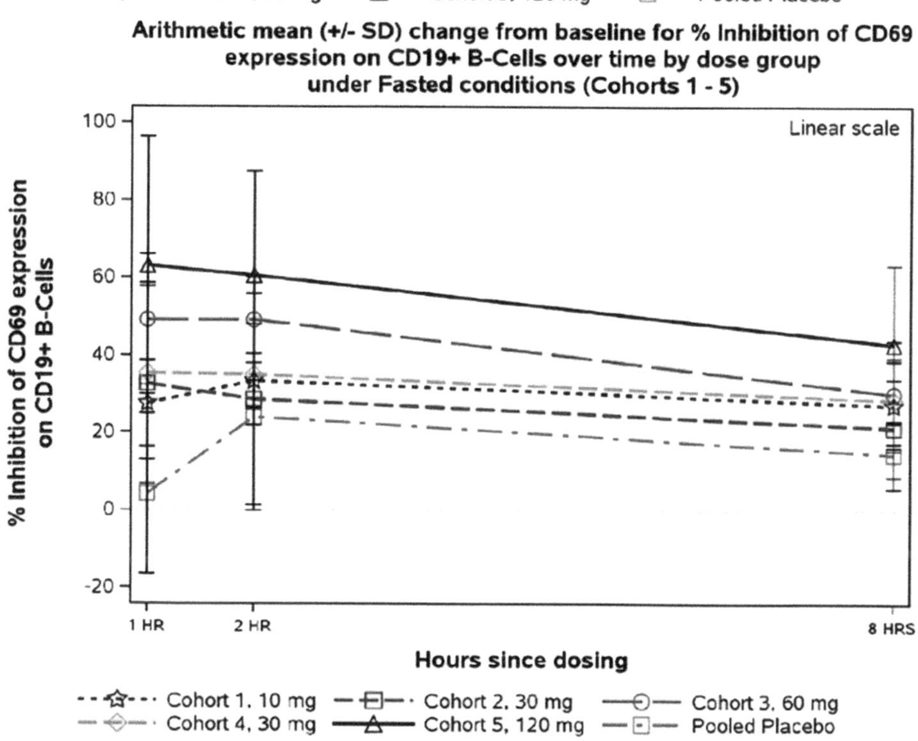

B.

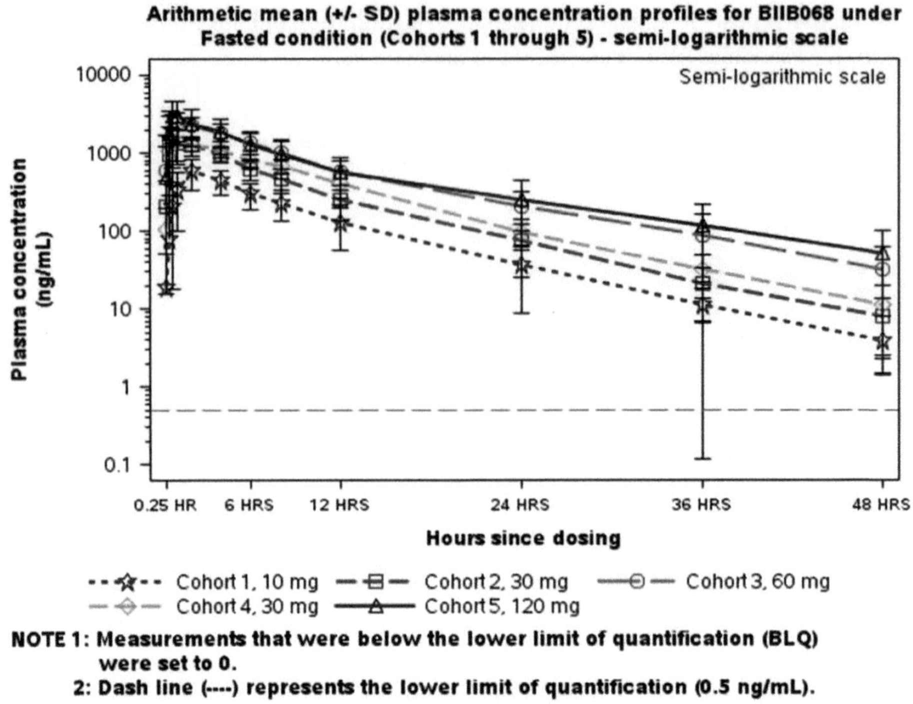

C.

*Figure 6. Dose response of BIIB068 (**1**) in healthy human cohorts (placebo, 10 mg, 30 mg, 60 mg, 120 mg). (A) pBTK inhibition. (B) CD69 inhibition. (C) Total drug concentration in plasma.*

Although full inhibition of pBTK was observed in healthy volunteers, this translated to only partial inhibition of BCR-mediated B cell activation as measured by CD69 expression when WB was stimulated with anti-IgD antibody. Due to apparent disconnect between target engagement (pBTK inhibition) and functional inhibition of B cell as measured using the anti-CD69 biomarker, additional studies may be required to determine if this has clinical significance for the treatment of autoimmune diseases.

Discovery Chemistry Synthesis of BIIB068

The discovery synthesis was designed to maximize access to the analogs of this chemical series. As illustrated in Scheme 1 by starting from commercially available nitrile **20**, reduction yielded amine **21** which was subsequently Boc-protected to give *N*-Boc **22**. After conversion to boronate **23**, Suzuki coupling led to the 4-phenylpyrimidine **24**. This was used extensively to achieve different analogs and remained to be a key building block into process development later.

Palladium catalyzed amination installed the 3-amino pyrazole moiety to yield the pyrazole-pyrimidine **25**, and deprotection yielded primary amine **26**. A CDI mediated urea formation led to the synthesis of BIIB068. Noticeably, two silica gel column chromatographic purifications were required at the pyrazole-pyrimidine **25** and BIIB068 stages. Tens of grams of BIIB068 were prepared through this route and supported the non-GLP studies. An extensive salt screening led to the identification of hemi-adipate salt form which was suitable for clinical use.

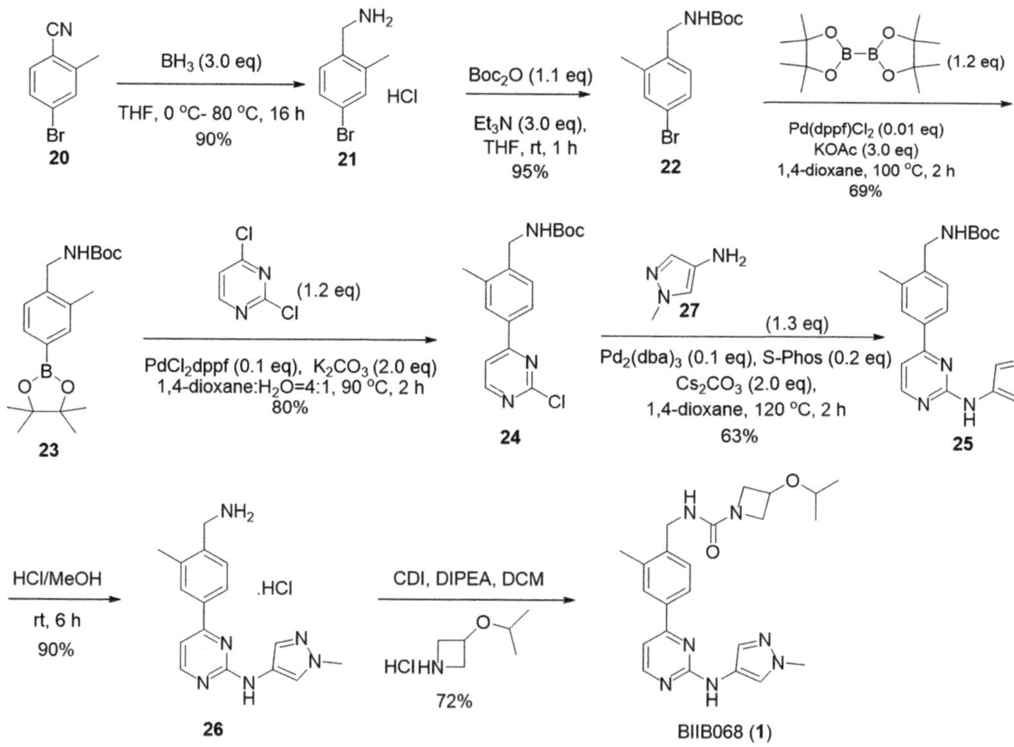

*Scheme 1. Synthesis of BIIB068 (**1**).*

Development of an Efficient Large-Scale Synthesis of BIIB068

General Strategy for Process Development

Although the discovery chemistry synthesis of BIIB068 (**1**) was largely a linear synthesis (*30*), we considered this route as suitable for the initial kilogram scale preparation of **1** for three reasons:

- The team was faced with tight timeline to produce enough API and support toxicology studies, formulation development and Phase I clinical trials
- Sufficient quantities (> 10 kg) of advanced intermediate **24** had been staged for GMP manufacturing
- Process improvement opportunities existed to a large extent considering the fact that the discovery synthesis steps were not thoroughly examined/optimized.

Therefore, we decided to focus our efforts on the development of final three steps (from **24** to **1**) and our goals were set to:

- Improve reaction profiles and yields of each transformation to enable large-scale production
- Closely monitor and appropriately control the impurities produced in the process to meet the product quality profile
- Ideally remove both chromatographic purifications (step **24** to **25** and step **26** to **1**) to facilitate large-scale synthesis of BIIB068.

Aminopyrazole Installation and In Situ Boc Deprotection

C-N coupling of 2-chloropyrimidine **24** with aminopyrazole **27** was achieved under Pd catalysis in the discovery chemistry route to BIIB068. However, as demonstrated in numerous literature reports, amination of 2-chloropyrimidines can be achieved under thermal conditions in the presence of either acids (*31*, *32*) or bases (*33*, *34*), in addition to transition metal-catalyzed processes (*35–37*).

a) Initial screening and optimization: An initial screening study was carried out and we discovered that transition metal (Pd) catalysis (*38*) or the undesirable 1,4-dioxane (*39*) as solvent were not prerequisites for a successful amination reaction (Table 5, entry 1). Although the reaction conditions using Cs_2CO_3 in the absence of Pd catalyst afforded only 10% desired product **25** (Table 5, entry 2), significant improvement (~ 50%) was realized when K_2CO_3 was used as an alternative base in 1-butanol or NMP (Table 5, entries 4 and 5). Satisfyingly, the reaction profiles were further improved to ~ 60% of **25** with ~ 30% of **24** remaining when the base was changed to organic bases such as DIPEA or Et_3N (Table 5, entries 6 and 7).

Gratifyingly, we discovered that the acidic conditions not only effected the desired C-N bonding formation, but also had the added benefit of removing the Boc protecting group in a one-pot process. In an initial experiment, 3 equiv of H_3PO_4 was used in the reaction (in 1-butanol) and 91% of the des-Boc product **26** was detected in the crude reaction mixture (Table 5, entry 8).

Next, we investigated the HCl salt of aminopyrazole **27** as an alternative coupling partner, expecting the deprotection reaction would proceed with the existing counter ion of HCl from the starting material and would not require additional acid, such as H_3PO_4. To our delight, the reaction proceeded cleanly to afford 92% of **26** (Table 5, entry 10). After further optimization of reaction conditions including solvent and temperature, the optimal reaction conditions were identified as heating **24** and **27.HCl** salt in 1-butanol to 85 °C. The reaction performed at 85 °C and 100 °C had similar reaction profile (Table 5, entry 11). Therefore, the HCl from the aminopyrazole **27.HCl** salt and that liberated from the displacement reaction was sufficient to trigger the Boc deprotection.

Close examination of reaction profiles revealed the formation of two butoxy impurities (**29** and **30**, Figure 7) (*40*) under various conditions (Table 5, entries 3 and 4) from solvent incorporation, therefore, we decided to switch from 1-butanol to the sterically more hindered 2-butanol. To our delight, we did not detect the formation of the corresponding 2-butanol impurities under these conditions. However, the reactivity in 2-butanol was lower in comparison with 1-butanol. For example, 15% of **25** and 9% of **28** remained after 16 h of reaction when 2-butanol was used as solvent (Table 5, entry 12).

After further optimization, the solvent mixture of 2-butanol/water was identified to give better reactivity than 2-butanol alone (Table 5, entry 13). The binary solvent system enabled a homogeneous reaction mixture compared with a heterogeneous reaction mixture when either 1-butanol or 2-butanol was used as a single solvent. This homogeneous binary solvent system also turned out to be advantageous for large-scale production. Our experiments indicated that the reactions proceeded well using 2-butanol / water mixture with volume ratios ranging from 5:2 to 2:3. However, the actual volume of 2-butanol had a significant impact on the product loss to the mother liquor. Two volumes of 2-butanol (with respect to the weight of **24**) gave maximal isolated yield. An excess of aminopyrazole **27.HCl** salt was required to drive the reaction to completion. The stoichiometry of **27.HCl** salt was initially set as 1.2 equiv, and later increased to 1.3 equiv in an effort to further accelerate the reaction. Under optimized reaction conditions, it was concluded that 1.3

equiv of aminopyrazole **27.HCl** salt gave optimal reaction as more did not give further improvement of reaction profile (Table 5, entries 13-15).

Table 5. Screening Results for the C-N Coupling of 24 with 27[a]

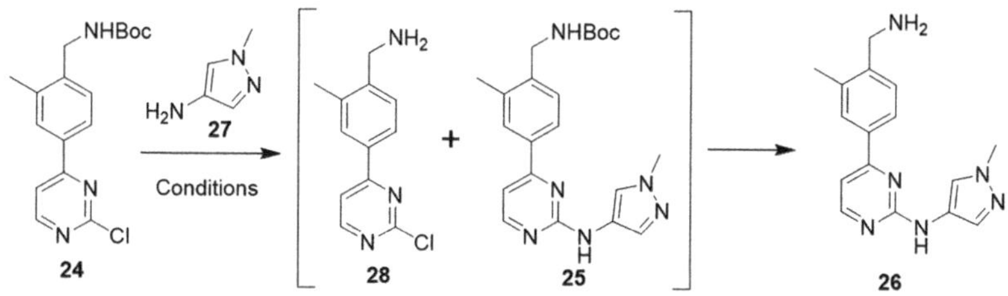

Entry	Reagents	Solvents	Temp	24 (HPLC A%)	25 (HPLC A%)	26 (HPLC A%)
1[b]	$Pd_2(dba)_3$(0.2 equiv), S-Phos(0.1 equiv), Cs_2CO_3 (2 equiv)	1,4-dioxane	100 °C	2	58	0
2	Cs_2CO_3 (2 equiv)	1,4-dioxane	100 °C	67	10	0
3[c]	Cs_2CO_3 (2 equiv)	1-butanol	100 °C	< 1	< 1	< 1
4[d]	K_2CO_3 (2 equiv)	1-butanol	100 °C	5	47	< 1
5	K_2CO_3 (2 equiv)	NMP	100 °C	25	52	< 1
6	DIPEA (2 equiv)	1-butanol	100 °C	31	61	< 1
7	Et_3N (2 equiv)	1-butanol	100 °C	32	60	< 1
8	H_3PO_4 (3 equiv)	1-butanol	100 °C	0	0	91
9	AcOH (3 equiv)	1-butanol	100 °C	0	43	54
10[e]	None	1-butanol	100 °C	0	0	92
11[e]	None	1-butanol	85 °C	0	0	91
12[e,f]	None	2-butanol	85 °C	< 1	15	71
13[e]	None	2-butanol/ water (2V/3V)	85 °C	0	0	95
14[e,g]	None	2-butanol/ water (2V/3V)	85 °C	0	0	94
15[e,h]	None	2-butanol/ water (2V/3V)	85 °C	0	0	93

[a]1.3 equiv of 27 or 27·HCl salt used for reaction conditions except for entries 14 and 15. All reactions were run for 16 h except for entries 1 and 2 for which the reactions aged for 2 h before analytical samples were taken. [b]Crude reaction contained 15% of 28. [c]Crude reaction contained 51% of 29 and 28% of 30. [d]Crude reaction contained 39% of 30. [e]27·HCl salt instead of 27 free base was used for the reactions. [f]Crude reaction mixture contained 9% of 28. [g]1.5 equiv of 27·HCl salt used. [h]2.0 equiv of 27·HCl salt used.

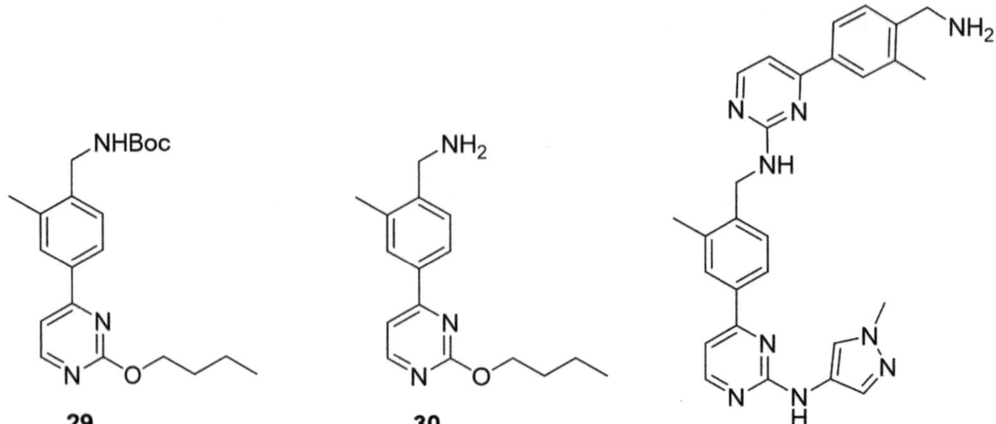

29
Table 5, entry 3

30
Table 5, enties 3 and 4

31

*Figure 7. Impurities generated in amination reaction of 2-chloropyrimidine **24** with aminopyrazole **27**.*

b) Reaction workup and product isolation: During process development, it was necessary to completely consume the 2-chloropyrimidine **24** prior to reaction workup, otherwise a dimer-like side-product **31** (Figure 7), which purged poorly in the subsequent step, would be generated as a result of displacing the chloride on unreacted **24** by the benzylic amino group of **26** under basic workup conditions. Therefore, an IPC specification of no more than 0.05% unreacted 2-chloropyrimidine **24** was set for reaction completion.

After exploring various workup conditions, we discovered that **26** can readily precipitate as the free base through the addition of aqueous ammonium hydroxide. Initial workup condition involved addition of ammonium hydroxide followed by water at 20 °C. However, this process was complicated as the free base of benzylamine **26** precipitated out quickly when the pH of the reaction mixture changed from acidic to basic, resulting in the formation of a very thick slurry that was difficult to stir. In order to alleviate this agitation issue, the workup temperature was increased to 70 °C which enabled mixing, but also resulted in an unidentified late-eluting impurity.

The optimal balance of workability and purity was accomplished by decreasing the workup temperature to 35 °C and diluting the reaction mixture with water followed by ammonium hydroxide addition to induce free base precipitation. Further process optimization resulted in the following workup protocol: addition of 60% of the total water amount to the reaction followed by addition of a mixture of ammonium hydroxide and the remaining 40% of the water amount. This protocol was applied in the large-scale synthesis and resulted in benzylamine **26** at 90% isolated yield and > 99% HPLC purity.

Urea Formation Leading to Synthesis of BIIB068 (1)

After optimization of the benzylamine **26** synthesis, our attention turned to the last chemical bond formation step: generating **1** via an urea synthesis (Scheme 2 and Table 6).

a) Choice of azetidine form: Selection of the appropriate azetidine salt form as starting material for the urea formation was deemed an important consideration in the chemical development process. 3-Isopropoxyazetidine (**32**) is readily available as either a liquid free base or a solid HCl salt. Both forms can be used in the current process, however, the free base was determined to be advantageous over the HCl salt based on the following:

- The HCl salt is extremely hygroscopic and required special handling
- The reaction was noticeably faster when the free base was used
- A dimethyl sulfoxide (DMSO) solution of the HCl salt had to be prepared prior to charging to the reaction, while the liquid free base could be conveniently charged into the reaction directly
- A higher level of the chloride ring-opening impurity **33** (Figure 8) occurred using the HCl salt for the reaction which suggests a side reaction of **1** with HCl in the reaction mixture.

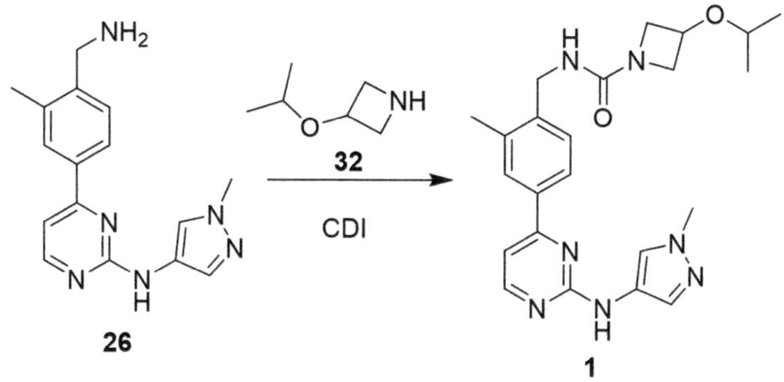

*Scheme 2. Urea synthesis to generate BIIB068 (**1**).*

Therefore, we selected the use of the free base 3-isopropoxyazetidine (**32**) instead of its HCl salt for further development.

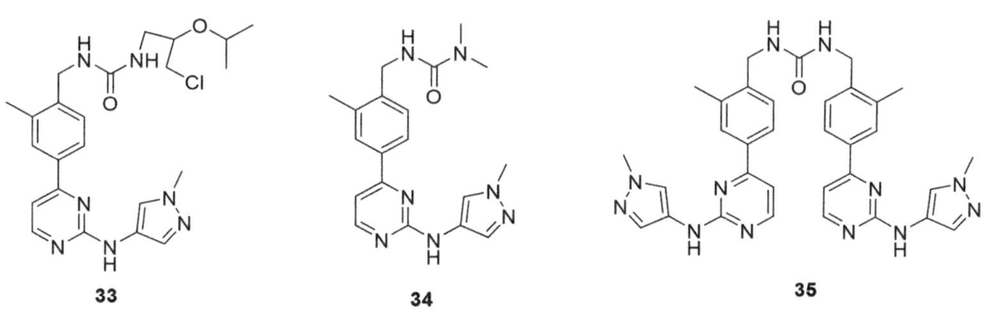

Figure 8. Impurities generated during urea formation step.

b) Reaction solvent selection: The urea formation reaction can be conducted in different solvents. To prevent the reaction stalling issue that was typical for some heterogeneous reactions, polar aprotic solvents were chosen for scale-up evaluation as they fully dissolved all materials and intermediates. *N,N*-Dimethylformamide (DMF) and dimethylacetamide (DMAc) were utilized for the synthesis of the early toxicology lots. After identification of an impurity associated with DMF (**34**, Figure 8; Table 6, entries 1-2) by LC-MS, we replaced it with DMAc. However, a bis-addition side-product **35** was formed in significant quantity (10~21%) when DMAc was used as the reaction solvent (Figure 8; Table 6, entries 3-4). After production of the early toxicology lots, DMSO, a class 3 solvent based on ICH guidelines, was identified as a better choice providing a cleaner reaction (Table 6, entry 5). Therefore, for all subsequent BIIB068 production campaigns supporting IND-enabling toxicology and clinical studies, DMSO was selected as the solvent for the urea formation step.

c) Bis-addition impurity 35: As previously discussed, formation of impurities **33** and **34** can be prevented by using alternative reactants or solvents. Unfortunately, various amounts (as low as close to 1% when DMSO was used as solvent) of dimeric impurity **35** occurred in all reactions. The impurity level of **35** was associated with the order of addition of **26** and CDI, as well as the CDI equivalents. When a solution of CDI was added to a solution of benzylamine **26**, impurity **35** was generated in a significantly larger amount (Table 6, entries 1-4). The level of dimeric urea **35** was reduced by adding the solution of **26** to the CDI solution over one hour at 30-40 °C, which was adopted for further optimization and large-scale production.

Table 6. Optimization of Urea Formation Step to Minimize Impurity 34 and 35[a]

Entry	*Solvent*	*Reaction conditions for mixing 26 with CDI*	*Product 1[b] (HPLC A%)*	*Impurity 34 (HPLC A%)*	*Impurity 35 (HPLC A%)*
1	DMF	1.2 equiv of CDI in DMF solution was added to **26** in DMF	92.7	1.8	1.3
2	DMF	**26** in DMF solution was added to 1.2 equiv of CDI in DMF	94.3	1.9	0.7
3	DMAc	0.8 equiv of CDI in DMAc solution was added to **26** in DMAc	73.6	0	21
4	DMAc	**26** in DMAc solution was added to 0.8 equiv of CDI in DMAc	86.5	0	10.2
5	DMSO	**26** in DMSO solution was added to 0.9 equiv of CDI in DMSO	95.8	0	1.4
6	DMSO	**26** in DMSO solution was added to 1 equiv of CDI in DMSO	96.5	0	0.8
7	DMSO	**26** in DMSO solution was added to 1.2 equiv of CDI in DMSO	96.6	0	0.6

[a] Reactions were all carried out at 30 °C first by mixing 26 and CDI for < 1 h and then followed by charging 32 and aging for 5 h. [b] < 0.5% HPLC A% of 26 was found in all of the crude reaction mixtures.

When CDI was undercharged, greater amounts of dimeric impurity **35** were also produced. For example, when 0.8 equiv of CDI was used in the process, the impurity level of **35** reached as high as ~20% (Table 6, entry 3). When the amount of CDI was optimized with DMSO as solvent, and it was found that 1.2 equiv of CDI were adequate to drive the reaction to completion without significant dimeric impurity **35** formation (Table 6, entries 5-7). Slight decomposition of the CDI reagent before use can also result in impurity **35**.

To mitigate this decomposition risk, exposure of CDI to air or water was alleviated by storing it with refrigeration under an inert atmosphere of nitrogen. In order to maintain reagent integrity in manufacturing practice, unopened, heat sealed, UV-protected bags were used to store CDI.

Removing the small amounts of dimeric impurity **35** from **1** was accomplished by dissolving crude **1** in ten volumes of ethanol at reflux. The insoluble impurity was removed by hot filtration. This procedure was incorporated in the final co-crystal formation step (see below).

The optimized urea formation procedure was demonstrated at > 100 g scale generating **1** in > 95% isolated yield and > 99% HPLC purity.

BIIB068 Hemi-Adipate Co-Crystal (36)

Initially, salt screening efforts resulted in the identification of two potential solid forms of **1**: the HCl salt and the hemi-adipate co-crystal **36** (*41*). The two forms were evaluated as possible suitable solid forms for further development. However, when **1** was treated with HCl, the formation of chloro-impurity **33** occurred, which was also flagged as a potential mutagenic impurity (PMI) by DEREK analysis (*42–44*). Therefore, we decided to focus our development efforts on the hemi-adipate co-crystal **36** (Scheme 3) (*45*).

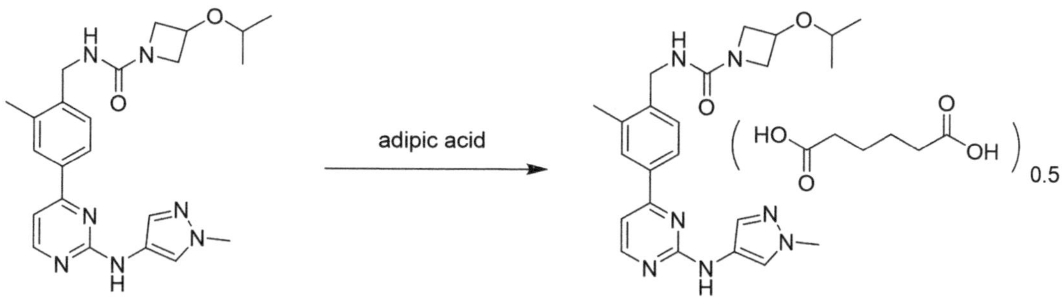

*Scheme 3. **1** Hemi-adipate co-crystal formation.*

Initial experiments suggested that both the hemi-adipate co-crystal and mono-adipate co-crystal can be isolated in salt formation attempts. To favor hemi-adipate co-crystal formation, 0.9-1.1 equiv of adipic acid was necessary, while for mono-adipate co-crystal formation, 1.9-2.1 equiv of adipic acid was used. The requirement of excess amounts of adipic acid equivalents to crystallize either one of the two co-crystal suggests potentially loosely bound molecules in both co-crystals. The hemi-adipate co-crystal was selected for further development.

In our early experiments, the hemi-adipate co-crystal was readily generated via primary nucleation without seeding. However, the process became challenging during development presumably because of the purity profile change of feed stream: the new impurity(ies) inhibited primary nucleation of hemi-adipate, which then could only be produced via seeding and secondary nucleation. To ensure a successful campaign, a seeded co-crystal protocol was therefore utilized.

THF or ethanol was originally used as the solvent for the hemi-adipate co-crystal formation, which gave similar yield, purity and processing in either solvent. In order to develop a practical co-crystal formation process, we conducted solubility screening. Both **1** and its hemi-adipate **36** exhibited high solubility in polar non-protic solvents such as DMSO, DMF and NMP, however, disproportionation was facilitated with these solvents. Moderate solubility (highly temperature-dependent) was observed in alcohol solvents (methanol, ethanol, 1-propanol, 2-propanol, 1-butanol) and in the general solvents THF, 1,4-dioxane, dichloromethane and methyl ethyl ketone. Overall, the solubility screen for **1** and **36** displayed alcohol systems (e.g.1-propanol, ethanol, 1-propanol/ethanol mixture) to be most promising for further crystallization development of **36**. The solubility of both **1** and **36** in alcohols was suitable for scale-up, with **36** exhibiting solubility of ca. 90-120 mg/mL at 70 °C and less than 10 mg/mL at 5 °C, rendering a cooling crystallization ramp feasible. At this stage of process development, ethanol, a Class 3 solvent, was selected for further development.

An ethyl carbamate impurity **37** arose in the hemi-adipate process when ethanol was used as solvent (Scheme 4). Subsequent control experiments suggested formation of **37** was the result of unreacted urea intermediate **38** (from urea formation step) in **1** reacting with ethanol in the co-crystal formation step. To minimize formation of **37**, the urea synthesis step was modified to include a residual limit of <0.1% of **38**. In order to achieve this limit, an increased urea step reaction time or an additional azetidine charge was implemented. Additional lab experiments suggested that **1** also reacted with ethanol to generate carbamate impurity **37** at ~ 0.002% per hour at 75 °C, which increased significantly with the addition of adipic acid. When adipic acid was present, impurity **37** formed at a rate of ~0.01% per hour at 75 °C. To minimize the formation of the ethyl carbamate impurity **37** from **1**, the GMP co-crystal formation process was modified such that instead of combining the materials and heating, **1** and the adipic acid were heated separately in ethanol to 75 °C and then combined to minimize the time both reactants would be heated together.

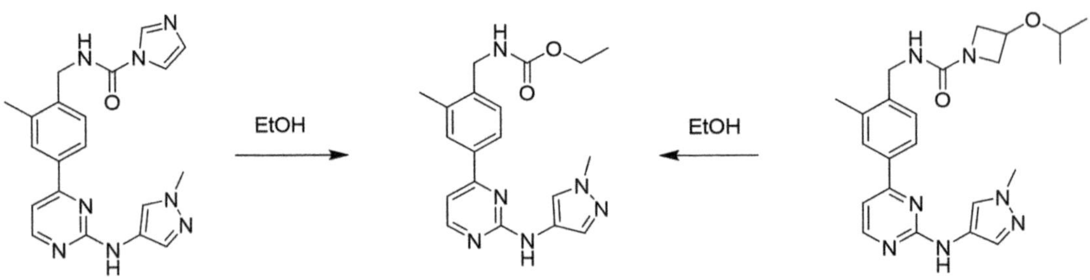

38 **37** **1**

Scheme 4. Formation pathway of ethyl carbamate impurity ***37****.*

A consistent and reproducible process was thus developed. The reaction was carried out at > 70 °C, followed by seeding and crystallization at 60 °C and ramp cooling to 0 °C. The product was isolated by filtration. Slow cooling was found to be essential as rapid cooling caused the mono-adipate co-crystal and/or **1** to precipitate regardless of the use of seeding. The optimized process generated hemi-adipate co-crystal **36** in 90% yield and > 99% HPLC purity.

Large-Scale Synthesis

The optimized conditions described above were utilized for multi-kilogram synthesis of BIIB068 hemi-adipate co-crystal **36**. The lab results were effectively reproduced in the large-scale manufacturing process generating 9.7 kg of **36** with high purity (99.4% by HPLC A%) in excellent overall yield (80%, Table 7). In comparison, the initial medicinal chemistry route generated **1** in 47% yield from the same starting material **24**. Gratifyingly, two silica gel column chromatography steps in the medicinal chemistry synthesis (**24** to **1**) were successfully removed in the large-scale process. In conclusion, process development efforts resulted in not only a robust chemical synthesis but also production of highly pure and sufficient quantities of BIIB068 (as its co-crystal form **36**) to support toxicology and clinical studies.

Table 7. Overview of Multi-kilogram Synthesis of BIIB068 Hemi-adipate Co-crystal 36

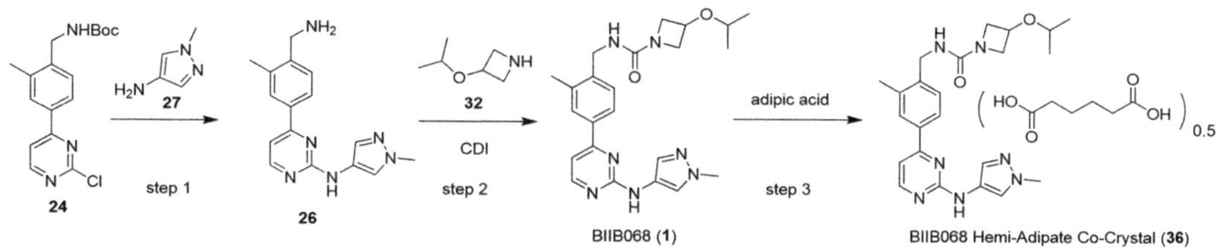

Step	Reaction	Starting weight (kg)	Product yield (kg)	Percentage Yield	Product Purity (HPLC A %)
1	Amination	9.6	7.7	90.0%	99.7%
2	Urea formation	6.6	9.7	99.0%	99.2%
3	Co-crystal formation	9.2	9.7	90.2%	99.4%

Summary

BIIB068 was discovered as a potent selective reversible BTK inhibitor through systematic rational drug design and optimization from a low micromolar screening hit. BIIB068 demonstrated favorable preclinical toxicology and DMPK properties across species which translated into good tolerability, PK, and target engagement in humans. The streamlined GMP chemical process led to multi- kilogram production of BIIB068 hemi-adipate which enabled GLP toxicology studies and early phase clinical trials. This represents a substantial progress of this series towards the successful use of a reversible BTK inhibitor for the treatment of autoimmune diseases.

References

1. Singh, S. P.; Dammeijer, F.; Hendriks, R. W. Role of Bruton's tyrosine kinase in B cells and malignancies. *Mol. Cancer* **2018**, *17*, 57, DOI: 10.1186/s12943-018-0779-z.
2. de Weers, M.; Verschuren, M. C.; Kraakman, M. E.; Mensink, R. G.; Schuurman, R. K.; van Dongen, J. J.; Hendriks, R. W. The Bruton's tyrosine kinase gene is expressed throughout B cell differentiation, from early precursor B cell stages preceding immunoglobulin gene rearrangement up to mature B cell stages. *Eur. J. Immunol.* **1993**, *23*, 3109–3114.
3. Ren, L.; Campbell, A.; Fang, H.; Gautam, S.; Elavazhagan, S.; Fatehchand, K.; Mehta, P.; Stiff, A.; Reader, B. F.; Mo, X.; Byrd, J. C.; Carson, W. E., III; Butchar, J. P.; Tridandapani, S. Analysis of the effects of the Bruton's tyrosine kinase (Btk) inhibitor Ibrutinib on monocyte Fcγ receptor (FcγR) function. *J. Biol. Chem.* **2016**, *291*, 3043–3052.
4. Florence, J. M.; Krupa, A.; Booshehri, L. M.; Davis, S. A.; Matthay, M. A.; Kurdowska, A. K. Inhibiting Bruton's tyrosine kinase rescues mice from lethal influenza-induced acute lung injury. *Am. J. Physiol. Lung Cell Mol. Physiol.* **2018**, *315*, L52–L58.
5. Li, X.; Kimberly, R. P. Targeting the Fc receptor in autoimmune disease. *Expert Opin. Ther. Targets* **2014**, *18* (3), 335–350.
6. Satoh, M.; Mizutani, A.; Behney, K. M.; Kuroda, Y.; Akaogi, J.; Yoshida, H.; Nacionales, D. C.; Hirakata, M.; Ono, N.; Reeves, W. H. X-linked immunodeficient mice spontaneously

produce lupus-related anti-RNA helicase A autoantibodies, but are resistant to pristane-induced lupus. *Int. Immunol.* **2003**, *5*, 1117–1124.

7. Honigberg, L. A.; Smith, A. M.; Sirisawad, M.; Verner, E.; Loury, D.; Chang, B.; Li, S.; Pan, Z.; Thamm, D. H.; Miller, R. A.; Buggy, J. J. The Bruton tyrosine kinase inhibitor PCI-32765 blocks B cell activation and is efficacious in models of autoimmune disease and B cell malignancy. *Proc. Natl. Acad. Sci. U. S. A.* **2010**, *107*, 13075–13080.
8. Katewa, A.; Wang, Y.; Hackney, J. A.; Huang, T.; Suto, E.; Ramamoorthi, N.; Austin, C. D.; Bremer, M.; Chen, J. Z.; Crawford, J. J.; Currie, K. S.; Blomgren, P.; DeVoss, J.; DiPaolo, J. A.; Hau, J.; Johnson, A.; Lesch, J.; DeForge, L. E.; Lin, Z.; Liimatta, M.; Lubach, J. W.; McVay, S.; Modrusan, Z.; Nguyen, A.; Poon, C.; Wang, J.; Liu, L.; Lee, W. P.; Wong, H.; Young, W. B.; Townsend, M. J.; Reif, K. Btk-specific inhibition blocks pathogenic plasma cell signatures and myeloid cell-associated damage in IFNα-driven lupus nephritis. *JCI Insight.* **2017**, *2*, e90111, DOI: 10.1172/jci.insight.90111.
9. U.S. Food and Drug Administration. *Approved drugs: Ibrutinib*. https://wayback.archive-it.org/7993/20170111231706/http://www.fda.gov/Drugs/InformationOnDrugs/ApprovedDrugs/ucm374857.htm.
10. Liang, C.; Tian, D.; Ren, X.; Ding, S.; Jia, M.; Xin, M.; Thareja, S. The development of Bruton's tyrosine kinase (BTK) inhibitors from 2012 to 2017: A mini-review. *Eur. J. Med. Chem.* **2018**, *151*, 315–326.
11. Wu, J.; Zhang, M.; Liu, D. Acalabrutinib (ACP-196): a selective second-generation BTK inhibitor. *J. Hematol. Oncol.* **2016**, *9*, 21, DOI: org/10.1186/s13045-016-0250-9.
12. Guo, Y.; Liu, Y.; Hu, N.; Yu, D.; Zhou, C.; Shi, G.; Zhang, B.; Wei, M.; Liu, J.; Luo, L.; Tang, Z.; Song, H.; Guo, Y.; Liu, X.; Su, D.; Zhang, S.; Song, X.; Zhou, X.; Hong, Y.; Chen, S.; Cheng, Z.; Young, S.; Wei, Q.; Wang, H.; Wang, Q.; Lv, L.; Wang, F.; Xu, H.; Sun, H.; Xing, H.; Li, N.; Zhang, W.; Wang, Z.; Liu, G.; Sun, Z.; Zhou, D.; Li, W.; Liu, L.; Wang, L.; Wang, Z. Discovery of Zanubrutinib (BGB-3111), a novel, potent, and selective covalent inhibitor of Bruton's tyrosine kinase. *J. Med. Chem.* **2019**, *62*, 7923–7940.
13. Murphy, G.; Isenberg, D. A. New therapies for systemic lupus erythematosus - past imperfect, future tense. *Nat. Rev. Rheumatol.* **2019**, *15*, 403–412.
14. Schafer, P. H.; Kivitz, A. J.; Ma, J.; Korish, S.; Sutherland, D.; Li, L.; Azaryan, A.; Kosek, J.; Adams, M.; Capone, L.; Hur, E. M.; Hough, D. R.; Ringheim, G. E. Spebrutinib (CC-292) affects markers of B cell activation, chemotaxis, and osteoclasts in patients with Rheumatoid Arthritis: Results from a mechanistic study. *Rheumatol. Ther.* **2020**, 7, 101–119.
15. Genovese, M. C.; Spindler, A.; Sagawa, A.; Park, W.; Dudek, A.; Kivitz, A.; Chao, J.; Chan, L.; Barchuk, W.; Nirula, A. SAT0089 Safety and efficacy of LY3337641, a Bruton's tyrosine kinase inhibitor in patients with Rheumatoid Arthritis: A randomized, double-blinded, placebo-controlled, 2-part phase 2 study. *Ann. Rheum. Dis.* **2019**, *78*, 1109–1110.
16. Montalban, X.; Arnold, D. L.; Weber, M. S.; Staikov, I.; Piasecka-Stryczynska, K.; Willmer, J.; Martin, E. C.; Dangond, F.; Syed, S.; Wolinsky, J. S. Placebo-controlled trial of an oral BTK inhibitor in Multiple Sclerosis. *N. Engl. J. Med.* **2019**, *380*, 2406–2417.
17. Ulrich, R. G. Idiosyncratic toxicity: A convergence of risk factors. *Annu. Rev. Med.* **2007**, *58*, 17–34.
18. Crawford, J. J.; Johnson, A. R.; Misner, D. L.; Belmont, L. D.; Castanedo, G.; Choy, R.; Coraggio, M.; Dong, L.; Eigenbrot, C.; Erickson, R.; Ghilardi, N.; Hau, J.; Katewa, A.; Kohli,

P. B.; Lee, W.; Lubach, J. W.; McKenzie, B. S.; Ortwine, D. F.; Schutt, L.; Tay, S.; Wei, B.; Reif, K.; Liu, L.; Wong, H.; Young, W. B. Discovery of GDC-0853: a potent, selective, and noncovalent Bruton's tyrosine kinase inhibitor in early clinical development. *J. Med. Chem.* **2018**, *61*, 2227–2245.

19. Watterson, S. H.; De Lucca, G. V.; Shi, Q.; Langevine, C. M.; Liu, Q.; Batt, D. G.; Bertrand, M. B.; Gong, H.; Dai, J.; Yip, S.; Li, P.; Sun, D.; Wu, D.-R.; Wang, C.; Zhang, Y.; Traeger, S. C.; Pattoli, M. A.; Skala, S.; Cheng, L.; Obermeier, M. T.; Vickery, R.; Discenza, L. N.; D'Arienzo, C. J.; Zhang, Y.; Heimrich, E.; Gillooly, K. M.; Taylor, T. L.; Pulicicchio, C.; McIntyre, K. W.; Galella, M. A.; Tebben, A. J.; Muckelbauer, J. K.; Chang, C.; Rampulla, R.; Mathur, A.; Salter-Cid, L.; Barrish, J. C.; Carter, P. H.; Fura, A.; Burke, J. R.; Tino, J. A. Discovery of 6-fluoro-5-(R)-(3-(S)-(8-fluoro-1-methyl-2,4-dioxo-1,2-dihydroquinazolin-3(4H)-yl)-2-methylphenyl)-2-(S)-(2-hydroxypropan-2-yl)-2,3,4,9-tetrahydro-1H-carbazole-8-carboxamide (BMS-986142): a reversible inhibitor of Bruton's tyrosine kinase (BTK) conformationally constrained by two locked atropisomers. *J. Med. Chem.* **2016**, *59*, 9173–9200.
20. Kawahata, W.; Asami, T.; Kiyoi, T.; Irie, T.; Taniguchi, H.; Asamitsu, Y.; Inoue, T.; Miyake, T.; Sawa, M. Design and synthesis of novel amino-triazine analogues as selective Bruton's tyrosine kinase inhibitors for treatment of Rheumatoid Arthritis. *J. Med. Chem.* **2018**, *61*, 8917–8933.
21. Chan, P.; Yu, J.; Chinn, L.; Prohn, M.; Huisman, J.; Matzuka, B.; Hanley, W.; Tuckwell, K.; Quartino, A. Population pharmacokinetics, efficacy exposure-response analysis, and model-based meta-analysis of Fenebrutinib in subjects with Rheumatoid Arthritis. *Pharm. Res.* **2020**, *37*, 1–13.
22. Isenberg, D.; Furie, R.; Jones, N.; Guibord, P.; Galanter, J.; Lee, C.; McGregor, A.; Toth, B.; Rae, J.; Hwang, O.; Miranda, P.; de Souza, V.; Jaller-Raad, J.; Maura Fernandes, A.; Garcia Salinas, R.; Chinn, L.; Townsend, M.; Morimoto, A.; Tuckwell, K. Efficacy, safety, and pharmacodynamic effects of the Bruton's tyrosine kinase Inhibitor, Fenebrutinib (GDC-0853), in moderate to severe systemic lupus erythematosus: Results of a phase 2 randomized controlled trial [abstract]. *Arthritis Rheumatol.* **2019**, *71* (suppl 10)
23. Ma, B.; Bohnert, T.; Otipoby, K. L.; Tien, E.; Arefayene, M.; Bai, J.; Bajrami, B.; Bame, E.; Chan, T. R.; Humora, M.; MacPhee, J. M.; Marcotte, D.; Mehta, D.; Metrick, C. M.; Moniz, G.; Polack, E.; Poreci, U.; Prefontaine, A.; Sheikh, S.; Schroeder, P.; Smirnakis, K.; Zhang, L.; Zheng, F.; Hopkins, B. T. Discovery of BIIB068: a selective, potent, reversible Bruton's tyrosine kinase inhibitor as an orally efficacious agent for autoimmune diseases. *J. Med. Chem.* **2020**ahead of print. doi:10.1021/acs.jmedchem.0c00702.
24. Li, C.; Franklin, L.; Chen, R.; Mack, T.; Humora, M.; Ma, B.; Hopkins, B. T.; Guzowski, J.; Zheng, F.; MacPhee, M.; Lin, Y.; Ferguson, S.; Patience, D.; Moniz, G. A.; Kiesman, W. F.; O'Brien, E. M. Process development and large-scale synthesis of BTK inhibitor BIIB068. *Org. Process Res. Dev.* **2020**, *24*, 1199–1206.
25. Di Paolo, J. A.; Huang, T.; Balazs, M.; Barbosa, J.; Barck, K. H.; Bravo, B. J.; Carano, R. A.; Darrow, J.; Davies, D. R.; DeForge, L. E.; Diehl, L.; Ferrando, R.; Gallion, S. L.; Giannetti, A. M.; Gribling, P.; Hurez, V.; Hymowitz, S. G.; Jones, R.; Kropf, J. E.; Lee, W. P.; Maciejewski, P. M.; Mitchell, S. A.; Rong, H.; Staker, B. L.; Whitney, J. A.; Yeh, S.; Young, W. B.; Yu,

C.; Zhang, J.; Reif, K.; Currie, K. S. Specific Btk inhibition suppresses B cell– and myeloid cell–mediated arthritis. *Nat. Chem. Biol.* **2011**, *7*, 41–50.

26. Bender, A. T.; Gardberg, A.; Pereira, A.; Johnson, T.; Wu, Y.; Grenningloh, R.; Head, J.; Morandi, F.; Haselmayer, P.; Liu-Bujalski, L. Ability of Bruton's tyrosine kinase inhibitors to sequester Y551 and prevent phosphorylation determines potency for inhibition of Fc receptor but not B-cell receptor signaling. *Mol. Pharmacol.* **2017**, *91*, 208–219.
27. Haselmayer, P.; Camps, M.; Liu-Bujalski, L.; Nguyen, N.; Morandi, F.; Head, J.; O'Mahony, A.; Zimmerli, S. C.; Bruns, L.; Bender, A. T.; Schroeder, P.; Grenningloh, P. Efficacy and pharmacodynamic modeling of the BTK inhibitor evobrutinib in autoimmune disease models. *J. Immunol.* **2019**, *202*, 2888–2906.
28. Goess, C.; Harris, C. M.; Murdock, S.; McCarthy, R. W.; Sampson, E.; Twomey, R.; Mathieu, S.; Mario, R.; Perham, M.; Goedken, E. R.; Long, A. J. ABBV-105, a selective and irreversible inhibitor of Bruton's tyrosine kinase, is efficacious in multiple preclinical models of inflammation. *Mod. Rheumatol.* **2018**, *29* (3), 510–522, DOI: 10.1080/14397595.2018.1484269.
29. SLE: a phase 1, single-ascending-dose, safety, tolerability, pharmacokinetic (PK), and pharmacodynamic (PD) study of BIIB068 in healthy participants; ClinicalTrials.gov Identifier: NCT02829541. https://clinicaltrials.gov/ct2/show/NCT02829541.
30. Dach, R.; Song, J.; Roschangar, F.; Samstag, W.; Senanayake, C. The eight criteria defining a good chemical manufacturing process. *Org. Process Res. Dev.* **2012**, *16*, 1697–1706.
31. Mishra, N. K.; Choi, M.; Jo, H.; Oh, Y.; Sharma, S.; Han, S. H.; Jeong, T.; Han, S.; Lee, S-Y; Kim, I. S. Direct C–H alkylation and indole formation of anilines with diazo compounds under rhodium catalysis. *Chem. Comm.* **2015**, *51*, 17229–17232.
32. Maddess, M.; Carter, R. SNAr reactions of 2-methylthio-4-pyrimidinones in pivalic acid: access to functionalized pyrimidinones and pyrimidines. *Synthesis* **2012**, *44*, 1109–1118.
33. Gill, D.; Iveson, M.; Collins, I.; Jones, A. A Mitsunobu reaction to functionalized cyclic and bicyclic N-arylamines. *Tetrahedron Lett.* **2018**, *59*, 238–242.
34. Ho, L.; Raston, C.; Stubbs, K. Transition-metal-free cross-coupling reactions in dynamic thin films to access pyrimidine and quinoxaline analogues. *Eur. J. Org. Chem.* **2016**, *36*, 5957–5963.
35. Zhang, G.; Wang, X.; Li, C.; Yin, D. Palladium-catalyzed cross-coupling of electron-deficient heteroaromatic amines with heteroaryl halides. *Synth. Commun.* **2013**, *43*, 456–463.
36. Shaw, J.; Grayson, D.; Rozas, I. Synthesis of cyclic guanidines: 2-arylamino-1,4,5,6-tetrahydropyrimidines. *ARKIVOC* **2014**, *2014*, 161–174.
37. Zhang, H.; Xia, Z.; Vasudevan, A.; Djuric, S. Efficient Pd-catalyzed synthesis of 2-arylaminopyrimidines via microwave irradiation. *Tetrahedron Lett.* **2006**, *47*, 4881–4884.
38. Bryan, M.; Dunn, P.; Entwistle, D.; Gallou, F.; Koenig, S.; Hayler, J.; Hickey, M.; Hughes, S.; Kopach, M.; Moine, G.; Richardson, P.; Roschangar, F.; Steven, A.; Weiberth, F. Key green chemistry research areas from a pharmaceutical manufacturers' perspective revisited. *Green Chem.* **2018**, *20*, 5082–5103.
39. 1,4-Dioxane is a class 2 solvent (solvents to be limited) based on ICH Q3C with permitted daily exposure of 3.8 mg/day.
40. ~ 0.4% of **30** was observed for entry 11 reaction in Table 5.

41. Initially the hemi-adipate form was named hemi-adipate salt, however, based on the calculated pKa of 2.75 it is very likely that the adipate is a co-crystal. Single crystal structure of hemi-adipate was later obtained and co-crystal form was therefore confirmed. See MacPhee, M.; Chen, R.; Ferguson, S.; Franklin, L.; Mack, T. Preparation of adipate forms and compositions of biaryl inhibitors of Bruton's tyrosine kinase. PCT Int. Appl. WO 2016201271 A1, December 15, 2016.
42. DEREK (Deductive Estimation of Risk from Existing Knowledge) is an expert rule-based computer modelling program developed by Lhasa (Oxford, UK) designed for computational toxicological assessments. Also see Greene, N.; Judson, P.; Langowski, J.; Marchant, C. Knowledge-based expert systems for toxicity and metabolism prediction: DEREK, StAR and METEOR. *SAR QSAR. Environ. Res.* **1999**, *10*, 299–314.
43. Dobo, K.; Greene, N.; Fred, C.; Glowienke, S.; Harvey, J.; Hasselgren, K.; Jolly, R.; Kenyon, M.; Munzner, J.; Muster, W.; Neft, R.; Reddy, V.; White, A.; Weiner, S. In silico methods combined with expert knowledge rule out mutagenic potential of pharmaceutical impurities: An industry survey. *Regul. Toxicol. Pharmacol.* **2012**, *62*, 449–455.
44. Galloway, S.; Reddy, V.; McGettigan, K.; Gealy, R.; Bercu, J. Potentially mutagenic impurities: Analysis of structural classes and carcinogenic potencies of chemical intermediates in pharmaceutical syntheses supports alternative methods to the default TTC for calculating safe levels of impurities. *Regul. Toxicol. Pharmacol.* **2013**, *66*, 326–335.
45. MacPhee, M.; Humora, M. Preparation of forms and compositions of biaryl inhibitors of Bruton's tyrosine kinase. PCT Int. Appl. WO 2016201280 A1, December 15, 2016.

Chapter 11

Discovery and Commercial Development of Cholesteryl Transfer Protein Inhibitor Evacetrapib

Nathan B. Mantlo[1] and Scott A. Frank[*,2]

[1]Discovery Chemistry Research and Technologies and Eli Lilly and Co., Lilly Research Laboratories, Indianapolis, Indiana 46285, United States

[2]Synthetic Molecule Design and Development, Eli Lilly and Co., Lilly Research Laboratories, Indianapolis, Indiana 46285, United States

[*]Email: frank_scott_a@lilly.com

Novel therapies which modulate HDL metabolism and thus serve to increase HDL-c levels have been proposed as an effective strategy in reducing overall residual cardiovascular risk. As an approximation, each mg/dL increase in plasma HDL cholesterol could result in a 2 – 3% decrease in cardiovascular risk. This chapter summarizes the discovery of cholesteryl ester transfer protein (CETP) inhibitor evacetrapib (LY2484595), a molecule capable of dramatically elevating plasma levels of HDL-c. Key themes described in this summary include the discovery and structure activity relationship of the scaffold, development of enabling synthetic methods to prepare preclinical quantities for early discovery efforts, and ultimately development of a commercially viable manufacturing process.

Introduction

Epidemiological studies show that both low HDL-cholesterol (HDL-c) levels and high LDL-cholesterol (LDL-c) levels are positive predictors of coronary artery disease and atherosclerotic cardiovascular (CV) events (*1*). Reducing LDL-c levels with statin treatment reduces cardiovascular event risk and has revolutionized CV disease prevention (*2*, *3*, *4*, *5*, *6*, *7*). However, CV risk is not eliminated even by aggressive statin treatment, because low HDL-c levels could still contribute to the residual patient risk (*8*, *9*). HDL-c promotes cholesterol efflux from peripheral tissues and transports it to the liver for excretion, a process known as reverse cholesterol transport. It is estimated that each 1 mg/dL increase in the plasma HDL-c decreases cardiovascular risk by about 2 - 3% (*1*). Therefore, a therapy that can modulate HDL metabolism to increase HDL-c levels may provide an effective strategy to reduce residual CV risk.

Niacin and fibrates are used to elevate HDL-c in humans, but their efficacy in reducing CV events is equivocal. While early studies are supportive, a recent study indicated that niacin was not effective in reducing cardiovascular event rates when dosed in combination with simvastatin, even though HDL-c was elevated by 15% (*10, 11*). Cholesteryl ester transfer protein (CETP) facilitates the transfer of cholesteryl ester and triglycerides between plasma lipoprotein particles HDL and LDL / VLDL, resulting in equilibration between these lipoprotein fractions. In humans, the deficiency of CETP produces an effect wherein cholesteryl ester predominately results in HDL particles as opposed to LDL particles (*12*).

Given the potential impact for CETP inhibition on overall cardiovascular health outcomes, four CETP inhibitors (Figure 1) were advanced to Phase 3 clinical trials in order to evaluate their safety and efficacy in large patient settings (*13, 14*). A fifth asset was also advanced into Phase 2 clinical studies (*15*). Within the group of CETP inhibitors that advanced into Phase 3 studies is evacetrapib (LY2484595), a molecule discovered at Eli Lilly and Company which evolved from an extensive SAR effort to identify a potent, orally bioavailable inhibitor of the CETP biological target.

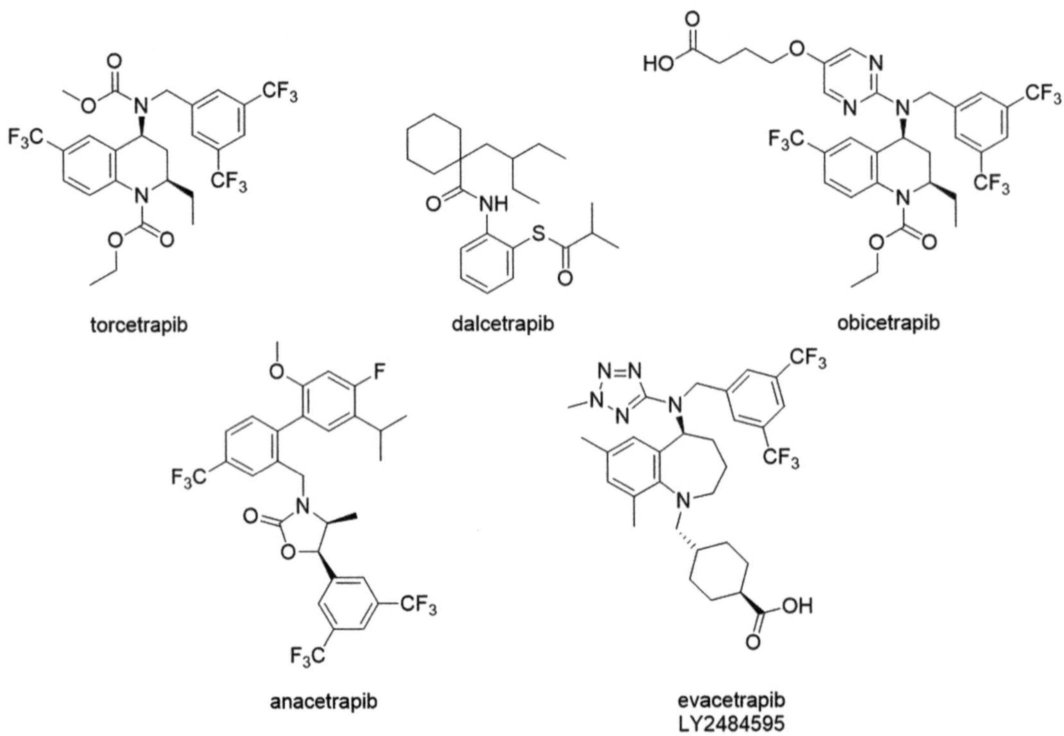

Figure 1. Structures of CETP Inhibitors which were investigated in clinical studies.

Medicinal Chemistry Effort and Preliminary Scaffold Evolution

Discovery efforts used in vitro assay formats that track CETP mediated cholesteryl ester (CE) and triglyceride (TG) movement between HDL and LDL particles. Early assays used tritiated cholesteryl ester (pre-packaged in donor particles) as the tracer (*16*). This assay worked acceptably in buffer but was difficult to adapt to plasma, which would have allowed us to gather enzyme inhibition in a system approximating circulating blood. Therefore, we used a fluorescence assay wherein a fluorescence labeled cholesteryl ester (CE-BODIPY) was pre-packaged in a donor particle. Fluorescence is observed as recombinant hCETP mediates the transfer of the labeled CE from the donor particle to an acceptor lipoprotein. This assay system was readily adaptable to run in both plasma and buffer and was further extended to monitor CETP activity ex vivo in blood samples from humans and in preclinical models, post-dose (*17*). In this section, we report in vitro CETP inhibition data for compounds as IC50's in both buffer and plasma.

In vivo data for CETP inhibitors were not generated in wild-type mice or rats because they are CETP deficient (*18*). While hamsters and non-human primates could be used, we chose double transgenic mice that co-express both human CETP and human apolipoprotein A-1 (hCETP/hApoAI double transgenic mice) (*19*). This transgenic mouse model readily delivered robust ex vivo CETP inhibition and HDL-c elevation data even after a single dose of a CETP inhibitor. Our early benzazepine racemic lead, (**1**, Table 1) was optimized beginning with a substituent scan at the C(7) position. Optimal substituents paralleled our previously described tetrahydronapthyridine CETP inhibitor SAR (*20*), with both favoring a methyl or trifluoromethyl group at the C(7) position (data not shown).

Table 1. Early Benzazepine SAR

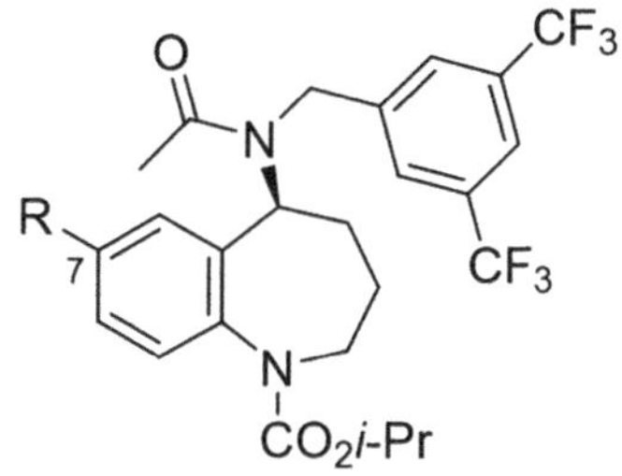

Compound	*R*	*CETP IC_{50} (nM)*	
		Buffer	*Plasma*
1	H	2900	5280
2	NH_2	13200	17200
3	$N(CH_3)_2$	579	1750
4	Ph	2900	5280
5	Me	470	1480
6	CF_3	280	640

Further substituent exploration at the C(8)- and C(9)-positions of the benzazepine is shown in Table 2. We observed that the 7,8-dimethyl substitution in **7** was superior to both 7-methyl analog **5**, and the 7,9-dimethyl analog **8**. The 7-methyl, 8-trifluoromethyl analog **9** produced further gains in potency, and the most potent substitution pattern in this set was the 7,8,9-trisubstituted benzazepine **11**. In the previously published work on the 6,6-fused tetrahydropyridine and napthyridine series (*20*), the majority of the CETP inhibition resided with a single enantiomer at the *bis*-trifluoromethylbenzylamine position. In this 6,7-fused benzazepine series we found a similar result with the 5-*S* enantiomer being preferred.

Table 2. Benzazepine SAR

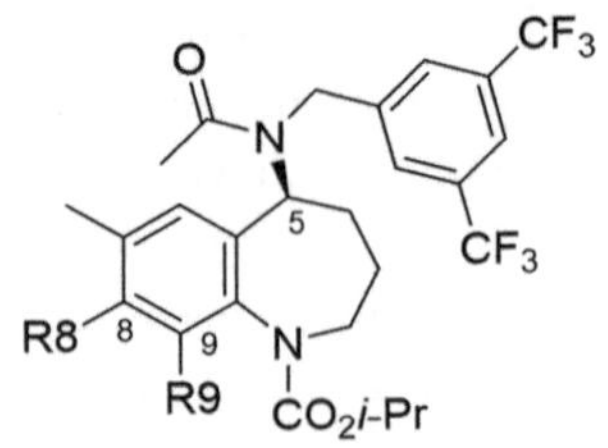

Compound	*C(5)*	*R8*	*R9*	*CETP IC_{50} (nM)*	
				Buffer	*Plasma*
5	*R/S*	H	H	470	1480
7	*R/S*	Me	H	217	432
8	*R/S*	H	Me	465	605
9	*R/S*	CF_3	H	136	199
10	*R/S*	Cl	H	80	307
11	*S*	CF_3	Me	30	42

Further Refinement of the Benzazepine Scaffold Series

By examining substituent effects on the benzene ring shown in Table 1 and 2, we had sufficient information to suggest that we could dial in the needed potency. But from a synthetic complexity and cost of goods perspective, we were unsatisfied that potency was consistent with extensive substitution or a trifluoromethyl moiety on the benzazepine portion (as exemplified in structure **11**). Further, these compounds were quite lipophilic and required oil vehicles for oral administration, thus we looked for areas of the molecule where we might introduce polarity while maintaining potency. Satisfactory alternatives for the northern *bis*-trifluoromethybenzyl group were not found (data not shown), and as a result, we extended our search for acetamide alternatives. We were aware that the acetamide could be replaced by a lower alkyl carbamate or any number of heterocycles but found the *N*-methyltetrazole **13** (Table 3) to be preferred with an improved in vitro potency profile.

Despite these improvements in activity, tetrazole **13** retained issues with high lipophilicity as previously mentioned. Unfortunately, polarity inducing modifications within that region were less active as exemplified by the des-methyl tetrazole analog (not shown). At this point we turned our SAR attention to polar modifications of the benzazepine nitrogen substituent. Simply modifying the isopropylcarbamate with more polar alkylcarbamate derivatives reduced in vitro potency. Thus, we investigated alkyl groups on the benzazepine nitrogen hoping to make hydrophilicity advances by eliminating the carbamate carbonyl and taking advantage of a mildly basic benzazepine nitrogen atom (Table 4).

Replacing the carbamate in **13** with a simple *N*-ethyl group (**14**) led to good in vitro activity (Table 4). Polar substituents such as hydroxyethyl **15** were also potent, but the more polar carboxymethyl derivative **16** led to a sharp decrease in activity. Potency returned with the benzoic acid analog **17** and the *trans*-1,4-cyclohexaneacetic acid derivative **18**. The maintenance of potency seemed consistent with a sufficiently long hydrocarbon spacer separating the carboxylic acid and the benzazepine nitrogen.

Table 3. Introduction of the *N*-Methyltetrazole Group

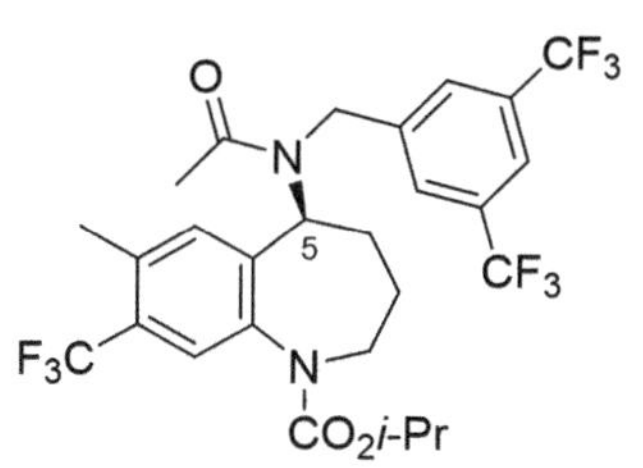

12

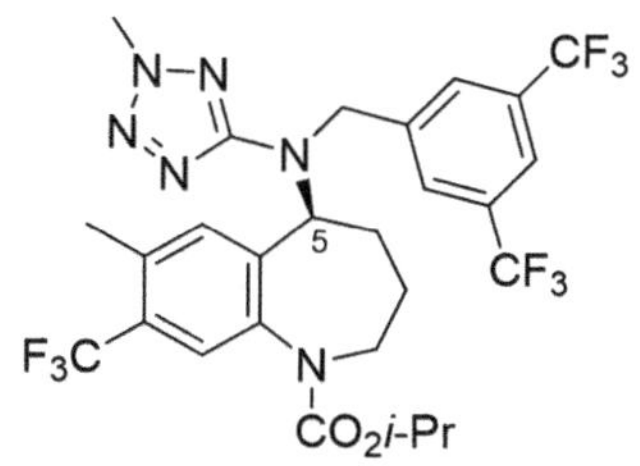

13

Compound[a]	*CETP IC₅₀ (nM)*	
	Buffer	*Plasma*
12	109	126
13	64	52

[a]Both compounds **12** and **13** possess C(5)-*S* stereochemistry.

Table 4. Effect of Substitution at the Benzazepine Nitrogen

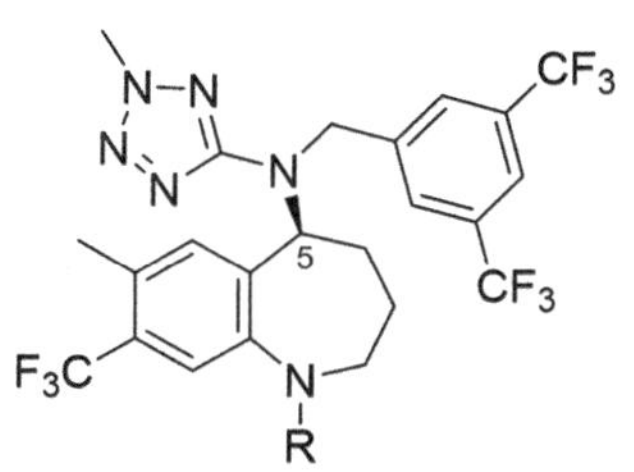

Compound[a]	*R*	*CETP IC_{50} (nM)*	
		Buffer	*Plasma*
13	$-CO_2$-iPr	64	52
14	$-CH_2CH_3$	27	58
15	$-CH_2CH_2OH$	43	37
16	$-CH_2CO_2H$	717	15500
17	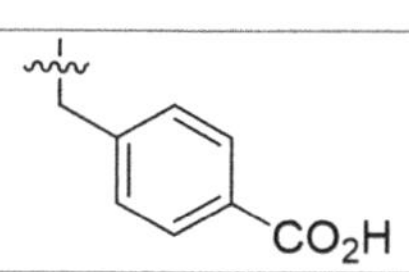	4	45
18	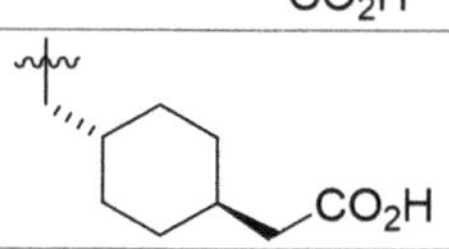	4	40

[a]Compounds **13** through **18** possess C(5)-*S* stereochemistry.

In addition to carboxylates, this distal portion of the scaffold tolerated mildly basic residues such as a 4-pyridylmethyl group. These analogs were not further pursued due to poor in vivo activity, despite retaining desirable absorption properties (data not shown). Nevertheless, we were nearing the endgame as these carboxylates were well absorbed in rodents when dosed in aqueous suspensions and were well behaved in early ADME and toxicology studies. We chose **18** for further studies because of its low IV clearance, potent ex vivo CETP inhibition, and finally HDL-c elevation in hCETP/hApoAI double transgenic mice. Unfortunately, further work revealed that **18** was unstable at stomach pH conditions, and the degradation products **19** and **20** indicated that expulsion of the benzylated aminotetrazole fragment **20** was readily occurring (Figure 2).

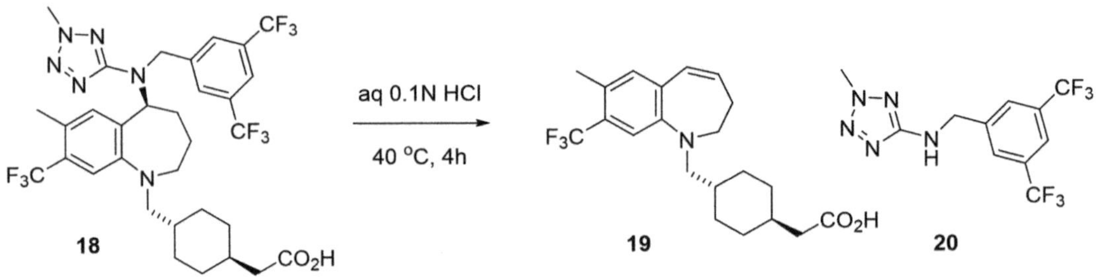

Figure 2. Gut pH instability of ***18****.*

Addressing the Acidic pH Instability Issues in the Benzazepine Scaffold Series

After a targeted screening directed towards stability of a number of compounds bearing a tetrazole moiety, we found the aforementioned carbamate analogues (*e.g.:* **13**, Table 3) and C(9)-methyl analogues (*e.g.:* **21**, Figure 3 below) devoid of this mode of degradation. In the case of the stable carbamate analogues, it is likely that electron delocalization to the carbonyl disfavors participation of the benzazepine nitrogen in the elimination process. However in the absence of the carbamate, stabilization from the elimination pathway may result from the steric interactions between the C(9)-methyl group and the N(1)-alkyl substituent. With the introduction of substitution at the C(9)-position, there is strain introduced by a conformational shift toward planarity of the benzazepine nitrogen and the arene pi-system essentially disfavors this shift. This hypothesis was based upon molecular mechanics simulations, and by comparison to the X-ray crystallographic structure of **21**, which revealed the N(1)-alkyl substituent in a pseudo-axial position significantly away from the C(9)-methyl group, as depicted in Figure 3. We envision that some overlap of the pi-system of the fused arene ring with the adjacent nitrogen lone pair may be necessary to facilitate the degradation pathway, which was observed with des-methyl analog **18**, but not in compound **21**.

With compound **21** in hand, we nevertheless desired options to retain the stability afforded by the C(9)-methyl group and also to eliminate the 8-trifluoromethyl group, thereby providing access to simplified benzazepine scaffolds. This search led us to inexpensive building blocks such as 2,4-dimethyl aniline, which in effect simplified the API synthesis toward the benzazepine scaffold. This modification along with a minor alteration in the cyclohexane substitution produced the candidate, evacetrapib (LY2484595, Figure 4) as the best performing compound from a potency, oral absorption, in vivo performance, toxicology, drug disposition and metabolism, and inherent stability perspective (*21*).

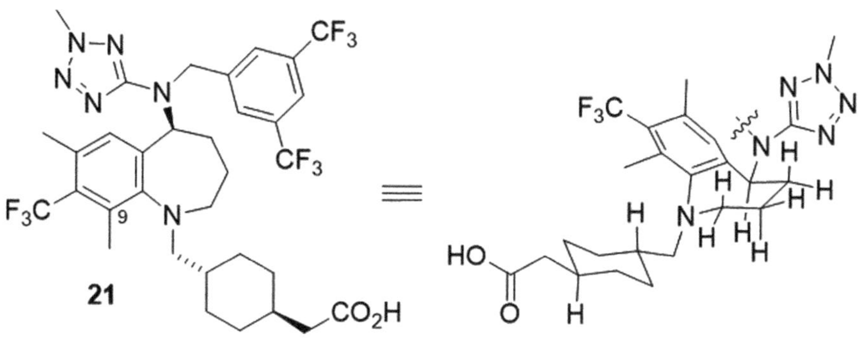

Figure 3. Proposed gut pH stabilization conferred by the C(9)-methyl moiety.

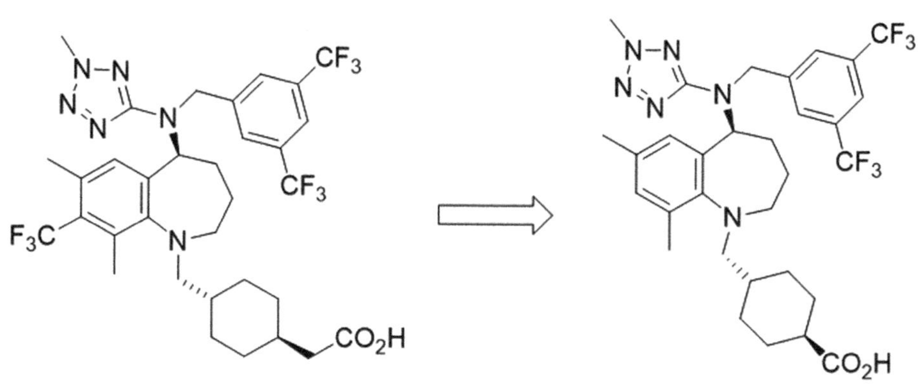

21

CETP IC_{50} (plasma) = 24 nM

evacetrapib (LY2484595)

CETP IC_{50} (plasma) = 24 nM
oral %F (vehicle: aqueous suspension):
rat (40%); monkey (29%)

*Figure 4. Optimization from **21** and discovery of evacetrapib (LY2484595).*

Preliminary Medicinal Chemistry Route Development toward Evacetrapib

The initial discovery efforts required a viable synthetic approach to the unique benzazepine ring core. While several routes are known for the assembly of benzazepine ring systems (22), a Dieckmann cyclization methodology was particularly attractive as we expected scalability and high yields to result (23). This approach also yields a carbonyl group at the 5-position, which could be readily elaborated into more complex intermediates, such as the aminotetrazole headpiece. Additionally, several commercially available anthranilic acid raw materials could be used for analog preparation. Analog work such as this was carried out in parallel to drive the SAR while continuing to develop evacetrapib preclinically.

To exemplify this attractive approach (Scheme 1), the anthranilic acid **22** was treated with sulfuric acid in methanol to yield methyl ester **23,** and subsequently allowed to react with di-*tert*-butyl dicarbonate thermally under solvent-free conditions. These more strenuous conditions were necessary to address reactivity issues with the largely unreactive aniline nitrogen.

After alkylation with ethyl 4-bromobutyrate in DMF using cesium carbonate as base, intermediate **25** was isolated in good yield. Dieckmann cyclization was then accomplished with potassium *tert*-butoxide which provided two crude intermediate esters: the expected product **26** as well as a co-product **27**, itself a result of transesterification with liberated methoxide. Nevertheless, both products were directly carried forward into the decarboxylation to yield the same product. This reaction also gave negligible amounts of undesired Boc removal which would be usually seen under more acidic conditions.

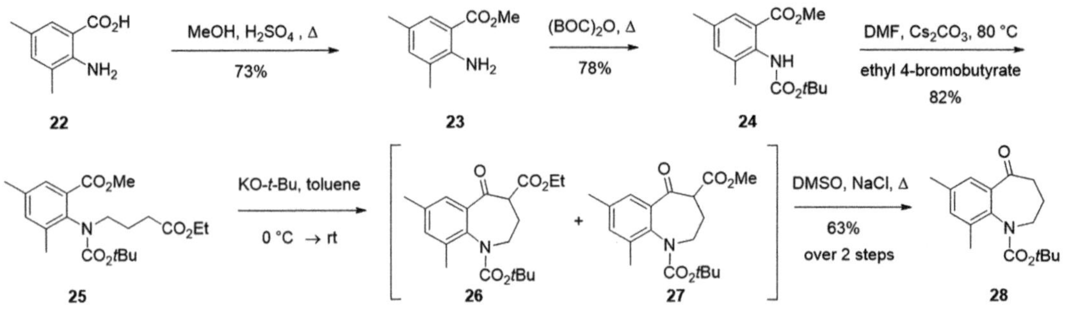

Scheme 1. Dieckmann approach to benzazepinone core ***28****.*

The initial chemistry approach for the asymmetric installation of the aminotetrazole headpiece proved to be a challenging enterprise. While alkylation of an installed secondary leaving group with amine-containing nucleophilic fragments yielded only elimination or other degradation products (not shown), attempts at asymmetric imine- or enamine-based hydrogenations were initially problematic. As such, a racemic route was rapidly developed and while requiring chiral separation, was facile for the preparation of multiple intermediates. (Scheme 2).

Condensation of benzazepinone **29** with 3,5-bistrifluoromethylbenzyl amine substrate **30** produced imine **31,** which was not rigorously characterized but subsequently reduced with sodium borohydride to yield racemic amine **32**. Treatment of **32** with cyanogen bromide in the presence of Hünig's base yielded cyanamide **33**. Thermal cycloaddition with tributyltin azide in toluene gave aminotetrazole **34**. Finally, subjection of the tetrazole with diethylazodicarboxylate and triphenylphosphine in the presence of methanol afforded methyltetrazole **13**, which could be resolved by chiral column chromatography.

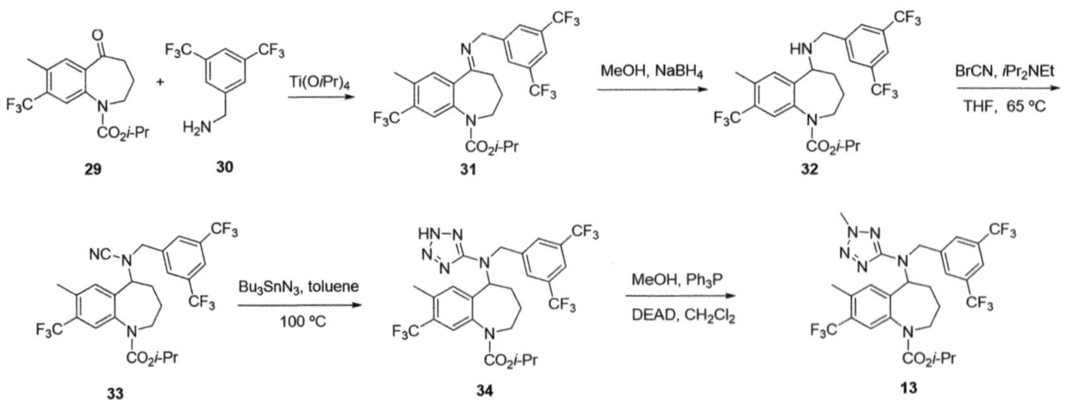

Scheme 2. Initial discovery racemic process to access the aminotetrazole scaffold.

It should be noted that the Mitsunobu conditions applied to the tetrazole provided excellent selectivity for methylation for the N(2)-nitrogen within the tetrazole ring (>10:1). This was preferred relative to other alkylation approaches which provided diminished selectivity. This synthesis approach was suitable to make multiple analogs which were evaluated in a variety of in vitro protocols as part of a parallel SAR evaluation.

Early Route Strategy for Installation of the Chiral Amine Fragment

While the Dieckmann sequence was scaled to prepare several hundred gram quantities of protected benzazepinone intermediates (i.e, **28**), the racemic sequence for installing the aminotetrazole fragment was deemed to not be acceptable for larger scale quantities that would be needed for preclinical toxicology studies. Given some of the challenges and timing required for late-stage chiral column chromatography, an asymmetric synthesis was sought which would alleviate these purification challenges.

An approach which yielded initial success is described in Scheme 3. Asymmetric reduction of ketone **28** with Corey-Bakshi-Shibata (CBS) oxazaborolidine protocol afforded chiral alcohol **35** in excellent yields and with desirable levels of enantioselectivity (>94% ee) (*24*). Whereas other nucleophilic displacement methods were unsuccessful, utilization of amine surrogate nucleophilic systems such as diphenylphosphoryl azide were successful in displacing the chiral alcohol in an S_N2 fashion, thereby providing the amine functionality necessary for elaboration to the aminotetrazole moiety. Reduction of azide **36** with standard palladium on carbon conditions under hydrogen provided primary amine **37**, which when treated with aldehyde **38** in the presence of sodium triacetoxyborohydride in dichloroethane gave secondary amine **39** in good yield over the three telescoped steps. However, it should be noted that at this stage of development, no crystalline intermediates were observed, and all intermediates were isolated by silica gel column chromatography either as foams or oils after concentration in vacuo.

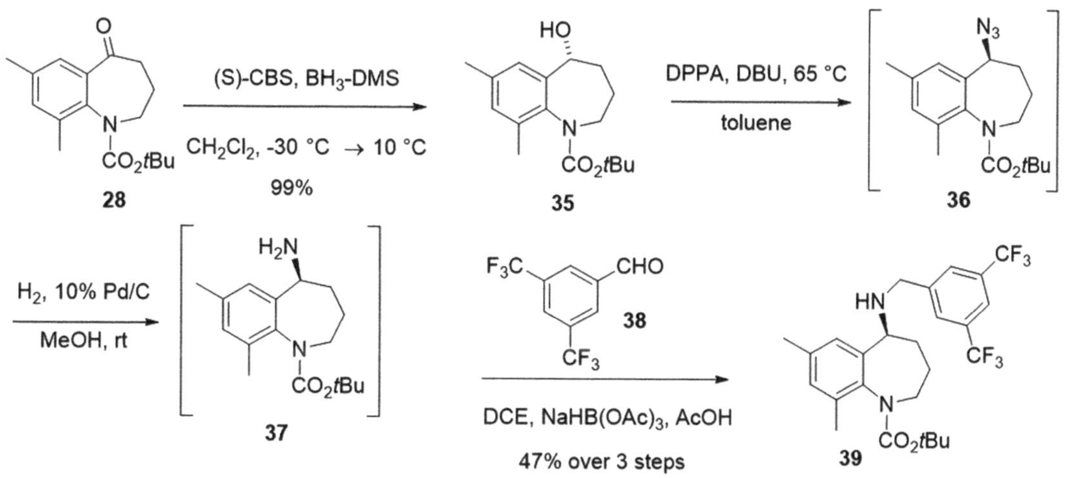

Scheme 3. Asymmetric synthetic approach to benzazepine scaffold.

Early Route Strategy for Construction of the Amino Tetrazole Fragment

While secondary amine **39** served as an excellent scaffold for further elaboration toward the SAR, it also served the primary purpose as a key intermediate in the ultimate manufacture of initial quantities of evacetrapib for advanced preclinical toxicology studies. As such, several hundred grams of this intermediate was prepared. However, on further examination of the enabling synthesis previously implemented for compounds like **13**, several improvements were sought prior to larger scale preparation. These desired improvements are:

- elimination of the use of tributylstannyl azide to mitigate removal of tributyltin waste streams later
- replacement of Mitsunobu methylation conditions to simplify purification
- development of reductive amination conditions which would install the *trans*-cyclohexylcarboxylate moiety within evacetrapib with minimal levels of the corresponding *cis*-isomer

The results of these enhancements to the originally disclosed route applied to the preparation of tetrazole **44** are summarized below (Scheme 4). We found that formation of cyanamide **40** could be conducted at a reduced temperature which afforded a slightly simplified reaction setup while retaining the reaction rate/product profile. With quantities of the cyanamide in hand, attention was then turned to the challenging tetrazole formation. Given some of the aforementioned challenges with using tributylstannyl azide on scale, the conditions (sodium azide, triethylamine hydrochloride, ethanol) for the preparation of tetrazole **41** worked well albeit somewhat hampered by an inability to drive the reaction to completion (roughly 5% remaining starting material). However, the reaction profile was otherwise quite clean and crude tetrazole **41** could be immediately processed forward without further purification or removal of any unreacted starting material.

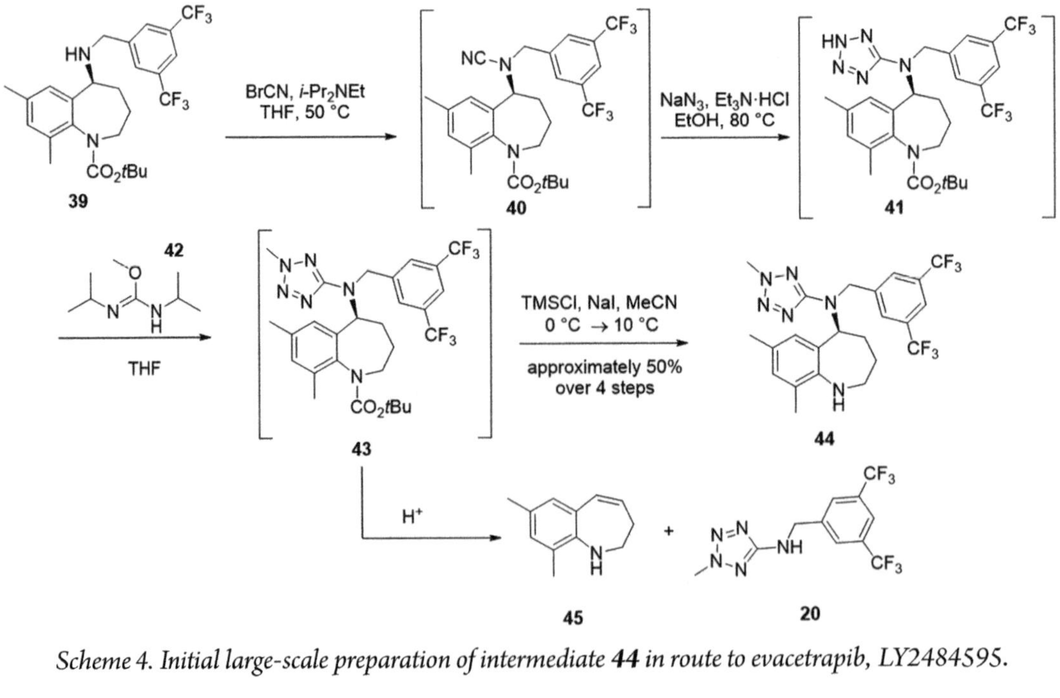

Scheme 4. Initial large-scale preparation of intermediate ***44*** *in route to evacetrapib, LY2484595.*

Regarding methylation of the tetrazole, we found that excellent N(2):N(1) selectivity could be achieved with methyl-transfer reagent **42** (*25*), which provided selectivity in ratios of 10:1, although in analogous systems selectivity of up to 30:1 had been observed (unreported). These conditions simplified purification in contrast to the Mitsunobu protocol previously described. Quite surprisingly, preparation of **44** from **43** proved to be more challenging than anticipated. More standard protic acid conditions provided the desired removal of the Boc group. But in addition, elimination of the benzylated aminotetrazole sidechain of **44** occurred, affording degradants benzazepine **45** and methyltetrazole **20**. This degradation is entirely analogous to that described for compound **18** (Figure 2). Fortunately, the desired Boc removal could be achieved by using trimethylsilyl chloride in the presence of sodium iodide. Under these conditions, the crude amine **44** could be isolated in about 50% yield over the four steps and was used directly in the subsequent reaction to prepare evacetrapib (*26*).

Observation of Benzazepine Conformational Isomer in the Route to Evacetrapib

One of the surprising observations during the initial development was the appearance of "conformational isomers" for substrates **35** through **37** and **39** through **41**, and **43**. This nature manifested itself in very complex ^{1}HNMR spectra, and in some intermediates, the conformational isomers could be separated by HPLC. While initially we attributed this property to slowly interconverting carbamate rotamers at the Boc-substituent, existing literature precedent indicated that this behavior was a result of extremely slow interconversion between benzazepine ring conformers, as previously reported by Hassner and co-workers on a related benzazepine system (*27*). A similar observation on analogous benzazepines had been also made by Hii and co-workers (*28*), and in this study they noted that the conformational interconversion was independent of the substitution at the 5-position of the benzazepine (for clarity, labeled below in Figure 5). In our series, interconversion between the benzazepine ring conformers apparently is slowed by the presence of the 9-methyl substituent, as similarly observed with a related analog previously reported by Hassner and co-workers (*29*).

A depiction of this behavior is shown below in Figure 5 for substrate **39**. As proposed by Hii and co-workers (*28*), with substitution of the N(1)-carbamate and the C(5)-benzylamine units, benzazepine conformational isomers may exist with those substituents in either a pseudo-diaxial or pseudo-diequatorial relationship. In converting between benzazepine ring conformations, apparently interactions between the C(9)-methyl group and neighboring Boc-protecting group are confounding enough to severely limit interconversion between these isomeric conformers. These energetic differences are sufficient to complicate ^{1}HNMR analysis and also afford separation by chromatographic methods. While we did not measure the barrier to inversion, we believe this to be well over 25 kcal·mol^{-1} which is consistent with the observations made in both of the previously reported benzazepine conformational analysis reports.

Not only did this phenomenon complicate the analysis of reaction completion and purity, it severely hampered any efforts at isolation of intermediates due to differences in crystallization behavior between the benzazepine ring conformers and varying rates of interconversion at intermediates with this substitution pattern. As such, the synthesis focused on telescoping these intermediates and where possible, minimizing chromatographic purification due to the complicating presence of multiple conformational isomers.

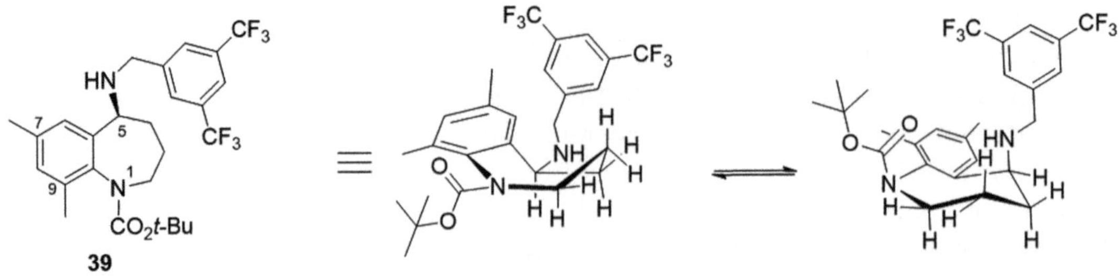

Figure 5. Potential conformational isomers observed in intermediate ***39****.*

Early Route Strategy for Appending the *trans*-Cyclohexylcarboxylate Fragment

With an acceptable route of synthesis to intermediate **44** in hand, the effort shifted focus on the attachment of the *trans*-cyclohexylcarboxylate fragment to the benzazepine nitrogen. Preliminary attempts at alkylation with activated leaving groups on the cyclohexylcarboxylate fragment proved to be entirely unsuccessful, likely due to the extreme steric environment in both the nucleophilic amine and the electrophilic components. A preliminary reductive amination was developed, called for a pre-mixing of the aldehyde species with the amine in the presence of catalytic acetic acid, followed by reduction with sodium triacetoxyborohydride.

However, while successful in terms of achieving high conversion in terms of coupling, the reaction was plagued by the formation of 15–20% of the corresponding *cis*-cyclohexylcarboxylate isomer which we attributed to reduction of the isomeric imine **49** (Figure 6). Control experiments performed in the absence of reducing agent established that in presence of secondary amine **44** alone, aldehyde **46** would not epimerize to the corresponding *cis*-isomer. However, when **46** was treated with **44** and acetic acid, rapid epimerization would take place.

Our working hypothesis for this epimerization behavior was that a rapid and reversible equilibrium was established, and that acid-catalyzed tautomerization of the initially formed iminium species **47** to the enamine species **48** would further isomerize and lead to a thermodynamic ratio of the *cis*- and *trans*-iminium species (**47** and **49**). In a fully equilibrated system, both *trans*- and *cis*-aldehydes would be anticipated to be in solution, along with **47** and **49**. Assuming the difference in the reduction rate between **47** and **49** is negligible under these conditions, attempting to preform iminium **47** would present a challenge toward a selective reductive amination.

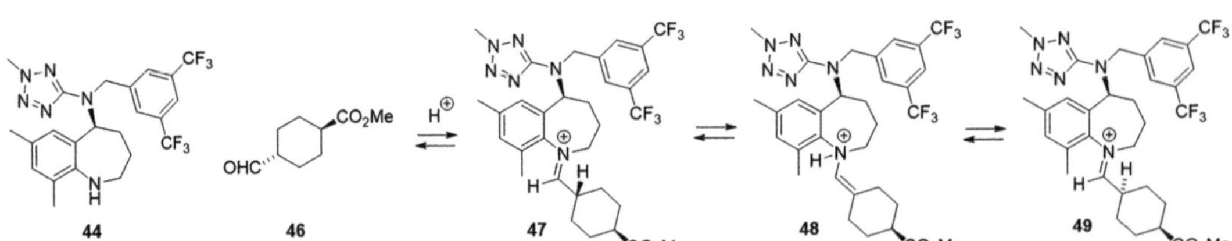

Figure 6. Rationale for observed epimerization of ***46*** *in the reductive amination of* ***44****.*

Nevertheless, the problem was largely solved by simply altering the sequence of reagent addition. Thus, addition of the aldehyde **46** and acetic acid to a mixture of the amine **44** and sodium triacetoxyborohydride in dichloroethane, resulted in an effective reductive amination and the formation of evacetrapib without substantial epimerization (a 97 : 3 observed ratio of isomers was obtained). Scheme 5 outlines the final assembly following this procedure, which was scaled up to produce >100 g of evacetrapib.

Executed as stated, the reductive amination product **50** was isolated in 48% yield after chromatography. This diminished yield is a result of both the incomplete reaction of **44** (due to competitive background reduction of aldehyde **46** to the corresponding alcohol) and the telescoping of several steps without purification. It should also be pointed out that compound **50** was devoid of any the observed benzazepine conformational isomerism. This observation is consistent with the reported behavior where the alkyl N-substituted analogs form a single benzazepine conformational isomer. Finally, hydrolysis was achieved under typical conditions, and after chromatographic purification of the acid product and manipulation to a more easily handled amorphous (L)-lysine salt, evacetrapib lysine was isolated in over 90% yield.

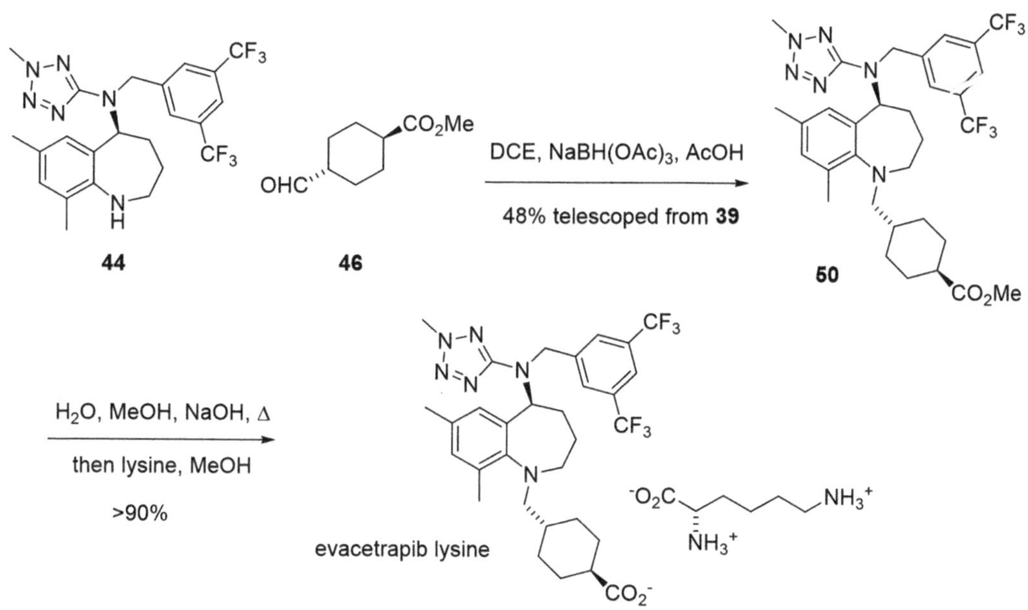

Scheme 5. Final sequence to form evacetrapib as its corresponding amorphous lysine salt.

Analysis of the Discovery Synthesis of Evacetrapib and Strategy for Route Redesign

While this synthesis yielded material of sufficient quality and quantity to evaluate in preclinical toxicology studies, there were several significant challenges which precluded viable commercial and late phase clinical supply preparation. Significantly, there was a general lack of crystallinity at almost all stages within the synthetic route, in no small part due to the conformational isomerism exemplified in intermediates such as **39**. This trend manifested itself in an inability to fully purge processing related impurities including minor levels of the undesired enantiomer (roughly 3%) introduced at chiral alcohol **35**, the methylation isomer of tetrazole **41**, and finally the *cis*-isomer produced at evacetrapib ester **50**. While suitable for testing in preclinical studies, it was anticipated that a more robust control strategy would be required to deliver material of higher purity for longer term clinical supply.

Secondly, with a longest linear sequence to prepare evacetrapib from anthranilic acid derivative **22** of roughly 15 steps (not including the preparation of **22**), a more convergent approach was envisioned to be desirable for larger scale manufacturing needs. While convergency can confer opportunities for higher yields and increased throughput, in a commercial manufacturing setting, prospects for greater flexibility in terms of capacity, and leveraging multiple manufacturing nodes can be of equivalent consideration.

Finally, more readily available raw materials were also desirable which would yield benefits in cost and overall cycle time reduction. While raw materials **22**, **30**, and **46** were available in several kilogram quantities for early deliveries of preclinical material, a more cost-attractive pool of raw materials was necessary to meet long term manufacturing targets. Emphatically, the potential for high volume annual demands drove the need for a truly innovative, scalable and novel approach to evacetrapib. In identifying the new synthetic route to evacetrapib, we relied heavily on the substantial experience gained in the preparation of preliminary quantities for preclinical toxicology evaluation. While we briefly explored scenarios for early installation of the *trans*-cyclohexylcarboxylic acid moiety, it became evident given the complexity of that subunit, a late installation that could provide excellent control for the desired isomer could be advantageous (Figure 7).

As such, we turned our attention to raw materials such as ester-acid **51** for constructing this fragment, itself available in high *trans* purity in commercial quantities via a well-known thermal equilibration of the mixture of diacid **52** (*30*). Amine **44** was envisioned to arrive via a convergent assembly of multiple fragments: primarily, amino-tetrazole **53** (derived from a material used extensively in the automotive industry); benzyl chloride **54**, generated from bromide **55** and widely available as a pharmaceutical starting material; and finally **56**, which could potentially be envisioned to arrive from aniline **57** and bromo-ester **58** (*31*), themselves commodity chemical items of ubiquitous applications. Certainly, while other raw materials and synthetic strategies can be envisioned, our previous route experience led us to the conclusion that this approach and key intermediates **44** and **46** could result in a vastly improved synthesis.

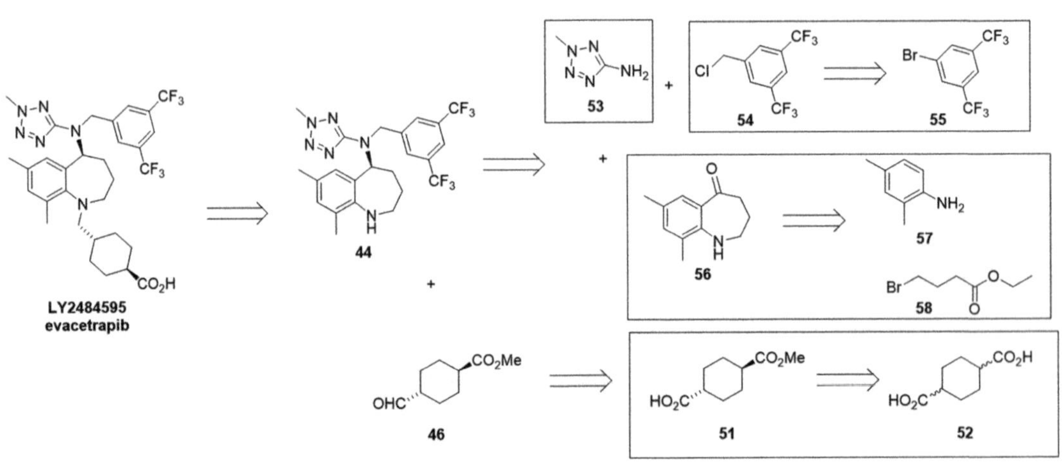

Figure 7. Retrosynthetic analysis of evacetrapib and identification of key raw materials.

Synthesis and Design of an Efficient Benzazepine Synthesis

In designing the synthesis of dimethylbenzazepinone **56**, we recognized that a substantially abbreviated sequence would be required. While the Dieckmann sequence depicted in Scheme 1 was used to manufacture kilogram quantities of **7**, it required five steps, and realistically, additional steps to prepare anthranilic acid **22** using traditional isatin chemistry (*32*). Inspiration for an alternative benzazepinone synthesis was derived from the work of Boekelheide (*33*). In this work, an intramolecular Friedel-Crafts reaction was envisioned that could rapidly access a benzazepinone scaffold. While this approach ultimately proved unsuccessful, yielding pyrollidinone and solvent derived byproducts (*34*), we thought a more judicious choice of solvent could address these issues. Additionally, we felt that the ortho-methyl group might influence the cyclization in a positive sense relative to the literature system. Thus, we set out on exploring this reaction platform as described in Scheme 6.

Alkylation of aniline **57** with ethyl 4-bromobutyrate (**58**) in toluene using triethylamine as base promoted the formation of ester **62**, which was isolated in 89% yield after an aqueous work up. While **62** could be isolated as a solid, it was determined more efficient to carry it forward without isolation into the subsequent transformations, contaminated with unreacted **57** and dialkylation byproduct. To avoid pyrrolidinone formation as observed in the original disclosures, we protected the monoalkylated aniline **62** as its corresponding methyl carbamate analog **63**. Preparation of **63** was accomplished in essentially quantitative yield by the treatment of **62** with methyl chloroformate and Na_2CO_3.

Selective hydrolysis of the ester was achieved with aqueous sodium hydroxide in MeOH followed by a series of aqueous work up manipulations to yield acid **64** in a quantitative yield. Initial attempts to cyclize the acid **64** directly were met with failure, However, we found that cyclization of the corresponding acid chloride **65** was successful upon treatment with aluminum trichloride. Thus, with the appropriate solvent selection and reagent stoichiometry, we obtained a high yield of the desired benzazepinone **66**. Specifically, treatment of the acid chloride **65** with excess aluminum trichloride (3.2 equiv) under moderately dilute conditions (roughly 0.25 M) drove the reaction to acceptable completion (> 85% conversion). After work up, **66** was isolated by crystallization from an ethanol water mixture in 76% yield (66% overall estimated from **57**). It is worthy to note that, there are no isolations between starting materials **57**/**58** and benzazepinone **66** and yields shown in the schematic are based upon single experiment isolations, or in situ chromatographic measurements (*35*).

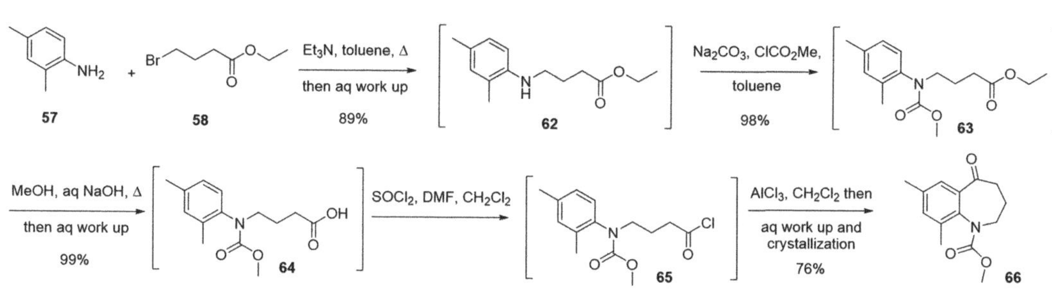

Scheme 6. Optimized intramolecular Friedel-Crafts reaction to form ***66****.*

In contrast to the historical reports on this cyclization approach to benzazepinones, we observed negligible formation of the pyrrolidinone byproduct. In fact, the conditions described in Scheme 6 worked successfully and the major byproducts of the Friedel-Crafts reaction primarily resulted from dimerization or other oligomerization events. We attribute the beneficial intramolecular Friedel-Crafts cyclization outcome to a more favorable solvent system (dichloromethane) as well as the *N*-protecting group, which may have served to coordinate with excess Lewis acid and thereby promote the desired cyclization through an undetermined affect (*36*). Additional studies are required to further elucidate the exact underlying factors which altered the reaction outcome. Nevertheless, the described sequence gave us ready access to key benzazepinone **66**, and in one campaign over three thousand kilograms was manufactured across seven batches, in an average batch size of 475 kg.

Strategy for the Asymmetric Installation of the Aminotetrazole Fragment

The previously used stepwise construction of the aminotetrazole fragment, while affording the desired stereo- and regiochemistry and supplying discovery efforts with necessary material, presented challenges for later clinical phase and larger-scale commercial supplies. Accordingly, we turned our attention toward more convergent strategies and determining desirable raw materials to improve synthetic efforts. As such, aminotetrazole **53** was identified, available from bulk commodity raw material **67** from within the automotive industry (Figure 8). While methylation of **67** had been reported previously (37, 38), ultimately we found that the desired 2-methyl-5-aminotetrazole **53** could be readily procured in high purity (ratio **53** : **68** >99.5 : 0.5) Indeed, with **53** available in large quantities, an approach for direct installation of the desired aminotetrazole fragment was prioritized.

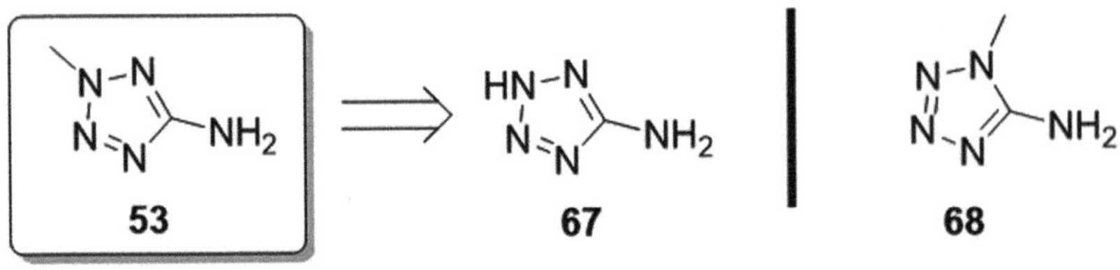

*Figure 8. 5-Aminotetrazole **53** and methylation from **67**.*

Several retrosynthetic strategies were envisioned to achieve a coupling approach to the benzazepine core using aminotetrazole **53**. While alkylation strategies using **53** as a nucleophile were entirely unsuccessful and yielded no evidence of any desired product (not shown), we turned our attention to asymmetric imine hydrogenation, a well-demonstrated approach for chiral amine formation (39). As shown in Scheme 7, we quickly determined analog **69** could be converted to a mixture of imine **70** and enamine **71** by treatment with an excess of **53** and an acid catalyst with azeotropic removal of water. On small scale, conversion to the **70**/**71** mixture required >24 h to complete to give a useful yield of the combined products. While perhaps generation of the imine/ enamine mixture could be deemed problematic, we envisioned that either **70** or **71** could be a ready substrate for asymmetric hydrogenation.

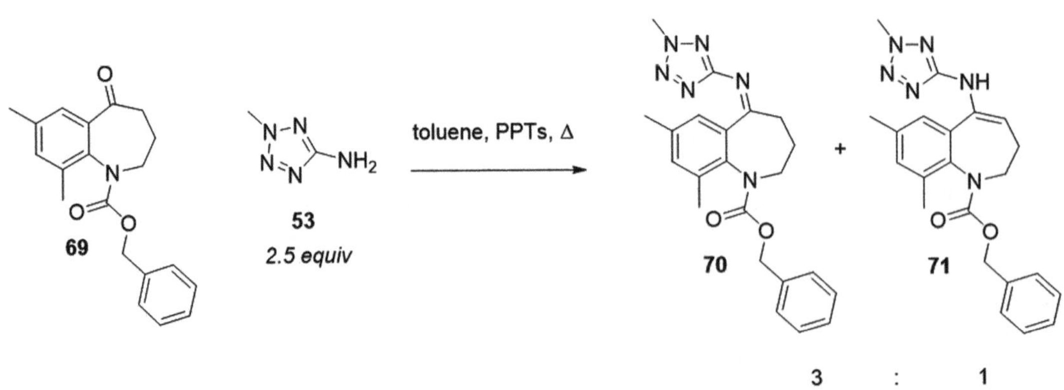

Scheme 7. Preliminary formation of imine/enamine asymmetric hydrogenation precursors.

To our delight, the mixture of **70** and **71** proved to work well in fairly-standard asymmetric hydrogenation conditions ($Ir_2Cl_2(COD)_2$, (*S*)-BINAP, and KI, at ~ 400–500 psig hydrogen) (*40*). Preliminary evidence suggested that imine **70** was the active species in the catalytic reduction, but **70** and **71** equilibrated under the reaction conditions at a satisfactory rate (*41*). We quickly transitioned our lab scale results to more meaningful larger scale demonstrations but observed a problematic trend in longer reaction times for formation of **70** and **71** from **69** and **53**. We attributed this not too surprising result to efficient water removal required to shift an equilibrium toward the desired reaction outcome. In large glassware equipped with a common Dean-Stark apparatus, the completion time shifted to approximately 48 hours. Further, as we scaled the from lab to pilot plant scale, reaction times for full conversion shifted to approximately two weeks to achieve the equivalent level of completion. This severe dependence on efficiency of water removal was attributed to inefficiencies within the specific reactor headspace. Additionally, for undetermined reasons, on pilot scale the asymmetric hydrogenation of the mixture of **70** and **71** required 4 days to reach completion departing from our lab experience of approximately 24 hours. To mitigate both complications, we began to explore improved alternatives to this preliminary asymmetric hydrogenation protocol.

Development of a Direct Asymmetric Reductive Amination (DARA) Reaction

The direct asymmetric reductive amination (DARA) is an ideal reaction to directly convert carbonyl compounds into the desired primary or secondary amine analogs while not necessarily requiring prior water removal from the system (*42*, *43*). As such, we investigated this approach with substrate **66**. Unfortunately, our excitement of such a reaction was tempered by the insignificant conversion to the desired product **75**, unless large excesses of **53** were used (Scheme 8). As well, only modest enantioselectivity was observed with a simple BINAP and iridium catalyst system.

Our prior observations on the criticality of water removal in the reaction of analog **69** inspired us to consider other means for removing water and facilitating imine formation. Utilization of 3Å molecular sieves provided one potential option for water removal in situ. Satisfactorily, an improved yield (81%) and enantioselectivity (91% ee) could be achieved with a xylyl-BINAP ligand and iridium catalyst. While this approach indeed appeared to be an ideal solution, ultimately concerns with scalability related to reactor compatibility from the use of sieves at higher temperatures made this approach undesirable (*44*).

We considered alternative approaches to remove the water prior to the DARA reaction, and considered ketal derivatives of **66**, which would in principle rapidly convert to the useful imine/enamine mixture via an oxonium species and the aminotetrazole. Formation of the ketal derivative of **66** gave crystalline solid **72**, which when treated with catalytic acid and stoichiometric levels of **53** (1.2 equiv), produced a mixture of imine **73** and enamine **74** (Scheme 9). Confirming this approach, when the mixture of **73** and **74** (prepared from **72**) was subjected to the asymmetric hydrogenation conditions, desired product **75** was formed in high conversion and enantioselectivity from both intermediates.

Even more satisfactorily, when ketal **72** was directly subjected to the aforementioned asymmetric hydrogenation conditions in the presence of **53**, the desired product could be isolated in 75% yield and with 94% enantiomeric excess utilizing the iridium xylyl-BINAP system. Furthermore, rather than the approximately 4-day reaction time required with substrates **70** and **71**, reaction times approached only 15 h (on a scale of 150 kg of **72**). For the first time, a streamlined DARA approach to installing the amino tetrazole moiety of the evacetrapib core was accomplished.

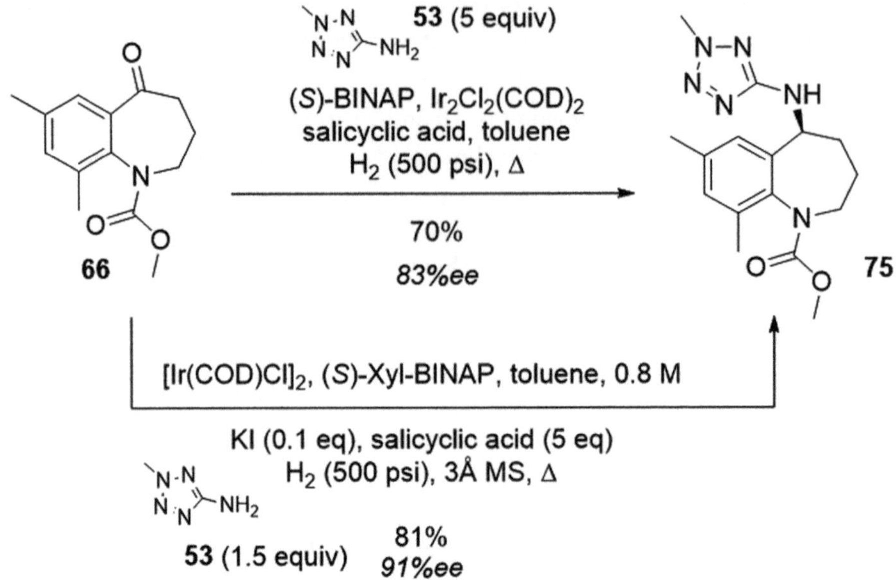

*Scheme 8. Initial DARA conditions used to prepare **75** from ketone **66** and aminotetrazole **53**.*

Nevertheless, we envisioned several opportunities for further improvement upon an overall commercialization approach toward **75** as shown in Scheme 9 given high-volume requirements for evacetrapib. First, while ketal formation and isolation of **72** was successful, there was some inherent inefficiency associated with its isolation prior to conversion to imine/enamine mixture **73**/**74**. A rough approximation of 10 – 15% material was lost during isolation of **72**. This yield loss was not associated with any improvement in purity but was mainly a reflection of the properties of **72** in terms of solubility and processing stability.

Additionally, isolation of intermediate **75** was complicated by a couple additional factors, While indeed **75** could be isolated in high purity, both chemical and chiral, there was a requirement to complete a thermal equilibration of a slurry of the benzazepine ring conformers of **75** in order to isolate a crystalline solid in practical yields. Apparently, one of the benzazepine ring conformers of **75** was crystalline while the alternative benzazepine conformer was non-crystalline, as observed with substrates from the early discovery route shown in Figure 5 (*45*). As well, we observed initial co-crystallization of minor levels of the wrong enantiomer with the desired product prior to the thermal equilibration used to isolate **75.** While this situation did provide an option for upgrading chiral purity, we were hesitant to rely on this pre-filtration as part of the final chiral control strategy for **75** given that we could not routinely remove all of the undesired enantiomer.

To address these modest inefficiencies, we sought an improved approach which maintained the overall architecture of this portion of the synthesis route while implementing some of the targeted development points. Primarily, eliminating the isolation of **72**, maintaining the in situ generation of **73** and **74**, and finally an alternative isolation to **75** were targeted as goals for optimization. Scheme 10 describes the outcome of these design criteria applied toward installing the aminotetrazole via the DARA approach. As a further part of the optimization, we sought a new catalyst system which could in principle harmonize some of the transformations as well provide improved cost factors.

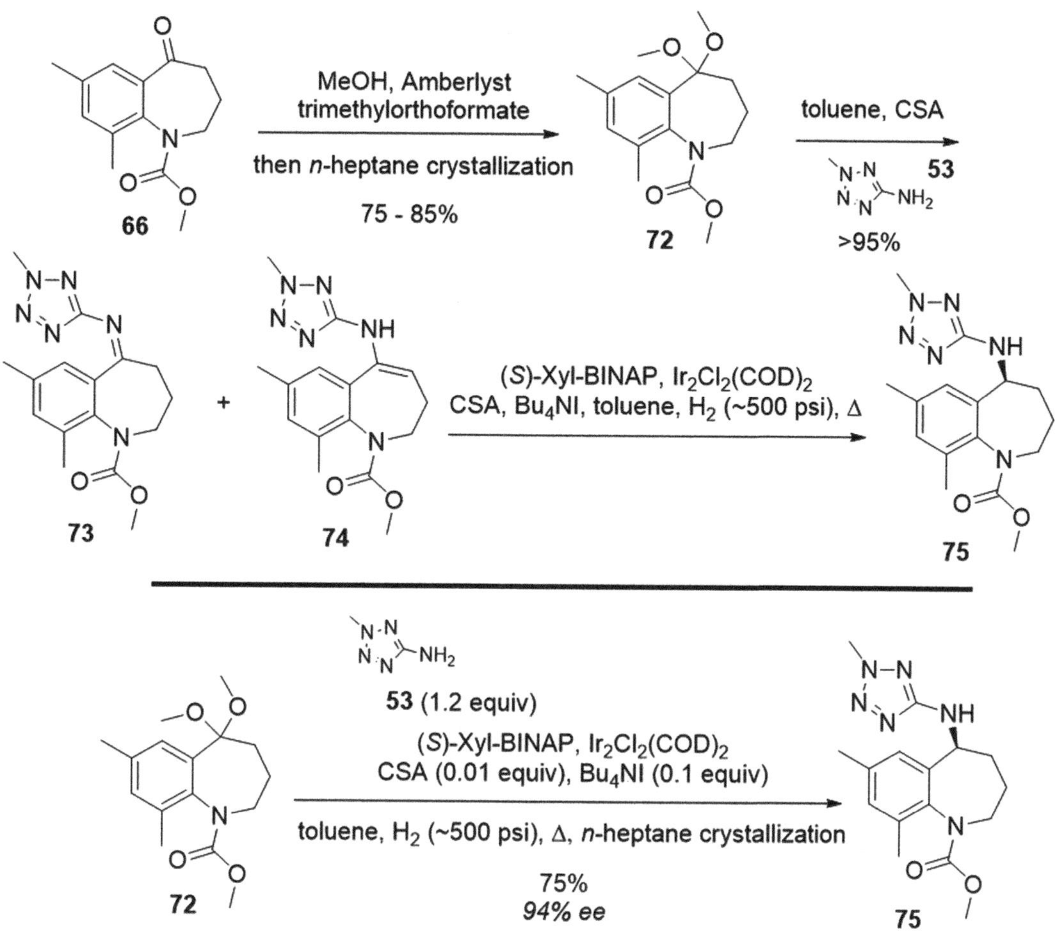

Scheme 9. Optimized DARA conditions utilizing ketal ***72*** *and aminotetrazole* ***53****.*

The use of ruthenium BINAP complexes is well documented to promote these types of catalytic transformations, in competition with iridium-based systems (*46, 47*), and we shortly discovered an alternative ruthenium catalyst system which accessed a similar BINAP ligand scaffold (*48*). This ruthenium based catalyst system was optimal relative to the iridium-catalyzed system for this DARA reaction as it allowed for a common solvent (methanol) to be used in the ketalization and DARA steps, in comparison to the iridium-based catalyst system which required toluene for the asymmetric hydrogenation portion. We also were pleased to discover that the newly discovered catalytic system maintained a practical substrate to catalyst loading (1000 : 1), with only slightly increased hydrogen pressure (roughly 750 psig). In addition, the same acid promoter could be used for the both the ketalization step as well as the in situ imine/enamine generation, and could be carried through as before without impact to the asymmetric hydrogenation.

One additional aspect of this sequence we desired to address was the isolation of ketal **72**, and potentially telescoping through that specific intermediate. We discovered that by subjecting **66** under slightly alternative conditions to form ketal **72** (*49*), a productive mixture of **72** and methyl enol ether **76** could be formed in situ (typically around 7 – 8 : 1). We were even more delighted to find that both species would further react in the presence of aminotetrazole **53**, and that the methanol solution, when subjected to the ruthenium xylyl-(*S*)-BINAP catalyzed asymmetric hydrogenation conditions, afforded desired **75** in >95% conversion and in approximately 94% ee. In isolation of the final product from this highly optimized sequence, ester **75** was not isolated but the solution cooled, at which point a small amount of a racemic mixture of **75** and its undesired enantiomer precipitated and was removed via an in-line filtration. While in the previously described stepwise process, this pre-filtration did not provide the robustness we desired, we chose to leverage it here in the telescoped sequence. This typically provided chiral purities of **75** in solution of greater than 98%. As such, we forward processed to remove the methyl carbamate protecting group after the removal of the racemic solid precipitate. The resulting methanolic filtrate was then treated with aqueous sodium hydroxide, which yielded surprisingly the intermediate sodium carbamic acid salt **77** (*50*). After dilution with toluene and treatment with aqueous acetic acid, decarboxylation readily ensued and chiral, enantiomerically enriched amino tetrazole benzazepine **78** was isolated after aqueous work up, crystallization, and filtration.

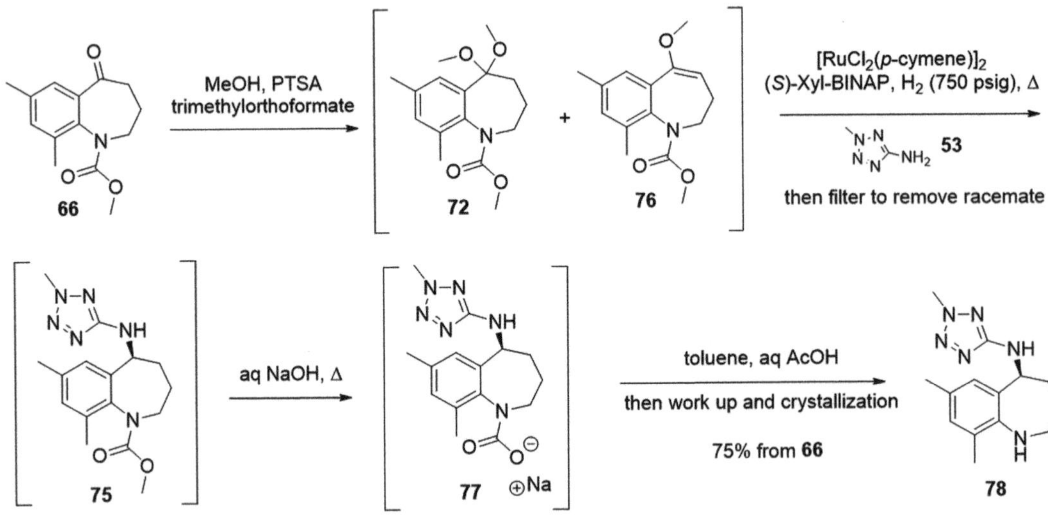

Scheme 10. Fully optimized DARA approach to intermediate ***78****.*

Given the telescoping of various unit operations and high pressures utilized in this sequence to manufacture **78**, we evaluated three types of continuous reactor platforms (a coiled tuber reactor, a vertically configured pipes-in series reactor, and a horizontally configured pipes-in-series reactor) in addition to the more standard batch autoclave reactor platform as part of the commercialization strategy (*51*). Ultimately, we selected the vertical pipes-in-series reactor (not shown), primarily for reasons of mass transfer efficiency with respect to hydrogen utilization (*52*), and typically full conversion could be achieved in less than five hours.

Moreover, we could fully integrate the continuous unit operations with the telescoped batch processing conditions to identify a truly hybrid continuous process as depicted in Scheme 10. To demonstrate, the ketal/enol ether **72**/**76** was conducted in batch mode with distillative removal of the methyl formate generated in the reaction. This solution was then treated with tetrazole **53**, and this solution was combined with the catalyst solution for the continuous asymmetric hydrogenation. This reaction stream was then collected in batch mode, and cooled to separate the racemic crystalline byproduct, and then forwarded through the hydrolysis of the carbamate after addition of aqueous sodium hydroxide. After dilution with toluene and neutralization with acetic acid, an aqueous work up, and a standard crystallization and filtration, product **78** was isolated in high purity. Most satisfactorily, the yield obtained from this fully telescoped, hybrid process was 75% (roughly 20%

yield enhancement over the stepwise process) and the enantiomeric purity of **78**, as a result of the elimination of racemate at intermediate **75**, was greater than 99%. Other impurities, residual metal and ligand in the isolated **78** were all well within specifications, demonstrating the robustness of this isolation protocol. This highly optimized process using the continuous DARA reaction was used to manufacture over 1000 kilograms of **78** as part of the supply chain for large, late phase clinical evaluation. Four batches were ultimately manufactured in this way, averaging over 250 kg in batch size, each with HPLC area purity ≥99%, chiral purity ≥99%, and residual metals present in <50 ppm routinely.

Selective Alkylation of the Amino Tetrazole Moiety

With a much-improved route and process to access the key chiral aminotetrazole fragment within the evacetrapib core, we turned our attention to the installation of the final two fragments to complete the synthesis. Preliminary attempts at reductive amination of tetrazolebenzazepine **78** with the cyclohexylcarboxaldehyde fragment **46** proved successful in terms of selectivity for the benzazepine ring nitrogen. However, the intermediate product unfortunately proved to be non-crystalline and as such, this sequence was abandoned.

Alternatively, we considered options for selective alkylation of the aminotetrazole nitrogen relative to the benzazepine ring nitrogen. Owing to the acidity differences between the two functionalized amines in amino tetrazole benzazepine **78**, we felt that a variety of conditions could drive selectivity toward benzylamine-substituted **44** over that of aniline **80**, and as such, set out on establishing some experiments to determine guiding parameters of the selectivity for **44** and byproduct **80**, as shown in Table 5.

In entry 1, our initial exploration revealed that under alkylation conditions utilizing Hünig's base and benzyl bromide **79**, indeed alkylation at the benzazepine ring nitrogen predominated. However, entry 2 demonstrated that preferential alkylation of the aminotetrazole N-H could be achieved via a selective deprotonation and treatment with **79**. Under these conditions, it was also found to be important to form the nitrogen anion prior to treatment with benzyl bromide **79** to avoid base-mediated degradation of benzyl bromide **79** (53).

In moving toward more cost-effective bases, treatment with sodium *tert*-butoxide also led to the desired product, albeit requiring a higher temperature for desired conversion (Entry 3). Entry 4 exemplified the subtle counter-ion effects of the reaction, as **44** could be prepared in high-selectivity at ambient temperatures (54). Nevertheless, these studies confirmed that indeed with the proper selection of conditions, selective alkylation could be achieved.

In optimization for scaleup of this step, we chose to utilize benzyl chloride **54**, derived from precursor **55** via a three-step sequence: Grignard formation and trapping with dimethylformamide, reduction with sodium borohydride, and treatment with thionyl chloride (55). As **54** could be made available in bulk as a toluene solution, we found that a mixed solvent system of toluene and tetrahydrofuran provided **44** in similar alkylation selectivity (~25:1) to the conditions with the bromide as described in Table 5 when treated with potassium *tert*-butoxide and roughly 1.1 equivalents of **54** (Scheme 11).

However, we observed during larger scale development runs of this chemistry, that over-alkylation of the free benzazepine ring nitrogen of product **44** would occur during solvent exchange (post work up) as we were exchanging solvents for the crystallization matrix. Further, presumably due to catalytic amounts of hydrochloric acid formed from the over-alkylation, we observed extensive degradation of the product **44**. As such, we instituted a scavenging step as part of the aqueous work up. By simply treating the crude reaction mixture prior to the solvent exchange with an excess of ethanolamine, the remaining traces **54** could be quenched and product **44** isolated routinely in 85% yield (56).

Table 5. Selective Alkylation of 78 to Prepare Key Intermediate 44

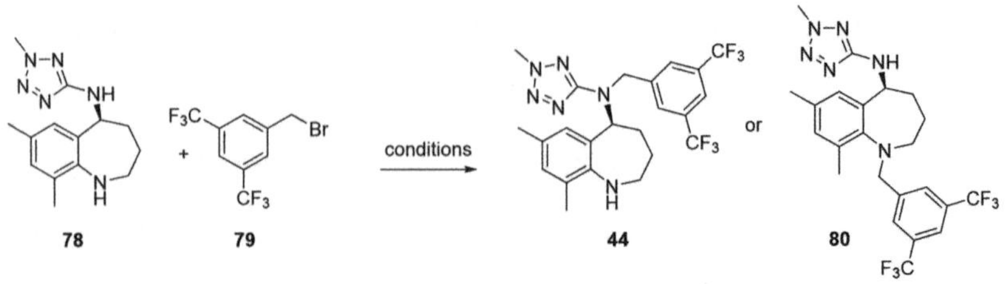

Entry	*Base*	*Solvent*	*Rxn Temp*	*Rxn Time (h)*	*Ratio* ***44 : 80***	*Yield*
1	*i*-Pr_2NEt	MeCN	80 °C	12	~ 1 : 12	90% (**80**)
2	KHMDS	THF	ambient	2	> 25 : 1	81%
3	NaO*t*-Bu	THF	60 °C	21	~ 20 : 1	82%
4	KO*t*-Bu	THF	ambient	6 – 8	>25 : 1	85%

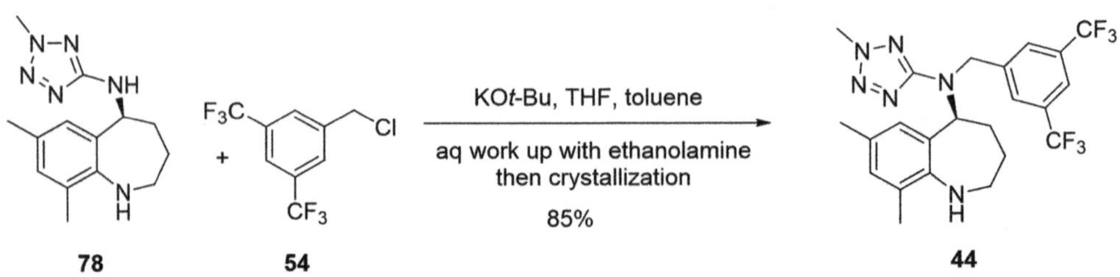

Scheme 11. Optimized conditions for alkylation to prepare ***44****.*

An additional benefit to the isolation of **44** was that remaining levels of the undesired (*R*)-enantiomer entering with **78** (approximately 1%) could be effectively purged to levels less than 0.1% routinely. Not only did this two-step sequence (described in Schemes 10 and 11) effectively establish the absolute stereochemistry within **44**, but also the methylation regioselectivity of the aminotetrazole moiety via entry as **53**. The latter was a marked improvement over the initial strategies for late stage methylation. In addition, the intermediate **44** proved to be an effective isolation control point within the large scale synthesis route to evacetrapib, and we manufactured a total of three batches in an average batch size of roughly 600 kg, each with HPLC area purity >99% and an enantiomeric excess ratio >99.8%.

Improved Process for the Installation of the *trans*-Cyclohexylcarboxylate Fragment

In designing the final major bond forming sequence for evacetrapib, we felt somewhat limited by available options for forming the nitrogen-carbon bond between the benzazepine core and the cyclohexylcarboxylate fragment given the challenges observed with the alkylation approaches attempted earlier in development. We felt that opportunities for optimizing the late stage coupling of **44** with aldehyde **46** utilized before were most appealing, provided that more suitable control over the aldehyde epimerization could be achieved.

As a starting point, aldehyde **46** from our preliminary medicinal chemistry synthetic route (Scheme 5) was prepared in an approximate ratio of 97:3 *trans* : *cis* isomers **46** : **81**. Initial studies also revealed that reactivity rates between the desired *trans*-aldehyde **46** and undesired *cis*-aldehyde **81** were indistinguishable in forming the corresponding reactive iminium species, and apparently both further reacted in the reductive amination (undesired leading to isomer **82**), as depicted in Figure 9. Secondly, equally concerning were reaction conditions employed in the reductive amination, as even stereochemically pure **46** could readily epimerize and thus lead to undesired **82**. Finally, a preponderance of evidence showed that crystallization of product **50** afforded minimal upgrade in terms of rejection of the unwanted *cis*-isomer **82**. As such, we set out on a dual-pronged strategy to design a route leading to improved purity of aldehyde **46**, and additionally identification of ideal reaction conditions to establish a scalable process for **50**.

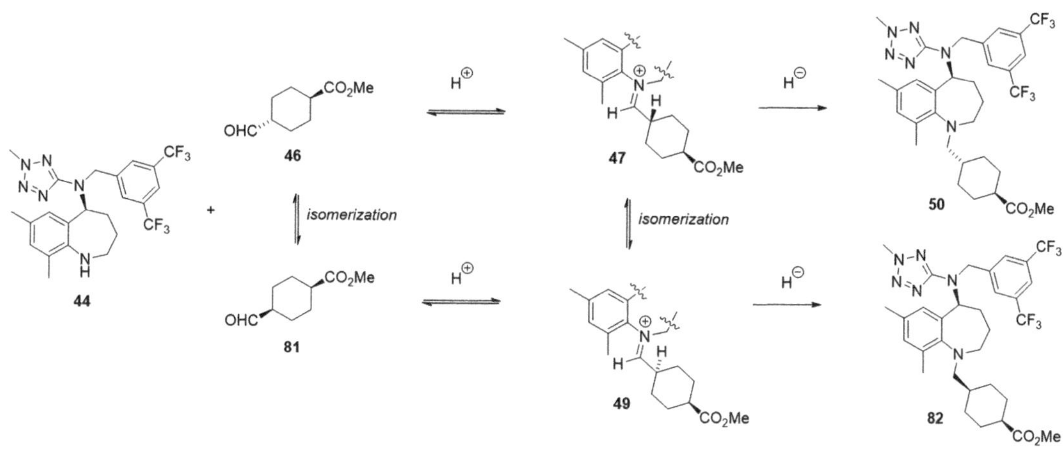

Figure 9. Possible reductive amination pathways generating trans- and cis-isomers.

As a starting point for toward a synthesis of **46**, we initiated our efforts employing commercially available *trans*-carboxylic acid ester **51**, with designs to implement a well-established Rosenmund reduction of an in situ generated acid chloride to afford desired *trans*-aldehyde **46** (Scheme 12) (57). Given that **51** was available in > 99.5% *trans*-isomeric purity, this approach seemed highly favorable of generating **46** with reduced levels of **81**. Formation of acid chloride **83** from **51** was accomplished via treatment with oxalyl chloride in the presence of catalytic 2,6-lutidine and the resulting product was directly treated with palladium on carbon in the presence of hydrogen and 2,6-lutidine in toluene as solvent (58). Presumably, addition of the 2,6-lutidine as an acid scavenger minimizes acid-mediated epimerization of *trans*-aldehyde **46**, and indeed this approach generated solutions of **46** which were of >99.5% *trans*-isomeric purity, retaining the isomeric purity derived from **51** (59). In further developing this overall strategy, an isolable form of **46** was sought, but given that *trans*-aldehyde **46** was a non-crystalline material and moderately unstable, this presented a significant challenge. The primary advantages for accessing an isolation point at **46** would be to add an additional control point, but also to potentially provide more on-demand access if **46** could be isolated and stored.

A common approach for isolation and purification of aldehydes is to use the analogous bisulfite adducts, which are well established as isolable intermediates (*60*). We quickly found that by treatment of aqueous tetrahydrofuran solutions of *trans*-aldehyde **46** with either aqueous sodium or potassium metabisulfite solutions, isolable crystalline solids could be readily obtained. In particular, potassium bisulfite adduct **84** could be isolated in roughly 90% yield, as a crystalline hemihydrate, in the two-step sequence from acid-ester **51**. Major benefits of the potassium analog relative to the sodium counter-ion resulted from decreased hygroscopicity (2% moisture sorption vs >25% for the sodium analog at 95% relative humidity) and markedly improved filtration rates due to its crystal morphology (*61*). Further, control experiments with samples of **46** enriched with elevated levels of the *cis*-isomer (>20%), established that residual amounts of the undesired *cis*-isomer could readily be purged in the isolation of bisulfite adduct **84**. However, absolute quantification of the *cis*-isomer in **84** was found to be difficult to levels <0.5% as previously described, due to limits of detection in either quantitative ^1H NMR methods or epimerization observed during sample preparation with other chromatographic methods.

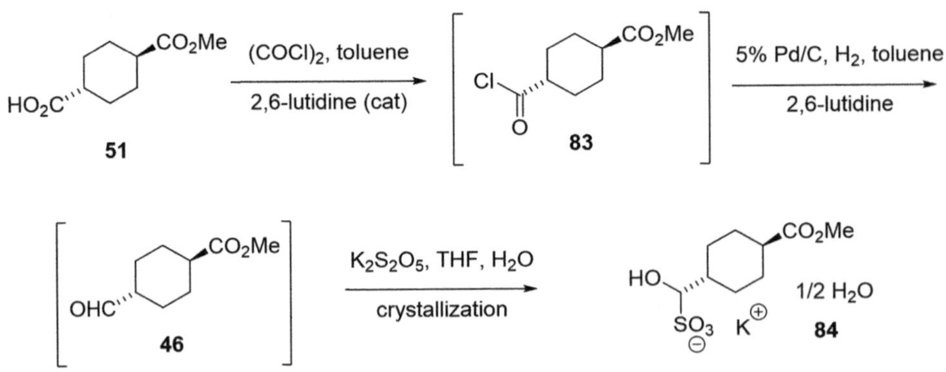

Scheme 12. Preparation of bisulfite adduct ***84*** *as a key isolable precursor.*

With a suitable route and isolation technique to aldehyde **46** via the bisulfite adduct **84**, our aspirations turned toward an optimized approach to the reductive amination of **44** which would be able to robustly deliver the penultimate ester **50** on the required scale. During the span of commercial development, both a sodium triacetoxyborohydride mediated reductive amination process (*62*) and a heterogeneous platinum-catalyzed process (*63*) were utilized to make material for clinical supply. These processes were both robust and manufactured hundreds of kilograms of **50**. However, we recognized that a substantial improvement might ultimately result from incorporating homogenous catalysis and continuous processing principles, as were utilized in the manufacture of intermediate **78**.

After some extensive screening of transition metals and conditions, we arrived at an analogous iridium-catalyzed homogenous reductive amination reaction with which to deliver penultimate **50** as shown in Scheme 13 (*64*). Thus, combination of aldehyde **46** with amine **44** in the presence of an in situ formed iridium catalyst (derived from commercially available iridium chloride cyclooctadiene dimer) with catalytic acetic acid and tetrabutylammonium iodide and high pressures of hydrogen (~750 psig) provided high yields of **50** in comparable purity to the previously described reductive amination conditions. These new conditions provided advantages over previously developed processes, either in terms of improved control strategies with respect to in-process generated impurities, such as **82**, or overall operational improvements. It is also interesting to note that in contrast to the ruthenium catalyzed process developed for **78**, iridium was the more preferred organometallic system for coupling of **44** and **46.**

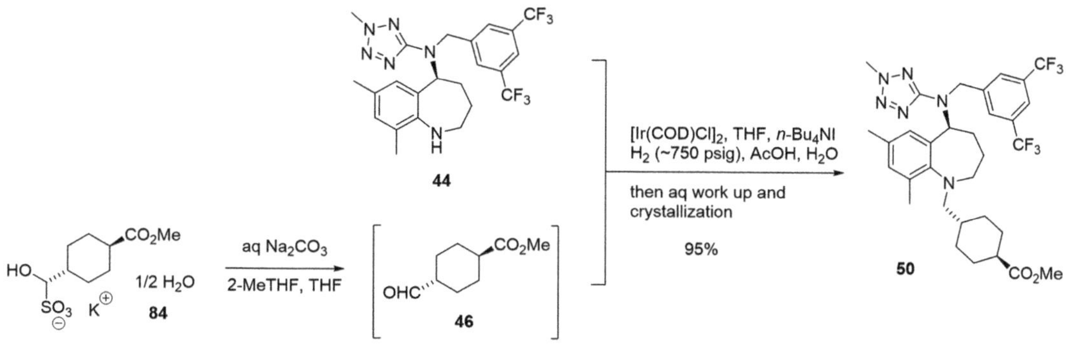

Scheme 13. Iridium-catalyzed transition metal catalyzed reductive amination to form ***50****.*

A proposed mechanistic analysis of the fairly complex overall catalytic process is depicted in Figure 10 . Under equilibrium conditions in the presence of acetic acid and water, a mixture of several intermediates, including iminium species **47** as well as hemiaminal **86** can be expected to result from combination of **44** and **46**, itself in equilibrium with hydrated form **85**. Coordination to the iridium center can occur, potentially to intermediate hemiaminal **86**, and subsequent delivery of hydride occurs assisted in some part by acetic acid (65). Within the iridium complex, ligand exchange and reductive elimination of acetic acid then regenerates an active catalytic species. A potential alternative option with the catalyst system reducing iminium **47** (unpictured) can be envisioned, but this pathway would arguably require complexation of the iridium system with the iminium species and thus seems less likely.

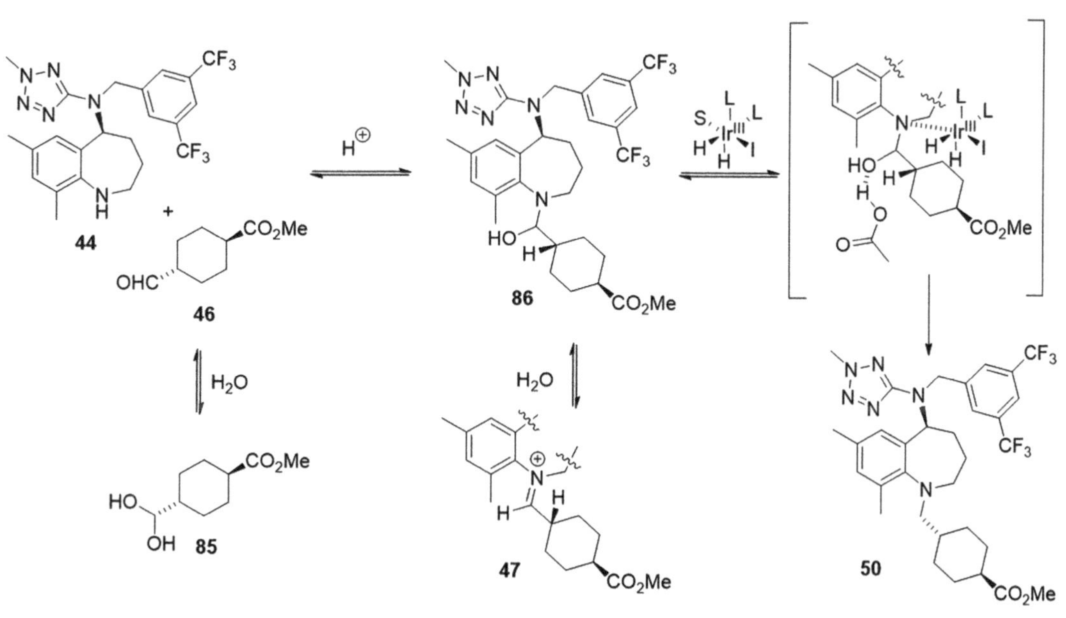

Figure 10. Potential iridium-catalyzed pathway to form ***50*** *via hemiaminal* ***86****.*

In developing the reaction to commercial-scale application, not only did the high-pressure conditions employed in the iridium-catalyzed reductive amination warrant themselves toward continuous processing principles, but the solution stability of aldehyde **46** provided additional rationale. Accordingly, the vertical pipes-in-series reactor that was implemented in the manufacture of **78** could also be applied in the manufacture of **50.** To provide further detail for the iridium-catalyzed reductive amination with **46**, regeneration of the aldehyde from the potassium bisulfite analog **84** was readily accomplished with aqueous sodium carbonate in a biphasic solvent mixture with tetrahydrofuran and 2-methyltetrahydrofuran, a mixture used to facilitate optimal layer separation during the aqueous work up. Deprotection of the bisulfite adduct under these conditions was quite rapid (< 1 h), and epimerization of **46** was minimized (typically at process scale, levels of the *cis*-isomer were below the limit of quantification of <0.5%). An additional aqueous wash with brine was required to remove trace amounts of bisulfite-related inorganic impurities, which tended to apparently poison the subsequent catalytic step. The solution of aldehyde **46** was adequately stable for immediate processing, however upon prolonged storage, amounts of the *cis*-isomer in solution would increase and thus it was best to use these solutions within 48 h.

To facilitate the continuous flow hydrogenation and to minimize isomerization of **46**, we developed the process utilizing multiple reagent streams. In addition to the tetrahydrofuran and 2-methyltetrahydrofuran solution of **46**, separate solutions of **44** in tetrahydrofuran, an aqueous solution of acetic acid, and finally a solution of tetrabutylammonium iodide with the iridium chloride cyclooctadiene dimer when combined under high pressures of hydrogen in the vertical pipes-in-series reactor yielded **50** in excellent yield and purity. While the catalytic system would work sufficiently well in the absence of the tetrabutylammonium iodide, further catalyst stability was provided by its addition. This added stability resulted, presumably, by preventing aggregation and plating of metallic iridium (*66*). Interestingly, during development we did determine that the plated iridium species would productively reduce aldehyde **46** to the corresponding alcohol, thus addition of the tetrabutylammonium iodide not only provided benefits for stabilization of the iridium catalyst required but also in preventing reduction of **46**. As such, roughly 1.4 equivalents of **46** could be used to achieve >99% conversion and practical amounts of the iridium chloride cyclooctadiene dimeric catalyst at a loading of 1000 : 1 substrate to catalyst were typically used.

A summary of the results of the continuous process manufacture for **50** is shown in Table 6, which was run at steady state for 24 days (with batch-mode work up and crystallization of four separate lots). The continuous reductive amination process routinely delivered material with in-process levels of the unwanted *cis*-isomer **82** well less than 0.5%, and upon work up and isolation by crystallization of **50**, less than 0.4%. As well, the yield for the coupling was a satisfactory 95% and well over 2000 kg of **50** were manufactured by this process.

Table 6. Summary of the Iridium-Catalyzed Reductive Amination to Manufacture 50

Batch	*Output kg*	*% Yield*	*Total Impurities*	**82** *(cis)*
1	469	86.8	0.24%	0.16%
2	546	95.1	0.19%	0.19%
3	549	99.1	0.22%	0.22%
4	490	97.3	0.21%	0.34%

Final Hydrolysis to Form Evacetrapib

At this point, all that remained for the synthesis was the seemingly simple hydrolysis of the methyl ester of **50** to reveal the carboxylic functionality of evacetrapib. While we expected that this transformation would be trivial, some unanticipated epimerization of the established *trans*-cyclohexylcarboxylate moiety occurred using conditions employed in Scheme 5. We suspected that the basic conditions used were responsible for the epimerization, but due to the aforementioned pathway for acid-promoted degradation, we were somewhat limited in conditions we could employ.

Thus, we began by screening basic reaction conditions for the hydrolysis, and quickly determined that proper choice of solvent in the sodium hydroxide mediated hydrolysis was key to minimization of epimerization forming **85** (Table 7). Surprisingly, significant levels of epimerization using methanol as solvent were observed, with slightly less in ethanol, and largely suppressed in 2-propanol. The epimerization leading to **85** was independent of counter ion as lithium and potassium hydroxide provided similar trends in terms of epimerization relative to sodium. While the 2-propanol conditions largely eliminated epimerization, hydrolysis in this solvent did lead to some level of transesterification, and so we were required to monitor for conversion of both esters to achieve the desired reaction completion. Nevertheless, suitable reaction completion could be achieved within a few hours.

Table 7. Reaction Conditions for the Hydrolysis of Methyl Ester 50

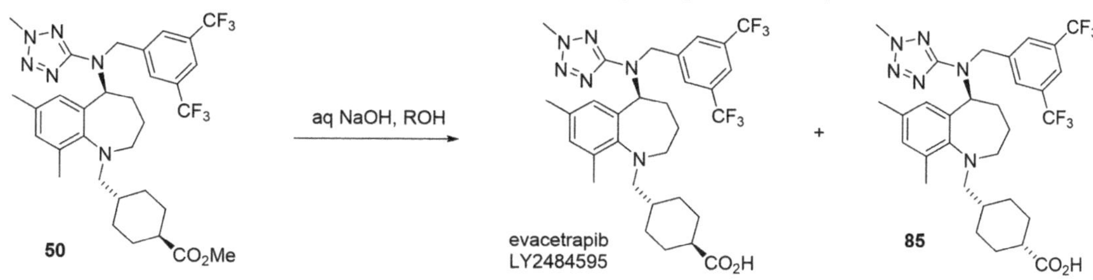

Solvent	*Temp*	*Rxn Time (h)*	***85** (cis)*
MeOH	ca. 65 °C	4 – 6	0.2 – 0.4%
EtOH	ca. 75 °C	2 – 3	0.1 – 0.2%
2-PrOH	ca. 82 °C	<2	<0.05%

Development and Isolation of Evacetrapib

Scheme 14 shows the final conditions for hydrolysis and isolation for evacetrapib incorporating the isolation of the anhydrate crystalline form. As directed by initial studies, hydrolysis was conducted in 2-propanol until completion, followed by neutralization, first with acetic acid and then final pH adjustment with aqueous hydrochloric acid. Ultimately, as the drug product formulation utilized an amorphous presentation of evacetrapib, crystalline form and particle size in the isolated material were less important parameters in the isolation and were not fully optimized. However, the hydrolysis and isolation conditions provided evacetrapib as a crystalline anhydrate in quantities necessary for large scale clinical demand, but also with desired purity (>99.5 HPLC area % and >99.8% enantiomeric excess) to support late phase purity specifications.

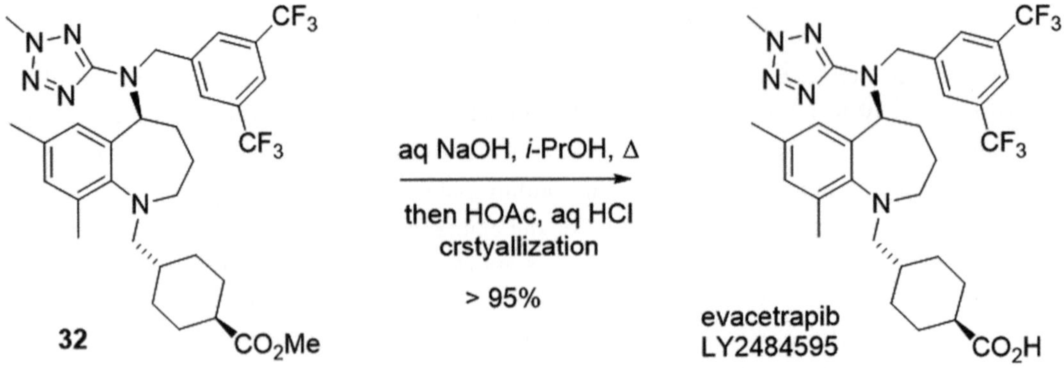

Scheme 14. Final hydrolysis conditions and isolation of evacetrapib.

Table 8 summarizes the large-scale manufacturing results achieved with the process described in Scheme 14. In total, 5 batches were manufactured across the campaign and all batches were isolated in high purity, where the only predominant impurity greater than 0.1% was *cis*-isomer **82**. As well, the average isolated yield across the campaign was >95%. While the first and final batch yields deviated from expectation, this is attributed to carryover of residual cake remaining in the dryer through the isolation.

Table 8. Summary of the Hydrolysis Step in the Manufacture of Evacetrapib

Batch	*Output (kg)*	*% Yield*	*Assay (%)*	*Total Impurities (area%)*	*cis* **82** *(area%)*
1	285	79.1	100.7	0.2	0.2
2	357	98.1	99.6	0.1	0.1
3	349	97.0	99.7	0.1	0.1
4	349	96.7	99.8	0.2	0.2
5	374	103.4	99.5	0.2	0.2

Summary of Route Development and Improvements in Evacetrapib Synthesis

Overall, the evolution of the route and process design resulted in a new manufacturing route for evacetrapib that was used to enable a large Phase 3 clinical study. This new route significantly reduced the overall number of steps in the longest linear sequence (from 15 steps to 11), dramatically improved the overall yield (from 6% to 38%), and resulted in the discovery of key crystalline intermediates which enabled the delivery of a robust impurity control strategy. Additionally, two continuous reductive aminations were discovered and developed that were key to the successful bond forming sequences and control strategy.

Perspective on Evacetrapib Future Development

This work highlights the significant team achievements for both the discovery of a novel CETP inhibitor, evacetrapib, as well as the evolution of the synthesis route: from the early discovery process which was pivotal in delivering hundreds of analogs for SAR fulfillment and quantities for early preclinical evaluation, to the putative commercial route which delivered quantities for large late-phase clinical studies. The route scouting and process development required significant time and investment in order to design a process capable of meeting commercial demands, from aspects of environmental, cost measures, and the ability of the manufacturing route to meet quantities for annual demand. Importantly, the development for evacetrapib was guided by key insights gained during a collaborative interface between discovery and development, especially with regards to the conformational aspects of the benzazepine ring core as well as the inherent stability challenges within the scaffold design. Key input from development allowed for input on optimization of the evacetrapib series, with respect to the design of simpler scaffolds. Additional knowledge gained about alternative routes investigated was also invaluable in informing the final route investigation activities.

Unfortunately, despite lengthy Phase 3 studies with demonstrated robust improvements in the biomarkers for HDL-c and LDL-c, evacetrapib did not demonstrate a sufficient reduction in overall cardiovascular health risk relative to standard of care. Unfortunately, all other inhibitors in this class also failed to achieve commercial regulatory approval. This was a result of either adverse effects revealed during large clinical studies (*67*), or a similar lack of reduction in the risk/benefit profile, or even other non-technical concerns despite improvements in the overall benefit/risk ratio (*68*). Nevertheless, the innovation and numerous learnings from evacetrapib obtained by the discovery and process development provided new capabilities and knowledge which can be applied to future projects.

Acknowledgments

The authors thank the significant numbers of colleagues across Lilly Research Laboratories who contributed to the development of evacetrapib, both from discovery as well as process development. In particular, we acknowledge the contributions of Craig Ruble, Todd Fields, Eric Moher, Scott May, Radhe Vaid, Mike Frederick, Chuck Alt, Chris Ley, Alison Campbell-Brewer, and in addition, Marty Johnson, Brian Haeberle, Phil Hoffman, Shujauddin Changi, Candice Wong, and Tim White for the development of the continuous reductive amination processes. Additionally, we also acknowledge the many colleagues from Lilly manufacturing, especially D. Declan Hurley and Marie Kissane who aided in the development of the iridium-catalyzed continuous reductive amination and manufacture of late phase clinical materials. Finally, we respectfully acknowledge the scientist staff at Albany Molecular Research Inc., Takasago and Evonik for their key contributions to the technical development and manufacture of intermediates of evacetrapib.

References

1. Di Angelantonio, E.; Sarwar, N.; Perry, P.; Kaptoge, S.; Ray, K. K.; Thompson, A.; Wood, A. M.; Lewington, S.; Sattar, N.; Packard, C. J.; Collins, R.; Thompson, S. G.; Danesh, J. Major Lipids, Apolipoproteins, and Risk of Vascular Disease. *JAMA* **2009**, *302*, 1993–2000.
2. Pedersen, T. R.; Kjekshus, J.; Berg, K.; Haghfelt, T.; Færgeman, O.; Thorgeirsson, G.; K Pyörälä, K.; Miettinen, T.; Wilhelmsen, L.; Olsson, A. G.; Wedel, H. Randomised Trial of Cholesterol Lowering in 4,444 Patients with Coronary Heart Disease: the Scandinavian Simvastatin Survival Study (4S). *Lancet* **1994**, *344*, 1383–1389.
3. Sharpe, N.; Hunt, D.; West, M.; Thompson, P.; White, H. Prevention of Cardiovascular Events and Death with Pravastatin in Patients with Coronary Heart Disease and a Broad Range of Initial Cholesterol Levels. *N. Engl. J. Med.* **1998**, *339*, 1349–1357.

4. Collins, R.; Armitage, J.; Parish, S.; Sleight, P.; Peto, R. MRC/BHF Heart Protection Study of Cholesterol Lowering with Simvastatin in 20,536 High-Risk Individuals: a Randomised Placebo Controlled Trial. *Lancet* **2002**, *360*, 7–22.
5. Downs, J. R.; Clearfield, M.; Weis, S.; Whitney, E.; Shapiro, D. R.; Beere, P. A.; Langendorfer, A.; Stein, E. A.; Kruyer, W.; Gotto, A. M., Jr. Primary Prevention of Acute Coronary Events With Lovastatin in Men and Women With Average Cholesterol Levels. *JAMA* **1998**, *279*, 1615–1622.
6. Shepherd, J.; Cobbe, S. M.; Ford, I.; Isles, C. G.; Lorimer, A. R.; MacFarlane, P. W.; McKillop, J. H.; Packard, C. J. Prevention of Coronary Heart Disease with Pravastatin in Men with Hypercholesterolemia. West of Scotland Coronary Prevention Study Group. *N. Engl. J. Med.* **1995**, *333*, 1301–1307.
7. Sacks, F. M.; Pfeffer, M. A.; Moye, L. A.; Rouleau, J. L.; Rutherford, J. D.; Cole, T. G.; Brown, L.; Warnica, J. W.; Arnold, J. M.; Wun, C. C.; Davis, B. R.; Braunwald, E. The Effect of Pravastatin on Coronary Events after Myocardial Infarction in Patients with Average Cholesterol Levels. *N. Engl. J. Med.* **1996**, *335*, 1001–1009.
8. Baigent, C.; Blackwell, L.; Emberson, J.; Holland, L. E.; Reith, C.; Bhala, N.; Peto, R.; Barnes, E. H.; Keech, A.; Simes, J.; Collins, R. Efficacy and Safety of More Intensive Lowering of LDL Cholesterol: a Meta-Analysis of Data from 170,000 Participants in 26 Randomised Trials. *Lancet* **2010**, *376*, 1670–1681.
9. Barter, P.; Gotto, A. M.; LaRosa, J. C.; Maroni, J.; Szarek, M.; Grundy, S. M.; Kastelein, J. J.; Bittner, V.; Fruchart, J. C. HDL Cholesterol, Very Low Levels of LDL Cholesterol, and Cardiovascular Events. *N. Engl. J. Med.* **2007**, *357*, 1301–1310.
10. Lavigne, P. M.; Karas, R. H. The Current State of Niacin in Cardiovascular Disease Prevention: a Systematic Review and Meta-Regression. *J. Am. Coll. Cardiol.* **2013**, *61*, 440–446.
11. Boden, W. E.; Probstfield, J. L.; Chaitman, B. R.; Weintraub, W.; Desvignes-Nickens, P.; Teo, K.; Anderson, T.; McBride, R.; Koprowicz, K. Niacin in Patients with Low HDL Cholesterol Levels Receiving Intensive Statin Therapy. *N. Engl. J. Med.* **2011**, *365*, 2255–2267.
12. Barkowski, R. S.; Frishman, W. H. HDL Metabolism and CETP Inhibition. *Cardiol. Rev.* **2008**, *16*, 154–162.
13. Bowman, L.; Hopewell, J. C.; Chen, F.; Wallenddszus, K.; Stevens, W.; Collins, R.; Wiviott, S. D.; Cannon, C. P.; Braunwald, E.; Sammons, E.; Landray, M. J. Effects of Anacetrapib in Patients with Atherosclerotic Vascular Disease. *N. Engl. J. Med.* **2017**, *377*, 1217–1227.
14. Rennings, A. J. M.; Stalenhoef, A. F. H. "JTT-705: Is There Still Future for a CETP inhibitor After Torcetrapib? *Expert Opin. Invest. Drugs* **2008**, *17*, 1589–1597.
15. Hovingh, G. K.; Kastelein, J. P. P.; Van Deventer, S. J. H.; Round, P.; Ford, J.; Saleheen, D.; Rader, D. J.; Brewer, H. B.; Barter, P. J. Cholesterol Ester Transfer Protein Inhibition by TA-8995 in Patients with Mild Dyslipidaemia (TULIP): a Randomised, Double-Blind, Placebo-Controlled Phase 2 Trial. *Lancet* **2015**, *386*, 452–460.
16. Loreau, L.; Gambert, P.; Lallemant, C. Immunospecific Scintillation Proximity Assay of Cholesteryl Ester Transfer Protein Activity. *Clin. Chem.* **1995**, *41*, 914–919.
17. Bisgaier, C. L.; Minton, L. L.; Essenburg, A. D.; White, A.; Homan, R. Use of Fluorescent Cholesteryl Ester Microemulsions in Cholesteryl Ester Transfer Protein Assays. *J. Lipid Res.* **1993**, *34*, 1625–1634.

18. Ha, Y. C.; Barter, P. J. Differences in Plasma Cholesteryl Ester Transfer Activity in Sixteen Vertebrate Species. *Comp. Biochem. Physiol. B* **1982**, *71*, 265–269.
19. Hayek, T.; Chajek-Shaul, T.; Walsh, A.; Agellon, L. B; Moulin, P.; Tall, A. R.; Breslow, J. L. An Interaction between the Human Cholesteryl Ester Transfer Protein (CETP) and Apolipoprotein A-I Genes in Transgenic Mice Results in a Profound CETP-Mediated Depression of High Density Lipoprotein Cholesterol Levels. *J. Clin. Invest.* **1992**, *90*, 505–510.
20. Fernandez, M-C.; Escribano, A.; Mateo, A. I.; Parthasarathy, S.; Martin de la Nava, E. M.; Wang, X.; Cockerham, S. L.; Beyer, T. P.; Schmidt, R. J.; Cao, G.; Zhang, Y.; Jones, Timothy M.; Borel, A.; Sweetana, S. A.; Cannady, E. A.; Stephenson, G.; Frank, S.; Mantlo, N. B. Design, Synthesis and Structure–Activity-Relationship of 1,5-Tetrahydronaphthyridines as CETP Inhibitors. *Bioorg. Med. Chem. Lett.* **2012**, *22*, 3056–3062.
21. Cao, G.; Beyer, T. P.; Zhang, Y.; Schmidt, R. J.; Chen, Y. Q.; Cockerham, S. L.; Zimmerman, K. M.; Karathanasis, S. K.; Cannady, E. A.; Fields, T.; Mantlo, N. B. Evacetrapib is a Novel, Potent, and Selective Inhibitor of Cholesteryl Ester Transfer Protein that Elevates HDL Cholesterol Without Inducing Aldosterone or Increasing Blood Pressure. *J. Lipid Res.* **2011**, *52*, 2169–2176.
22. Boger, D. L.; Turnbull, P. Synthesis and Evaluation of CC-1065 and Duocarmycin Analogs Incorporating the 1,2,3,4,11,11a-Hexahydrocyclopropa[c]naphtho[2,1-b]azepin-6-one (CNA) Alkylation Subunit: Structural Features that Govern Reactivity and Reaction Regioselectivity. *J. Org. Chem.* **1997**, *62*, 5849–5863.
23. Koebel, R. F.; Needham, L. L.; Blanton, C. D., Jr. Synthesis of Thieno[2,3-6]azepin-4-ones as Potential Antineoplastic Agents. *J. Med. Chem.* **1975**, *18*, 192–194.
24. Corey, E. J.; Bakshi, R. K.; Shibata, S.; Chen, C. P.; Singh, V. K. A Stable and Easily Prepared Catalyst for the Enantioselective Reduction of Ketones. Applications to Multistep Syntheses. *J. Am. Chem. Soc.* **1987**, *109*, 7925–7926.
25. Henry, R. H. *t*-Butylation of 5-Substituted Tetrazoles. *J. Het. Chem.* **1976**, *13*, 391–392.
26. The yield information provided is inferred from a chromatographed sample of the crude material. In practice, it was found to best to carry the material forward unpurified.
27. Hassner, A.; Amit, B.; Marks, V.; Gottlieb, H. E. Dynamic and Static Conformational Analysis of Acylated Tetrahydrobenzazepines. *J. Org. Chem.* **2003**, *68*, 6853–6858.
28. Qadir, M.; Cobb, J.; Sheldrake, P. W.; Whittall, N.; White, A. J. P.; Hii, K. K.; Horton, P. N.; Hursthouse, M. B. Conformation Analyses, Dynamic Behavior, and Amide Bond Distortions of Medium-sized Heterocycles. 1. Partially and Fully Reduced 1- Benzazepines. *J. Org. Chem.* **2005**, *70*, 1545–1551.
29. 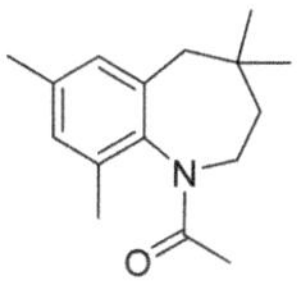

The analogous substrate is shown to the left and as reported in the original work, the ring conformers can be separated by chiral chromatography.
30. Endou, K.; Nakamura, H.; Tanaka, S. U.S. Patent 7,595,423, 2007.
31. Lavety, J.; Proctor, G. R. Ethyl γ-Bromobutyrate. *Org. Synth.* **1965**, *55*, 42–43.
32. For a variation on the classic Sandmeyer isatin process, see: Klein, L. L.; Tufano, M. D. Synthesis of Substituted Isatins. *Tetrahedron Lett.* **2013**, *54*, 1008–1011.
33. Astill, B. D.; Boekelheide, V. The Synthesis of 1-Benzazepine Derivatives as Model Compounds Related to Apo-b-Erythroidine. *J. Am. Chem. Soc.* **1955**, *77*, 4079–4084.

34. Proctor, G. R.; Thomson, R. H. Azabenzocycloheptenones. Part I. The Friedel–Crafts Reaction with Aryl-Amino-Acids. *J. Chem. Soc.* **1957**, 2302–2311.

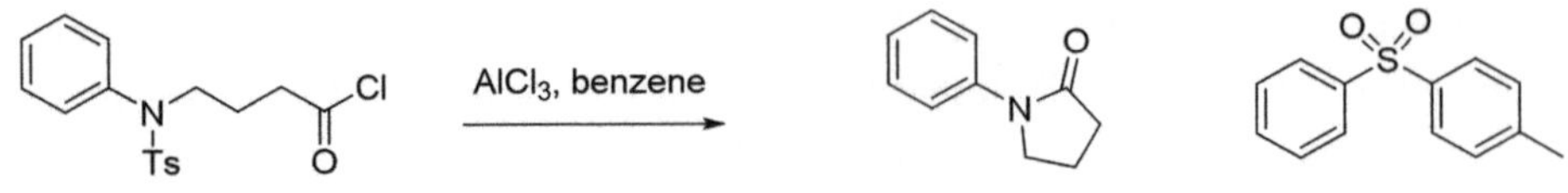

59 **60** **61** The main products as described in the reference are shown above, resulting from side reaction of the solvent benzene with the tosyl protecting group and intramolecular lactam formation. .

35. Vaid, R. K.; Boini, S. K.; Alt, C. A.; Spitler, J. T.; Hadden, C. E.; Frank, S. A.; Moher, E. D. Synthesis of Methyl 7,9-Dimethyl-5-oxo-2,3,4,5-tetrahydro-1H-benzo[b]azepine-1-carboxylate and Its Analogues. *Synthesis* **2014**, *46*, 2463–2470.

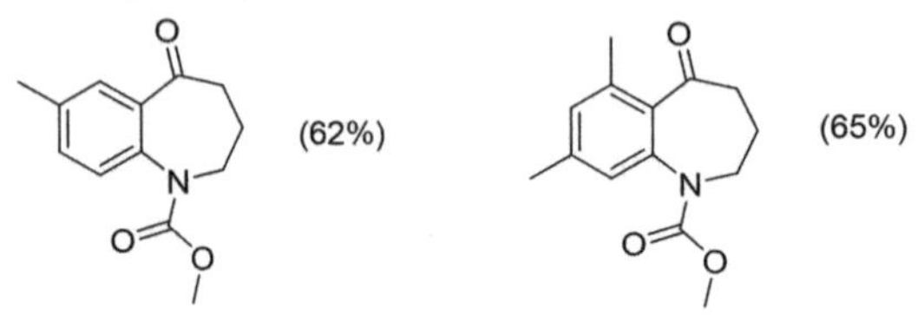

36. The ortho-methyl group (shown to the left) has partial influence on the outcome of the reaction, however additional substrates tested in the Friedel-Crafts cyclization not possessing this substitution pattern also cyclized productively.

37. Henry, R. A.; Finnegan, W. G. Mono-Alkylation of Sodium 5-Aminotetrazole in Aqueous Medium. *J. Am. Chem. Soc.* **1954**, *76*, 923–926.
38. Spear, R. J. Positional Selectivity of the Methylation of 5-Substituted Tetrazolate Anions. *Aust. J. Chem.* **1984**, 37, 2453–2468Presumably the undesired 1-methyl isomer **68** can be readily separated due to facile separation of the regioisomers via crystallization.

39. For an industrial example of a high commercial volume agricultural product, see: Blaser, H.-U. The Chiral Switch of (S)-Metolachlor: A Personal Account of an Industrial Odyssey in Asymmetric Catalysis. *Adv. Synth Catal.* **2002**, *344*, 17–31.
40. Chen, X.; Frank, S. A.; Remick, D. M.; Pedersen, S. W. U.S. Patent 8,299,060, 2010.

41.

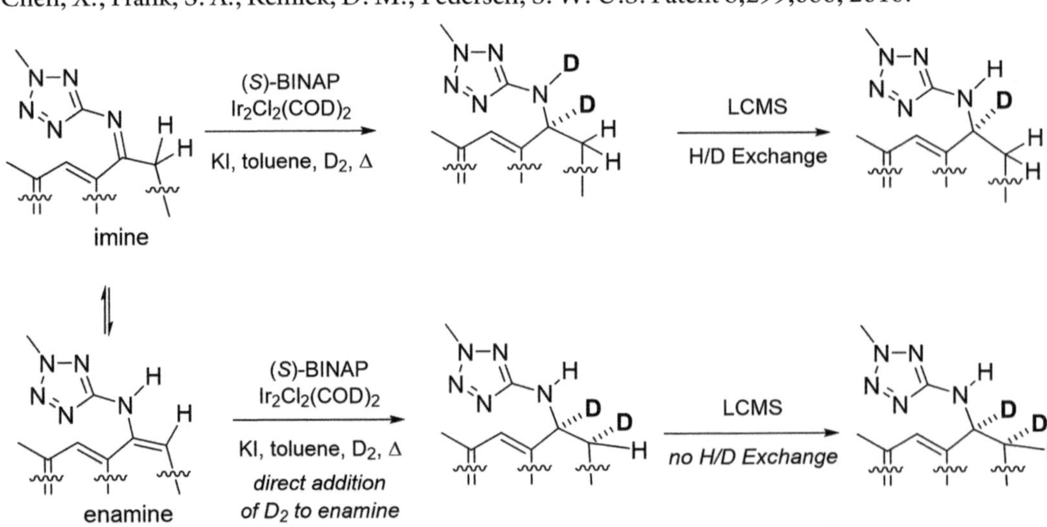

A sample of enamine (shown above) was obtained via chromatography and subjected to the asymmetric hydrogenation conditions under an atmosphere of deuterium gas. LC-MS analysis of the hydrogenation products indicated single deuterium incorporation. H/D exchange of the

aminotetrazole N-H moiety is sufficient in the aqueous analytical sample mixture to allow for exchange of an intermediate doubly deuterated species.
42. Steinhubel, D.; Sun, Y.; Matsumura, K.; Sayo, N.; Saito, T. Direct Asymmetric Reductive Amination. *J. Am. Chem. Soc.* **2009**, *131*, 11316–11317.
43. Villa-Marcos, B.; Li, C.; Mulholland, K. R.; Hogan, P. J.; Xiao, J. Bifunctional Catalysis: Direct Reductive Amination of Aliphatic Ketones with an Iridium-Phosphate Catalyst. *Molecules* **2010**, *15*, 2453–2472.
44. The combination of heat and the abrasive aspects of molecular sieves were of immediate concern with regards to use in fixed large-scale high-pressure autoclaves, due to potential for etching of internal reactor walls. However, more systematic material of construction studies were not conducted as we sought an alternative solution.
45. The typical equilibration time required to achieve interconversion and full crystallization was >8 h.
46. Tan, X.; Gao, S.; Zeng, W.; Xin, S.; Yin, Q.; Zhang, X. Asymmetric Synthesis of Chiral Primary Amines by Ruthenium-Catalyzed Direct Reductive Amination of Alkyl Aryl Ketones with Ammonium Salts and Molecular H_2. *J. Am. Chem. Soc.* **2018**, *140*, 2024–2027.
47. Gallardo-Donaire, J.; Hermsen, M.; Wyscoki, J.; Ernst, M.; Rominger, F.; Trapp, O.; Hashmi, A. S. K.; Schaefer, A.; Comba, P.; Schaub, T. Direct Asymmetric Ruthenium-Catalyzed Reductive Amination of Alkyl–Aryl Ketones with Ammonia and Hydrogen. *J. Am. Chem. Soc.* **2018**, *140*, 355–361.
48. Saito, T.; Yokozawa, T.; Ishizaki, T.; Moroi, T.; Sayo, N.; Miura, T.; Kumobayashi, H. New Chiral Diphosphine Ligands Designed to have a Narrow Dihedral Angle in the Biaryl Backbone. *Adv. Synth. Catal.* **2001**, *343*, 264–267.
49. A distillation unit operation was required to remove levels of methyl formate which complicated the subsequent DARA reaction.
50. Caplow, M. Kinetics of Carbamate Formation and Breakdown. *J. Am. Chem. Soc.* **1968**, *90*, 6795–6803.
51. Changi, S. M.; Yokozawa, T.; Yamamoto, T.; Nakajima, H.; Embry, M.; Vaid, R.; Luciana, C. V; Wong, S.-W.; Johnson, M.; Moher, E. D. Mechanistic Investigation of a Ru-Catalyzed Direct Asymmetric Reductive Amination Reaction for a Batch or Continuous Process Scale-up: An Industrial Perspective. *React. Chem. Eng.* **2017**, *2*, 720–739.
52. We found that the vertically configured pipes-in-series had much more efficient hydrogen incorporation and was optimal in terms of hydrogen stoichiometry requirements.
53. Brunner, M.; Saarenketo, P.; Straub, T.; Rissanen, K.; Koskinen, A. M. P. Stereocontrolled α-Alkylation of Fully Protected L-Serine. *Eur. J. Org. Chem.* **2004**, 3879–3883.
54. Our working hypothesis for the differences in reactivity between the sodium and potassium amide anions was rationalized to be a result in aggregation effects of the different counterions. Some preliminary spectroscopic evidence supported this assertion, but given the complexities of the system we were unable to definitively show this as the key contributing factor for the reactivity differences.
55. Nardello, A.; Pretto, M.; Faccin, A. E. P. Patent 1,673,329, 2006.
56. The ethanol amine serves to scavenge both the excess chloride **54** as well as to buffer the resulting solvent mixture during the exchange. Byproducts from this scavenging step were readily rejected in the crystallization of **44**.
57. Sawada, N.; Uehara, R.; Kabuto, S.; Kyotani, Y.; Kawamura, K. Japan Patent 49,048,639, 1974.

58. Houpis, I. N.; Molina, A.; Reamer, R. A.; Lynch, J. E.; Volante, R. P.; Reider, P. J. Towards the Synthesis of HIV-Protease Inhibitors. Synthesis of Optically Pure 3-Carboxyl-Decahydroisoquinolines. *Tetrahedron Lett.* **1993**, *34*, 2593–2596.
59. Absolute quantification of the levels of the *cis*-aldehyde present in **46** was complicated by epimerization related to the method and at best, a limit of quantification of 0.5% was typically assumed.
60. Kjell, D. P.; Slattery, B. J.; Semo, M. J. A Novel, Nonaqueous Method for Regeneration of Aldehydes from Bisulfite Adducts. *J. Org. Chem.* **1999**, *64*, 5722–5724.
61. Kissane, M. G.; Frank, S. A.; Rener, G. A.; Ley, C. P.; Alt, C. A.; Stroud, P. A.; Vaid, R. K.; Boini, S. K.; McKee, L. A.; Vicenzi, J. P.; Stephenson, G. A. Counterion Effects in the Preparation of Aldehyde–Bisulfite Adducts. *Tetrahedron Lett.* **2013**, *54*, 6587–6591.
62. The sodium triacetoxyborohydride conditions used to prepare **50** are described in reference 40. For a review on the use of this reagent, see: Abdel-Magid, A. F.; Mehrman, S. J. A Review on the Use of Sodium Triacetoxyborohydride in the Reductive Amination of Ketones and Aldehydes. *Org. Process Res. Dev.* **2006**, *10*, 971–1031.
63. Frederick, M. O.; Frank, S. A.; Vicenzi, J. T.; LeTourneau, M. E.; Berglund, K. D.; Edward, A. W.; Alt, C. A. Development of a Hydrogenative Reductive Amination for the Synthesis of Evacetrapib: Unexpected Benefits of Water. *Org. Process Res. Dev.* **2014**, *18*, 546–551.
64. May, S. A.; Johnson, M. D.; Buser, J. Y.; Campbell, A. N.; Frank, S. A.; Haeberle, B. D.; Hoffman, P. C.; Lambertus, G. R.; McFarland, A. D.; Moher, E. D.; White, T. D.; Hurley, D. D.; Corrigan, A. P.; Gowran, O.; Kerrigan, N. G.; Kissane, M. G.; Lynch, R. R.; Sheehan, P.; Spencer, R. D.; Pulley, S. R.; Stout, J. R. Development and Manufacturing GMP Scale-Up of a Continuous Ir-Catalyzed Homogeneous Reductive Amination Reaction. *Org. Process Res. Dev.* **2016**, *20*, 1870–1898.
65. For an example of a hemiaminal intermediate reductive amination, see: Berdini, V.; Cesta, M. C.; Curti, R.; D'Annaballe, G.; Di Bello, N.; Nano, G.; Nicolini, L.; Topai, A.; Allegretti, M. A Modified Palladium Catalysed Reductive Amination Procedure. *Tetrahedron* **2002**, *58*, 5669–5674.
66. Hopman, K. H.; Bayer, A. On the Mechanism of Iridium-Catalyzed Asymmetric Hydrogenation of Imines and Alkenes: A Theoretical Study. *Organometallics* **2011**, *30*, 2483–2497.
67. Joy, T. R.; Hegele, R. A. The Failure of Torcetrapib: What Have we Learned? *Br. J. Pharmacol.* **2008**, *154*, 1379–1381.
68. Tall, A. R.; Rader, D. J. Trials and Tribulations of CETP Inhibitors. *Circ. Res.* **2017**, *122*, 106–112.

Editors' Biographies

Jaan A. Pesti

Jaan A. Pesti was born on Long Island, NY in 1954. He received his Bachelor's degree from Long Island University, Brooklyn, NY in 1976 and his PhD in organic chemistry from Columbia University in the laboratories of Prof. Thomas Katz in 1981. Following postdoctoral research in synthetic organic chemistry at University of California, Berkeley, CA from 1981 to 1983, he joined the staff of Dupont's Medical Products Division in Wilmington, DE and Deepwater, NJ, conducting research in process development from bench scale to pilot plant manufacture. He later transitioned to Dupont Merck Pharmaceuticals, Dupont Pharmaceuticals, and joined Bristol-Myers Squibb Pharmaceuticals (BMS) in 2001. He retired from BMS in 2013 after 30 years of service credited, but has worked for NAL Pharma and Gelest Inc. and currently is part of PRG Inc. and EnginZyme AB.

His career has focused on the process development of potential drugs and medical devices, including contributing to the processes to manufacture the marketed drugs losartan and efavirenz. He was the Associate Editor for the journal *Organic Process Research & Development* from 2008 to 2016. He cofounded the conference series Balticum Organicum Syntheticum (BOS), held biennially in the Baltics since 2000, and the symposium series Process Chemistry: New Developments in Pharmaceutical Process Development held biennially held at the ACS National meetings.

Ahmed F. Abdel-Magid

Ahmed F. Abdel-Magid was born in Giza-Egypt in 1947. He received his Bachelor's degree (1969) and Master's degree (1973) from Cairo University (Cairo-Egypt) and PhD from Temple University, Philadelphia, PA USA (1980). After postdoctoral appointments at the University of South Carolina, Columbia, SC (1981-1982) and SmithKline Beckman Pharmaceuticals, Philadelphia, PA (1983-1984), he joined Drexel University, Philadelphia, PA as Teaching/Research Associate (1985-1986) and then as Adjunct Professor (1986-2006). He started his industrial career at the Process Chemistry Department of Wyeth Laboratories as Senior Chemist (1986-1987). He then moved to the Process Chemistry Group at Johnson & Johnson Pharmaceutical Research and Development (1987-2008) as a Senior Scientist and worked his way through the ranks to Senior Research Fellow. He joined Therachem Research Medilab in 2009 where he currently holds the titles of Executive VP and Chief Scientific Officer.

Dr. Abdel-Magid's career was focused on Process Chemistry, application of hydride reductions in organic synthesis, and teaching. His process Chemistry career has spanned more than 33 years. He contributed to the development of more than 60 processes including several commercial manufacturing processes for marketed drugs. He is an inventor of 28 issued patents and two pending applications, editor of 7 books and author of 40 publications and book chapters. His contributions to the field of hydride reductions in organic synthesis include the development of the frequently used and vastly cited procedure for the reductive amination of aldehydes and ketones by using sodium triacetoxyborohydride. He taught several undergraduate organic chemistry courses as an Adjunct Professor at Drexel University. Dr. Abdel-Magid served as the Executive Officer and Chair of the 39th National Organic Chemistry Symposium (NOS) held at the University of Utah in June 2005. He also served as the National Program Chair and member of the Executive Committee of the Division of Organic Chemistry of the ACS (2008-2014). He is currently serving as a Patent Panelist for the ACS Medicinal Chemistry Letters (2012-Present) where he contributed 142 patent highlights to date. He was recognized by several awards and honors including: the Philip B. Hofmann Award for

Outstanding Research from Johnson & Johnson (1994), the Philadelphia Organic Chemists' Club Award for Outstanding Contributions to Organic Chemistry (2007), the ACS-Philadelphia Section Award for Conspicuous Scientific Achievement Through Research (2013) and he was selected as a Fellow of the American Chemical Society (Inaugural class, 2009).

Rajappa Vaidyanathan

Rajappa Vaidyanathan was born in Madras, India. He obtained his Bachelor's degree from Loyola College, Madras, and his Master's from the Indian Institute of Technology, Madras. He completed his PhD in 1998 from the University of California, Irvine, CA, working in the laboratories of Prof. Scott Rychnovsky. After a postdoctoral appointment at Eli Lilly and Company, he joined the Chemical Process R&D group of Pharmacia Corporation, Kalamazoo, MI, and subsequently Pfizer in Groton, CT. During this period, he led several interdisciplinary teams in the discovery and development of practical, environmentally responsible processes for New Chemical Entities, three of which were commercialized as approved drugs (sunitinib, tofacitinib, and bosutinib). In 2012, he moved to India as Group Director and Head of Chemical Development and API Supply at Bristol-Myers Squibb. In 2020, he took over as the Chief Operating Officer at the Medicines for All Institute in Richmond, VA.

He has authored several peer-reviewed articles and book chapters, and has co-edited five books as part of the ACS Symposium Series. Since 2012, he has been co-chairing the biennial symposium series, Process Chemistry: The Role of Organic Synthesis in Early Clinical Drug Development at the ACS National Meetings. In addition, he is an Adjunct Professor in the Department of Chemistry at the Indian Institute of Technology, Madras. He has won numerous awards, most notably, the American Chemical Society's "Heroes of Chemistry" Award (2015).

Indexes

Author Index

Subject Index

C

G

M

R

S

T

U

Printed in the USA/Agawam, MA
April 26, 2022

792224.111